AF805538

Systems Biology of Marine Ecosystems

Manoj Kumar • Peter Ralph
Editors

Systems Biology of Marine Ecosystems

Springer

Editors
Manoj Kumar, PhD
Climate Change Cluster
Faculty of Science
University of Technology Sydney (UTS)
Sydney, NSW, Australia

Peter Ralph, PhD
Climate Change Cluster
Faculty of Science
University of Technology Sydney (UTS)
Sydney, NSW, Australia

ISBN 978-3-319-62092-3 ISBN 978-3-319-62094-7 (eBook)
DOI 10.1007/978-3-319-62094-7

Library of Congress Control Number: 2017953394

© Springer International Publishing AG 2017
This work is subject to copyright. All rights are reserved by the Publisher, whether the whole or part of the material is concerned, specifically the rights of translation, reprinting, reuse of illustrations, recitation, broadcasting, reproduction on microfilms or in any other physical way, and transmission or information storage and retrieval, electronic adaptation, computer software, or by similar or dissimilar methodology now known or hereafter developed.
The use of general descriptive names, registered names, trademarks, service marks, etc. in this publication does not imply, even in the absence of a specific statement, that such names are exempt from the relevant protective laws and regulations and therefore free for general use.
The publisher, the authors and the editors are safe to assume that the advice and information in this book are believed to be true and accurate at the date of publication. Neither the publisher nor the authors or the editors give a warranty, express or implied, with respect to the material contained herein or for any errors or omissions that may have been made. The publisher remains neutral with regard to jurisdictional claims in published maps and institutional affiliations.

Printed on acid-free paper

This Springer imprint is published by Springer Nature
The registered company is Springer International Publishing AG
The registered company address is: Gewerbestrasse 11, 6330 Cham, Switzerland

Preface

Marine organisms are exposed to diverse environmental fluctuations, anthropogenic stresses, and threats from invasive species and pathogens. Increasing ocean temperature and acidification in a changing climate continually alter the structure and function of marine ecological systems, thereby forcing the marine organisms to either tolerate or adapt to new ocean conditions. Ecophysiology-based approaches in studies of ecological adaptation to altered environmental conditions, such as measuring photosynthesis, growth, and morphological changes, have generally been unable to precisely predict future changes in the performance and persistence of marine organisms under a scenario of global climate change and increased anthropogenic activities. In terrestrial ecosystems, however, approaches in systems biology have played a major role in elucidating the functional adaptation of land plants to biotic and abiotic stress conditions.

Systems biology integrates data from various disciplines, such as physiology, genomics, transcriptomics, proteomics, and metabolomics, into numerical models in order to simulate the physiology of a whole organism. It not only analyzes the topology of biochemical and signaling networks in response to stress but also captures the dynamics of these responses. Systems biology has been used extensively to study terrestrial vegetation and their ecological adaptation to future climate change scenarios, but to a lesser extent in the study of marine organisms.

This book, *Systems Biology in Marine Ecosystem*, describes current advances in the biological and functional interplay within four marine ecosystems: seaweed (Chaps. 1–5), seagrasses (Chaps. 6–9), microorganisms and microalgae (Chaps. 10–13), and their bacterial interactions (Chaps. 14–16). It describes how systems biology has been applied to advance knowledge of the stress response in these important marine ecosystems to climatic and anthropogenic perturbations. This knowledge is linked to mechanisms of resilience and persistence under varying environmental scenarios, which have important implications for the conservation and management of these ecosystems. In addition, the book describes how systems biology has been used in research and discovery to benefit the marine biotechnology sector.

Seaweeds are the focus of Chaps. 1–5. Seaweeds, also known as macroalgae, are the dominant flora of coastal ecosystems globally. Among the most important primary producers, seaweeds are of great importance both ecologically and economically. Seaweeds are exposed to a variety of stressors which affect their physiological and ecological performance. Integrated "omics" is a powerful technique to identify the genes, proteins, and metabolic pathways that respond to biotic and abiotic stresses in plants. Chapter 1 focuses on the application of functional genomics to study stress physiology in seaweeds and the challenges associated with this approach. It also describes how functional genomics has been used to identify the mechanism of biosynthesis of secondary metabolites that comprise the cell wall of seaweeds. This knowledge is highly beneficial when genetically altering the cell wall composition of seaweeds; such alterations can facilitate oil extraction for biofuel production and the production high-value bio-products from diverse seaweeds. In recent years, with the availability of genomic and transcriptomic (expressed sequence tags or ESTs) resources, several studies have integrated physiological, transcriptomic, and/or proteomic approaches to determine stress tolerance mechanisms in seaweeds. Chapters 2 and 3 summarize how integrated omics approaches, when coupled with physiological observations, have led to new mechanistic understandings of stress tolerance in seaweeds. For example, it was discovered that biochemical pathways/networks coordinate within the cell to scavenge reactive oxygen species in order to increase tolerance in seaweeds to desiccation and to detoxify heavy metals. These pathways/networks included the upregulation of antioxidant machinery, phycobilisomes (light-harvesting complexes), vesicular trafficking, heat shock proteins, polyamines, phytochelatins, lipoxygenases, and ATP-binding cassette transporter proteins. These findings may explain the permanence of stress-tolerant algal species in the upper intertidal zone, compared with sensitive algal species located in the lower intertidal zone.

In recent years, another allied omics platform, "lipidomics," has gained momentum in marine science to reveal the role of diverse lipids and fatty acids and their oxidized counterparts (commonly known as oxylipins) in biological systems. These studies have shown how lipid metabolites influence membrane architecture and the modulation of transcription and translation and thus provide tolerance and acclimation to marine organisms in altered environmental conditions. In Chap. 4, current knowledge of lipidomics, advanced analytical tools, and techniques to examine lipids and their derivatives are given. The integration of lipidomics with allied sister omics branches to identify unknown gene/protein functions and the development of systems biology networks to advance knowledge of lipid biochemistry in seaweed development and acclimation to stress conditions are also discussed. Chapter 5 describes recent advances in understanding how volatile compounds emitted from seaweeds, such as ethylene and DMSP, affect seaweeds' physiology, reproduction, and developmental biology.

Seagrasses are the focus of Chaps. 6–9. Seagrasses are monocotyledonous flowering plants that have adapted to the marine environment for over 130 million years. Despite their immense ecological (carbon sink) and commercial value, they are declining at an alarming rate due to climate change and anthropogenic activities attributed directly (e.g., dredging) or indirectly (e.g., eutrophication) to light stress. Chapter 6 provides an overview of the development of high-throughput molecular technologies (e.g., omics) to bridge the gap between the genome and phenotype in order to elucidate the molecular mechanisms that underpin tolerance to abiotic stress in seagrass. Seagrasses live in dynamic coastal aquatic environments and experience complex photosynthetic and respiratory responses. Chapter 7 describes systems biology approaches to the understanding of photosynthetic processes in seagrasses and the accurate estimation of the carbon budgets of seagrass meadows. Furthermore, Chapter 8 summarizes how system-based approaches are crucial in predicting the fitness and response of seagrasses to the combined impacts of environmental constraints and how their interactions with other organisms in their ecological niche at different trophic levels affect the marine system dynamics at numerous points in the network. The adaptive fitness of seagrasses to any environment requires a mechanistic understanding of environmental influence on metabolic networks that eventually control energy assimilation, growth, and reproduction. Chapter 9 explores nontargeted metabolite profiling and how metabolomic information in seagrasses is integral to linking genotype to phenotype in the context of global climate change.

Marine microalgae and microorganisms are the focus of Chaps. 10–13. Chapter 10 describes the availability of complete marine microalgal genome sequences, meta-transcriptomic data, and other omics-based datasets. These molecular resources have enabled precise molecular descriptions of complete biological systems and have enabled rigorous hypothesis testing to study the connections between genotype and phenotype, phenotype and the environment, species and ecosystems, and the interspecies evolution and adaptation of microalgae. A discussion of the potential of meta-barcoding and meta-genomics to characterize ocean microbial communities rapidly and effectively is given in Chap. 11. In addition, this chapter describes the potential of cultivation-independent omics approaches to understand how microbial taxa adjust their molecular and physiological machinery to take advantage of changing environmental conditions and, in turn, shape microbial community structure. Chapter 12 summarizes the individual and combined effects of ocean acidification and ultraviolet radiations on marine photosynthetic carbon fixation. Chapter 13 provides a comprehensive overview of bioprospecting of microalgae while culturing under stress conditions to enhance secondary metabolite production and biofuel potential. This chapter further highlights how the integration of multiple omics is effective for discovering new metabolic pathways that are integral for the use of microalgae as biofactories.

Chemical communications between host and microbial community are the focus of Chaps. 14–16. The host (marine macro- and microalgae/corals)-microbial interaction and its significance within a hostile marine environment are described. Furthermore, the interactions that are essential to regulate the host defense system, their morphology and development, quorum sensing, and exchange of info-chemicals informed by systems biology approaches, together with meta-genomics and meta-transcriptomics, are discussed.

This book describes the latest advances in systems biology in four pillars of the marine ecosystems: seaweed, seagrasses, microalgae, and corals. This knowledge will not only benefit marine biology students and researchers but also resource managers and marine biotechnologists. We thank all the authors for their generous contribution to this book and collaboration in revising the manuscript. We are also indebted to the consistent support from the reviewers for providing their critical inputs to improve the articles and eventually this book. We are extremely thankful to the entire team of Springer for their support and effort in producing this book.

Sydney, New South Wales, Australia

Manoj Kumar
Peter Ralph

Contents

Part I

1 **Macroalgal Functional Genomics: A Missing Area** 3
Vishal Gupta, Mukesh Jain, and C.R.K. Reddy

2 **Tolerance Pathways to Desiccation Stress in Seaweeds** 13
Loretto Contreras-Porcia, Camilo López-Cristoffanini, Andrés Meynard, and Manoj Kumar

3 **Marine Metal Pollution and Effects on Seaweed Species** 35
Loretto Contreras-Porcia, Andrés Meynard, Camilo López-Cristoffanini, Nicolas Latorre, and Manoj Kumar

4 **Seaweed Lipidomics in the Era of 'Omics' Biology: A Contemporary Perspective** 49
Puja Kumari

5 **Volatiles in the Aquatic Marine Ecosystem: Ethylene and Related Plant Hormones and Sporulation in Red Seaweeds** 99
P. Garcia-Jimenez and R.R. Robaina

Part II

6 **Abiotic Stress of Seagrasses: Recent Advances in Transcriptomics, Genomics, and Systems Biology** 119
E.E. Malandrakis, T. Danis, A. Iona, and A. Exadactylos

7 **Photobiology of Seagrasses: A Systems Biology Perspective** 133
Pimchanok Buapet

8 **Systems Biology and the Seagrass Paradox: Adaptation, Acclimation, and Survival of Marine Angiosperms in a Changing Ocean Climate** 167
Richard C. Zimmerman

9 Gas and Liquid Chromatography-Mass Spectrometry-Based Metabolic Profiling of Marine Angiosperm *Zostera muelleri* (Alismatales, Zosteraceae) 189
Unnikrishnan Kuzhiumparambil, Manoj Kumar, and Peter Ralph

Part III

10 Marine Microalgae: Systems Biology from 'Omics' 207
Justin Ashworth

11 Application of 'Omics' Approaches to Microbial Oceanography 223
Deepa R. Varkey and Martina A. Doblin

12 Effects of Ocean Acidification and UV Radiation on Marine Photosynthetic Carbon Fixation 235
Kunshan Gao and Donat-P. Häder

13 Oxidative Stress-Induced Bioprospecting of Microalgae 251
Kaumeel Chokshi, Imran Pancha, Arup Ghosh, and Sandhya Mishra

Part IV

14 Bioactive Small Molecules Mediate Microalgal-Bacterial Interactions 279
Leen Labeeuw, Anna R. Bramucci, and Rebecca J. Case

15 Exploring the Complexity of Macroalgal-Bacterial Interactions Through Interkingdom Signalling System 301
Ravindra Pal Singh, Ramesh Kothari, and Suhelen Egan

16 Role of Bacteria in Coral Ecosystem 317
Neha P. Patel, Sweta B. Kumar, and S. Haldar

Index 343

Contributors

Justin Ashworth Climate Change Cluster, University of Technology Sydney, Sydney, NSW, Australia

Anna R. Bramucci Department of Biological Sciences, University of Alberta, Edmonton, AB, Canada

Pimchanok Buapet Department of Biology, Faculty of Science, Prince of Songkla University, Songkhla, Thailand

Coastal Oceanography and Climate Change Research Center, Prince of Songkla University, Hat Yai, Songkhla, Thailand

Rebecca J. Case Department of Biological Sciences, University of Alberta, Edmonton, AB, Canada

Kaumeel Chokshi Discipline of Salt and Marine Chemicals, CSIR—Central Salt and Marine Chemicals Research Institute, Bhavnagar, India

Academy of Scientific and Innovative Research (AcSIR), CSIR—Central Salt and Marine Chemicals Research Institute, Bhavnagar, India

Loretto Contreras-Porcia Departamento de Ecología y Biodiversidad, Facultad de Ecología y Recursos Naturales, Universidad Andres Bello, Santiago, Chile

Center of Applied Ecology and Sustainability (CAPES), Pontificia Universidad Católica de Chile, Santiago, Chile

Centro de Investigación Marina Quintay (CIMARQ), Facultad de Ecología y Recursos Naturales, Universidad Andres Bello, Quintay, Chile

T. Danis Department of Ichthyology and Aquatic Environment, School of Agricultural Sciences, University of Thessaly, Volos, Greece

Martina A. Doblin Climate Change Cluster, University of Technology Sydney, Sydney, NSW, Australia

Suhelen Egan School of Biological, Earth and Environmental Science, Centre for Marine Bio-Innovation, The University of New South Wales, Sydney, NSW, Australia

A. Exadactylos Department of Ichthyology and Aquatic Environment, School of Agricultural Sciences, University of Thessaly, Volos, Greece

Kunshan Gao State Key Laboratory of Marine Environmental Science, College of Ocean and Earth Sciences, Xiamen University, Xiamen, China

P. Garcia-Jimenez Department Biology, Facultad de Ciencias del Mar, Universidad de Las Palmas de Gran Canaria, Las Palmas de Gran Canaria, Spain

Arup Ghosh Academy of Scientific and Innovative Research (AcSIR), CSIR—Central Salt and Marine Chemicals Research Institute, Bhavnagar, India

Discipline of Plants Omics, CSIR—Central Salt and Marine Chemicals Research Institute, Bhavnagar, India

Vishal Gupta Biological Oceanography Division, CSIR-National Institute of Oceanography, Panaji, Goa, India

Donat-P. Häder Department of Biology, Friedrich-Alexander University Erlangen-Nürnberg, Erlangen, Germany

S. Haldar Marine Biotechnology and Ecology Division, CSIR- Central Salt and Marine, Chemicals Research Institute (CSIR-CSMCRI), Bhavnagar, Gujarat, India

Academy of Scientific and Innovative Research, CSIR-Central Salt and Marine Chemicals Research Institute (CSIR-CSMCRI), Council of Scientific and Industrial Research (CSIR), Bhavnagar, Gujarat, India

A. Iona Hellenic National Oceanographic Data Centre-HCMR, Aghios Kosmas, Attiki, Greece

Mukesh Jain School of Computational and Integrative Sciences, Jawaharlal Nehru University, New Delhi, India

Ramesh Kothari Department of Biosciences, Saurashtra University, Rajkot, Gujarat, India

Manoj Kumar Climate Change Cluster (C3), Faculty of Science, University of Technology Sydney (UTS), Sydney, NSW, Australia

Sweta B. Kumar Marine Biotechnology and Ecology Division, CSIR- Central Salt and Marine, Chemicals Research Institute (CSIR-CSMCRI), Bhavnagar, Gujarat, India

Academy of Scientific and Innovative Research, CSIR-Central Salt and Marine Chemicals Research Institute (CSIR-CSMCRI), Council of Scientific and Industrial Research (CSIR), Bhavnagar, Gujarat, India

Puja Kumari Institute of Plant Sciences, Agricultural Research Organization, Volcani Centre, Rishon LeTsiyon, Israel

Unnikrishnan Kuzhiumparambil Climate Change Cluster, Faculty of Science, University of Technology Sydney (UTS), Sydney, NSW, Australia

Leen Labeeuw Department of Biological Sciences, University of Alberta, Edmonton, AB, Canada

Plant Functional Biology and Climate Change Cluster (C3), University of Technology, Sydney, NSW, Australia

Nicolas Latorre Programa de Doctorado en Medicina de la Conservación, Facultad de Ecología y Recursos Naturales, Universidad Andres Bello, Santiago, Chile

Camilo López-Cristoffanini Departamento de Biología Vegetal, Facultad de Biología, Universitat de Barcelona, Barcelona, Spain

E.E. Malandrakis Department of Ichthyology and Aquatic Environment, School of Agricultural Sciences, University of Thessaly, Volos, Greece

Andrés Meynard Departamento de Biología Vegetal, Facultad de Biología, Universitat de Barcelona, Barcelona, Spain

Centro de Investigación Marina Quintay (CIMARQ), Facultad de Ecología y Recursos Naturales, Universidad Andres Bello, Quintay, Chile

Sandhya Mishra Discipline of Salt and Marine Chemicals, CSIR—Central Salt and Marine Chemicals Research Institute, Bhavnagar, India

Academy of Scientific and Innovative Research (AcSIR), CSIR—Central Salt and Marine Chemicals Research Institute, Bhavnagar, India

Imran Pancha Discipline of Salt and Marine Chemicals, CSIR—Central Salt and Marine Chemicals Research Institute, Bhavnagar, India

Neha P. Patel Marine Biotechnology and Ecology Division, CSIR- Central Salt and Marine, Chemicals Research Institute (CSIR-CSMCRI), Bhavnagar, Gujarat, India

Academy of Scientific and Innovative Research, CSIR-Central Salt and Marine Chemicals Research Institute (CSIR-CSMCRI), Council of Scientific and Industrial Research (CSIR), Bhavnagar, Gujarat, India

Peter Ralph Climate Change Cluster, Faculty of Science, University of Technology Sydney (UTS), Sydney, NSW, Australia

C.R.K. Reddy Division of Marine Biotechnology and Ecology, CSIR-Central Salt and Marine Chemicals Research Institute, Bhavnagar, Gujarat, India

R.R. Robaina Department Biology, Facultad de Ciencias del Mar, Universidad de Las Palmas de Gran Canaria, Las Palmas de Gran Canaria, Spain

Ravindra Pal Singh Department of Biosciences, Saurashtra University, Rajkot, Gujarat, India

Deepa R. Varkey Climate Change Cluster, University of Technology Sydney, Sydney, NSW, Australia

Richard C. Zimmerman Department of Ocean, Earth and Atmospheric Sciences, Old Dominion University, Norfolk, VA, USA

Authors' Biography

Dr. Manoj Kumar is a research scientist (ARC-DECRA fellow) in the Climate Change Cluster (C3), University of Technology Sydney (UTS), Australia. He obtained a PhD in marine biotechnology at the Central Salt and Marine Chemicals Research Institute (CSMCRI-CSIR), Gujarat, India. His major research activities have been focused on the ecophysiology and stress tolerance mechanisms of benthic marine macrophytes (seaweeds and seagrasses) to diverse environmental and anthropogenic perturbations under the scenario of global climate change through omics approaches (proteomics and metabolomics). His research interests have also included the use of seaweeds in food and fuel, quality seed stock generation of seaweeds using tissue culture, and protoplast isolation techniques. He has also researched the epigenetic regulation of marine and terrestrial plants under environmental stress conditions. Dr. Kumar was a visiting scientist at the Agricultural Research Organization (ARO), Volcani Center, Israel, where he expanded his research activities in the area of phytohormone signaling that regulates root and shoot architecture in the *Arabidopsis* model plant. Currently, he is studying the early signals of seagrass loss using advanced "omics" approaches in order to understand eutrophication- and light attenuation-related phenotypic responses in seagrasses.

Dr. Kumar is the recipient of the Blaustein Fellowship (Israel), Japan Society for the Promotion of Science Fellowship (JSPS, Japan), National Fund for Scientific and Technological Development (FONDECYT, Chile), and Australian Research Council—Discovery Early Career Researcher Award (ARC-DECRA, Australia). He has published several research papers and book chapters (over 35 in total) in high-impact scientific journals and books. He is a member of the editorial board for the international journal *Frontiers in Plant Science* (Switzerland) as a review editor for Plant Abiotic Stress. He is a member of scientific societies such as the Australasian Society for Phycology and Aquatic Botany (ASPAB), International Society for Applied Phycology (ISAP), and Australian Society of Plant Scientists (ASPS) and a general committee member of ASPAB. He is also an invited reviewer for a number of international scientific journals including *New Phytologist*, *Scientific Reports*, *Algal Research*, *Frontiers in Plant Science*, and research grant proposals including ARC-DECRA and ARC-Discovery.

Prof. Peter Ralph is the executive director of the Climate Change Cluster (C3) at the University of Technology Sydney. He leads a dynamic, multidisciplinary group of researchers dedicated to improving our predictions about the impacts of climate change. Over the past 8 years, he has grown this research institute to support over 100 staff and students. He is a member of the Blue Carbon International Scientific Working Group, formed under the auspices of the International Union for Conservation of Nature, United Nations Environmental Programme, and Intergovernmental Oceanographic Commission. He is the former leader of the CSIRO Marine and Coastal Carbon Biogeochemistry Cluster, which brought together the expertise of eight Australian universities in Australia's largest coastal blue carbon accounting, mapping, and measurement study. He has recently been appointed to the Scientific Advisory Board for CzechGlobe (Global Change Research Institute—Czech Academy of Science).

Prof. Ralph's diverse research background combines his expertise in photophysiology with molecular biology, bioinformatics, modeling, bio-optics, and ecology. He undertakes multiscale analyses—organismic, cellular, and molecular—of seagrasses, freshwater macrophytes, macroalgae, microalgae/phytoplankton, and terrestrial plants to understand their response to climate impacts, with a vision to preserve these ecosystems in future climates.

Currently, Prof. Ralph is leading research in the areas of seagrass health and algae biotechnology. Within seagrass health, he is developing diagnostic tools to assess seagrass health when seagrasses are exposed to anthropogenic stress. Within algae biotechnology, his research is aimed at improving algal growth productivity, optimizing algal growth conditions, and developing a library of algal species best suited for the production of high-value bio-products, such as functional foods, food additives, aquaculture feedstocks, as well as proteins, nutraceuticals, and pharmaceuticals. He established the NSW Deep Green Biotech Hub which brings together industry, academics, and entrepreneurs to develop the algae biotechnology sector in NSW, Australia.

Prof. Ralph has published over 200 research papers, is an invited member of a number of international scientific advisory boards, was an associate editor for the journal *Marine Biology*, and is a review editor for the journal *Marine Ecology Progress Series*. He is a recipient of the UTS Vice-Chancellor's Award for Research Leadership.

Part I

Chapter 1
Macroalgal Functional Genomics: A Missing Area

Vishal Gupta, Mukesh Jain, and C.R.K. Reddy

Abstract Functional genomics may be defined as the study of deciphering the function and regulation of genes for various traits. Functional genomics has made significant advances in decoding functionality of gene(s) and their regulations furthering our knowledge of systems biology of an organism. The true benefits of such studies have widely been realized in terrestrial plants by understanding their bio-architecture, physiology, regulation and metabolic activations. The functional genomics studies for marine macroalgae (seaweeds) are underdetermined despite their proven economic value. Seaweeds are generally found growing in the intertidal region and experience diverse chronic stresses, including the desiccation, intense irradiance, ultraviolet radiation, salinity and submergence/exposure arising from periodic regular tidal rhythms. The molecular basis of genetic regulations involved in physiological adaptation of seaweeds is limited. The whole genome sequences available for *Ectocarpus siliculosus* and *Chondrus crispus* remained largely functionally unannotated. On the other hand, the seaweed improvement programmes also retarded due to limitation of mapping of functional traits over genetic loci. So far, only a few verities were developed using time-consuming conventional breeding approach. The advancement in functional genomics in seaweeds can significantly contribute to these gap areas. Moreover, the functional genomics will facilitate decoding of the mechanisms regulating biosynthesis of species-specific valuable products. This will support the genetic manipulation research for improvements of desired traits in seaweeds. This review, therefore, highlights the potentials of functional genomics in understanding and resolving the unexplored facts about seaweed physiology and trait characterization for developing strategies towards crop improvement.

V. Gupta (✉)
Biological Oceanography Division, CSIR-National Institute of Oceanography, Goa, India
e-mail: vishalg@nio.org

M. Jain
School of Computational and Integrative Sciences, Jawaharlal Nehru University, New Delhi, India

C.R.K. Reddy (✉)
Division of Marine Biotechnology and Ecology, CSIR-Central Salt and Marine Chemicals Research Institute, Bhavnagar, Gujarat, India
e-mail: crk@csmcri.res.in

© Springer International Publishing AG 2017
M. Kumar, P. Ralph (eds.), *Systems Biology of Marine Ecosystems*,
DOI 10.1007/978-3-319-62094-7_1

Keywords Genomics • Metabolomics • Seaweed • Systems biology • Transcriptome

Contents

1.1 Introduction 4
1.2 Need for Functional Genomics 4
1.3 Functional Genomics from the Context of Biofuel 7
1.4 Elite Variety Development 8
1.5 Seaweed Genetic Resource for Translational Research 8
1.6 Conclusion 9
References 10

1.1 Introduction

Marine macroalgae or seaweeds are the assemblage of the macroscopic, multicellular photosynthetic organisms classified in three groups based on their pigmentation and cell architecture as green, red and brown. As an integral component of marine macrophytes (including seagrass and mangroves), seaweeds occupy the basal position in aquatic food web. The exceptional chemical diversity of seaweeds offers a very unique opportunity for their use in food, feed, fertilizer, cosmetics, pharmaceutical, nutraceuticals, biofuel and agro-based applications (Radulovich et al. 2015). Seaweed with other macrophytes acts as blue carbon sink as they capture CO_2, sequester and store it in live tissue or sediment for longer durations. Macrophytic blue carbon system is considered more efficient than terrestrial systems (Nellemann et al. 2009; McLeod et al. 2011). The 2009 UNEP blue carbon report states that coastal macrophytes account for less than 0.5% of the seabed community structure and sequester between 114 and 328 teragrams of carbon per year or 1.6 to 4.6% of total anthropogenic emissions (7200 Tg per year) (http://www.unep.org/ecosystemmanagement/Portals/7/Documents/factsheets/BlueCarbonInitiativeFactSheet.pdf). Seaweeds are now globally considered as key drivers for attaining food and fuel security (Baghel et al. 2015; Trivedi et al. 2016).

1.2 Need for Functional Genomics

The functional diversity among seaweed species can be noted from their diverse phylogeographic distribution from marine water to brackish water and freshwater and from the tropical islands to polar regions (Hu et al. 2016). Despite a huge functional diversity, seaweeds do not have morphological or anatomical complexity. Thus the survival strategies adopted by seaweeds to sustain the highly fluctuating oceanic conditions, such as desiccation and submergence cycle and associated multiple stressors arising from periodic tidal rhythms, for which there is no equivalent terrestrial counterpart (Kumar et al. 2011), are mostly at genetic levels. The research on ecophysiology and economic trait mapping in seaweeds is underdetermined due to the lack of genetic markers. In order to illustrate this statement, a few examples are the following : the 1940s witnessing the continued expansion in the applications of in vitro techniques of plant cell and tissue culture and the same in seaweed gaining momentum in the early 1990s (Reddy et al. 2008a,b, 2010) and unavailability of genome editing technique in seaweed except a recent report of Oertel et al. (2015) on developing suitable vectors and stable transformation in *Ulva mutabilis*; however, this line of research gained gear in the 1990s in terrestrial plants and now advanced to specific genome editing tools like TALEN and CRISPR/Cas9 (Kumar and Jain 2015). The attempts for genetic transformation in seaweeds and associated challenges have been reviewed by Mikami (2013, 2014).

Nevertheless, scientific community made attempts to answer the seaweed high diversity and adaptation to hypervariable environmental conditions through generation of expression sequence tag (EST) libraries and a few transcriptome profiles (Collén et al. 2007; Dittami et al. 2009; Gravot et al. 2010; Pearson et al. 2010; Dittami et al. 2011; Heinrich et al. 2012; Coelho et al. 2013). The genome level understanding in seaweeds gained momentum after release of whole genome sequence data of a brown alga *Ectocarpus siliculosus* (Dillwyn), Lyngbya (Cock et al. 2010), a red alga *Chondrus crispus* (Irish moss) (Collen et al. 2013a, b) and another brown alga *Saccharina japonica* (Ye et al. 2015). The whole genome sequence revealed genomic features that evolved in these groups of organisms for their successful propagation and proliferation in coastal environment. For example, the *Ectocarpus* genome revealed the presence of a complex photosynthetic system facilitating its propagation even in highly variable light conditions and flavonoid pathway genes homologous to plants synthesizing high phenolic contents protecting the alga from ultraviolet radiations. An uncommon halide metabolism has been deciphered based on the presence of 21 putative dehalogenases and 2 haloalkane dehalogenases. Moreover, the genes and gene families associated with the development of multicellularity and evolution of the brown algal lineage have been identified (Cock et al. 2010). Similarly, the genome sequence of red alga has also elucidated metabolic adaptations pertaining to halogen metabolism, synthesis of oxylipins, microRNA and transcription factors for the development of multicellularity (Collen et al. 2013a, b). This study also revealed unique metabolic features that are otherwise part of bacterial and fungal metabolism (cellulose synthesis and cell wall remodelling) and are absent in genome of brown alga. The whole genome analysis of *S. japonica* revealed gene families for iodine synthesis and concentration mechanism, biogenesis and remodelling of cell wall polysaccharides. The *S. japonica* is quite different from *E. siliculosus* in terms of more complex differentiation, large blade size, higher polysaccharide content and high iodine accumulation. The genome analysis could identify some of the striking features for these differences. The phase-dependent Imm gene families were identified upregulated in gametophytic phase of *S. japonica* similar to *E. siliculosus* (Ye et al. 2015).

Though whole genome sequence studies explored some uncommon genomic features related to primary and secondary metabolism in seaweeds, the large part of the genomic information remained unannotated. Independent transcriptomics studies with de novo assembly and annotation of data define the regulations seaweed possess for habiting the coastal environmental conditions. For example, the transcriptome for *Ulva linza* revealed the existence of land-specific genes, special photoprotective mechanism based on both LhcSR and PsbS, evidence of C4-like carbon-concentrating mechanisms, multi-origin transporters for essential inorganic nutrients, multiple and complex P450s for its colonization on coastal waters and bloom formation (Zhang et al. 2012). A few more transcriptomic studies further advance our knowledge towards the physiology and regulations in *Sargassum* (Liu et al. 2014), *Pyropia* (Im et al. 2015) and *Undaria* (Shan et al. 2015). These studies have revealed a few unknown pathways and highlighted the existence of specialized mechanisms in seaweeds, which are quite distinct from terrestrial counterparts. However, more than 40% of transcriptomic information generated from these studies remained unassigned as there is no BLAST hit for those regions. Thus unique regulatory elements of seaweeds either at functional or physiological aspects could not be defined. Missing such valuable information is a major setback towards understanding of seaweed unique physiology and functioning. Thus, integration of datasets from functional and comparative genomics is mandatory which can generate a more accurate and holistic view of genes of unknown function (Xu et al. 2014). A few studies have been carried out to understand the metabolic processes in response to salinity and oxidative stress in a brown alga and were correlated with transcriptome data (Gravot et al. 2010; Dittami et al. 2011; Konotchick et al. 2013). A most recent study by Konotchick et al. (2013) conceptualized a depth-dependent physiology of seaweed by transcriptome analysis. The study with *Ectocarpus* aimed at understanding the response to salinity showed the active function of γ-aminobutyric acid (GABA) synthesized through a salt stress-induced putrescine degradation pathway (Dittami et al. 2011). Another study by the same group showed impregnation of genomic alterations at metabolite level to stabilize the transition of evolutionary colonization of alga from fresh water to marine habitat (Dittami et al. 2012). Sun et al. (2015) developed a comparative transcriptome profile in *Pyropia* in response to temperature stress indicating the complicated and diverse regulation mechanism, such as upregulation of FAD in low-temperature stress and HSP in heat stress. Similar results were obtained from the study on *Fucus* from the Arctic and subarctic regions experiencing thermal stress (Smolina et al. 2016). These few comparative studies thereby could define the ecophysiology of seaweeds but suffer from inevitable issue of tipping down the seaweed-specific regulatory elements in want of assays precisely defining their roles by overexpression or knockdown approach.

As an example of breakthrough research on seaweed functional genomics, the work published by Wang et al. (2015) and Oliveira et al. (2015) can be cited. The former revealed the role of auxin polar transport, auxin signal transduction, cross-talk with other endogenous plant hormones and antioxidant systems, for adventitious buds formation in *G. lichenoides* explants in vitro. This work represented a foundation stone for defining the important genomic factors that could facilitate the process of regeneration of agarophytic macroalgae. The improvement in the regeneration capacity of agarophytes may induce agar yield and hence provide economic support to this industry. Oliveira et al. (2015) defined 20 different genes involved in the biosynthesis of terpenoid precursors and 21 different genes coding for terpene synthases in *Laurencia dendroidea*. The authors could tap the mevalonate pathway involved in the biosynthesis of terpenes in *L. dendroidea*. This therefore opens opportunity for possible heterologous biosynthesis of terpenes from *L. dendroidea* exhibiting ecological or biotechnological interest. These studies represent the advantages of functional genomics in seaweeds and thus attribute to expand the seaweed research in the direction of genetic manipulations to improve the traits. With the understanding of the importance of functional genome mapping, the seaweed-specific functional traits of industrial relevance may be determined in due course.

1.3 Functional Genomics from the Context of Biofuel

Seaweeds with wide species-specific architectural and physiological variations indicate huge array of information which needs to be explored. For example, the cell wall biosynthesis in seaweeds is a complex phenomenon as it composed of heteropolysaccharide agar and carrageenan in seaweeds belonging to red, alginates in brown and ulvan in seaweed species of Ulvophyceae. A better understanding of biosynthesis mechanism of these hydrocolloids may help to initiate genetic manipulation studies to improve the seaweed for these traits, to be more specific, cell wall architectural engineering and remodelling. Understanding the *Ulva* cell wall biosynthesis is of paramount importance as this alga is considered as biofuel crop (Trivedi et al. 2015, 2016). This is composed of microfibrillar cellulose in the matrix of heteropolysaccharide ulvan. Upregulation of cellulose synthesis machinery with concomitant decrease in ulvan biosynthesis will make the biofuel production from this alga more feasible. Likewise, the oxylipin biosynthetic machinery in brown seaweeds showed gene cluster homologous to both plant and animal. The sulfinic acid compounds and their derivatives reported by Gupta et al. (2013) in seaweeds are of great interest, and their biosynthetic pathways need to be explored for understating their role in free radical detoxification.

With the practical demonstration of biorefinery concept in seaweeds (Trivedi et al. 2016; Baghel et al. 2015), the biomass can be utilized in various streams and the fibrillary polysaccharide component, i.e. cellulose, can be processed for the production of ethanol or other ancillary industrial components. The introduction of other heteropolysaccharides (agar, alginate and carrageenan), which already constitute an essential commodity product from seaweeds to biofuel production, will be a critical risk to existing market. In the given case scenario, research on cell wall architecture and remodelling is required. Thus understanding of cell wall biosynthesis in particular the cellulose biosynthesis and trafficking in seaweeds is most essential. The genome sequence of a few seaweeds though revealed the existence of land plant orthologs for cellulose biosynthesis (Cock et al. 2010; Collen et al. 2013a, b; Ye et al. 2015), but their functions are not yet explained.

1.4 Elite Variety Development

Macroalgae (seaweeds) represent the second largest aquaculture production with 27 million tonnes fresh weight (FAO 2014). Also, the market for marine macroalgal products is steadily growing at 9% per annum (FAO 2014). Further, research is continued on the development of newer low-volume high-value products for nutraceutical and pharmaceutical applications (Hafting et al. 2015). The volumetric growth of existing commodity products industry in the last one decade and also by newly developed sectors has shown the overwhelming demand for seaweed biomass. Though 94% of seaweed biomass is generated through cultivation, the genetic resource improvement strategies are barely tapped (Loureiro et al. 2015). Thus it is impractical to sustain the market demand for seaweed biomass without developing elite varieties with improved traits along with developing more efficient and economic cultivation and harvesting technologies. Serial subculturing of a seaweed germplasm, as conventionally performed, results into loss of vigour, decline in production and susceptibility to multifaceted diseases and pests (Barrento et al. 2016). There are growing evidences of loss in productivity and multifaceted diseases on major commercial crops of *Kappaphycus*, *Pyropia* and *Saccharina* which were grown after multiple inbreeding and/or serial subculturing (Robinson et al. 2013; Loureiro et al. 2015; Barrento et al. 2016). The huge research efforts implied for sustainability of agriculture, livestock and aquaculture remained underutilized in seaweeds (Barrento et al. 2016). Developing molecular markers for functional trait analysis followed by marker-assisted breeding and selection technologies is needed for the time.

1.5 Seaweed Genetic Resource for Translational Research

Seaweeds are habitants of a dynamic environment experience highly fluctuating abiotic and biotic conditions. Thus, seaweeds could be a genetic resource for understanding and translating the adaptations for such variable environments. The investigated genetic regulations may then be translated to agriculture crops for improving their tolerance to some abiotic stresses. For example, Kishimoto et al. (2013) transformed rice with an animal type-Na^+ – ATPase gene from a marine red seaweed, *Porphyra yezoensis*, which conferred salinity tolerance in rice. Similarly, gaseous flux tolerance in intertidal seaweeds arising from submergence/desiccation cycle due to tidal rhythms is unique. Understanding the genetic regulations for gaseous flux mitigation may aid in designing strategies for improving waterlogging tolerance in agriculture crops. Functional genomics in seaweeds may support in presenting the seaweed as genetic resource for improving traits in terrestrial crops.

Another aspect which is to be looked upon is developing the assay to test the effect of seaweed-specific genes with unknown functions. The candidate gene can be overexpressed now with the availability of suitable vectors or can be knocked out by RNAi, and the concomitant effects can be determined. This will define the seaweed-specific regulations not found in any other domain of life. Gupta et al. (2014) highlighted the need for a coexpression match between mRNA (gene) and targeted metabolite and their homologous expression. Further, Kumar et al. (2016) also reviewed the importance of metabolomics in seaweeds (and other marine plants) to investigate their unique metabolism. Reverse genetic approach knocking out the gene then leads to identify responsible metabolites and its function. This leads to a discovery of relationship between gene regulatory networks with specialized metabolic pathway opening new avenues for metabolic engineering for the production of targeted specialized metabolites in seaweeds.

1.6 Conclusion

In conclusion, the expansion of functional genomics is essential to understand the newer aspects of seaweed physiological and metabolic regulations. This will support in improvement and diversification of seaweed resource and their subsequent commercial utilization. The functional genomics advancement in seaweeds will decode the function and regulation of genes specific to seaweed traits, which is only possible by developing assays determining the effect after overexpression or knockout. Functional genomics in seaweeds must advance with more of tools developed to assay the effects of genes with no match with other life forms. Functional genomics of seaweeds may open new avenues for translational research of utilizing seaweed genetic resource for improving agriculture productivity by conferring tolerance in them against various abiotic stresses mainly the salt and waterlogging.

Acknowledgements The authors would like to thank all the researchers contributing in seaweed genomics. The first author (Vishal Gupta) would like to thank the Department of Science and Technology, India, for INSPIRE Faculty award to initiate seaweed functional genomics. VG also thanks the International Centre for Genetic Engineering and Biotechnology, India, for providing the research facility. VG is thankful to Dr. N Ramaiah, Chief Scientist, CSIR-National Institute of Oceanography for providing research support. The financial support received from PSC0206 is acknowledged to extend the seaweed functional genomics research.

References

Baghel RS, Trivedi N, Gupta V, Neori A, Reddy CRK, Lali A, Jha B (2015) Biorefining of marine macroalgal biomass for production of biofuel and commodity chemicals. Green Chem 17:2443–2443

Barrento S, Camus C, Sousa-Pinto I, Buschmann A (2016) Germplasm banking of the giant kelp: Our biological insurance in a changing environment. Algal Res 13:134–140

Cock JM, Sterck L, Rouze P, Scornet D, Allen AE, Amoutzias G et al (2010) The *Ectocarpus* genome and the independent evolution of multicellularity in brown algae. Nature 465:617–621

Coelho SM, Simon N, Ahmed S, Cock JM, Partensky F (2013) Ecological and evolutionary genomics of marine photosynthetic organisms. Mol Ecol 22:867–907

Collén J, Guisle-Marsollier I, Léger JJ, Boyen C (2007) Response of the transcriptome of the intertidal red seaweed Chondrus crispus to controlled and natural stresses. New Phytol 176:45–55

Collen J, Porcel B, Carre W, Ball SG, Chaparro C, Tonon T et al (2013a) Genome structure and metabolic features in the red seaweed *Chondrus crispus* shed light on evolution of the Archaeplastida. Proc Natl Acad Sci U S A 110:5247–5252

Collen J, Porcel B, Carre W, Ball SG, Chaparro C, Tonon T et al (2013b) Genome structure and metabolic features in the red seaweed *Chondrus crispus* shed light on evolution of the Archaeplastida. Proc Natl Acad Sci U S A 110:5247–5252

Dittami SM, Gravot A, Goulitquer S, Rousvoal S, Peters AF, Bouchereau A et al (2012) Towards deciphering dynamic changes and evolutionary mechanisms involved in the adaptation to low salinities in *Ectocarpus* (brown algae). Plant J 71:366–377

Dittami SM, Gravot A, Renault D, Goulitquer S, Eggert A, Bouchereau A et al (2011) Integrative analysis of metabolite and transcript abundance during the shortterm response to saline and oxidative stress in the brown alga *Ectocarpus siliculosus*. Plant Cell Environ 34:629–642

Dittami SM, Scornet D, Petit JL, Segurens B, Da Silva C, Corre E et al (2009) Global expression analysis of the brown alga *Ectocarpus siliculosus* (Phaeophyceae) reveals largescale reprogramming of the transcriptome in response to abiotic stress. Genome Biol 10

Food and Agriculture Organization of the United Nations (2014) Fisheries and Aquaculture Information and Statistics Services. [WWW document] URL http://www.fao.org/figis/

Gravot A, Dittami SM, Rousvoal S, Lugan R, Eggert A, Collen J et al (2010) Diurnal oscillations of metabolite abundances and gene analysis provide new insights into central metabolic processes of the brown alga *Ectocarpus siliculosus*. New Phytol 188:98–110

Gupta V, Thakur RS, Baghel RS, Reddy CRK, Jha B (2014) Seaweed metabolomics: a new facet of functional genomics. In: Jacquot JP, Gadal P (Serial eds) and Bourgougnon N (Serial vol edn), Advances in botanical research vol 71. Sea plants, pp 31–52

Gupta V, Thakur RS, Reddy CRK, Jha B (2013) Central metabolic processes of marine macrophytic algae revealed from NMR based metabolome analysis. RSC Adv 3:7037–7047

Hafting JT, Craigie JS, Stengel DB, Loureiro RR, Buschmann AH, Yarish C et al (2015) Prospects and challenges for industrial production of seaweed bioactives. J Phycol 51:821–837

Heinrich S, Valentin K, Frickenhaus S, John U, Wiencke C (2012) Transcriptomic analysis of acclimation to temperature and light stress in *Saccharina latissima* (Phaeophyceae). PLoS One 7:e44342

Hu ZM, Duan DL, Lopez-Bautista J (2016) Seaweed phylogeography from 1994 to 2014: an overview. In: Hu ZM, Fraser CI (eds) Seaweed phylogeography: adaptation and evolution of seaweeds under environmental change. Springer, Heidelberg, pp 3–22

Im S, Choi S, Hwang MS, Park EJ, Jeong WJ, Choi DW (2015) De novo assembly of transcriptome from the gametophyte of the marine red algae *Pyropia seriata* and identification of abiotic stress response genes. J Appl Phycol 27:1343–1353

Kishimoto M, Shimajiri Y, Oshima A, Hase A, Mikami K, Akama K (2013) Functional expression of an animal type-Na+-ATPase gene from a marine red seaweed *Porphyra yezoensis* increases salinity tolerance in rice plants. Plant Biotechnol 30:417–422

Konotchick T, Dupont CL, Valas RE, Badger JH, Allen AE (2013) Transcriptomic analysis of metabolic function in the giant kelp, *Macrocystis pyrifera*, across depth and season. New Phytol 198:398–407

Kumar V, Jain M (2015) The CRISPR-Cas system for plant genome editing: advances and opportunities. J Exp Bot 66:47–57

Kumar M, Kuzhiumparambil U, Pernice M, Jiang Z, Ralph PJ (2016) Metabolomics: an emerging frontier of systems biology in marine macrophytes. Algal Res 16:76–92

Kumar M, Gupta V, Trivedi N, Kumari P, Bijo AJ, Reddy CRK, Jha B (2011) Desiccation induced oxidative stress and its biochemical responses in intertidal red alga *Gracilaria corticata* (Gracilariales, Rhodophyta). Environ Exp Bot 72:194–201

Liu FL, Sun XT, Wang WJ, Liang ZR, Wang FJ (2014) De novo transcriptome analysis-gained insights into physiological and metabolic characteristics of *Sargassum thunbergii* (Fucales, Phaeophyceae). J Appl Phycol 26:1519–1526

Loureiro R, Gachon CM, Rebours C (2015) Seaweed cultivation: potential and challenges of crop domestication at an unprecedented pace. New Phytol 206:489–492

Mcleod E, Chumra GL, Bouillon S, Salm R, Björk M, Duarte CM et al (2011) A blueprint for blue carbon: toward an improved understanding of the role of vegetated coastal habitats in sequestering CO_2. Front Ecol Environ 9:552–560

Mikami K (2013) Current advances in seaweed transformation. In: Baptista GR (ed) An integrated view of the molecular recognition and toxinology—from analytical procedures to biomedical applications. InTech Open Access Publisher, Rijeka, pp 323–347

Mikami K (2014) A technical breakthrough close at hand: feasible approaches toward establishing a gene-targeting genetic transformation system in seaweeds. Front Plant Sci 5:498

Nellemann C, Corcoran E, Duarte CM, Valdés L. De Young C, Fonseca L, Grimsditch G (2009) Blue Carbon. A Rapid Response Assessment. United Nations Environment Programme, GRID-Arendal, www.grida.no

Oertel W, Wichard T, Weissgerber A (2015) Transformation of *Ulva mutabilis* (Chlorophyta) by vector plasmids integrating into the genome. J Phycol 51:963–979

Oliveira LS, Tschoeke DA, Oliveira AS, Hill LJ, Paradas WC, Salgado LT et al (2015) New insights on the terpenome of the red seaweed *Laurencia dendroidea* (Florideophyceae, Rhodophyta). Mar Drugs 13:879–902

Pearson GA, Hoarau G, Lago-Leston A, Coyer JA, Kube M, Reinhardt R et al (2010) An expressed sequence tag analysis of the intertidal brown seaweeds *Fucus serratus* (L.) and *F. vesiculosus* (L.) (Heterokontophyta, Phaeophyceae) in response to abiotic stressors. Mar Biotechnol 12:195–213

Reddy CRK, Jha B, Fujita Y, Ohno M (2008a) Seaweed micropropagation techniques and their potentials: an overview. J Appl Phycol 20:609–617

Reddy CRK, Gupta MK, Mantri VA, Jha B (2008b) Seaweed protoplasts: status, biotechnological perspectives and needs. J Appl Phycol 20:619–632

Reddy CRK, Gupta V, Jha B (2010) Developments in biotechnology of red algae. In: Seckbach J, Chapman DJ (eds) Red algae in the genomic age. Springer, Dordrecht, pp 307–341

Radulovich R, Neori A, Valderrama D, Reddy CRK, Cronin H, Forster J (2015) Farming of seaweeds. In: Tiwari BK, Troy DJ (eds) Seaweed sustainability—food and non-food applications. Elsevier, Amsterdam

Robinson N, Winberg P, Kirkendale L (2013) Genetic improvement of macroalgae: status to date and needs for the future. J Appl Phycol 25:703–716

Shan T, Pang S, Li J, Li X (2015) De novo transcriptome analysis of the gametophyte of *Undaria pinnatifida* (Phaeophyceae). J Appl Phycol 27:1011–1019

Smolina I, Kollias S, Jueterbock A, Coyer JA, Hoarau G (2016) Variation in thermal stress response in two populations of the brown seaweed, *Fucus distichus*, from the Arctic and subarctic intertidal. R Soc Open Sci 3:150429

Sun P, Mao Y, Li G, Cao M, Kong F, Wang L, Bi G (2015) Comparative transcriptome profiling of *Pyropia yezoensis* (Ueda) M.S. Hwang & H.G. Choi in response to temperature stresses. BMC Genomics 16:463

Trivedi N, Baghel RS, Bothwell JH, Gupta V, Reddy CRK, Jha B, Lali A (2016) An integrated process for the extraction of fuel and chemicals from marine macroalgal biomass. Sci Rep 6:30728

Trivedi N, Gupta V, Reddy CRK, Jha B (2015) Marine macroalgal biomass as a renewable source of bioethanol. In: Kim S-K, Lee C-G (eds) Marine bioenergy: trends and developments. CRC Publisher, Taylor and Francis Group, USA, pp 197–216

Wang W, Li H, Lin X, Yang S, Wang Z, Fang B (2015) Transcriptome analysis identifies genes involved in adventitious branches formation of *Gracilaria lichenoides in vitro*. Sci Rep 5:17099

Xu J, Sun J, Yin J et al (2014) Comparative analysis of four essential Gracilariaceae species in China based on whole transcriptomic sequencing. Acta Ocean Sin 33:54–62

Ye N, Zhang X, Miao M, Fan X, Zheng Y, Xu D et al (2015) Saccharina genomes provide novel insight into kelp biology. Nat Comm 6:6986

Zhang X, Ye N, Liang C, Mou S, Fan Y, Xu J, Xu D, Zhuang Z (2012) De novo sequencing and analysis of the *Ulva linza* transcriptome to discover putative mechanisms associated with its successful colonization of coastal ecosystems. BMC Genomics 13:565

Chapter 2
Tolerance Pathways to Desiccation Stress in Seaweeds

Loretto Contreras-Porcia, Camilo López-Cristoffanini, Andrés Meynard, and Manoj Kumar

Abstract Seaweeds are sessile organisms that inhabit coastal benthic systems and are key species for the equilibrium of marine communities. Rocky intertidal zone seaweeds are distributed in marked patterns determined by interactions between biotic and abiotic factors influenced by tide levels. It has been proposed that the distribution and abundance of organisms in the upper intertidal zones, with longer emersions, are mostly regulated by abiotic factors. Desiccation is a particularly noteworthy abiotic factor since, during low tide, algae of the upper intertidal zones can lose more than 90% of cellular water content, which can ultimately induce oxidative stress. Considering the necessary activation of several desiccation tolerance mechanisms, these algal species are ideal research models in ecophysiology. In fact, several studies using physiological, transcriptomic, and proteomic approaches have determined that desiccation tolerance mechanisms are expressed within a well-coordinated network that includes morphological and cell wall changes, photosynthetic activity

L. Contreras-Porcia (✉)
Departamento de Ecología y Biodiversidad, Facultad de Ecología y Recursos Naturales, Universidad Andres Bello, República 440, Santiago, Chile

Center of Applied Ecology and Sustainability (CAPES), Pontificia Universidad Católica de Chile, Avda. Libertador Bernardo O'Higgins 340, Santiago, Chile

Centro de Investigación Marina Quintay (CIMARQ), Facultad de Ecología y Recursos Naturales, Universidad Andres Bello, Quintay, Chile
e-mail: lorettocontreras@unab.cl

C. López-Cristoffanini
Departamento de Biología Vegetal, Facultad de Biología, Universitat de Barcelona, Barcelona, 08028, Spain

A. Meynard
Departamento de Biología Vegetal, Facultad de Biología, Universitat de Barcelona, 08028 Barcelona, Spain

Centro de Investigación Marina Quintay (CIMARQ), Facultad de Ecología y Recursos Naturales, Universidad Andres Bello, Quintay, Chile

M. Kumar
Climate Change Cluster (C3), Faculty of Science, University of Technology Sydney (UTS), Sydney, NSW 2007, Australia

© Springer International Publishing AG 2017
M. Kumar, P. Ralph (eds.), *Systems Biology of Marine Ecosystems*,
DOI 10.1007/978-3-319-62094-7_2

diminishment, increased expression of desiccation-associated proteins, hormone accumulation, ROS scavenging by antioxidant enzymes and compounds, and osmolyte and protein synthesis. These mechanisms explain the permanence of tolerant algae species in the upper intertidal zone in comparison with lower intertidal species. Therefore, this chapter focuses on identifying tolerant algal species, and explaining the mechanisms underlying the high capacity of these species to cope with desiccation-induced oxidative stress.

Keywords Desiccation • Seaweeds • Tolerance pathways • Ecophysiology • Intertidal distribution

Contents

2.1 Desiccation Stress Tolerance in Seaweeds: Ecological and Physiological Aspects 14
2.1.1 Early Studies Relating Desiccation Stress to Intertidal Zonation Patterns 15
2.1.2 Differential Buffering Capacities of Seaweeds Against the Overproduction of ROS and Cellular Damage During Desiccation Stress 17
2.1.3 Species of the Bangiales Order as Models for Studying Desiccation Stress Tolerance and Seaweed Distribution in the Intertidal Zone 18
2.1.4 Differential Tolerances to Emersion Stressors and the Geographic Distribution of Seaweeds Across Intertidal Shores 19
2.1.5 Perspectives 20
2.2 Molecular Mechanisms of Desiccation Tolerance in Seaweeds 21
2.2.1 Decrease of Photosynthetic Activity 22
2.2.2 Morphological Changes and the Accumulation of Compatible Solutes 23
2.2.3 Increased Expression of Desiccation-Associated Proteins 24
2.2.4 Increased Antioxidant Activity 25
2.2.5 Perspectives 26
References 28

2.1 Desiccation Stress Tolerance in Seaweeds: Ecological and Physiological Aspects

Macro- and microalgae are the main primary producers in coastal benthic ecosystems, with other organisms in the food web directly or indirectly reliant on algae (Hurd et al. 2014). Macroalgae provide habitats, refuges against predators or other physical threats, and recruitment sites, particularly in stressful environments for the juvenile stages of other algae, fish, and invertebrates (e.g., Wright et al. 2006; Hay 2009; Watt and Scrosati 2013). Therefore, fluctuations in seaweed abundances and distributions by natural or anthropogenic events can impact higher trophic levels, affecting the equilibrium and persistence of whole communities. In this context, understanding the physiological and biochemical mechanisms underlying the local and regional distribution and abundances of seaweeds in rocky intertidal communities is of great relevance. Furthermore, such information could provide significant contributions toward the sustainable economic exploitation of algae, in addition to expanding upon existing scientific knowledge.

Intertidal zonation patterns along rocky coastlines depend on interactions between several biotic and abiotic factors and the relative contribution of these at the different tidal levels. The seminal study by Stephenson and Stephenson (1949) established at least three intertidal zones between widely occurring tidal marks, with each zone characterized by specific distributions and abundances of invertebrate and algal species. Posterior experimental investigations of rocky coasts around the world have shown that the distribution and abundance of organisms in the upper tidal zones, with longer emersion times, are usually regulated by abiotic factors such as UV/PAR radiation, light, salinity, temperature fluctuations, nutrient availability, and desiccation, whereas the presence and abundance of organisms in the lower tidal zones, with longer immersion times, are mainly regulated by biological interactions such as herbivory, predation, competition and/or facilitation (e.g., Connell 1961, 1972; Paine 1974). Additional authors have further established that these tidal zones, or intertidal fringes, vary from shore to shore and expand or shrink depending on wave exposure, climate variables, tidal amplitude, and topographic conditions (Mislan et al. 2009). More recently, Bird et al. (2013) also postulated that vertical zonation patterns and shoreline water levels primarily depend not only on tide patterns but also on wave height.

Tolerant marine species are able to cope with these very stressful conditions as a result of phenotypical plasticity (Fierro et al. 2017). As a consequence of extended emersion periods characterized principally by water depletion, seaweeds exhibit the intracellular production and accumulation of reactive oxygen species (ROS), as in other environmental conditions, which can subsequently result in a condition of oxidative stress (e.g., Rijstenbil 2001; Lee and Shin 2003; Contreras et al. 2005, 2007, 2009; Liu et al. 2007; Kumar et al. 2010, 2011).

In this chapter, we mainly focus on desiccation stress-tolerant algae species, and particularly on two genera of the Bangiales order, *Porphyra* and *Pyropia*, which have extraordinary capacities to withstand the harsh physical and chemical stresses of the upper intertidal levels. These genera are ideal research models in ecophysiology, and facilitate relating high tolerances to specific abiotic stressors with local and regional algae abundances and distributions. This chapter will also present algae species that thrive in the lower intertidal zones and that show varying degrees of tolerance to abiotic stressors. With this discussion, it will shed light onto some of the physiological mechanisms accounting for the differential distribution patterns of intertidal seaweed communities at rocky shores.

2.1.1 Early Studies Relating Desiccation Stress to Intertidal Zonation Patterns

In intertidal zones of the northern Atlantic, species of the genus *Fucus* (Fucales, Phaeophyceae) are distributed within different, but overlapping, vertical distributions (Billard et al. 2010). Indeed, *Fucus* species from the upper shore to the midlittoral zone show varied tolerances to desiccation (Dring and Brown 1982) and temperature extremes (Chapman 1995; Davison and Pearson 1996), ranging from the highly tolerant *F. spiralis* to the less tolerant *F. vesiculosus* and least tolerant *F. serratus*. Chapman (1990) experimentally demonstrated that *F. spiralis* is excluded from the midlittoral zone by the competitively superior *F. vesiculosus*. This suggests that the range of emersion time tolerated by *F. spiralis* depends more on tolerance ranges to abiotic stressors rather than on a specialization and circumscribed exploitation of the drier intertidal zones.

Indeed, most Fucales species do not require periodic drying out, with the exception of *Pelvetia canaliculata* (Rugg and Norton 1987), a candidate for studying algal specialization to upper intertidal abiotic stressors. On the other hand, *F. vesiculosus* from the Baltic Sea, which were introduced to this non-tidal sea approximately 7,500 years ago, evidence lesser desiccation stress tolerance to experimental emersion than close relatives from the North Sea. As demonstrated by Pearson et al. (2000), the photoprotective processes of Baltic Sea algae are rapidly impaired during desiccation stress at or below 10% water content, and photosynthetic activity is not as rapidly or as completely recovered as North Sea *F. vesiculosus* specimens. This differentiated tolerance could be the result of interactions between desiccation and temperature or between desiccation tolerance and salinity.

Abe et al. (2001) measured changes in water potential and photosynthetic activity during dehydration and rehydration for 18 algal species from the upper, mid, and lower intertidal zones along the southwest Pacific shoreline of Shimoda, Japan. These researchers evaluated the water potential, in addition to water content, as a reliable indicator of the thermodynamic state of cellular water content and degree of desiccation. It was found that *Porphyra dentata* (Bangiales), an upper intertidal species, is able to tolerate a water potential as low as -158 MPa (i.e., 2% water content with 30% relative air humidity), with the photosynthetic apparatus of this alga still in motion. Moreover, the photosynthetic activity of *Po. dentata* completely recovered after rehydration, showing a superior desiccation tolerance capacity than species occurring in lower intertidal zones. In contrast, *Chondrus verrucosus* (Gigartinales), *Petalonia fascia* (Ectocarpales), and *Gelidium elegans* (Gelidiales), the least tolerant species of the lower intertidal zone, were injured after reaching their lowest water potential threshold (−14 MPa). Finally, mid-intertidal species showed intermediate levels of cellular desiccation tolerance, with water potentials ranging from −14 to −158 MPa.

Early studies on desiccation stress in macroalgae mainly focused on determining the minimum cellular water content tolerated by different intertidal species and the capability of the photosynthetic systems to recover after long exposures to dry air (e.g., Schonbeck and Norton 1980; Smith et al. 1986; Lipkin et al. 1993; Abe et al. 2001). All of these studies agree that the tolerance or non-tolerance of algae species to desiccation and, generally to emersion, can be determined by the buffering capacities of each species to the damage produced during this environmentally stressful condition. Therefore, it is the ability to withstand desiccation stress and quickly recover during rehydration through buffering mechanisms, and not the ability to retain cellular water, that characterizes algae able to flourish in the drier intertidal zones.

2.1.2 *Differential Buffering Capacities of Seaweeds Against the Overproduction of ROS and Cellular Damage During Desiccation Stress*

The main source of injury during emersion (low tide) for sessile marine organisms such as seaweeds is osmotic imbalance with the subsequent loss of cellular water content. Moreover, depending on climatic and oceanographic conditions, other types of stressors could also play important roles in algal injury. During air exposure and water depletion, ROS are produced in excess, surpassing the buffering capacities of cells and resulting in the oxidation of macromolecules such as lipids and proteins (Foyer and Noctor 2009). ROS are directly produced by O_2 excitation and the subsequent formation of singlet oxygen, or by the transfer of one, two, or three electrons to O_2, which results in the formation of superoxide radicals, hydrogen peroxide, or hydroxyl radicals, respectively. It is worth mentioning that ROS are a normal by-product of the metabolism and function, for example, as signaling molecules, in addition to participating in local immune responses to parasitic and herbivorous attacks (Baker and Orlandi 1995). In non-tolerant algae species, water loss through an osmotic imbalance can also cause tearing of the plasmalemma from the cell wall, thus threatening cell integrity (Flores-Molina et al. 2014). In fact, cellular dehydration resulting from desiccation or emersion stress in general increases electrolyte concentrations in the cell, altering membrane structures such as thylakoids (Kim and Garbary 2007).

Desiccation stress-tolerant species, such as *Pyropia orbicularis* [formerly *Pyropia columbina* (Bangiales, Rhodophyta)] (Ramírez et al. 2014) collected from the upper intertidal zone of a southeast Pacific shoreline of Central Chile (18°–53°S), can tolerate water loss near 96% (Contreras-Porcia et al. 2011). Despite an overproduction of ROS such as hydrogen peroxide (H_2O_2), this species is able to increase the activity of diverse antioxidant enzymes and lower the production of oxidized proteins during desiccation. Contreras-Porcia et al. (2011) also demonstrated that *Py. orbicularis* decreases photosynthetic activity and chlorophyll content, but maintains high levels of phycocyanin and phycoerythrin, during rehydration, as well as evidencing rapid photosynthetic apparatus activity recovery during rehydration.

In addition to the activation of antioxidant enzymes, a low production of oxidized macromolecules can result from the production of lipid- and water-soluble compounds, which also regulate ROS levels during dehydration (Noctor and Foyer 1998; Asada 1999; Wang et al. 2009). In fact, macroalgae collected from lower intertidal zones of Central Chile and treated with algal extracts from desiccated *Py. orbicularis* can complete post-germination development under desiccation stress treatments (Contreras-Porcia et al. 2012). This extract-mediated adaptation contrasts with the normal response of *Lessonia spicata* (Laminariales), the least tolerant algal species studied. Specifically, without the extract treatment, when *L. spicata* is exposed to desiccation stress (2 h/day), it is able to germinate but cannot complete post-germination development.

A number of comparisons have been conducted between dehydrated and hydrated states in the most desiccation-tolerant seaweed *Py. orbicularis* (Contreras-Porcia et al. 2013; López-Cristoffanini et al. 2015; Fierro et al. 2017). The dehydrated plants overexpress various tolerance genes related to antioxidant enzymes, heat shock proteins, cytochrome P450, cell wall proteins, and the ATP-binding cassette transporter proteins (see sections below for molecular description of tolerance mechanisms). A study by Kumar et al. (2011) in *Gracilaria corticata* (Gracilariales), collected from the Veraval coast (Gujarat, India) during low tide, revealed that this seaweed possess mechanisms similar to *Py. orbicularis* for regulating desiccation stress-induced damage, in addition to having specific mechanisms for effectively responding to desiccation stress. The adaptive response of *G. corticata* to desiccation is associated with the activation of antioxidant enzymes and phycobiliproteins, as well as increased amounts of specific polyunsaturated fatty acids and of specific soluble and bound insoluble compounds, such as putrescine and spermine. Moreover, Kumar et al. (2011) observed the activation of specific isoforms of the antioxidant enzymes superoxide dismutase (SOD), ascorbate peroxidase (AP), and glutathione peroxidase during desiccation-induced stress. The activation of particular antioxidant compounds and specific antioxidant enzyme isoforms, as well as an increased presence of polyunsaturated fatty acids, are desiccation-related responses that differ from responses to other induced abiotic stressors, such as salinity (Kumar et al. 2010).

2.1.3 Species of the Bangiales Order as Models for Studying Desiccation Stress Tolerance and Seaweed Distribution in the Intertidal Zone

Taxa of the Bangiales order are an ideal model for studying tolerance mechanisms to environmental stressors due to worldwide distribution that extends from tropical to polar seas and from lacustrine to fully marine habitats (Butterfield 2000; Sutherland et al. 2011). This order is evolutionarily significant as it contains *Bangiomorpha pubescens* NJ Butterfield, the oldest taxonomically resolved eukaryote fossil record and the first known record of a sexual and multicellular organism (Butterfield 2000). While traditional taxonomic studies of the Bangiales order are based only on very simple and plastic morphologies, recent phylogenetic analyses based on molecular markers have redefined the frontiers of numerous taxa over the last 10 years, where several cryptic genera have been discovered and various classical genera have been redefined (e.g., Broom et al. 2002; Brodie et al. 2007; Neefus et al. 2008; Nelson 2013). Indeed, the worldwide molecular study by Sutherland et al. (2011) proposed the existence of eight genera of bladed Bangiales, including *Boreophyllum*, *Clymene*, *Fuscifolium*, *Miuraea*, *Lysithea*, *Porphyra*, *Pyropia*, and *Wildemania*. At a lower taxonomic level, cryptic genetic diversity has also been found, as particularly evidenced by the redefined genera *Porphyra* and *Pyropia* (Niwa et al. 2009; Broom et al. 2010; Guillemin et al. 2016). Related to this, representative *Porphyra* and *Pyropia* specimens have been intensively studied to uncover physiological ecology patterns of rocky intertidal seaweed communities, the origin of sexual reproduction, the diversity of reproductive strategies, and the evolutionary biology of plastids (Blouin et al. 2011; Wang et al. 2013).

Global ecological research of intertidal zonation patterns for *Pyropia* and *Porphyra* suggests that these species seasonally or annually change in regard to relative abundances and distributions along intertidal levels within and across sites. These distributional patterns would be caused not by adaptations to intertidal fringes per se, but by adaptations to abiotic factors inducing physiological stress during high tide emersion, which are highly variable over space and time. For example, West et al. (2005) used molecular tools to identify *Pyropia* and *Porphyra* species and described the intertidal distributions of these genera at two sites from the New England coast (USA). At the first site, Fort Stark, the upper intertidal zone was dominated over the year by *Porphyra umbilicalis*, with the mean biomass of this alga peaking during the summer. In the lower intertidal zone, *Po. umbilicalis* and *Pyropia leucosticta* were most conspicuous and showed maximum biomasses and coverage percentages in the summer. In contrast, no species occurred during winter in the lower intertidal zone at Fort Stark. At the second site, Dover Point, *Po. umbilicalis* was restricted to the upper intertidal zone where, together with *Po. purpurea*, this was the dominant species for the majority of seasons. However, during winter and early spring *Pyropia yezoensis* was the most abundant algal species in the upper intertidal zone of Dover Point. Using the above examples, it is possible to infer that: (1) the range of tidal levels occupied by a specific species, such as *Po. umbilicalis*, can change across sites (Fort Stark vs Dover Point), and (2) across seasons (summer vs winter) the same species, such as *Po. umbilicalis* and *Py. leucosticta* at Fort Stark, can show contrasting patterns of occurrence, biomass, and coverage percentage within the same intertidal level.

As a result of varied specializations, research on the ecophysiological performance and distribution of different foliose Bangiales species requires significant changes in the sampling design. Modifications are needed since environmental factors, such as temperature, UV and PAR radiation, and relative humidity, do not always coincide with tidal fringes and usually fluctuate on smaller spatial scales due in part to topographical irregularities of the intertidal landscape and, on larger scales, due to climatic and oceanographic variations. Therefore, updated sampling models will need to (1) more finely measure important stress factors to define intertidal areas with similar abiotic conditions and (2) compare the likely expansion or shrinkage of algal patches seasonally and yearly.

2.1.4 Differential Tolerances to Emersion Stressors and the Geographic Distribution of Seaweeds Across Intertidal Shores

Depending on the seaweed species, tolerances to different abiotic stressors can likely explain distribution ranges across biogeographic zones. For example, the endemic Antarctic brown alga *Desmarestia anceps* (Desmarestiales) flourishes most of the year in the more stable and nutrient-rich lower subtidal zone (>5 m deep). Nevertheless, this alga is also adapted to grow under the high UV radiation characteristic of the spring and early summer, when sea ice breakage causes a deeper penetration of irradiance into the highly transparent Antarctic coastal waters. Flores-Molina et al. (2016) showed that tolerance to UV radiation and temperature is dependent on high concentrations of constitutive soluble phlorotannins with UV-absorbing properties (Gómez and Huovinen 2015). Increased soluble/non-soluble and constitutive/inducible phlorotannin concentrations under high UV stress lessen the normal inhibition of the photosynthetic system and DNA damage, as has been demonstrated in several brown algae (e.g., Swanson and Druehl 2002).

Two sibling brown algae species, *Lessonia berteroana* and *L. spicata*, are distributed along the Chilean coast in two contiguous biogeographic regions, between 18°S and 30°S and between 29°S and 41°S, respectively. These species evidence contrasting regimes of air exposure during high tide and differential stress tolerance capacities. For example, post-germination development in *L. spicata*, which thrives in the southern biogeographic region and is exposed to milder emersion regimes, abruptly stopped after 1 or 2 h of daily desiccation treatments (López-Cristoffanini et al. 2013). In contrast, 75% of germinated *L. berteroana* spores developed into gametophytes despite 1 h of daily desiccation treatment, and 50% continued development even with 2 h of daily desiccation. First, this indicates that desiccation stress is a selective force able to explain the disjunctive distribution between the less tolerant species of the lower intertidal zone along the Chilean coast. Secondly, this also supports that *L. berteroana* evolved greater desiccation tolerance than *L. spicata* to successfully recruit and develop along the drier intertidal shores of northern Chile.

Recent research by Contreras-Porcia et al. (unpublished) evaluated if tolerance mechanisms to desiccation stress present different capacities in *Porphyra* and *Pyropia* spp. along Chilean shores spanning the previously mentioned biogeographic zone (18°S–41°S). Higher levels of protein and lipid oxidation were measured in *Porphyra* and *Pyropia* spp. from northern Chilean shores, confirming that more extreme emersion regimes with higher desiccation stress prevail in this region. However, no significant differences were found between species and latitude in regard to the activation of antioxidant enzymes under desiccation stress. Indeed, all *Porphyra* and *Pyropia* spp. showed a similarly adaptive and plastic regulation of these enzymes under natural and in vitro desiccation (Fig. 2.1). Therefore, for algae species thriving in the upper intertidal zone, different environmental factors could elucidate disjunctive distributional ranges along the Chilean shoreline.

2.1.5 Perspectives

Considering the prior information, we can conclude that there is a general lack of studies investigating causal relationships between geographical distribution and the physiological tolerances of seaweeds to diverse environmental abiotic factors. Moreover, despite the discoveries of specific tolerance mechanisms under a number of stressful conditions, few researchers have synchronously compared the roles of these mechanisms with specific regulations under different abiotic stressors. This highlights the necessity to further investigate and compare the ecophysiological similarities and

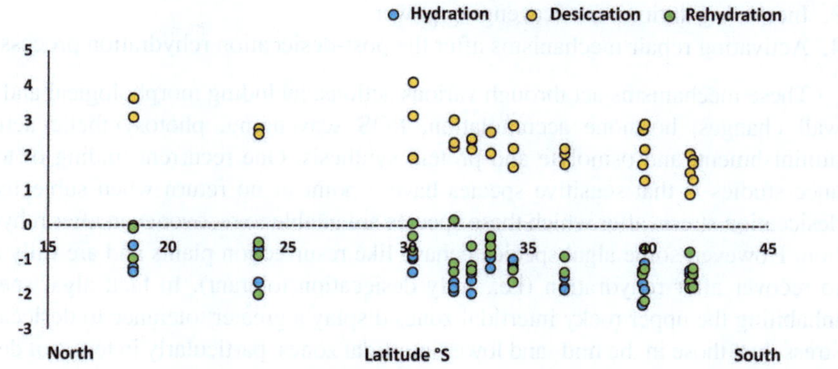

Fig. 2.1 Principal components analysis plot (constructed in PAST 3.11) for the activities of five antioxidant enzymes (ascorbate peroxidase, catalase, thiol-dependent peroxidase, pyruvate dehydrogenase, and thioredoxin) measured under natural hydration, desiccation, and rehydration conditions in *Porphyra* and *Pyropia* spp. sampled along the Chilean coast (18°–41°S). Individual points correspond to the ordination of principal components. Most points within the same latitude correspond to a single *Porphyra* or *Pyropia* species (three replicates per species and treatment), excepting the two southernmost groups of points, where pairs of very close sites, with different species, were indistinguishable due to the plotting scale

differences between the strategies employed by species to cope with different stressful conditions. Further research will facilitate a better understanding of seaweed abundance and distribution in intertidal communities, both locally and regionally.

2.2 Molecular Mechanisms of Desiccation Tolerance in Seaweeds

When faced with detrimental biotic and/or abiotic factors, sessile organisms are more severely affected than mobile counterparts due to highly limited or fixed positions within the environment (Shelford 1914; Huey et al. 2002; Hohmann 2004). The fixed environmental position of sessile organisms such as plants and seaweeds necessitates the development of adaptations to cope with the fluctuations produced by desiccation stress. Some of these strategies include changes in form, structure, or relative limb position, as well as in subcellular mechanisms (Shelford 1914; Bewley 1979; Ingram and Bartels 1996; Ramanjulu and Bartels 2002). Due to the availability of different -omics techniques and several physiological analyses, desiccation tolerance mechanisms have been found to occur in several species rather than being species specific (Oliver and Bewley 1997; Oliver et al. 1998; Contreras-Porcia et al. 2011, 2013; López-Cristoffanini et al. 2015; Moore and Farrant 2015; Fierro et al. 2017). Desiccation tolerance mechanisms can be grouped according to the following four action criteria:

1. Limiting damage to a reparable level
2. Maintaining physiological integrity under desiccation

3. Increasing antioxidant scavenging power
4. Activating repair mechanisms after the post-desiccation rehydration process

These mechanisms act through various actions, including morphological and cell wall changes, hormone accumulation, ROS scavenging, photosynthetic activity diminishment, and osmolyte and protein synthesis. One recurrent finding of tolerance studies is that sensitive species have a point of no return when subjected to desiccation stress, after which these species are unable to recover even after rehydration. However, some algal species behave like resurrection plants and are fully able to recover after rehydration (i.e., fully desiccation tolerant). In fact, algal species inhabiting the upper rocky intertidal zones display a greater tolerance to desiccation stress than those in the mid- and lower intertidal zones, particularly in terms of development arrest and physiological and molecular alterations. The following sections review desiccation stress in algae according to tolerance mechanisms, with emphasis on how they help explain the vertical distribution of algae within the intertidal zone.

2.2.1 Decrease of Photosynthetic Activity

To our knowledge, Schonbeck and Norton (1978) were the first to address desiccation stress as a key factor controlling the vertical distribution of brown algae within the rocky intertidal zone. This investigation demonstrated that algae subjected to desiccation display diminished photosynthetic activity that is recovered after rehydration. Moreover, the recovery process was found to be faster in species inhabiting the upper, rather than lower, intertidal zones, an effect since confirmed by other researchers (Dring and Brown 1982; Leuschner et al. 1998; Contreras-Porcia et al. 2011; Flores-Molina et al. 2014; Guajardo et al. 2016; Fierro et al. 2016, 2017). Dring and Brown (1982) studied intertidal brown algae photosynthesis during desiccation and recovery periods, and although all of the assessed species evidenced decreased photosynthesis due to tissue water loss, the extent of recovery was greater in upper shore species. In relation to water loss, Leuschner et al. (1998) determined that for *Zostera noltii* (Alismatales) inhabiting the German Wadden Sea, carbon dioxide assimilation is dependent on the water content in leaves, with higher water content translating into higher CO_2 assimilation. Importantly, leaf water content is a factor directly impacted by the duration of air exposure.

Algal photosynthetic activity declines during air exposure to reduce damages associated with the photosynthetic by-products of desiccation stress and excessive UV light exposure (Gómez et al. 2004; Chaves et al. 2009; Contreras-Porcia et al. 2011; Dinakar et al. 2011). Several studies have assessed the underlying mechanisms of decreased algal photosynthetic activity. For example, Wiltens et al. (1978) analyzed the chlorophyll fluorescence of the desiccation-tolerant red algae *Porphyra sanjuanensis* and established that rehydration resulted in rapid recovery from severe desiccation. In particular, the fluorescence changes observed during the desiccation/rehydration cycle suggested that (1) electron transport between photosystems I and II and water splitting are partial desiccation-sensitive reactions; (2) intersystem electron transport becomes blocked at ~25% water content;

(3) further desiccation leads to a loss of water splitting activity and, eventually, the complete loss of fluorescence photosystem II reaction centers; and (4) intersystem electron transport begins almost immediately after rehydration while the recovery of water splitting requires several minutes. This final point was confirmed in a detailed study performed by Contreras-Porcia et al. (2011) on the desiccation-tolerant red algae *Py. orbicularis*. Specifically, the photosynthetic efficiency and electron transport flux per cross of *Py. orbicularis* returned to the basal levels only 5 min after rehydration.

An extensive study on various algal species sampled from numerous locations within the Chilean rocky intertidal zone (Flores-Molina et al. 2014) demonstrated that the mid-intertidal algae species *Mazzaella laminarioides* and *Scytosiphon lomentaria* (Ectocarpales) recover basal photosynthetic activity after rehydration. However, algae from the lower intertidal zone, including *Ulva compressa*, *L. spicata*, and *Gelidium rex*, do not fully recover from desiccation stress. This would partly explain the positions of these algal species within the rocky intertidal zone of the Chilean coast. Furthermore, researchers have posited that the poorer photosynthetic performance of sensitive species during desiccation/rehydration cycles is generated by disarrayed photosynthetic machinery and not by a photoinhibition mechanism.

Transcriptomic and proteomic studies performed on *Py. orbicularis* indicate a downregulation of photosynthesis-related RNAs and proteins such as RubisCO, photosystem I and II proteins, and ferredoxin-$NADP^+$ during stressful periods (Contreras-Porcia et al. 2013; López-Cristoffanini et al. 2015). These downregulated expressions during desiccation, together with rapid expression recoveries during rehydration, suggest homoiochlorophyllous-like behavior, which is the storing of proteins by dismantling under stressful conditions.

2.2.2 Morphological Changes and the Accumulation of Compatible Solutes

Resurrection plants are known for leaf curling when subjected to desiccation stress, a mechanism also observed in algal species suffering a cellular water deficit. Like resurrection plants, desiccated algal tissue contains a low cellular water content and, as observed through optical microscopy, a reduced cellular volume (~30–60% of the non-stressed volume). Sensitive species from the Chilean coast show slightly reduced cellular volumes (25–43%), as well as ultrastructural damage, under air exposure. Specifically, *L. spicata* does not recover the fine structure of organelles and cellular components after rehydration (Flores-Molina et al. 2014). In *Py. orbicularis*, the normal greenish-red color of cells is replaced by dark purple, and cells become tightly folded, stiff, and brittle (Contreras-Porcia et al. 2011). This reduction and folding of cells may occur due to reduced contents of microtubule and microfilament proteins, as previously observed in resurrection plants (Pressel et al. 2006; Oliver et al. 2011; Cruz de Carvalho et al. 2014).

As in *Py. orbicularis*, desiccation stress drastically alters the actin cytoskeleton of *Klebsormidium crenulatum* (Charophyta) (Holzinger et al. 2011). Proteomic and transcriptomic analyses found that the reduction in actin quantity was due to an upregulation of a cofilin/actin depolymerizing factor (Holzinger et al. 2011; López-Cristoffanini et al. 2015; Fierro et al. 2017). This factor disassembles actin filaments and prevents denaturation, thereby maintaining an actin pool in the cell (Maciver and Hussey 2002). Further ultrastructural analyses of both desiccated *Py. orbicularis* and *K. crenulatum* evidenced tissues with irregular contours, plasma membrane folding, and a notable accumulation of electron-dense bodies inside the chloroplast. These observations support that cytoplasmic components, such as compatible solutes, agglomerate, and cytoplasm viscosity, increase when the cellular water content is low. Both effects decrease the probability of molecular interactions that may cause protein denaturation and membrane fusion.

Proteins associated with the synthesis of compatible solutes, including triose-phosphate isomerase, *S*-adenosylmethionine synthetase, GDP-D-mannose-3′,5′-epimerase, and phosphomannomutase, occur in desiccated *Py. orbicularis*. Due to the inability of red algae such as *Py. orbicularis* to produce sucrose, triose-phosphate isomerase could be important in glycerol synthesis, which couples with D-galactose to form floridoside and isofloridoside, the primary photosynthates. *S*-adenosylmethionine synthetase, which is upregulated in plants and algae exposed to stress, synthesizes adenosylmethionine, a cofactor in the synthesis of compatible solutes such as glycine betaine. Thus, the recovery of cellular volume and conformation could be due to the intricate coordination of mechanisms, such as compatible solutes production, channel activation, and an efficient dismantling and assembly of the cytoskeleton present in desiccation-tolerant species.

2.2.3 Increased Expression of Desiccation-Associated Proteins

In resurrection plants, desiccation stress increases the production of heat shock and late embryogenesis abundant proteins (Scott 2000; Alpert 2005; Leprince and Buitink 2010). These protein groups are possible desiccation tolerance chaperones that avert macromolecule aggregation and maintain the proper conformations and activities of other proteins (Timperio et al. 2008; Toldi et al. 2009). These proteins have also been associated with internal cellular structures, such as cytoskeleton filaments that maintain cell stability. Furthermore, these proteins act as water substitute given a high glycine content (ca. >6%) (Oliver and Bewley 1997; Hoekstra et al. 2001). Desiccated *Py. orbicularis* overproduces heat shock proteins and chaperones (e.g., chaperonin 60, GRP78, BiP, and KAR2) (Contreras-Porcia et al. 2013; López-Cristoffanini et al. 2015). Likewise, the green alga *Asterochloris erici* (Trebouxiales) overproduces five heat shock protein 90 transcripts under desiccation (Gasulla et al. 2013; Holzinger and Karsten 2013). In addition, increased peptidylprolyl isomerase production was detected in desiccated *Py. orbicularis* using a proteomic approach (López-Cristoffanini et al. 2015). This enzyme rotates the peptide bonds prior to proline to induce protein folding and prevent water loss damage (Baniwal et al. 2004). Thus, although the available literature regarding algae desiccation-associated proteins is not as extensive as that of resurrection plants, a careful review of the existing research indicates that similar mechanisms occur in algae subjected to desiccation stress, information that contributes toward explaining the vertical distribution of intertidal seaweed species.

2.2.4 *Increased Antioxidant Activity*

Along with the modifications and effects mentioned in the previous subsections, desiccation can also cause oxidative stress . This condition occurs when there is an uncontrolled production of free oxygen radicals that consequently increase the levels of ROS. Algae have successfully developed scavenging mechanisms to prevent harmful ROS overproduction, and the first study on antioxidant responses to desiccation stress in algae analyzed *Stictosiphonia arbuscula* (Ceramiales) individuals collected from the upper and lower rocky intertidal zones of a New Zealand beach (Burritt et al. 2002). The authors showed that desiccation stress triggered an organism-wide antioxidant response, although glutathione reductase and AP enzyme activities were greater in individuals sampled from the upper rocky intertidal zones. This also correlates with the finding that low-band specimens produce more hydrogen peroxide than high-band ones.

An in-depth analysis of antioxidant responses in *Py. orbicularis* established that this alga increases the production of glutathione reductase, AP, catalase (CAT), thioredoxin (TRX), dehydroascorbate reductase, peroxiredoxin (PRX), and *CAT*, *PRX*, and *TRX* transcripts, among other factors, under desiccation (Contreras-Porcia et al. 2011; Fierro et al. 2017). Interestingly, Western-blot analyses of desiccated *Py. orbicularis* fronds display low PRX protein levels, and in rehydrated specimens, PRX activity drops to undetectable levels. PRXs belong to the thiol-dependent peroxidase family and are known to attenuate oxidative stress and modulate redox-dependent signaling cascades (Rouhier and Jacquot 2002; Tripathi et al. 2009; Dietz 2011). Therefore, the observed differential inductions of PRX could be involved in the transduction of signals by means of ROS, which are needed for the genetic activation of tolerance factors during air exposure.

Recently, Guajardo et al. (2016) showed that under desiccation stress, free abscisic acid (ABA) levels in *Py. orbicularis* were four- to sevenfold higher than sensitive species. Using the ABA inhibitors sodium tungstate and ancymidol, ABA was found to regulate the activation of antioxidant enzyme activities during desiccation (AP, CAT, and PRX), concomitant with low lipid peroxidation and high cell viability. These results demonstrate the participation of ABA in the regulation of desiccation tolerance in seaweeds and suggest that regulatory mechanisms with ABA signaling could be of great importance for the adaptation of these organisms to environmental stress.

López-Cristoffanini et al. (2015) detected the proteins SOD and lactoylglutathione lyase, also known as glyoxalase I (GlyI), in *Py. orbicularis*. GlyI is key to detoxifying the methylglyoxal that may be overproduced during stressful conditions, including desiccation (Blomstedt et al. 1998; Hossain et al. 2009). A recent study by Fierro et al. (2016) assessed methylglyoxal production and GlyI activity in tolerant (*Py. orbicularis*) and sensitive (*L. spicata*) algae species, finding that *Py. orbicularis* increases GlyI activity and maintains reduced methylglyoxal levels during desiccation. In contrast, *L. spicata* GlyI activity does not sufficiently increase, resulting in increased methylglyoxal levels and cellular alterations.

In addition to antioxidant enzyme activity, resurrection plants produce several antioxidant compounds to reduce ROS levels, including anthocyanin and accessory pigments. Just as in resurrection plants, algae also produce these compounds during desiccation stress. An increase in phycobiliproteins such as phycoerythrin, phycocyanin, and allophycocyanin occurs in *Py. orbicularis* and *Gracilaria corticata* during desiccation stress, as evidenced using direct extraction and pigment profiling in acrylamide gels (Kumar et al. 2011; López-Cristoffanini et al. 2015). These proteins can both canalize light energy to diminish and neutralize the amount of ROS produced by light excess (Bhat and Madyastha 2001; Romay et al. 2003; Cano-Europa et al. 2010). Moreover, desiccation in *Py. orbicularis* results in an overproduction of the enzyme adenosine 5′-phosphosulfate kinase. This catalyzes the production of 3′-phosphoadenosine 5′-phosphosulfate (Kopriva and Koprivova 2004), which is involved in the biosynthesis of sulfated polysaccharides (McCandless and Craigie 1979), molecules that present high antioxidant potential (Tannin-Spitz et al. 2005; Rocha de Souza et al. 2007). Kumar et al. (2011) determined that desiccation-stressed *G. corticata* increase ascorbate production, an important antioxidant. This correlates with the findings of López-Cristoffanini et al. (2015), who showed increases in phosphomannomutase and GDP-D-mannose-3′,5′-epimerase, enzymes involved in ascorbate biosynthesis (Oesterhelt et al. 1996; Valpuesta and Botella 2004). Additionally, *Py. orbicularis* also overproduces the stress-inducible pyridoxine biosynthesis protein involved in the synthesis of pyridoxal 5′-phosphate, a form of vitamin B6 with potent antioxidant functions (Mittenhuber 2001; Raschke et al. 2011).

2.2.5 Perspectives

Mechanisms such as increased antioxidant enzyme activity coupled with an overproduction of antioxidant compounds further explain the dominance displayed by some algal species in the upper rocky intertidal zones. These mechanisms have also have been registered in resurrection plants, which evidence conserved mechanisms of tolerance to water deficit. Specifically, *Py. orbicularis* demonstrates a well-coordinated network of mechanisms that successfully scavenge ROS (Fig. 2.2),

Fig. 2.2 (continued) *OEE1* plastid oxygen-evolving enhancer 1, *P-Gal* GDP-L-galactose, *PC* phycocyanin beta subunit, *PE* phycoerythrin beta subunit, *PGK* phosphoglycerate kinase, *PMM* phosphomannomutase, *PPI* peptidylprolyl isomerase, *Rab11* rab GTPase family 11, *RNP* RNP domain-containing protein, *RNApol* RNA polymerase, *RPL35* ribosomal protein L35, *RubisCO* ribulose-1,5-bisphosphate carboxylase/oxygenase large subunit (chloroplast), *SAM* *S*-adenosylmethionine synthetase, *Sar1* small GTP-binding protein Sar1, *Sec 7* Sec 7 domain, *SLG* *S*-D-lactoylglutathione, *SOR* stress-inducible pyridoxine biosynthesis protein SOR, *TPI* triosephosphate isomerase, *TRR* thioredoxin reductase, *TRX* thioredoxin, *VDAC* voltage-dependent anion channel

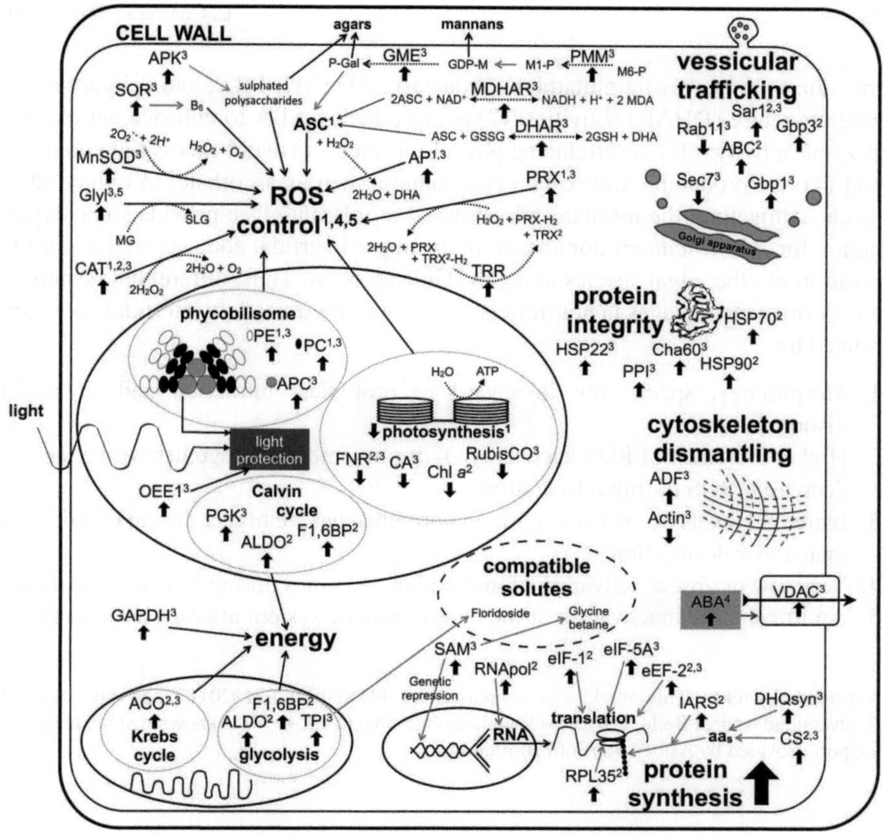

Fig. 2.2 Cell diagram of *Pyropia orbicularis* mechanisms involved in tolerance to desiccation stress induced during natural low tide periods. Adapted from López-Cristoffanini et al. (2015); numbers next to protein, enzyme, or molecule indicate evidence from [1]Contreras-Porcia et al. (2011) and Fierro et al. (2017), [2]Contreras-Porcia et al. (2013), [3]López-Cristoffanini et al. (2015), [4]Guajardo et al. (2016), and [5]Fierro et al. (2016). Processes and enzymes with decreased activities as well as downregulated proteins and transcripts during desiccation are followed by↓, whereas those with increased activity and upregulated proteins and transcripts are followed by ↑. Organelles within the cell are depicted by a circle with a continuous line, whereas cycles or structures within these are denoted by a circle with a dotted line. The direct action of an enzyme/protein/molecule on a mechanism is indicated with *black arrows*. An enzyme/protein/molecule is indicated with a gray arrow when part of an enzymatic pathway. Dotted arrows indicate the enzyme's mechanism. Proteins are abbreviated as follows: *ABA* abscisic acid, *ABC* ATP-binding cassette transporters, *ACO* aconitate hydratase, *ADF* cofilin/actin depolymerizing factor, *ALDO* fructose-bisphosphate aldolase, *AP* ascorbate peroxidase, *APC* allophycocyanin alpha subunit, *APK* adenosine 5′-phosphosulfate kinase, *ASC* ascorbate, *B6* vitamin B6, *CA* carbonic anhydrase, *CAT* catalase, *Cha60* chaperonin 60, *Chl a* chlorophyll a, *CS* cysteine synthase, *DHA* dehydroascorbate, *DHAR* dehydroascorbate reductase, *DHQsyn* 3-dehydroquinate synthase/*O*-methyltransferase fusion, *eEF-2* elongation factor 2, *eIF-1* eukaryotic initiation factor 1, *eIF-5A* translation initiation factor eIF5A, *F1,6BP* fructose 1,6-bisphosphatase, *FNR* ferredoxin-NADP+ reductase, *GAPDH* glyceraldehyde 3-phosphate dehydrogenase, *Gbp1* guanylate binding protein 1, *Gbp3* guanylate binding protein 3, *GDP-M* GDP-d-mannose, *GlyI* lactoylglutathione lyase, *GME* GDP-D-mannose-3′0.5′-epimerase, *GR* glutathione reductase, *GSH* glutathione, *GSSG* glutathione disulfide, *HSP22* heat shock protein 22, *HSP70* heat shock protein 70, *HSP90* heat shock protein 90, *IARS* isoleucyl-tRNA synthetase, *M1-P* mannose 1-phosphate, *M6-P* mannose 6-phosphate, *MDA* monodehydroascorbate, *MDHAR* monodehydroascorbate reductase, *MG* methylglyoxal, *MnSOD* manganese superoxide dismutase,

including (1) increasing glutathione reductase, AP, CAT, PRX, and dehydroascorbate reductase (DHAR) activities; (2) overproducing ABA to enhance antioxidant enzyme activity; (3) overproducing phycobiliproteins to avoid excess light damage; and (4) overproducing antioxidant compounds, such as ascorbate, to reduce ROS levels. Altogether, the mechanisms observed in tolerant algae provide some explanation for *Py. orbicularis* dominance in the upper intertidal zone, as well as for the position of other algal species along the Chilean coast. Thus, differential responses and permanent damages in non-tolerant species restricted to low intertidal zones are related to:

1. Morphology, specifically in regard to protoplast retraction and thylakoid disorganization
2. Higher increases of ROS such as H_2O_2 during emersion, together with a lack of decreased ROS during rehydration
3. Increased levels of oxidized lipids and proteins that are higher during rehydration than under desiccation
4. Very low or absent activation of antioxidant enzymes during desiccation stress
5. An irreversible inactivation of the photosynthetic system after desiccation stress

Acknowledgments This work was supported by FONDECYT 1120117 and DI-501-14/R (Universidad Andrés Bello, Proyectos Regulares Internos) to LC-P. We acknowledge the language support provided by Ashley VanCott, BioPub Ltd.

References

Abe S, Kurashima A, Yokohama Y, Tanaka K (2001) The cellular ability of desiccation tolerance in Japanese intertidal seaweeds. Bot Mar 44(4):125–131

Alpert P (2005) The limits and frontiers of desiccation-tolerant life. Integr Comp Biol 45:685–695

Asada K (1999) The water–water cycle in chloroplasts: scavenging of active oxygen and dissipation of excess photons. Annu Rev Plant Physiol 50:601–639

Baker CJ, Orlandi EW (1995) Active oxygen in plant pathogenesis. Annu Rev Phytopathol 33:299–321

Baniwal SK, Bharti K, Chan KY et al (2004) Heat stress response in plants: a complex game with chaperones and more than twenty heat stress transcription factors. J Biosci 29:471–487

Bewley JD (1979) Physiological aspects of desiccation tolerance. Annu Rev Plant Physiol 30:195–238

Bhat VB, Madyastha KM (2001) Scavenging of peroxynitrite by phycocyanin and phycocyanobilin from *Spirulina platensis*: protection against oxidative damage to DNA. Biochem Biophys Res Commun 285:262–266

Billard E, Serrão E, Pearson G, Destombe CH, Valero M (2010) *Fucus vesiculosus* and *spiralis* species complex: a nested model of local adaptation at the shore level. Mar Ecol Prog Ser 405:163–174

Bird CE, Franklin EC, Smith CM, Toonen RJ (2013) Between tide and wave marks: a unifying model of physical zonation on littoral shores. PeerJ 1:e154. doi:10.7717/peerj.154

Blomstedt CK, Gianello RD, Hamill JD, Neale AD, Gaff DF (1998) Drought-stimulated genes correlated with desiccation tolerance of the resurrection grass *Sporobolus stapfianus*. Plant Growth Regul 24:153–161

Blouin NA, Brodie JA, Grossman AC, Xu P, Brawley SH (2011) *Porphyra*: a marine crop shaped by stress. Trends Plant Sci 16:29–37

Brodie J, Bartsch I, Neefus C, Orfanidis S, Bray T, Mathieson AC (2007) New insights into the cryptic diversity of the North Atlantic-Mediterranean *Porphyra leucosticta* complex: *P. olivii* sp. nov. and *P. rosengurttii* (Bangiales, Rhodophyta). Eur J Phycol 42:3–28

Broom JE, Nelson WA, Farr TJ, Jones WA, Aguilar-Rosas R, Aguilar-Rosas LE (2002) A reassessment of the taxonomic status of *Porphyra suborbiculata*, *Porphyra carolinensis* and *Porphyra lilliputiana* (Bangiales, Rhodophyta) based on molecular and morphological data. Eur J Phycol 37:227–235

Broom JE, Nelson WA, Yarish C, Phillips LE, Clayton MN (2010) A molecular survey of the summer *Porphyra* (Bangiales, Rhodophyta) flora of the Falkland Islands using rbcL, and nSSU sequence data. Aust Syst Bot 23:27–37

Butterfield NJ (2000) *Bangiomorpha pubescens* n. gen., n. sp.: implications for the evolution of sex, multicellularity, and the Mesoproterozoic/Neoproterozoic radiation of eukaryotes. Paleobiology 26:386–404

Burritt DJ, Larkindale J, Hurd CL (2002) Antioxidant metabolism in the intertidal red seaweed *Stictosiphonia arbuscula* following desiccation. Planta 215:829–838

Cano-Europa E, Ortiz-Butrón R, Gallardo-Casas CA, Blas-Valdivia V, Pineda-Reynoso M, Olvera-Ramírez R, Franco-Colin M (2010) Phycobiliproteins from *Pseudanabaena tenuis* rich in c-phycoerythrin protect against $HgCl_2$-caused oxidative stress and cellular damage in the kidney. J Appl Phycol 22:495–501

Chapman ARO (1990) Competitive interactions between *Fucus spiralis* L and *Fucus vesiculosus* L (Fucales, Phaeophyta). Hydrobiologia 204:205–209

Chapman ARO (1995) Functional ecology of fucoid algae: twenty-three years of progress. Phycologia 34:1–32

Chaves MM, Flexas J, Pinheiro C (2009) Photosynthesis under drought and salt stress: regulation mechanisms from whole plant to cell. Ann Bot 103:551–560

Connell JH (1961) Effects of competition, predation by *Thais lapillus*, and other factors on natural populations of the barnacle *Balanus balanoides*. Ecol Monogr 31:61–104

Connell JH (1972) Community interactions on marine rocky intertidal shores. Annu Rev Ecol Syst 3:169–192

Contreras L, Moenne A, Correa JA (2005) Antioxidant responses in *Scytosiphon lomentaria* (Phaeophyceae) inhabiting copper-enriched coastal environments. J Phycol 41:1184–1195

Contreras L, Medina MH, Andrade S, Oppliger V, Correa JA (2007) Effects of copper on early developmental stages of *Lessonia nigrescens* Bory (Phaeophyceae). Environ Pollut 145:75–83

Contreras L, Mella D, Moenne A, Correa JA (2009) Differential responses to copper-induced oxidative stress in the marine macroalgae *Lessonia nigrescens* and *Scytosiphon lomentaria* (Phaeophyceae). Aquat Toxicol 94:94–102

Contreras-Porcia L, Thomas D, Flores V, Correa JA (2011) Tolerance to oxidative stress induced by desiccation in *Porphyra columbina* (Bangiales, Rhodophyta). J Exp Bot 62:1815–1829

Contreras-Porcia L, Callejas S, Thomas D, Sordet C, Pohnert G, Contreras A, Lafuente A, Flores-Molina MR, Correa JA (2012) Seaweeds early development: detrimental effects of desiccation and attenuation by algal extracts. Planta 235:337–348

Contreras-Porcia L, López-Cristoffanini C, Lovazzano C, Flores-Molina MR, Thomas D, Núñez A, Fierro C, Guajardo E, Correa JA, Kube M, Reinhardt R (2013) Differential gene expression in *Pyropia columbina* (Bangiales, Rhodophyta) under natural hydration and desiccation conditions. Lat Am J Aquat Res 41:933–958

Cruz de Carvalho R, Bernardes Da Silva A, Soares R, Almeida AM, Coelho AV, Marques da Silva J, Branquinho C (2014) Differential proteomics of dehydration and rehydration in bryophytes: evidence towards a common desiccation tolerance mechanism. Plant Cell Environ 37:1499–1515

Davison IR, Pearson GA (1996) Stress tolerance in intertidal seaweeds. J Phycol 32:197–211

Dietz KJ (2011) Peroxiredoxins in plants and cyanobacteria. Antioxid Redox Signal 15:1129–1159

Dinakar C, Djilianov D, Bartels D (2011) Photosynthesis in desiccation tolerant plants: energy metabolism and antioxidative stress defense. Plant Sci 182:29–41

Dring MJ, Brown FA (1982) Photosynthesis of intertidal brown algae during and after periods of emersion: a renewed search for physiological causes of zonation. Mar Ecol Prog Ser 8:301–308

Fierro C, López-Cristoffanini C, Latorre N, Rivas J, Contreras-Porcia L (2016) Methylglyoxal metabolism in seaweeds during desiccation. Rev Biol Mar Oceanogr 51:187–191

Fierro C, López-Cristoffanini C, Meynard A, Lovazzano C, Castañeda F, Guajardo E, Contreras-Porcia L (2017) Expression profile of desiccation tolerance factors in intertidal seaweed species during the natural tidal cycle. Planta 245(6):1149–1164

Flores-Molina MR, Thomas D, Lovazzano C, Núñez A, Zapata J, Kumar M, Correa JA, Contreras-Porcia L (2014) Desiccation stress in intertidal seaweeds: effects on morphology, photosynthetic performance and antioxidant responses. Aquat Bot 113:90–99

Flores-Molina MR, Rautenberger R, Muñoz P, Huovinen P, Gomez I (2016) Stress tolerance of the endemic antarctic brown alga *Desmarestia anceps* to UV radiation and temperature is mediated by high concentrations of phlorotannins. Photochem Photobiol 92:455–466

Foyer CH, Noctor G (2009) Redox regulation in photosynthetic organisms: signaling, acclimation, and practical implications. Antiox Redox Signal 11:861–905

Gasulla F, Jain R, Barreno E, Guéra A, Balbuena TS, Thelen JJ, Oliver MJ (2013) The response of *Asterochloris erici* (Ahmadjian) Skaloud et Peksa to desiccation: a proteomic approach. Plant Cell Environ 36:1363–1378

Gómez I, López-Figueroa F, Ulloa N, Morales V, Lovengreen C, Huovinen P, Hess S (2004) Patterns of photosynthesis in 18 species of intertidal macroalgae from southern Chile. Mar Ecol Prog Ser 270:103–116

Gómez I, Huovinen P (2015) Lack of physiological depth patterns in conspecifics of endemic Antarctic brown algae: a trade-off between UV stress tolerance and shade adaption? PLoS One 10(8):e0134440

Guajardo E, Correa JA, Contreras-Porcia L (2016) Role of abscisic acid (ABA) in activating antioxidant tolerance responses to desiccation stress in intertidal seaweed species. Planta 243:1–15

Guillemin ML, Contreras-Porcia L, Ramírez ME, Macaya EC, Bulboa Contador C, Woods H, Wyatt C, Brodie J (2016) The bladed Bangiales (Rhodophyta) of the South Eastern Pacific: molecular species delimitation reveals extensive diversity. Mol Phylogenet Evol 94:814–826

Hay ME (2009) Marine chemical ecology: chemical signals and cues structure marine populations, communities, and ecosystems. Annu Rev Mar Sci 1:193–212

Hoekstra FA, Golovina EA, Buitink J (2001) Mechanisms of plant desiccation tolerance. Trends Plant Sci 6:431–438

Hohmann S (2004) Plant responses to abiotic stress, 1st edn. Springer-Verlag, Berlin, Heidelberg

Holzinger A, Lütz C, Karsten U (2011) Desiccation stress causes structural and ultrastructural alterations in the aeroterrestrial green alga *Klebsormidium crenulatum* (Klebsormidiophyceae, Streptophyta) isolated from an alpine soil crust. J Phycol 47:591–602

Holzinger A, Karsten U (2013) Desiccation stress and tolerance in green algae: consequences for ultrastructure, physiological and molecular mechanisms. Front Plant Sci 4:327

Hossain MA, Hossain MZ, Fujita M (2009) Stress-induced changes of methylglyoxal level and glyoxalase I activity in pumpkin seedlings and cDNA cloning of glyoxalase I gene. Aust J Crop Sci 3:53–64

Huey RB, Carlson M, Crozier L, Frazier M, Hamilton H, Harley C, Hoang A, Kingsolver JG (2002) Plants versus animals: do they deal with stress in different ways? Integr Comp Biol 42:415–423

Hurd CL, Harrison PJ, Bischof K, Lobban SC (eds) (2014) Seaweed ecology and physiology, 2nd edn. Cambridge University Press, Cambridge, UK

Ingram J, Bartels D (1996) The molecular basis of dehydration tolerance in plants. Annu Rev Plant Physiol Plant Mol Biol 47:377–403

Kim KY, Garbary DJ (2007) Photosynthesis in *Codium fragile* (Chlorophyta) from a Nova Scotia estuary: responses to desiccation and hyposalinity. Mar Biol 151:99–107

Kopriva S, Koprivova A (2004) Plant adenosine 5′–phosphosulphate reductase: the past, the present, and the future. J Exp Bot 55:1775–1183

Kumar M, Kumari P, Gupta V, Reddy CRK, Jha B (2010) Biochemical responses of red alga *Gracilaria corticata* (Gracilariales, Rhodophyta) to salinity induced oxidative stress. J Exp Mar Biol Ecol 391:27–34

Kumar M, Gupta V, Trivedi N, Kumari P, Bijo AJ, Reddy CRK, Jha B (2011) Desiccation induced oxidative stress and its biochemical responses in intertidal red alga *Gracilaria corticata* (Gracilariales, Rhodophyta). Environ Exp Bot 72:194–201

Lee MY, Shin HW (2003) Cadmium-induced changes in antioxidant enzymes from the marine alga *Nannochloropsis oculata*. J Appl Phycol 15:13–19

Leprince O, Buitink J (2010) Desiccation tolerance: from genomics to the field. Plant Sci 179:554–564

Leuschner C, Landwehr S, Mehlig U (1998) Limitation of carbon assimilation of intertidal *Zostera noltii* and *Z. marina* by desiccation at low tide. Aquat Bot 62:171–176

Lipkin Y, Beer S, Eshel A (1993) The ability of *Porphyra linearis* (Rhodophyta) to tolerate prolonged periods of desiccation. Bot Mar 36(6):517–524

Liu W, Au DWT, Anderson DM, Lam PKS, Wu RSS (2007) Effects of nutrients, salinity, pH and light: dark cycle on the production of reactive oxygen species in the alga *Chattonella marina*. J Exp Mar Biol Ecol 346:76–86

López-Cristoffanini C, Tellier F, Otaíza R, Correa JA, Contreras-Porcia L (2013) Tolerance to air exposure: a feature driving the latitudinal distribution of two sibling kelp species. Bot Mar 56:431–440

López-Cristoffanini C, Zapata J, Gaillard F, Potin P, Correa JA, Contreras-Porcia L (2015) Identification of proteins involved in the tolerance responses to desiccation stress in the red seaweed *Pyropia orbicularis* (Rhodophyta. Bangiales). Proteomics 15:3954–3968

Maciver SK, Hussey PJ (2002) The ADF/cofilin family: actin-remodeling proteins. Genome Biol 3:1–12

McCandless EL, Craigie JS (1979) Sulfated polysaccharides in red and brown algae. Annu Rev Plant Physiol 30:41–53

Mislan KAS, Wethey DS, Helmuth B (2009) When to worry about the weather: role of tidal cycle in determining patterns of risk in intertidal ecosystems. Glob Change Biol 15:3056–3065

Mittenhuber G (2001) Phylogenetic analyses and comparative genomics of vitamin B6 (pyridoxine) and pyridoxal phosphate biosynthesis pathways. J Mol Microbiol Biotechnol 3:1–20

Moore JP, Farrant JM (2015) Editorial: current advances and challenges in understanding plant desiccation tolerance. Front Plant Sci 6:1–3

Niwa K, Iida S, Kato A, Kawai H, Kikuchi N, Kobiyama A, Aruga Y (2009) Genetic diversity and introgression in two cultivated species (*Porphyra yezoensis* and *Porphyra tenera*) and closely related wild species of *Porphyra* (Bangiales, Rhodophyta). J Phycol 45(2):493–502

Neefus CD, Mathieson AC, Bray TL (2008) The distribution, morphology, and ecology of three introduced Asiatic species of *Porphyra* (Bangiales, Rhodophyta) in the Northwestern Atlantic. J Phycol 44:1399–1414

Nelson WA (2013) *Pyropia plicata* sp. nov. (Bangiales, Rhodophyta): naming a common intertidal alga from New Zealand. PhytoKeys 21:17–28

Noctor G, Foyer CH (1998) Ascorbate and glutathione: keeping active oxygen under control. Annu Rev Plant Physiol 49:249–279

Oesterhelt C, Schnarrenberger C, Gross W (1996) Phosphomannomutase and phosphoglucomutase in the red alga *Galdieria sulphuraria*. Plant Sci 121:19–27

Oliver MJ, Bewley JD (1997) Desiccation tolerance of plant tissues: a mechanistic overview. In: Janick J (ed) Horticultural reviews, vol 18, 1st edn. Wiley, New York, pp 171–213

Oliver MJ, Wood AJ, Mahony PO (1998) "To dryness and beyond "—preparation for the dried state and rehydration in vegetative desiccation-tolerant plants. Plant Growth Regul 24(3):193–201

Oliver MJ, Jain R, Balbuena TS, Agrawal G, Gasulla F, Thelen JJ (2011) Proteome analysis of leaves of the desiccation-tolerant grass, *Sporobolus stapfianus*, in response to dehydration. Phytochemistry 72:1273–1284

Paine RT (1974) Intertidal community structure. Experimental studies in the relation between a dominant competitor and its principle predator. Oecologia 15:93–120

Pearson GA, Kautsky L, Serrao E (2000) Recent evolution in Baltic *Fucus vesiculosus*: reduced tolerance to emersion stresses compared to intertidal (North Sea) populations. Mar Ecol Prog Ser 202:67–79

Pressel S, Ligrone R, Duckett JG (2006) Effects of de- and rehydration on food-conducting cells in the moss *Polytrichum formosum*: a cytological study. Ann Bot 98:67–76

Ramanjulu S, Bartels D (2002) Drought- and desiccation-induced modulation of gene. Plant Cell Environ 25:141–151

Ramírez ME, Contreras-Porcia L, Guillemin ML, Brodie J, Valdivia C, Flores-Molina MR, Nuñez A, Bulboa Contador C, Lovazzano C (2014) *Pyropia orbicularis* sp. nov. (Rhodophyta, Bangiaceae) based on a population previously known as *Porphyra columbina* from the central coast of Chile. Phytotaxa 158(2):133–153

Raschke M, Boycheva S, Crèvecoeur M, Nunes-Nesi A, Witt S, Fernie AR, Amrhein N, Fitzpatrick TB (2011) Enhanced levels of vitamin B (6) increase aerial organ size and positively affect stress tolerance in *Arabidopsis*. Plant J 66:414–432

Rijstenbil JW (2001) Effects of periodic, low UVA radiation on cell characteristics and oxidative stress in the marine planktonic diatom *Ditylum brightwellii*. Eur J Phycol 36:1–8

Rocha de Souza MC, Marques CT, Guerra Dore CM, Ferreira da Silva FR, Oliveira Rocha HA, Leite EL (2007) Antioxidant activities of sulfated polysaccharides from brown and red seaweeds. J Appl Phycol 19:153–160

Rouhier N, Jacquot JP (2002) Plant peroxiredoxins: alternative hydroperoxide scavenging enzymes. Photosynth Res 74:259–268

Romay C, González R, Ledón N, Remirez D, Rimbau V (2003) C-phycocyanin: a biliprotein with antioxidant, anti-inflammatory and neuroprotective effects. Curr Protein Pept Sci 4:207–216

Rugg DA, Norton TA (1987) *Pelvetia canaliculata*, a seaweed that shuns the sea. In: Crawford RMM (ed) Plant life in aquatic and amphibious environments. Blackwells, Oxford, pp 347–358

Scott P (2000) Resurrection plants and the secrets of eternal leaf. Ann Bot 85:159–166

Shelford VE (1914) A comparison of the responses of sessile and motile plants and animals. Am Soc Nat 48:641–674

Schonbeck M, Norton TA (1978) Factors controlling the upper limits of fucoid algae on the shore. J Exp Mar Biol Ecol 31:303–313

Schonbeck MW, Norton TA (1980) The effects on intertidal fucoid algae of exposure to air under various conditions. Bot Mar 23(3):141–148

Smith CM, Satoh K, Fork DC (1986) The effects of osmotic tissue dehydration and air drying on morphology and energy transfer in two species of *Porphyra*. Plant Physiol 80:843–847

Stephenson TA, Stephenson A (1949) The universal features of zonation between tide-marks on rocky coasts. J Ecol 2:289–305

Sutherland JE, Lindstrom S, Nelson W, Brodie J, Lynch M, Hwang MS, Choi HG, Miyata M, Kikuchi N, Oliveira M, Farr T, Neefus C, Mortensen A, Milstein D, Müller K (2011) A new look at an ancient order: generic revision of the Bangiales. J Phycol 47:1131–1151

Swanson AK, Druehl LD (2002) Induction, exudation and the UV protective role of kelp phlorotannins. Aquat Bot 73:241–253

Tannin-Spitz T, Bergman M, van-Moppes D, Grossman S, Arad S (Malis) (2005) Antioxidant activity of the polysaccharide of the red microalga *Porphyridium* sp. J Appl Phycol 17:215–222

Timperio AM, Egidi MG, Zolla L (2008) Proteomics applied on plant abiotic stresses: role of heat shock proteins (HSP). J Proteome 71:391–411

Toldi O, Tuba Z, Scott P (2009) Vegetative desiccation tolerance: is it a goldmine for bioengineering crops? Plant Sci 176:187–199

Tripathi BN, Bhatt I, Dietz KJ (2009) Peroxiredoxins: a less studied component of hydrogen peroxide detoxification in photosynthetic organisms. Protoplasma 235:3–15

Valpuesta V, Botella MA (2004) Biosynthesis of L-ascorbic acid in plants: new pathways for an old antioxidant. Trends Plant Sci 9:573–577

Wang T, Jónsdóttir R, Ólafsdóttir G (2009) Total phenolic compounds, radical scavenging and metal chelation of extracts from Icelandic seaweeds. Food Chem 116:240–248

Wang L, Mao Y, Kong F, Li G, Ma F, Zhang B, Sun P, Bi G, Zhang F, Xue H, Cao M (2013) Complete sequence and analysis of plastid genomes of two economically important red algae: *Pyropia haitanensis* and *Pyropia yezoensis*. PLoS One 8(5):e65902

Watt CA, Scrosati RA (2013) Bioengineer effects on understory species richness, diversity, and composition change along an environmental stress gradient: experimental and mensurative evidence. Estuar Coast Shelf Sci 123:10–18

West AL, Mathieson AC, Klein AS, Neefus CD, Bray TL (2005) Molecular ecological studies of New England species of *Porphyra* (Rhodophyta, Bangiales). Nova Hedwigia 80:1–24

Wiltens J, Schreiber U, Vidaver W (1978) Chlorophyll fluorescence induction: an indicator of photosynthetic activity in marine algae undergoing desiccation. Can J Bot 56:2787–2794

Wright JP, Jones CG, Boeken B, Shachak M (2006) Predictability of ecosystem engineering effects on species richness across environmental variability and spatial scales. J Ecol 94:815–824

Chapter 3
Marine Metal Pollution and Effects on Seaweed Species

Loretto Contreras-Porcia, Andrés Meynard, Camilo López-Cristoffanini, Nicolas Latorre, and Manoj Kumar

Abstract Heavy metals are significant pollutants continuously released into the biosphere, both naturally and anthropogenically. Conceptually, metal speciation, bioavailability, and associated toxicity in marine organisms depend on a wide array of abiotic and biotic factors. Among these, pH variation is one of the most important environmental factors influencing metal speciation and toxicity. Due to this, ocean acidification is expected to modify metal speciation, altering the effects these nondegradable contaminants have on marine organisms, such as seaweeds. One clear effect of heavy metals on seaweeds is the rapid formation of reactive oxygen species (ROS). The production of ROS beyond the physiological tolerance threshold causes an oxidative stress condition that, if not attenuated, can irreversibly damage cellular

L. Contreras-Porcia (✉)
Departamento de Ecología y Biodiversidad, Facultad de Ecología y Recursos Naturales, Universidad Andres Bello, República 440, Santiago, Chile

Center of Applied Ecology and Sustainability (CAPES), Pontificia Universidad Católica de Chile, Avda. Libertador Bernardo O'Higgins 340, Santiago, Chile

Centro de Investigación Marina Quintay (CIMARQ), Facultad de Ecologíay Recursos Naturales, Universidad Andres Bello, Quintay, Chile
e-mail: lorettocontreras@unab.cl

A. Meynard
Departamento de Ecología y Biodiversidad, Facultad de Ecología y Recursos Naturales, Universidad Andres Bello, República 440, Santiago, Chile

Centro de Investigación Marina Quintay (CIMARQ), Facultad de Ecologíay Recursos Naturales, Universidad Andres Bello, Quintay, Chile

C. López-Cristoffanini
Departament de Biologia Evolutiva, Ecologia i Ciencies Ambientals, Secció de Fisiologia Vegetal, Universitat de Barcelona, 08028 Barcelona, Spain

N. Latorre
Programa de Doctorado en Medicina de la Conservación, Facultad de Ecología y Recursos Naturales, Universidad Andres Bello, República 440, Santiago, Chile

M. Kumar
Climate Change Cluster (C3), Faculty of Science, University of Technology Sydney (UTS), Sydney, NSW 2007, Australia

© Springer International Publishing AG 2017

M. Kumar, P. Ralph (eds.), *Systems Biology of Marine Ecosystems*,
DOI 10.1007/978-3-319-62094-7_3

constituents such as DNA/RNA, proteins, and lipids. To cope with heavy metal excess, several mechanisms exist in tolerant seaweed species, including the activation of an efficient ROS-scavenging system constituted by metal-binding compounds, antioxidant enzymes, and oxygenated polyunsaturated fatty acids, among others. Another adaptive mechanism involves the participation of ATP-binding cassette (ABC) transporter proteins in translocating a wide variety of compounds across cell membranes, including heavy metals. In contrast, an excessive heavy metal presence can inhibit photosynthesis, reduce pigment concentration and growth, induce cation losses, and disrupt gametophyte development in non-tolerant seaweed species. In a scenario of lowered ocean pH and increased heavy metal toxicity, the important roles played by non-tolerant seaweed species in structuring communities could be severely compromised, with unknown consequences for associated organisms. Therefore, in the upcoming decades, marine pollution could majorly shift and rearrange community compositions and the distributional ranges of species.

Keywords Seaweeds • Heavy metal stress • Tolerance mechanisms • Ocean acidification

Contents

3.1 Heavy Metal Toxicity in Marine Ecosystems 36
3.2 Tolerance Mechanisms in Seaweeds to Heavy Metal Toxicity 38
3.3 Influence of Abiotic Factors on Metal Toxicity: The Case of the Ocean Acidification 40
3.4 Seaweeds as a Study Model for Heavy Metal Toxicity and Ocean Acidification 42
3.5 Conclusion 44
References 44

3.1 Heavy Metal Toxicity in Marine Ecosystems

Heavy metals are significant toxic pollutants, and extensive literature details the accumulation of heavy metals in coastal marine ecosystems (Walker et al. 2012). Metals are continuously released into the biosphere by volcanoes and the natural weathering of rocks, in addition to release through numerous anthropogenic activities such as mining, fuel combustion, and industrial, urban, and agricultural activities.

The input of metals into the sea, natural or anthropogenic in origin, is mainly via the atmosphere, whereas the anthropogenic contribution of some metals (e.g., copper, cadmium, and lead) is greater than from natural sources (Duce et al. 1991). Metal deposition occurs primarily through a gas exchange at the sea surface, through either particle fallout (dry deposition) or through dragging from the air by rain (wet deposition) (Walker et al. 2012).

Unlike organic chemicals, metals are neither created nor destroyed by biological or chemical processes. However, these processes can transform metals from one species to another (valence states, metal speciation) and can switch metals back and forth from inorganic to organic forms. Although metals such as copper (Cu) and zinc (Zn) are essential to life, others such as lead (Pb), cadmium (Cd), and mercury (Hg) do not have known useful biochemical functions (Allan 1997). Additionally, excess intake of any of these metals can be highly toxic when exceeding certain concentration thresholds. It is important to mention that although Cd is not essential, and is potentially toxic even in very low doses, it enhances phytoplankton photosynthesis and growth rate by increasing carbonic anhydrase activity in cells (Cullen et al. 1999).

Seaweeds are fundamental components of coastal benthic ecosystems and are responsible for much of the coastal primary production, community structure, and ecosystem function (Hurd et al. 2014; Schiel and Foster 2015). Natural or anthropogenic phenomena, such as heavy metal deposition, can adversely affect macroalgae, which in turn can directly or indirectly affect organisms at higher trophic levels and, ultimately, the integrity of entire ecosystems. For example, the deposition of copper into Chañaral Bay (26°S, 71°W) in northern Chile through the transport of mining wastes to the coast, occurring since 1938, negatively influenced algal biodiversity and eliminated several benthic herbivores and all benthic carnivores (Correa et al. 2000; Medina et al. 2005). This waste deposition has resulted in persistently high levels of total copper in the seawater, with 20 μg L^{-1} currently and 200 μg L^{-1} at the discharge point in the past (Castilla 1996; Henríquez-Castillo et al. 2015).

Particularly, the brown kelp *Lessonia berteroana* (Phaeophyceae) (formerly *L. nigrescens*), a keystone species dominating the lower intertidal rocky shores along the northern Chilean coast, is absent from this copper-enriched area (Medina et al. 2005; Lovazzano et al. 2013). Since *Lessonia* spp. regulate community structure (Cancino and Santelices 1984), the absence of *L. berteroana* has negatively impacted local biodiversity. It is important to emphasize that the high heavy metal concentration in seawater is not restricted to northern Chile. According to POAL (Coastal Monitoring Program, Chilean Navy DIRECTEMAR: www.directemar.cl), over the last 5 years, certain central coastal zones of Chile have registered close to 14 μg L^{-1} of total dissolved Cu, 5.5 μg L^{-1} of Cd, and 7 μg L^{-1} of Pb in the seawater, levels considered toxic for certain marine organisms by the Environmental Protection Agency (EPA) of the United States and as low-quality seawater by Chilean standards.

The excess presence of heavy metals can inhibit photosynthesis (Nielsen et al. 2003; Xia et al. 2004), reduce pigment concentration and growth (Contreras et al. 2007), and induce a loss of cations (Overnell 1975; Brown and Newman 2003). Particularly, copper and cadmium can disrupt gametophyte development in the brown kelp *Macrocystis pyrifera* (Phaeophyceae) and *Lessonia spicata* (Anderson and Hunt 1988; Garman et al. 1994; Contreras et al. 2007). These heavy metals also affect the distribution of other cellular compounds, such as free fatty acids (Ritter et al. 2008), as well as competitively inhibiting other metals (Franklin et al. 2002; Andrade et al. 2006). Therefore, it is clear that attenuation responses in seaweeds are necessary for controlling the adverse and environmentally stressful conditions that result from heavy metal over-enrichment.

3.2 Tolerance Mechanisms in Seaweeds to Heavy Metal Toxicity

Seaweeds accumulate metals through a two-stage process that begins with a rapid and reversible physicochemical adsorption on the algal surface, followed by a slower, metabolically arranged intracellular uptake (Garnham et al. 1992). Therefore, heavy metal concentration is generally dependent both on external factors (pH, salinity, and complex inorganic and organic molecules) and on physicochemical parameters that control metabolic rate (temperature, light, oxygen, and nutrients).

One clear effect of heavy metals on marine organisms, including seaweeds, is the rapid formation of reactive oxygen species (ROS) through the Haber-Weiss reaction, which is characterized by a heavy metal-catalyzed production of hydroxyl radicals from hydrogen peroxide (Gledhill et al. 1997; Pinto et al. 2003) (Fig. 3.1). ROS are directly produced by O_2 excitation and the subsequent formation of singlet oxygen, or by the transfer of one, two, or three electrons to O_2, which results in the formation of superoxide radicals, hydrogen peroxide, or hydroxyl radicals, respectively. The production of ROS beyond the physiological tolerance range of an organism can negatively affect physiological maintenance due to oxidative damage to cellular constituents such as DNA/RNA, proteins, and lipids (Vranová et al. 2002; Hung et al. 2005; Contreras et al. 2009; Lovazzano et al. 2013) (Fig. 3.1). This overproduction can lead to a state of oxidative stress, but a coordinated attenuation system can be activated by the affected organism to eliminate ROS excess (Collén and Davison 1999; Sordet et al. 2014). To cope with heavy metal excess, several mechanisms exist in tolerant seaweed species such as *Scytosiphon* spp. (Phaeophyceae) or *Ulva* spp. (Ulvophyceae), including the activation of an efficient ROS-scavenging system constituted by compounds and enzymes, including catalase (CAT), ascorbate peroxidase (AP), peroxiredoxin (PRX), and lipoxygenase, among others (Ratkevicius et al. 2003; Contreras et al. 2005; Contreras et al. 2010; Lovazzano et al. 2013) (Fig. 3.1). In contrast, species sensitive to heavy metal excess have low activities of tolerance response enzymes, which would explain the inability to flourish in heavy metal-enriched environments. For example, activities of diverse antioxidant enzymes are lower in the sensitive species *L. spicata* than in the tolerant *Scytosiphon lomentaria*, and others as glutathione peroxidase (GP) and dehydroascorbate reductase (DHAR) are completely inhibited at higher copper concentrations (Contreras et al. 2009). Copper toxicity also induces uncontrolled lipoperoxide accumulation in *L. spicata*, which leads to cell damage and dysfunction.

Using proteomic analysis, a PRX was identified in *Scytosiphon gracilis* exposed to copper excess (Contreras et al. 2010). PRXs belong to the thiol-dependent peroxidase family and are known to attenuate oxidative stress; reduce hydrogen peroxide, alkyl hydroperoxides, and peroxynitrite; and modulate redox-dependent signaling cascades (Dayer et al. 2008; Foyer and Noctor 2009; Tripathi et al. 2009; Dietz 2011). In algae, the protective function of PRXs is scarcely documented. For example, the *prx* gene, which presents homology with the 2-Cys PRX of higher plants (Baier and Dietz 1997), was first identified in *Porphyra purpurea* (Bangiophyceae) (Reith and Munholland 1993). By studying the responses to desiccation stress in

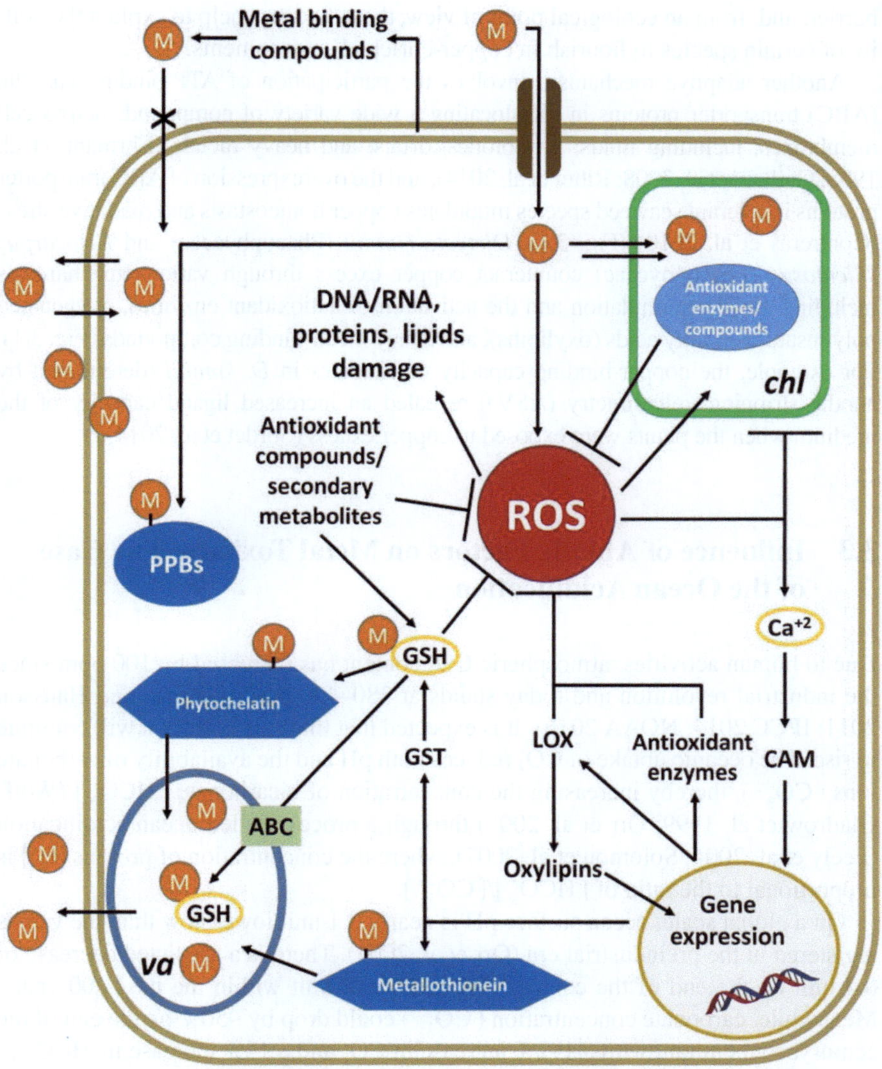

Fig. 3.1 Mechanisms involved in the tolerance to heavy metal stress in seaweeds. ABC transporter (ABC); calmodulin (CAM); chloroplast (*chl*); glutathione (GSH); glutathione S-transferase (GST); heavy metals like Cd, Cu, Pb, and others (M); polyphosphate bodies (PPBs); reactive oxygen species (ROS); vacuole (*va*). Some enzymes involved in these mechanisms are catalase (CAT), ascorbate peroxidase (AP), peroxiredoxin (PRX), and lipoxygenase (LOX), among others

Pyropia orbicularis (Bangiophyceae), Contreras-Porcia et al. (2011a) found that PRX activity and gene expression significantly increased. Similarly, when the green alga *U. compressa* was exposed to copper excess for 7 days, genes encoding for PRX and its reducing agent, thioredoxin, significantly increased (Contreras-Porcia et al. 2011b), and in *S. gracilis* PRX protein levels increased under copper excess (Lovazzano et al. 2013). These results suggest that PRX acts as an active antioxidant barrier, and, from an ecological point of view, these findings help to explain the ability of certain species to flourish in copper-enriched environments.

Another adaptive mechanism involves the participation of ATP-binding cassette (ABC) transporter proteins in translocating a wide variety of compounds across cell membranes, including lipids, xenobiotics, drugs, and heavy metals (Ehrmann et al. 1998; Gaillard et al. 2008; Ritter et al. 2014), and the overexpression of ABC transporter proteins in tolerant seaweed species modulates copper homeostasis and oxidative stress (Contreras et al. 2010) (Fig. 3.1). *Dictyota kunthii* (Phaeophyceae) and *Ectocarpus siliculosus* (Phaeophyceae) counteract copper excess through various mechanisms including metal accumulation and the activation of antioxidant enzymes, oxygenated polyunsaturated fatty acids (oxylipins), and heavy metal-binding compounds (Fig. 3.1). For example, the copper-binding capacity of exudates in *D. kunthii* (determined by anodic stripping voltammetry (ASV)) revealed an increased ligand capacity of the medium when the plants were exposed to copper excess (Sordet et al. 2014).

3.3 Influence of Abiotic Factors on Metal Toxicity: The Case of the Ocean Acidification

Due to human activities, atmospheric CO_2 content has increased by 100 ppm since the industrial revolution and today stands at 380–400 ppm (Gattuso and Hansson 2011; IPCC 2014; NOAA 2016). It is expected that this concentration will continue to rise. The oceanic uptake of CO_2 reduces both pH and the availability of carbonate ions (CO_3^{2-}), thereby increasing the concentration of bicarbonate [HCO_3^-] (Wolf-Gladrow et al. 1999; Orr et al. 2005) through a process called ocean acidification (Feely et al. 2004; Solomon et al. 2007), where the concentration of protons [H^+] is proportional to the ratio of [HCO_3^-]/[CO_3^{2-}].

On a global scale, ocean surface pH is nearly 0.1 unit lower now than the values registered in the preindustrial era (Orr et al. 2005). There is a predicted decrease of 0.4 unit by the end of the century and nearly 0.8 unit within the next 300 years. Meanwhile, carbonate concentration (CO_3^{2-}) could drop by ~50% by the end of the century, concomitant with a 192% increase in CO_2 and a 14% increase in HCO_3^- .

Metal speciation, or the form of the metal in terms of chemical species, influences its bioaccessibility, bioavailability fate, and toxic effects. Metal bioavailability and associated toxicity vary widely according to the physical, chemical, and biological conditions to which an organism is exposed; and sensitivity to metals varies with age, sex, nutritional status, and genetic polymorphisms. Due to this, the toxic action of metals on one particular organism conceptually depends on a wide array of abiotic and biotic factors (Fig. 3.2). In this context and because pH variation is probably one of the most important environmental factors influencing metal speciation and behavior (Byrne 1988), ocean acidification is expected to modify metal speciation, thereby altering the effects that these nondegradable contaminants have on marine organisms, including seaweeds.

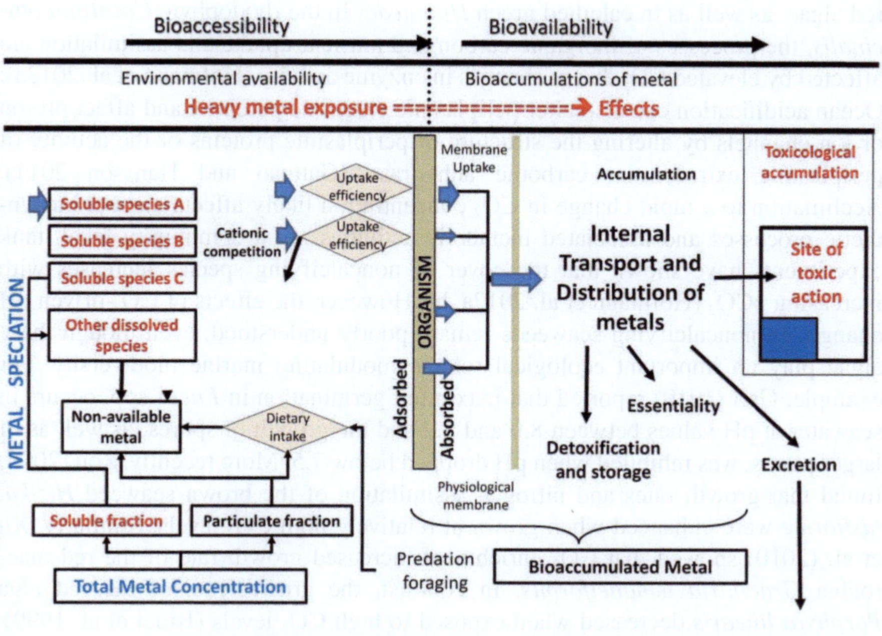

Fig. 3.2 Conceptual diagram for uptake mechanisms and speciation of metals in marine organisms. Modified from US EPA (2007)

Metal complexes with sulfate, fluoride, chloride, and phosphate are the most stable and, importantly, present at pH levels below 7, whereas metal carbonate (CO_3^{2-}) and hydroxide (OH^-) complexes are notably more unstable and present at pH levels above pH 6–8 (Millero et al. 2009). Metals that form complexes with CO_3^{2-} and OH^- have a higher fraction in their free form at a lower pH. This directly affects metal speciation and, thus, metal bioavailability and toxic effects. For example, at the current pH of seawater (ca. 8.1), Fe(III) is at its minimum solubility, and solubility is further influenced by organic matter (Liu and Millero 2002). A decrease in pH would increase Fe(III) solubility by ~40%, which could have a large impact on biogeochemical cycles. Increased solubility would make iron more available to marine autotrophs, leading to an increased primary production (Martin 1990). Likewise, copper is an important nutrient for all organisms, and its bioavailability influences several cellular functions. A low pH decreases the strength of natural organic matter/Cu^{2+} interactions, driving affinities to negligible levels at a lower pH. Thus, the proportion of non-complexed free Cu^{2+} ranges from less than 0.1% above pH 8 or higher to 30% at pH 7.0 (Zirino and Yamamoto 1972; Millero et al. 2009). Like copper, cadmium is strongly associated with chloride ions, but at a low pH, metal carbonate complexes increase, therefore reducing the bioavailability of carbonate and bicarbonate molecules.

Ocean acidification is consistently related to reduced growth rates in calcareous algae (e.g., planktonic coccolithophores and benthic calcifying macroalgae), and reduced calcification rates have been recorded in crustose and articulated coralline red algae, as well as in calcified green *Halimeda*. In the rhodophyte *Corallina officinalis*, the processes of inorganic carbon and nutrient uptake and assimilation are affected by elevated CO_2 due to changes in enzyme activity (Hofmann et al. 2012a). Ocean acidification can also alter periplasmic electrode potential and affect proton or ion channels by altering the structure of periplasmic proteins or the activity of periplasmic extracellular carbonic anhydrase (Gattuso and Hansson 2011). Acclimation to a rapid change in CO_2 concentration likely affects most photosynthetic processes and associated metabolic activities. At a community level, tank experiments have shown that the cover of noncalcifying species increases with increasing pCO_2 (Hofmann et al. 2012a, b). However, the effects of CO_2-driven pH changes in noncalcifying seaweeds remain poorly understood, even though these algae play an important ecological role in modulating marine biodiversity. For example, Gail (1919) reported that maximum germination in *Fucus* spp. occurs in seawater at pH values between 8.0 and 8.2 and the growth of spores, as well as of larger plants, was inhibited when pH dropped below 7.5. More recently, Zou (2005) found that growth rates and nitrogen assimilation of the brown seaweed *Hizikia fusiforme* were enhanced when grown at relatively high CO_2 levels. Similarly, Xu et al. (2010) showed that CO_2 enrichment increased growth rate of the red macroalga *Gracilaria lemaneiformis*. In contrast, the growth rate of the red alga *Porphyra linearis* decreased when exposed to high CO_2 levels (Israel et al. 1999). Zou and Gao (2002) found that some intertidal macroalgae increased the photosynthetic fixation of carbon at high CO_2 levels. At a biochemical level, the cellular composition of the red alga *Porphyra leucosticta* was significantly affected by increased pCO_2; while soluble proteins decreased, carbohydrates increased threefold (Mercado et al. 1999). García-Sánchez et al. (1994) reported similar results while studying the red alga *Gracilaria tenuistipitata*, and experiments conducted on *Ulva rigida* showed that under CO_2 enrichment, this alga has enhanced growth and net photosynthesis rates but decreased soluble protein and internal carbon contents, thus maintaining a constant C:N ratio (Gordillo et al. 2001). Recently, Duarte et al. (2016) registered a decrease in protein and organic matter contents in the tissues of *Durvillaea antarctica* (Phaeophyceae) exposed to high pCO_2 levels. Altogether, the published evidence suggests that algal responses to elevated CO_2 levels are species dependent and the ways in which CO_2 enrichment affects the physiology of a species are poorly understood.

3.4 Seaweeds as a Study Model for Heavy Metal Toxicity and Ocean Acidification

Multiple environmental factors are predicted to shift throughout the twenty-first century, and there is an urgent need to examine the interactive effect of these stressors to estimate potential impacts on the marine environment. Heavy metal enrichment and ocean acidification may have a particularly wide range of impacts on habitat-forming seaweed species such as *L. spicata* (Phaeophyceae), *M. pyrifera* (Phaeophyceae), and

P. orbicularis (Bangiophyceae). In Chile, these species promote kelp fisheries and mariculture, but some are sensitive to environmental stressors such as heavy metals, temperature, and desiccation (Contreras et al. 2007; Oppliger et al. 2012; Flores-Molina et al. 2014; Schiel and Foster 2015). Moreover, plasticity, acclimation, and tolerance responses may vary under different local environmental stress conditions since these seaweed species inhabit different intertidal zones (i.e., upper, mid, or lower) along a wide latitudinal gradient. Particularly along the Chilean coastline, these species are most abundant in the upper to shallow sub-intertidal zones (Macaya and Zuccarello 2010; Ramírez et al. 2008, 2014; Guillemin et al. 2016). These seaweeds facilitate algal and invertebrate recruitment, modulate local biological diversity and community structure, and are commercially exploited by seaweed-based industries for alginate, bioethanol, and organic fertilizer production, abalone feed, and human consumption (e.g., Almanza and Buschmann 2013; Aitken et al. 2014). Notably, exports reached 530,000 tons in 2013. However, ecophysiological studies in these organisms are scarce; and this lack of information impedes predicting potential outcomes of global stress factors.

The impacts of heavy metal toxicity and ocean acidification enrichment on the microscopic life stages of seaweeds have been largely ignored despite that algal recruitment depends mainly on the survival of these early stages. The combined results of heavy metals and pCO_2 might include shifts in species diversity and ecosystem composition due to reduced habitat range. Growth, development, and reproduction are among the life history traits of aquatic organisms known to be affected by copper and cadmium toxicity, among other heavy metals. In *L. spicata*, concentrations higher than 20 μg L^{-1} of dissolved copper interrupt spore development after settling. This leads to failure in the formation of male and female gametophytes and, as a consequence, results in a complete disruption of the normal life cycle (Contreras et al. 2007). The absence of *Lessonia* from copper-enriched environments is due to high sensitivity in the early life cycle stages, which limits growth and maturation of the gametophytic (*n*) microscopic phase and, subsequently, prevents development of the macroscopic sporophytic phase (*2n*).

Recently, the effects of pH shifts between 7.59 and 8.50 on meiospore germination and sex determination in the noncalcifying kelp *M. pyrifera* were determined, with lower pH resulting in significantly reduced germination and kelp spore mortality (Gaitán-Espitia et al. 2014). Additionally, Roleda et al. (2012) evidenced that the proportion of male to female *M. pyrifera* gametophytes was not significantly affected by reduced pH, and inhibition of meiospore germination under low pH could be counteracted by the availability of dissolved inorganic carbon.

In natural habitats, algae species often experience severe environmental and anthropogenic stresses as a result of periodic exposure to a wide range of atmospheric conditions. Therefore, the relative abundance, survivability, and distribution of seaweeds are determined mainly by specific tolerance levels to diverse environmental stressors. In a scenario of ocean acidification, it is tempting to predict that seaweeds will benefit from the increase in inorganic carbon concentration (Beardall et al. 1998). However, CO_2-driven effects on photosynthesis and growth depend on the degree to which carbon is limiting, which in turn varies between habitat types and among taxa

(Harley et al. 2012). Moreover, pCO_2 enrichment could modify the behavior of other environmental factors, such as heavy metal excess and, as a result, negatively affect the physiological performance of a given species. Therefore, comprehensive studies that include cellular, biochemical, and molecular biological approaches are required for determining the responses of marine seaweeds to interacting stressors.

3.5 Conclusion

According to the International Council for Science, five priorities must be considered among the 17 Sustainable Development Goals that were formally launched in September 2015 by the United Nations (70th session, General Assembly). Notable among these is the initiative to significantly reduce marine pollution and minimize the impacts of ocean acidification, as well as the effects of these on human and ecosystem health. In this context, it is important to elucidate the interactive effect of both factors on community performance and ecosystem function, especially considering that human dependence on the oceans places many socioeconomic practices associated with marine organisms at risk under a scenario of global climate change. Although it is very clear that many environmental stressors can act in combination to synergistically, antagonistically, or additively affect many physiological processes in marine organisms (Schiedek et al. 2007; Darling and Côté 2008; Gooding et al. 2009), studies that evaluate these interactive effects are still scarce. Heavy metal enrichment from anthropogenic sources and ocean acidification due to increasing greenhouse gas concentrations at the end of the twenty-first century may have a particularly wide range of effects on seaweed species and community structure. Although many seaweed species are known to be vulnerable to physical and chemical changes in the marine environment, the impacts of ongoing and future anthropogenic climate change remain poorly understood. In this context, physiological studies on seaweed communities will enhance our ability to predict future changes in the performance and persistence of marine organisms under global change.

Acknowledgments This work was supported by FIC-R Gobierno Regional de Valparaíso BIP n° 30397482-0, DI-501-14/R (Universidad Andrés Bello, Proyectos Regulares Internos) and FONDECYT N° 1170881 to LC-P. We appreciate the valuable assistance provided by Prof. Juan A. Correa and Prof. Cristian Duarte in reviewing the heavy metal and ocean acidification section. We also acknowledge the language support provided by Ashley VanCott, BioPub Ltd.

References

Aitken D, Bulboa C, Godoy-Faundez A et al (2014) Life cycle of macroalgae cultivation and processing for biofuel production. J Clean Prod 75:45–56

Allan R (1997) Introduction: mining and metals in the environment. J Geochem Explor 58(2–3):95–100

Almanza V, Buschmann A (2013) The ecological importance of *Macrocystis pyrifera* (Phaeophyta) forests towards a sustainable management and exploitation of Chilean coastal benthic co-management areas. Int J Environ Sus Develop 12:341–360

Anderson BS, Hunt JW (1988) Bioassay methods for evaluating the toxicity of heavy metals, biocides and sewage effluent using microscopic stages of giant kelp *Macrocystis pyrifera* (Agardh): a preliminary report. Mar Environ Res 26(2):113–134

Andrade S, Medina MH, Moffett JW, Correa JA (2006) Cadmium–copper antagonism in seaweeds inhabiting coastal areas affected by copper mine waste disposals. Environ Sci Technol 40(14):4382–4387

Baier M, Dietz KJ (1997) The plant 2-Cys peroxiredoxin BAS1 is a nuclear-encoded chloroplast protein: its expressional regulation, phylogenetic origin, and implications for its specific physiological function in plants. Plant J 12(1):179–190

Beardall J, Beer S, Raven JA (1998) Biodiversity of marine plants in an era of climate change: some predictions based on physiological performance. Bot Mar 41(1–6):113–123

Brown MT, Newman JE (2003) Physiological responses of *Gracilariopsis longissima* (S.G. Gmelin) Steentoft, L.M. Irvine and Farnham (Rhodophyceae) to sub-lethal copper concentrations. Aquat Toxicol 64(2):201–213

Byrne RH (1988) The influence of temperature and pH on trace metal speciation in seawater. Mar Chem 25(2):163–181

Cancino J, Santelices B (1984) Importancia ecológica de los discos adhesivos de *Lessonia nigrescens* Bory (Phaeophyta) en Chile central. Rev Chil Hist Nat 57:23–33

Castilla JC (1996) Copper mine tailing disposal in Northern Chile rocky shores: *Enteromorpha compressa* (Chlorophyta) as a sentinel species. Environ Monit Assess 40(2):171–184

Contreras L, Moenne A, Correa JA (2005) Antioxidant responses in *Scytosiphon lomentaria* (Phaeophyceae) inhabiting copper-enriched coastal environments. J Phycol 41(6):1184–1195

Collén J, Davison IR (1999) Reactive oxygen production and damage in intertidal *Fucus*spp. (Phaeophyceae). J Phycol 35(1): 54–61

Contreras L, Medina MH, Andrade S, Oppliger V, Correa JA (2007) Effects of copper on early developmental stages of *Lessonia nigrescens* Bory (Phaeophyceae). Environ Pollut 145(1):75–83

Contreras L, Mella D, Moenne A, Correa JA (2009) Differential responses to copper-induced oxidative stress in the marine macroalgae *Lessonia nigrescens* and *Scytosiphon lomentaria* (Phaeophyceae). Aquat Toxicol 94(2):94–102

Contreras L, Moenne A, Gaillard F, Potin P, Correa JA (2010) Proteomic analysis and identification of copper stress-regulated proteins in the marine alga *Scytosiphon gracilis* (Phaeophyceae). Aquat Toxicol 96(2):85–89

Contreras-Porcia L, Thomas D, Flores V, Correa JA (2011a) Tolerance to oxidative stress induced by desiccation in *Porphyra columbina* (Bangiales, Rhodophyta). J Exp Bot 62(6):1815–1829

Contreras-Porcia L, Dennett G, González A, Vergara E, Medina C, Correa JA, Moenne A (2011b) Identification of copper-induced genes in the marine alga *Ulva compressa* (Chlorophyta). Mar Biotechnol 13:544–556

Correa JA, Ramírez MA, de la Harpe JP, Román D, Rivera L (2000) Copper mining effluents and grazing as potential determinants of algal abundance and diversity in Northern Chile. Environ Monit Assess 61(2):265–281

Cullen JT, Lane TW, Morel FMM, Sherrell RM (1999) Modulation of cadmium uptake in phytoplankton by seawater CO_2 concentration. Nature 402:165–167

Darling ES, Côté IM (2008) Quantifying the evidence for ecological synergies. Ecol Lett 11(12):1278–1286

Dayer R, Fischer BB, Eggen RIL, Lemaire SD (2008) The peroxiredoxin and glutathione peroxidase families in *Chlamydomonas reinhardtii*. Genetics 179(1):41–57

Dietz KJ (2011) Peroxiredoxins in plants and cyanobacteria. Antioxid Redox Signal 15(4):1129–1159

Duarte C, López J, Benítez S, Manríquez PH, Navarro JM, Bonta CC, Torres R, Quijón P (2016) Ocean acidification induces changes in algal palatability and herbivore feeding behavior and performance. Oecologia 180(2):453–462

Duce RA, Liss PS, Merrill JT, Atlas EL, Buat-Menard P, Hicks BB, Miller JM, Prospero JM, Arimoto R, Church TM, Ellis W, Galloway JN et al (1991) The atmospheric input of trace species to the world ocean. Global Biogeochem Cycles 5(3):193–259

Ehrmann M, Ehrle R, Hofmann E, Boos W, Schlösser A (1998) The ABC maltose transporter. Mol Microbiol 29(3):685–694

Feely RA, Sabine CL, Lee K et al (2004) The impact of anthropogenic CO_2 on the $CaCO_3$ system in the oceans. Science 305(5682):362–366

Flores-Molina MR, Thomas D, Lovazzano C, Núñez A, Zapata J, Kumar M, Correa JA, Contreras-Porcia L (2014) Desiccation stress in intertidal seaweeds: effects on morphology, photosynthetic performance and antioxidant responses. Aquat Bot 113:90–99

Foyer CH, Noctor G (2009) Redox regulation in photosynthetic organisms: signaling, acclimation, and practical implications. Antioxid Redox Signal 11(4):861–905

Franklin NM, Stauber JL, Lim RP, Petocz P (2002) Toxicity of metal mixtures to a tropical freshwater alga (*Chlorella* sp): the effect of interactions between copper, cadmium, and zinc on metal cell binding and uptake. Environ Toxicol Chem 21(11):2412–2422

Gail FW (1919) Hydrogen ion concentration and other factors affecting the distribution of *Fucus*. Puget Sound Biol Sta 2:287–306

Gaillard S, Jacquet H, Vavasseur A, Leonhardt N, Forestier C (2008) AtMRP6/AtABCC6, an ATP-Binding Cassette transporter gene expressed during early steps of seedling development and up-regulated by cadmium in *Arabidopsis thaliana*. BMC Plant Biol 8:22

Gaitán-Espitia JD, Hancock JR, Padilla-Gamiño JL, Rivest EB, Blanchette CA, Reed DC, Hofmann GE (2014) Interactive effects of elevated temperature and pCO_2 on early-life-history stages of the giant kelp *Macrocystis pyrifera*. J Exp Mar Biol Ecol 457:51–58

García-Sánchez MJ, Fernández JA, Niell X (1994) Effect of inorganic carbon supply on the photosynthetic physiology of *Gracilaria tenuistipitata*. Planta 194(1):55–61

Garman GD, Pillai MC, Cherr GN (1994) Inhibition of cellular events during algal gametophyte development: effects of select metals and an aqueous petroleum waste. Aquat Toxicol 28(1–2):127–144

Garnham GW, Codd GA, Gadd GM (1992) Kinetics of uptake and intracellular location of cobalt, manganese and zinc in the estuarine green alga, *Chlorella salina*. Appl Microbiol Biotechnol 37(2):270–276

Gattuso JP, Hansson L (eds) (2011) Ocean acidification. Oxford University Press, Oxford, p 326

Gledhill M, Nimmo M, Hill SJ et al (1997) The toxicity of copper(II) species to marine algae, with particular reference to macroalgae. J Phycol 33(1):2–11

Gooding RA, Harley CDG, Tang E (2009) Elevated water temperature and carbon dioxide concentration increase the growth of a keystone echinoderm. PNAS 106(23):9316–9321

Gordillo FJL, Niell FX, Figueroa FL (2001) Non-photosynthetic enhancement of growth by high CO_2 level in the nitrophilic seaweed *Ulva rigida* C. Agardh (Chlorophyta). Planta 213(1):64–70

Guillemin ML, Contreras-Porcia L, Ramírez ME, Macaya EC, Contador CB, Woods H, Wyatt C, Brodie J (2016) The bladed Bangiales (Rhodophyta) of the South Eastern Pacific: molecular species delimitation reveals extensive diversity. Mol Phylogenet Evol 94(Pt B):814–826

Harley CDG, Anderson KM, Demes KW, Jorve JP, Kordas RL, Coyle TA, Graham MH (2012) Effects of climate change on global seaweed communities. J Phycol 48(5):1064–1078

Henríquez-Castillo C, Rodríguez-Marconi S, Rubio F, Trefault N, Andrade S, De la Iglesia R (2015) Eukaryotic picophytoplankton community response to copper enrichment in a metal-perturbed coastal environment. Phycol Res 63(3):189–196

Hofmann LC, Yildiz G, Hanelt D, Bischof K (2012a) Physiological responses of the calcifying rhodophyte, *Corallina officinalis* (L.), to future CO_2 levels. Mar Biol 159(4):783–792

Hofmann LC, Straub S, Bischof K (2012b) Competition between calcifying and noncalcifying temperate marine macroalgae under elevated CO_2 levels. Mar Ecol Prog Ser 464:89–105

Hung SH, Yu CW, Lin CH (2005) Hydrogen peroxide functions as a stress signal in plants. Bot Bull Acad Sin 46(1):1–10

Hurd CL, Harrison PJ, Bischof K, Lobban SC (eds) (2014) Seaweed ecology and physiology, 2nd edn. Cambridge University Press, Cambridge, UK

IPCC (2014) Cambio climático 2014: impactos, adaptación y vulnerabilidad – Resumen para responsables de políticas. Contribución del Grupo de trabajo II al Quinto Informe de Evaluación del Grupo Intergubernamental de Expertos sobre el Cambio Climático [Field CB, Barros VR, Dokken DJ, Mach KJ, Mastrandrea MD, Bilir TE, Chatterjee M, Ebi KL, Estrada YO, Genova RC, Girma B, Kissel ES, Levy AN, MacCracken S, Mastrandrea PR, White LL (eds)]. Organización Meteorológica Mundial, Ginebra, Suiza, p 39

Israel A, Katz S, Dubinsky Z, Merrill JE, Friedlander M (1999) Photosynthetic inorganic carbon utilization and growth of *Porphyra linearis* (Rhodophyta). J Appl Phycol 11:447–453

Liu X, Millero FJ (2002) The solubility of iron in seawater. Mar Chem 77(1):43–54

Lovazzano C, Serrano C, Correa JA, Contreras-Porcia L (2013) Comparative analysis of peroxiredoxin activation in the brown macroalgae *Scytosiphon gracilis* and *Lessonia nigrescens* (Phaeophyceae) under copper stress. Physiol Plant 149(3):378–388

Macaya EC, Zuccarello GC (2010) Genetic structure of the giant kelp *Macrocystis pyrifera* along the southeastern Pacific. Mar Ecol Prog Ser 420:103–112

Martin JH (1990) Glacial-interglacial CO_2 change: the iron hypothesis. Paleoceanography 5(1):1–13

Medina M, Andrade S, Faugeron S, Lagos N, Mella D, Correa JA (2005) Biodiversity of rocky intertidal benthic communities associated with copper mine tailing discharges in northern Chile. Mar Pollut Bull 50(4):396–409

Mercado JM, Javier F, Gordillo L, Niell FX, Figueroa FL (1999) Effects of different levels of CO_2 on photosynthesis and cell components of the red alga *Porphyra leucosticta*. J Appl Phycol 11(5):455–461

Millero FJ, Woosley R, Ditrolio B, Waters J (2009) Effect of ocean acidification on the speciation of metals in seawater. Oceanography 22(4):72–85

Nielsen HD, Brownlee C, Coelho SM, Brwon MT (2003) Inter-population differences in inherited copper tolerance involve photosynthetic adaptation and exclusion mechanisms in *Fucus serratus*. New Phytol 160(1):157–165

NOAA (2016) National Oceanic and Atmospheric Administration. http://www.noaa.gov/

Oppliger LV, Correa JA, Engelen AH, Tellier F, Vieira V, Faugeron S, Valero M, Gomez G, Destombe C (2012) Temperature effects on gametophyte life-history traits and geographic distribution of two cryptic kelp species. PLoS One 7(6):e39289

Orr J, Fabry VJ, Aumont O, Bopp L, Doney SC et al (2005) Anthropogenic ocean acidification over the twenty-first century and its impact on calcifying organisms. Nature 437:681–686

Overnell J (1975) The effect of heavy metals on photosynthesis and loss of cell potassium in two species of marine algae, *Dunaliella tertiolecta* and *Phaeodactylum tricornutum*. Mar Biol 29(1):99–103

Pinto E, Sigaud-Kutner TCS, Leitão MAS, Okamoto OK, Morse D, Colepicolo P (2003) Heavy metal-induced oxidative stress in algae. J Phycol 39(6):1008–1018

Ramírez ME, García-Huidobro MR, Orellana AN (2008) Flora marina bentónica de Caleta Quintay (Región de Valparaíso), Litoral central de Chile. Bol Mus Nac Hist Nat Chile 57:9–19

Ramírez ME, Contreras-Porcia L, Guillemin ML, Brodie J, Valdivia C, Flores-Molina MR, Nuñez A, Bulboa Contador C, Lovazzano C (2014) *Pyropia orbicularis* sp. nov. (Rhodophyta, Bangiaceae) based on a population previously known as *Porphyra columbina* from the central coast of Chile. Phytotaxa 158(2):133–153

Ratkevicius N, Correa JA, Moenne A (2003) Copper accumulation, synthesis of ascorbate and activation of ascorbate peroxidase in *Enteromorpha compressa* (L.) Grev. (Chlorophyta) from heavy metal-enriched environments in Northern Chile. Plant Cell Environ 26(10):1599–1608

Reith M, Munholland J (1993) A high-resolution gene map of the chloroplast genome of the red alga *Porphyra purpurea*. Plant Cell 5(4):465–475

Ritter A, Goulitquer S, Salaün J, Tonon T, Correa JA, Potin P (2008) Copper stress induces biosynthesis of octadecanoid and eicosanoid oxygenated derivatives in the brown algal kelp *Laminaria digitata*. New Phytol 180(4):809–821

Ritter A, Dittami SM, Goulitquer S, Correa JA, Boyen C, Potin P, Tonon T (2014) Transcriptomic and metabolomic analysis of copper stress acclimation in *Ectocarpus siliculosus* highlights signaling and tolerance mechanisms in brown algae. BMC Plant Biol 14(116):1–17

Roleda MY, Morris JN, McGraw CM, Hurd CL (2012) Ocean acidification and seaweed reproduction: increased CO_2 ameliorates the negative effect of lowered pH on meiospore germination in the giant kelp *Macrocystis pyrifera* (Laminariales, Phaeophyceae). Glob Chang Biol 18(3):854–864

Schiedek D, Sundelin B, Readman JW, Macdonald RW (2007) Interactions between climate change and contaminants. Mar Pollut Bull 54(12):1845–1856

Schiel DR, Foster MS (eds) (2015) The biology and ecology of giant kelp forests. University of California Press, Oakland, CA, p 416

Solomon S, Qin D, Manning M et al (eds) (2007) Climate change 2007: the physical science basis. Cambridge University Press, Cambridge, UK, p 996

Sordet C, Contreras-Porcia L, Lovazzano C, Goulitquer S, Andrade S, Potin P, Correa JA (2014) Physiological plasticity of *Dictyota kunthii* (Phaeophyceae) to copper excess. Aquat Toxicol 150:220–228

Tripathi BN, Bhatt I, Dietz KJ (2009) Peroxiredoxins: a less studied component of hydrogen peroxide detoxification in photosynthetic organisms. Protoplasma 235(1–4):3–15

US EPA (Environmental Protection Agency) (2007) Framework for metals risk assessment. EPA 120/R–07/001.2007

Vranová E, Inzé D, Van Breusegen F (2002) Signal transduction during oxidative stress. J Exp Bot 53(372):1227–1236

Walker CH, Sibly RM, Hopkins SP, Peakall DB (2012) Principles of ecotoxicology, 4th edn. CRC Press (Taylor and Francis), p 386. ISBN 9781439862667

Wolf-Gladrow DA, Bijma J, Zeebe RE (1999) Model simulation of the carbonate chemistry in the microenvironment of symbiont bearing foraminifera. Mar Chem 64(3):181–198

Xia JR, Li YJ, Lu J, Chen B (2004) Effects of copper and cadmium on growth, photosynthesis, and pigment content in *Gracilaria lemaneiformis*. Bull Environ Contam Toxicol 73:979–986

Xu Z, Zou D, Gao K (2010) Effects of elevated CO_2 and phosphorus supply on growth, photosynthesis and nutrient uptake in the marine macroalga *Gracilaria lemaneiformis* (Rhodophyta). Bot Mar 53(2):123–129

Zirino A, Yamamoto S (1972) A pH-dependent model for the chemical speciation of copper, zinc, cadmium and lead in sea water. Limnol Oceanogr 17(5):661–671

Zou D, Gao K (2002) Effects of desiccation and CO_2 concentrations on emersed photosynthesis in *Porphyra haitanensis* (Bangiales, Rhodophyta), a species farmed in China. Eur J Phycol 37(4):587–592

Zou DH (2005) Effects of elevated atmospheric CO_2 on growth, photosynthesis and nitrogen metabolism in the economic brown seaweed, *Hizikia fusiforme* (Sargassaceae, Phaeophyta). Aquaculture 250:726–735

Chapter 4
Seaweed Lipidomics in the Era of 'Omics' Biology: A Contemporary Perspective

Puja Kumari

Abstract Lipidomics, a branch of "omic" sciences, refers to the analysis of lipids on the systems level together with their interacting factors including expression of proteins involved in lipid metabolism and function and gene regulation. Seaweeds are one of the important ecosystem drivers that inhabit an unique aquatic environment being exposed to a diverse range of environmental fluctuations (salinity, light, desiccation, and temperature), pathogens, invasive species, and anthropogenic factors that affect their phenotype as well as acclimatory strategies. As a result of thriving in such diverse and extreme environments, they produce an array of unique bioactive, complex, exotic acyl lipids and fatty acids that are not generally present in terrestrial plants. Seaweeds have been extensively studied for their bioactive lipids mainly for their nutritionally important polyunsaturated fatty acids, oxylipins, and their pharmaceutical and biotechnological utilization. Most of the studies have been limited to the elucidation of lipid and fatty acids composition, their metabolic pathways, the genes and enzymes involved, as well as their roles in stress responses, innate immunity, and defense against pathogens. Although lipidomics has been extensively used in terrestrial plants and even microalgae to unravel their lipidome, functional annotation of unknown genes involved in lipid metabolism, correlation of genotype-phenotype, and understanding of the pleiotropic roles of lipids in cell development and biotic/abiotic stresses, only a few seaweeds have been studied with lipidomic approach. This chapter presents updated information on lipidomics, advanced analytical tools and techniques, and their applicability in seaweed studies along with its limitations. Further, an overview of how integrating lipidomics with allied sister branches of metabolomics, transcriptomics, and proteomics can help in the identification of unknown gene/protein functions and development of systems biology networks advancing our knowledge of lipid biochemistry in seaweed development and acclimation to stress conditions are discussed.

Keywords Seaweeds • Lipids • Lipidomics • Systems biology • Oxylipins • Stress

---

P. Kumari (✉)
Institute of Plant Sciences, Agricultural Research Organization, Volcani Centre, HaMaccabim, Road, P.O.B 15159, Rishon LeTsiyon 7528809, Israel
e-mail: pujamashal@gmail.com

© Springer International Publishing AG 2017

M. Kumar, P. Ralph (eds.), *Systems Biology of Marine Ecosystems*,
DOI 10.1007/978-3-319-62094-7_4

Contents

4.1 Introduction 50
4.2 Seaweed Lipids 52
4.2.1 Glycerophospholipids 52
4.2.2 Saccharolipids 54
4.2.3 Glycerolipids 55
4.2.4 Betaine Lipids 55
4.2.5 Sphingolipids 56
4.2.6 Fatty Acyls 56
4.2.7 Sterols 59
4.3 Seaweed Lipidomics: An Update 60
4.4 Tools and Techniques in Seaweed Lipidomics 61
4.4.1 Sampling and Lipid Extraction 61
4.4.2 Analytical Platforms for Lipidomics 63
4.4.3 Lipidomic Data Analysis 66
4.5 Application of Lipidomics to the Seaweed Systems Biology 76
4.5.1 Elucidation of Lipidomic Profiling and Identification of Novel Lipids 80
4.5.2 Lipidomic Changes in Seaweed Acclimation Strategies to Abiotic/Biotic Stress 83
4.6 Integration of Lipidomics with Allied Omics Platforms 85
4.7 Conclusion 89
References 90

4.1 Introduction

Lipids comprise a very diverse group of compounds that not only forms the membrane bilayer but also plays multiple critical roles in cellular functions. They provide hydrophobic environment for membrane protein functions and interactions, and fatty acids and triglycerides serve as energy reservoirs. They also play prominent roles in the regulation of cellular bioenergetics through integrating oxidative metabolism (Michalik et al. 2006), modulating systemic energy balance through eicosanoid and lysolipid production (Vegiopoulos et al. 2010), and regulating mitochondrial electron transport chain flux and coupling efficiency (Breen et al. 2005). Moreover, many classes of lipids such as eicosanoids, lysolipids, diacylglycerols, phosphatidic acids, and ceramides serve as secondary messengers in cellular signaling pathways (Gross and Han 2011).

Lipids are defined as amphiphilic biological substances consisting of fatty acids and their derivatives and the substances that are biosynthetically or functionally related to these compounds (Christie 1993). Lipid molecules exhibit high structural diversity due to variable chain length; a multitude of oxidative, reductive, substitutional, and ring-forming biochemical transformations; modification with sugar residues; and other functional groups of different biosynthetic origin (Fahy et al. 2011). The International Lipid Classification and the Nomenclature Committee, together with the Lipid Metabolites and Pathways Strategy (LIPID MAPS) Consortium, defined eight categories of lipids and divided them into classes and subclasses (Fahy et al. 2005, 2009, 2011) by their chemically functional backbones and biochemical principles in (1) fatty acyls, (2) glycerolipids, (3) glycerophospholipids (also known as phospholipids), (4) saccharolipids (also known as galactolipids), (5) sphingolipids, (6) sterol lipids, (7) prenol lipids, and (8) polyketides. The entire collection of chemically distinct lipid species in a cell is referred to as a lipidome (Gross and Han 2011). According to different estimates, an eukaryotic lipidome contains 9000–100,000 individual lipid molecular species (van Meer 2005; Han and Jiang 2009). Moreover, the estimated number of individual molecular species varies from cell to cell due to the intrinsic programmed genetic information, status of the cellular activation, metabolic signaling, and nutritional history (Han et al. 2012). Such large number and diversity of lipid species complicate their separation and identification thereby making the lipid analysis a challenge to accomplish. However, the advent of omic sciences has been a stimulus to determine the lipid molecular profile in biological systems, paving the way for "lipidomics." The term lipidomics was first used by Kishimoto et al. in 2001 (Kishimoto et al. 2001) and later defined by Han and Gross in 2003 (Han and Gross 2003). Currently, it is defined as the study of full complement of lipid molecules (lipidome) on the systems level together with their interacting factors including expression of proteins involved in lipid metabolism and function and gene regulation. Lipidomics enables to understand the role that lipids play in biological systems such as how lipids influence membrane architecture, the modulation of transcription and translation, and answers to environmental changes due to physiological processes (Rolim et al. 2015). This field has undergone rapid progress, mainly because of technology-driven transformation in instrumentation, most notably in mass spectrometry and its ancillary techniques (such as liquid chromatography and ionization sources) that have enabled quantitative lipid analyses with an unprecedented level of sensitivity and precision (Brügger 2014). Recent studies in lipidomics have largely focused on the identification of novel lipid classes and molecular species, development of quantitative methods for lipid analysis at attomole to femtomole levels per mg of protein, tissue mapping of altered lipid distribution present in different organs or in response to external cues, and bioinformatic approaches for the automated high-throughput processing and molecular modeling with lipidomic data (Wang et al. 2016). Now, the researchers have realized that metabolism of lipid molecular species or between individual lipid classes is interwoven and the metabolism of the entire lipidome should be investigated in a systems biology approach to better understand the functions of lipids in biological systems (Dennis 2009).

Seaweeds are benthic marine macroalgae comprising a diverse group of fascinating multicellular photosynthetic forms growing mostly attached to rocks in coastal waters. They are harvested and commercially utilized for food, feed, phycocolloids, fertilizer, energy, medicines, cosmetics and nutraceuticals, and biotechnological, bioremediation, and aquaculture applications with current market value of USD 7 billion (FAO 2014). Seaweeds are one of the important ecosystem drivers that inhabit an unique aquatic environment being exposed to a diverse range of environmental fluctuations (salinity, light, desiccation, and temperature), pathogens, invasive species, and anthropogenic factors that affect their phenotype as well as acclimatory strategies. As a result of thriving in such diverse and extreme environments, they produce an array of unique bioactive, complex, exotic acyl lipids and fatty acids that are not generally present in terrestrial plants. Seaweeds have been extensively studied for their bioactive lipids mainly for their nutritionally important polyunsaturated fatty acids (PUFAs), oxylipins, and their pharmaceutical and biotechnological utilization. Most of the studies have been limited to the elucidation of lipid and fatty acid composition, their metabolic pathways, the genes and enzymes involved, as well as their roles in stress responses, innate immunity, and defense against pathogens. Although lipidomics has been extensively used in terrestrial plants and even microalgae to unravel their lipidome, functional annotation of unknown genes involved in lipid metabolism, correlation of genotype-phenotype, and understanding of the pleiotropic roles of lipids in cell development and biotic/abiotic stresses, only a few seaweeds have been studied with lipidomic approach. This chapter presents updated information on lipidomics, advanced analytical tools and techniques, and their applicability in seaweed studies along with its limitations. Further, a comprehensive overview of how integrating lipidomics with allied sister omics branches can help in the identification of unknown gene/protein functions and development of systems biology networks advancing our knowledge of lipid biochemistry in seaweed development and acclimation to stress conditions are discussed.

4.2 Seaweed Lipids

Seaweed lipids mainly consist of glycerophospholipids, saccharolipids, glycerolipids, sphingolipids, fatty acyls (including oxylipins), and sterols analogous to higher plants along with betaine and some unusual lipids that may be characteristic of a particular genus or species. Their chain length and degree of unsaturation are also significantly higher than those of higher plants. The detailed structure, occurrence, functionalities, and bioactivities of seaweed lipids have been discussed by Kumari et al. (2013a) and recently reviewed by Maciel et al. (2016). A brief overview is presented here.

4.2.1 Glycerophospholipids

Glycerophospholipids or phospholipids (PLs) consist of a glycerol molecule linked with two fatty acyl chains at *sn*-1 and *sn*-2 positions and a phosphate group at *sn*-3 position, which is further linked to a hydrophilic head group that classifies

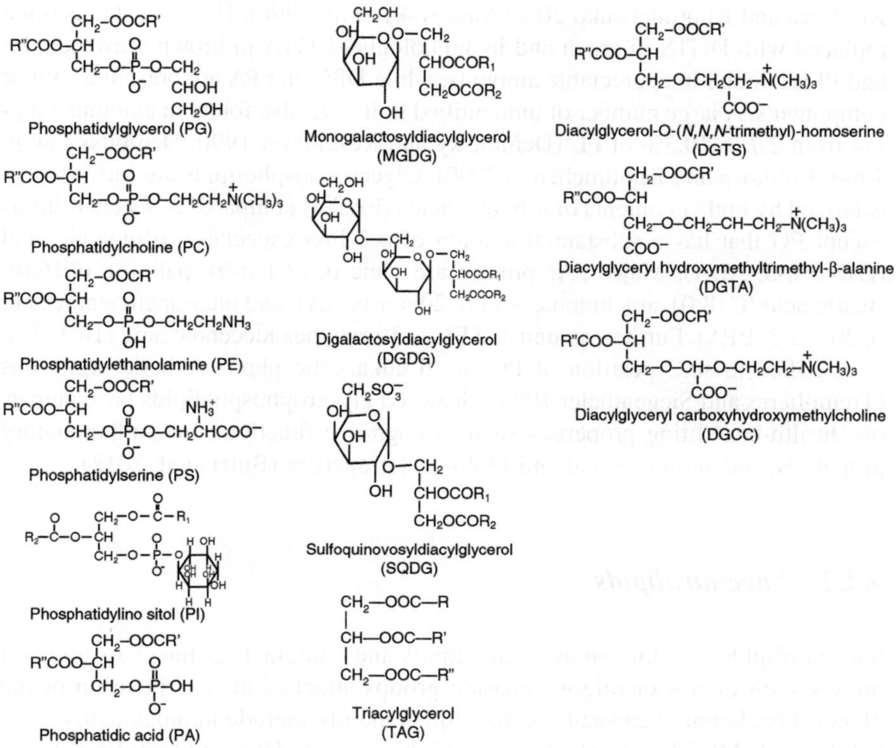

Fig. 4.1 Structures of common lipid classes found in seaweeds (Source: Kumari et al. 2013a) [Reproduced from Functional ingredients from algae for food and nutraceuticals, Domínguez H (ed)., Kumari P, Kumar M, Reddy CRK et al (2013) Algal lipids, fatty acids and sterols, Woodhead Publishing Ltd, Cambridge, UK, pp 87–134, with permissions from Elsevier]

individual glycerophospholipid molecules. The major glycerophospholipids found in seaweeds are phosphatidylglycerol (PG), phosphatidylcholine (PC), phosphatidylethanolamine (PE), phosphatidylserine (PS), phosphatidylinositol (PI), and phosphatidic acid (PA) containing glycerol, choline, ethanolamine, serine, myo-inositol, and phosphomonoester as their characteristic head groups, respectively (Fig. 4.1). Glycerophospholipids represent 10–20% of total lipids in seaweeds (Dembitsky and Rozentsvet 1990; Dembitsky and Rozentsvet 1996). They are located in extra-chloroplast membranes with the exception of PG, which occurs in significant amounts in thylakoid membranes. Cell membranes utilize the amphiphilic nature of phospholipids to maintain its structural integrity and selective permeability, while PG aids saccharolipids in maintaining the stability of photosynthetic apparatus. PG is the dominant phospholipid in Chlorophyta and accounts for 20–47% of PL, while PC in red represents >60% of PL and both PC and PE in brown algae account for 11.3–29.3% of PL (Dembitsky et al. 1990; Dembitsky and Rozentsvet 1996; Illijas et al. 2009; Khotimchenko et al. 1990;

Kulikova and Khotimchenko 2000; Vaśkovsky et al. 1996). However, PC is often replaced with DGTS in green and its homologue, DGTA in brown seaweeds. PS and PI are found in appreciable amounts, while DPG and PA are present as minor components. A large number of unidentified lipids are also found in amounts ranging from 2.7 to 10.3% of PL (Dembitsky and Rozentsvet 1990; Dembitsky et al. 1990; Kulikova and Khotimchenko 2000). Glycerophospholipids are further characterized by higher contents of *n*-6 fatty acids (FAs) as compared to saccharolipids except PG that has a substantial amount of *n*-3 FAs especially α-linolenic acid (C18:3 *n*-3, ALA). Major FAs present are oleic (C18:1 *n*-9), palmitic (C16:0), stearic acid (C18:0), arachidonic acid (C20:4 *n*-6, AA), and eicosapentaenoic acid (C20:5 *n*-3, EPA). Further, an unusual FA, Δ3-trans-hexadecenoic acid (16:1, 3 *t*), is esterified to *sn*-2 position of PG in all eukaryotic photosynthetic organisms (Tremolieres and Siegenthaler 1998). Seaweed glycerophospholipids have numerous health-benefitting properties such as cognitive functions, anti-inflammatory properties, and antitumor and anti-biological properties (Burri et al. 2012).

4.2.2 Saccharolipids

Saccharolipids also known as galactolipids and contain 1, 2-diacyl-*sn*-glycerol moiety with mono- or oligosaccharide groups attached at *sn*-3 position of the glycerol backbone. Seaweed saccharolipids mainly include monogalactosyldiacylglycerol (MGDG), digalactosyldiacylglycerol (DGDG) and sulfolipid sulfoquinovosyldiacylglycerol (SQDG). MGDG and DGDG contain one and two galactose molecules, respectively, and are uncharged at physiological pH, while SQDG carries a negative charge due to its sulfonic acid residue at position 6 of the monosaccharide moiety (Fig. 4.1). Saccharolipids are predominantly located in photosynthetic membranes with MGDG and SQDG strictly restricted to the thylakoid membranes of the chloroplast, while DGDG is also found in extraplastidial membranes. They constitute more than half of the lipids with MGDG representing 31–56% (Sanina et al. 2004; Kumari et al. 2015; Yan et al. 2011), with the exception of a few red algae (such as *Palmaria stenogona* Perestenko, *Ceramium kondoi* Yendo, *Laurencia nipponica* Yamada, *Ahnfeltia tobuchiensis* Kanno & Matsubara, *and Exophyllum wentii* Weber-van Bosse) where DGDG content was higher (35.7–64% of polar lipids) and the members of Fucales (brown seaweeds) contained higher SQDG (36.8–48.8%) (Khotimchenko 2002; Sanina et al. 2004). A unique feature of these lipids is their high *n*-3 PUFA contents similar to higher plants. MGDG is the most unsaturated saccharolipid in green and red seaweeds with DGDG in brown algae, while SQDG was the most saturated one. MGDG and DGDG contain hexadecatetraenoic acid (C16:4 *n*-3), ALA, stearidonic acid (C18:4 *n*-3, STA), and linoleic acid (C18:2 *n*-6, LA) in green seaweeds, AA and EPA in red, and both in brown algae, while SQDG contains palmitic and oleic acid as major FAs (Hofmann and Eichenberger 1997; Illijas et al. 2009; Khotimchenko 2002, 2003; Sanina et al. 2004).

Saccharolipids are indispensable for assembly and functional regulation of PSII (Mizusawa and Wada 2012 and the references therein). They play crucial roles as markers for cellular recognition and stabilization of membrane bilayers and during phosphate limitation in seaweeds (and in microalgae and plants) by replacing phospholipids to combat the stress condition. Further, seaweed saccharolipids have notable anti-inflammatory, antimicrobial, antitumor, and antiviral properties, and a large number of saccharolipids have been isolated from the species of *Ulva*, *Chondria*, *Laurencia*, *Palmaria*, *Fucus*, *Sargassum*, and others (discussed in Maciel et al. 2016 and references therein).

4.2.3 Glycerolipids

Glycerolipids are characterized by glycerol backbone esterified with hydrophobic acyl chains that may be saturated or unsaturated either at one, two, or all the three positions (*sn*-1, *sn*-2, and *sn*-3) forming monoacylglycerol, diacylglycerol, and triacylglycerol, respectively. Triacylglycerol is the most prevalent glycerolipid accumulated in seaweeds as storage product and energy reservoirs. Its level is highly plastic and ranges between 1% and 59.3% (Dembitsky et al. 1993; Dembitsky and Rozentsvet 1996; Hofmann and Eichenberger 1997; Illijas et al. 2009; Khotimchenko and Kulikova 1999; Kim et al. 1996; Kulikova and Khotimchenko 2000; Rozentsvet et al. 1995).

4.2.4 Betaine Lipids

Betaine lipids are acylglycerolipids characterized by a betaine moiety (a quaternary amine alcohol) instead of phosphorus or carbohydrate as a polar group linked to *sn*-3 position of glycerol by an ether bond with fatty acids esterified in *sn*-1 and *sn*-2 positions. These betaine lipids are all zwitterionic at neutral pH due to their positively charged trimethylammonium group and a negatively charged carboxyl group. The betaine lipids present in seaweeds are 1,2-diacylglyceryl-3-O-4′-(*N*,*N*,*N*-trimethyl)-homoserine (DGTS) and 1,2-diacylglyceryl-3-*O*-2′-(hydroxymethyl)-(*N*,*N*,*N*-trimethyl)-β-alanine (DGTA) (Fig. 4.1). Betaine lipids are widely distributed in seaweeds and extensively reviewed by Dembitsky (1996) and Kato et al. (1996). These two betaine lipids resemble PC due to their quaternary ammonium group and hence replace PC in most of the seaweeds, even to traces such as in *Ulotrichales*, *Scytosiphonales*, *Desmarestiales*, and others. DGTS abundantly occurs in Chlorophyta with 5.2–56.5% of polar lipids and DGTA in brown algae with 7.3–96.8% of polar lipids (Dembitsky and Rozentsvet 1996; Eichenberger et al. 1993; Kulikova and Khotimchenko 2000; Makewicz et al. 1997; Muller and Eichenberger 1994). DGTS in seaweeds contain long-chain PUFAs at both the *sn*-1 and *sn*-2 positions, while DGTA contain palmitic, myristic, oleic, LA, ALA, AA, and EPA as major FAs (Hofmann and Eichenberger 1997; Makewicz et al. 1997). Monoacylglyceryl-*N,N,N*-trimethylhomoserine (MGTS) molecular species have also been reported recently in one of the green seaweed of taxa *Codium tomentosum* Stackhouse (da Costa et al. 2015). Betaine lipids constitute minor fractions or have been unidentified in most of the red seaweeds except for DGTS (Künzler and Eichenberger 1997) until recently when Melo et al. (2015) identified 36 DGTS molecular species using advanced lipidomic approaches in *Chondrus crispus* Stackhouse, containing DGTS 16:0/16:1 as the most abundant DGTS molecular species followed by 16:0/16:0 and 14:0/18:0. This study shows the unparalleled strength of advanced lipidomic approaches that may lead to re-characterization of seaweed lipid profiles in the near future. DGTA is considered to play an important role in the redistribution of acyl chains and the biosynthesis of saccharolipids and DGTS in lipid-linked desaturation of fatty acids (Hofmann and Eichenberger 1998).

4.2.5 Sphingolipids

Sphingolipids in seaweeds are mainly present in red seaweeds in small amounts. These include cerebrosides and ceramides and inositolphosphoceramide (IPC) (Bano et al. 1990; Khotimchenko et al. 2000; Khotimchenko and Vaśkovsky 2004; Lo et al. 2001; Vaśkovsky et al. 1996). Only IPC was fully characterized from *Gracilaria verrucosa* (Hudson) Papenfuss (Khotimchenko et al. 2000; Khotimchenko and Vaśkovsky 2004) that contained palmitic (51.7%), stearic (23.2%), myristic (9.8%), oleic (9.8%), and palmitoleic acids. Recently, Melo et al. (2015) characterized 15 molecular species of IPC from *C. crispus* containing C18:0/C26:0 as the most abundant molecular species.

4.2.6 Fatty Acyls

Fatty acyls include fatty acids, their conjugates, and their oxidized products (oxylipins).

4.2.6.1 Fatty Acids

Fatty acids (FAs) are carboxylic acids with long aliphatic chains that may be straight or branched, saturated or unsaturated. Seaweeds mostly contain even carbon fatty acids (C4-C28); however, odd-chain FAs are also present in minor amounts. FAs are classified as monounsaturated FAs (MUFAs, with 1 double bond) and polyunsaturated FAs (PUFAs, with $\geq$2 double bonds) on the basis of the number of double bonds present. Further, PUFAs are classified as *n*-3 or *n*-6 FAs depending on the position of the first double bond from the methyl end. Seaweeds have been extensively explored for their fatty acids, especially PUFAs (representing10–70% of total fatty acids; TFAs) due to their chemotaxonomic and nutritional importance, with their compositions varying even within the same phyla (Galloway et al. 2012; Khotimchenko et al. 2002; Kumari et al. 2010; Kumari et al. 2013b; Li et al. 2002). The characteristic chemotaxonomic biomarkers are established in seaweeds, with green seaweeds containing higher contents of C18 PUFAs (ALA, STA, and LA), red seaweeds containing C20 PUFAs (AA and EPA), and brown seaweeds containing both C18 and C20 PUFAs in appreciable amounts. These long-chain PUFAs, particularly *n*-3 PUFAs (ALA, STA, and EPA) which cannot be synthesized by humans and thus obtained through diet, are indispensable for proper growth and development of organisms, prevention of cardiovascular and other chronic diseases such as diabetes, hypertension, and autoimmune diseases, and DHA for visual and neurological health, while AA and EPA are precursors of bioregulators prostaglandins, thromboxanes, and other eicosanoids, which influence inflammation processes and immune reactions (Calder and Grimble 2002).

4.2.6.2 Oxylipins

Oxylipins are lipid signaling oxygenated derivatives of PUFAs formed enzymatically either by lipoxygenases (LOX) or α-dioxygenases (α-DOX) or by chemical (auto) oxidation that mediates intra- and intercellular processes such as development, inflammation, and stress responses. These compounds are widely distributed in seaweeds with considerable species-specific differences due to the variability of both FAs and enzymatic transformations. Seaweeds possess both plant- and animal-type oxylipins, i.e., octadecanoid and eicosanoid pathways emanating from C18 and C20 PUFAs, respectively, as well as docosanoid pathway emanating from C22PUFAs, recently reviewed by Kumari et al. (2013a, 2014a) and Barbosa et al. (2016). C18 PUFAs are metabolized either at C-9 or C-13 via 9- and 13-LOX, respectively; C20 PUFAs are transformed at C-5, C-8, C-9, C-11, C-12, and C-15 via 5-, 8-, 9-, 11-, 12-, and 15-LOX, respectively; and C22 PUFAs are mainly transformed at C-14 by 14-LOX, forming their respective hydroperoxides (Barbosa et al. 2016 and Kumari et al. 2013a, b for detailed structures and occurrence of seaweed oxylipins). Further, these hydroperoxides are transformed into hydroxy-, oxo-, and epoxy-fatty acids and polyunsaturated aldehydes (PUAs) by the action of peroxidases, oxygenases, epoxygenases, and hydroperoxide lyases (HPL), respectively (Figs. 4.2 and 4.3) (Andreou and Feussner 2009; Bouarab et al. 2004; Gerwick et al. 1993; Kumari et al. 2014a; Lion et al. 2006; Ritter et al. 2008, 2014). Moreover, some red algae also form prostaglandins and leukotrienes either nonenzymatically or by the enzymatic action of allele oxide synthase/cyclase (AOS/AOC) or cyclooxygenase (COX) analogous to animals (Andreou and Feussner 2009). Recently, Kanamoto et al. (2011) identified COX gene in *Gracilaria vermiculophylla* (Ohni) Papenfuss and cloned it in *Escherichia coli* for the production of $PGF_{2\alpha}$. Apart from these simple oxylipins, macroalgae also contain various

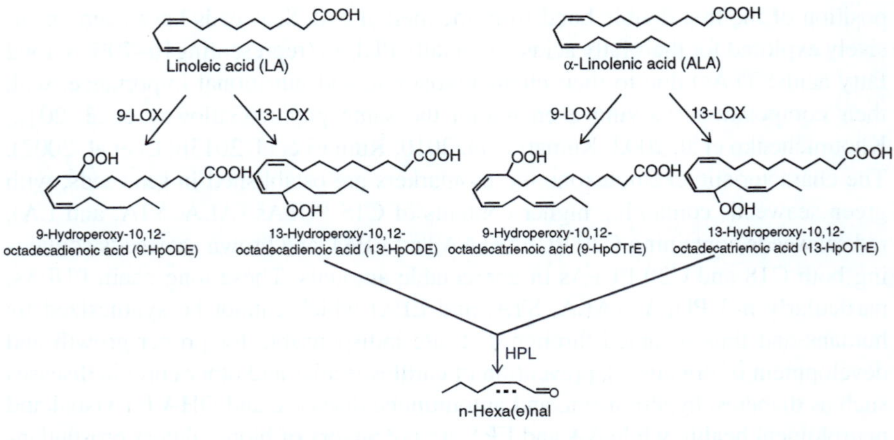

Fig. 4.2 Octadecanoid pathway in seaweeds. Modified from Andreou and Feussner (2009)

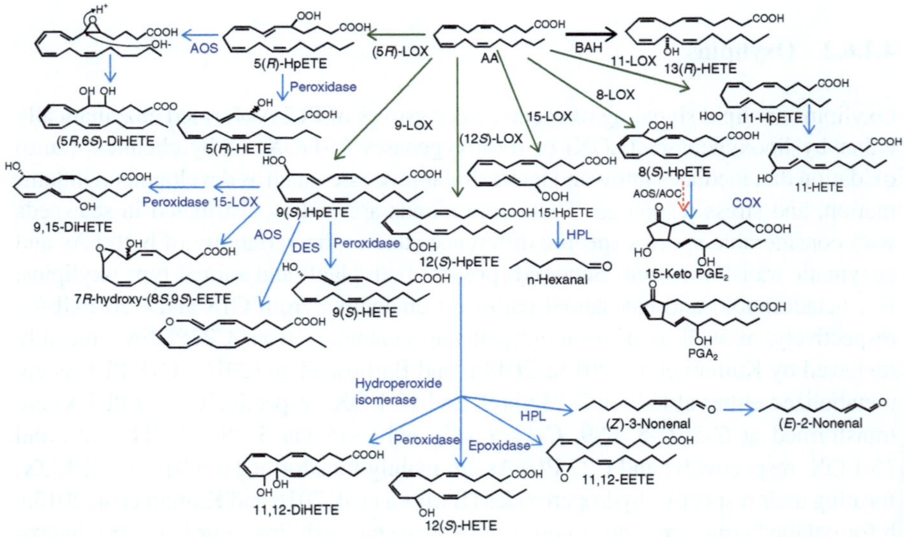

Fig. 4.3 Eicosanoid pathway in seaweeds. *Dashed line* in *red* shows putative reactions (*LOX* lipoxygenase, *AOS* allene oxide synthase, *AOC* allene oxide cyclase, *COX* cyclooxygenase, *DES* divinyl ether synthase). Modified from Andreou and Feussner (2009)

complex oxylipins such as polycyclic oxylipins, cyclopropyl hydroxyeicosanoids, egregialactones, ecklonialactones, hybridialactones, bicyclic cymathere ethers, cymatherelactones, and cymatherols, most of which are formed from intramolecular rearrangements of hydroperoxides of either ALA (C18:3, *n*-3) or stearidonic acid (C18:4, *n*-3) (Choi et al. 2012; Gerwick et al. 1990; Kousaka et al. 2003; Lion et al.

2006; Nagle and Gerwick 1990; Proteau and Gerwick 1993; Rempt et al. 2012; Weinberger et al. 2011). Phytoprostanes nonenzymatically derived from ALA have also been identified in seaweeds (Barbosa et al. 2015; Ritter et al. 2014). The occurrence and distribution of naturally occurring free phytoprostanes has been found to be highly unpredictable differing between species and as a consequence of the surrounding growth conditions, with F_{1t}-phytoprostanes (including both 9-F_{1t}-phytoprostane and 9-epi-9-F_{1t}-phytoprostane) being the dominant and L1-phytoprostanes the minor class (Barbosa et al. 2015). Recently, methyl jasmonate (MeJA) enzymatically derived from ALA by LOX pathway has been detected and quantified for the first time, from the cystocarps of red seaweed *Grateloupia imbricata* Holmes (Pilar et al. 2016) that releases this volatile compound in significant amounts (1.27 ± 0.20 mM • mg fw^{-1} • h^{-1} in fertile thalli and 0.95 ± 0.12 mM • mg fw $^{-1}$ • h^{-1} in infertile thalli).

Seaweed oxylipins play various important roles in growth regulation and defense and confer innate immunity in response to biotic and abiotic stresses such as pathogenic bacteria, herbivores, wounding, and metal toxicity in seaweeds (Bouarab et al. 2004; Gaquerel et al. 2007; Küpper et al. 2006, 2009; Lion et al. 2006; Nylund et al. 2011; Rempt et al. 2012; Ritter et al. 2008; Weinberger et al. 2011). However, most of the information available regarding seaweed oxylipins has come from the metabolic studies rather than the genomic studies due to limited number of available seaweed genome sequences as compared to higher plants and microalgae. Consequently, only four putative LOX sequences are available in NCBI database isolated from *C. crispus* (accession number XM_005718216.1), *Ectocarpus siliculosus* (Dillwyn) Lyngbye (Cock et al. 2010), *Gracilaria chilensis* C. J. Bird, McLachlan & E. C. Oliveira (accession number JF896804), *Porphyra purpurea* (Roth) C. Agardh (Liu and Reith 1994), *Pyropia haitanensis* (T. J. Chang & B. F. Zheng) N. Kikuchi & M. Miyata (accession number JX188386), and one AOC sequence in *E. siliculosus* (Cock et al. 2010). Besides the ecophysiological role of these oxidized lipid derivatives and their relevance in seaweeds, the exact mechanisms of stress tolerance are not known. Moreover, and because metabolites of this class also play a crucial role in both mammalian physiology and disease, interest in the structural chemistry, biosynthesis, and pharmacological activities of these marine products has increased. With present context, there is a need to improve our knowledge of the pathways of oxylipin biosynthesis, their individual role in cellular responses, and the target elements involved in gene regulation, which can only be achieved using the systems biology approach of combined genomics, transcriptomics, and metabolomics/lipidomics tools.

4.2.7 *Sterols*

Sterols are important structural components of cell membranes that regulate membrane fluidity and permeability. They are amphipathic compounds that originate in isoprenoid biosynthesis forming a group of triterpenes with a tetracyclic cyclopenta(α)phenanthrene structure and a side chain at C17 (please refer to Kumari et al. 2013a for detailed description of seaweed sterols structures and occurrences). Seaweed sterols are extremely diverse with both the mevalonate (MVA) and methyl-D-erythritol-4-phosphate (MEP) pathways of isoprenoid biosynthesis existing in seaweeds. Cholesterol is the dominant sterol in red seaweeds and fucosterol in brown seaweeds, while the dominant sterol seems to vary within the orders in green seaweeds (Al Easa et al. 1995) such as isofucosterol in Ulvales and clionasterol in Bryopsidales and Siphonocladales. These seaweed sterols also possess beneficial health-promoting effects such as hypercholesterolemic, antioxidant, anticancer, antidiabetic, antihypertensive, and anti-inflammatory responses (Kim and Ta 2011 and references therein).

4.3 Seaweed Lipidomics: An Update

Seaweeds have been studied for decades for their health-benefitting and bioactive lipids. Multiple approaches have been employed for the separation, quantification, and characterization of lipids including thin-layer chromatography (TLC), gas chromatography (GC), gas chromatography mass spectrometry (GC-MS), nuclear magnetic resonance (NMR), liquid chromatography (LC) or high-pressure liquid chromatography (HPLC), and mass spectrometry with or without conjunction of a variety of complementary procedures such as chemical hydrolysis, regiospecific enzymatic cleavage, and spectrophotometric assays mainly to unravel the lipid diversity in seaweeds, specifically fatty acids, oxylipins, and sterols, and to quantitate their abundance or for their isolation and structural characterization. However, no method has yet been developed that can decipher the complete lipidome. Recently, Kumari et al. (2015) elucidated the first polar lipidome of *Gracilaria dura* under methyl jasmonate stress; Melo et al. (2015) revealed whole polar lipidome of *C. crispus* and Chen et al. (2016) of *Pyropia haitanensis* under high-temperature stress (discussed in detail in latter section). Moreover, "lipidomics" is relatively a new avenue in seaweed lipid research as compared to microalgae where the impetus of biodiesel production has driven the development of lipidomic field faster leading to the elucidation of lipid biosynthetic pathways and the advancement in lipid and FA extraction techniques, strategies to manipulate lipid metabolic pathways to obtain higher yields of lipid and FAs. Numerous desaturases and elongases have been cloned and characterized from microalgae, which are extensively and timely reviewed by Harwood and Guschina (2009), Khozin-Goldberg and Zvi (2011), and Khozin-Goldberg (2016). On contrary, our knowledge on metabolic pathways of lipid and FA metabolism of seaweeds and the genes involved is mainly based on those of higher plants and microalgae, and is believed to be similar to them in one or more aspects. Recently, Chan et al. (2012) identified the enzymes involved in FA biosynthesis such as acetyl CoA carboxylase, FAS I/II, desaturases, and elongases and studied the FA desaturation patterns in transcriptomes of *Pyropia* spp. These authors identified all the four genes encoding the subunits of acetyl CoA carboxylase complex (accA through accD) on the plastid genome of *Pyropia* sp., except for the biotin carboxylase gene (accC) that was located on the nuclear transcriptome data of *Pyropia* sp. Moreover, no KAS II gene was identified in *Pyropia* sp., suggesting that 16:0-ACP (rather than 18:0-ACP) is the final product of fatty acid synthesis, or this last elongation step can alternatively be accomplished by KAS I. This indicated that 16:0-ACP is the main fatty acid conjugate exported from the plastid and/or the elongation rate of C18-fatty acids is high. Besides this "plant-type" FAS complex, orthologs of the fungal enzymes were also identified in *Pyropia* spp. (Chan et al. 2012). Furthermore, these authors reported that *Pyropia* spp. lack plastid desaturation pathway including the soluble acyl-ACP-desaturase FAB2. Therefore, they hypothesized that possibly, saturated FAs (16:0 and possibly 18:0) are exported from the plastid to the ER for desaturation in contrast to higher plants, where oleic acid (18:1) is the major fatty acid that is synthesized in chloroplasts and exported to the ER. It has been found that KAS II, one of the key enzymes involved in FA biosynthesis in higher plants and microalgae, is not found in *Pyropia* sp., and its role is played by another isoform, KAS I. The whole genomes of brown seaweed *Ectocarpus* (Cock et al. 2010) and red seaweed *Chondrus* (Collén et al. 2013), *Pyropia* (Nakamura et al. 2013), *Saccharina* (Ye et al. 2015), and *Cladosiphon* (Nishitsuji et al. 2016) have identified the genes/enzymes involved in seaweed lipid metabolism but have also mentioned many unidentified loops such as in fatty acid, oxylipin (especially methyl jasmonate), and sphingolipid pathways. The need is to apply integrated approach of genomics, proteomics, and lipidomics to completely understand the unresolved riddle of seaweed lipid metabolism.

4.4 Tools and Techniques in Seaweed Lipidomics

The detection, identification, and precise quantification of lipid compounds are prerequisite for their potential utilization and exploration. Lipidomics aims at characterization and functional annotation of a broad range of lipid molecular species in organisms. As seaweed lipids are highly complex and diverse, full characterization of all of its structural diversity and quantification is quite a challenge. An overview of key factors involved in seaweed lipidomics is presented here.

4.4.1 Sampling and Lipid Extraction

For lipidomics, sample preparation is one of the crucial steps, and utmost care should be taken while sample harvesting, and the samples should be immediately quenched after harvest by snap freezing in liquid nitrogen to disrupt the cell metabolism and inactivate all the endogenous hydrolytic enzymes. The frozen samples should be stored at −80 °C until analysis. Cell disruption (homogenization by mortar pestle or automated techniques) is also important to separate the biomass from the extracellular matrix of intracellular metabolites and to make them accessible for different solvents for lipid extraction. After homogenization lipids can be extracted with appropriate solvent(s) to extract a wide range of lipid metabolites exhibiting the utmost structural diversity as possible. The liquid-liquid extraction, organic solvent precipitation, and solid-phase extraction have been used for lipid extraction in seaweeds, recently reviewed by Maciel et al. (2016). The conventional organic solvent extraction methods based on chloroform/methanol/water, both Bligh and Dyer (chloroform/methanol; 1/2, v/v) (Bligh and Dyer, 1959) and Folch method (chloroform/methanol; 2/1, v/v) (Folch et al. 1957), have been invariably used as standard methods for lipid extraction from seaweeds (Galloway et al. 2012; Kumari et al. 2010, 2011, 2013a, b). Other solvents such as dichloromethane methanol (Graeve et al. 2002), n-butanol (Kim et al. 2007), and diethyl ether (El-Shoubaky et al. 2008) have also been used but were not so effective. Kumari et al. (2011) reported that the solvent system comprising of chloroform/methanol/phosphate buffer is a better combination for extracting lipids because chloroform and methanol in combination exhibit strong dissolving power for the entire range of polarity found in lipids, as well as the ability to break up membrane and denature (-lipo)proteins (Schreiner 2006), and the addition of buffer helped to overcome the ionic adsorption effects of salt that hinder lipid extraction from seaweeds. An extraction procedure using methanol/methyl-tert-butyl-ether (MTBE)/water has also been successfully employed for polar lipidomics of seaweeds (Matyash et al. 2008; Melo et al. 2015). The advantage of this procedure is that MTBE is nontoxic and noncarcinogenic, so a green solvent for lipid extraction and the lipid-containing phase forms the upper layer during phase separation. Some novel green techniques have also been developed such as supercritical fluid extraction, microwave-assisted extraction, ultrasound-assisted extraction, and pressurized solvent extraction pulsed electric field-assisted extraction and enzyme-assisted extraction, mostly being tested in microalgae reviewed by Kumari et al. (2013a) and Maciel et al. (2016) and references therein. The expanding lipidomic field will surely develop a more suitable environment-friendly, cost-effective, and reproducible green lipid extraction procedures.

Further, it should be taken into account that complete recovery of every lipid class is difficult to achieve with any known method of lipid extraction. Any incomplete recovery may lead to an incomplete measurement of lipid content in a sample and inconsistencies in the results between inter-laboratory experiments, if analytical methods are based on external calibration. It is advisable to always add internal standards (one or more stable isotope labeled compounds) of each lipid type/class during the extraction procedure, for reliable, accurate, and reproducible quantification of lipid class of interest (Vaz et al. 2015; Wang et al. 2016). At last, there are a few precautions to be taken while preparing lipid extracts for lipidomic analysis. Care should be taken during sample preparation and storage of lipid extracts from being chemically or enzymatically modified as it negatively influences the lipidomic data analysis. Lipids in tissues and cells are relatively protected by natural antioxidant systems and compartmentalization. But after sample homogenization, the cellular content is mixed and unavoidably diluted, which renders lipids more prone to chemical or enzymatic modification. Enzymatic modifications can be minimized by extracting lipids at temperatures close to 0 °C or by adding a small percentage of organic solvent to the homogenization buffer. After extraction, lipids become more prone to chemical oxidation, and thus the lipid extracts should preferably be stored in glass vials and solubilized in sufficient organic solvent at −80 °C, and air/oxygen should be eliminated by flushing with inert gasses (Vaz et al. 2015).

4.4.2 Analytical Platforms for Lipidomics

The lipid metabolite analysis of any sample can be achieved by several methods such as those for separation (TLC, LC, GC) and those methods used for detection (MS, ESI, NMR), stand-alone or in combination such as LC-MS, GC-MS, ESI-MS, and TLC with GC-MS/LC-MS/NMR. Traditionally, TLC (both one and two dimensional) has been utilized for decades to separate lipid classes in seaweeds using silica as the stationary phase and different elution solvents depending on the polarity of the lipid classes to be isolated. Nonpolar lipids (glycolipids, free fatty acyls, and sterols) have been separated using hexane and diethyl ether and polar lipids (glycerophospholipids, saccharolipids, and sphingolipids) using chloroform, methanol, acetone, benzene, water, and triethylamine in different combinations in seaweeds (Khotimchenko and Vaśkovsky 2004; Kulikova and Khotimchenko 2000; Sanina et al. 2008). The different lipid classes have been detected by observing the intensity of the spots after spraying with a solution of primuline in acetone and visualizing under a UV lamp, or by placing the plate in iodine vapor, and identified based on the comparison with migration of pure lipid standards applied to the same TLC plate and the relative quantification achieved by densitometry. The analysis of the molecular species has been achieved by scraping the spots of each lipid class, with organic solvents and then analyzing by GC-MS (Dembitsky and Rozentsvet 1990; Khotimchenko and Vaśkovsky 2004; Sanina et al. 2004, 2008; Vaśkovsky et al. 1996). However, TLC is not a suitable method for comprehensive lipidomic studies, as it is a time-consuming method limited to detection of lipid classes, requires a large sample size, and has low resolution and sensitivity.

GC has been used for analyzing subsets of lipid compounds such as fatty acids, oxylipins, and sterols in a large number of seaweeds and has been typically coupled with MS (GC-MS) or flame-ionization detection (GC-FID) (Barbosa et al. 2015; Bouarab et al. 2004; Galloway et al. 2012; Gaquerel et al. 2007; Kamenarska et al. 2004; Kumari et al. 2010, 2011, 2013a, b; Küpper et al. 2006; Lion et al. 2006; Wiesemeier et al. 2008). One of its prerequisites is the capacity of lipid compound to enter vapor phase under conditions that will alter their molecular structure. Thus, this method is sensitive to the polarity of the compound and requires derivatization steps to improve volatility (Christie 1993). Fatty acids have been analyzed as fatty acid methyl esters (FAMEs) using GC/GC-MS generated either by acid-catalyzed trans-esterification (using BF_3/HCl/H_2SO_4) or alkali-catalyzed transmethylation using 2 M ethanolic/methanolic KOH of lipid extracts. Oxylipins and sterols have been mostly analyzed by methylation with ethereal diazomethane and silylation with a mixture of BSTFA (*N*,*N*-bistrimethylsilyl-trifluoroacetamide)/TMCS (trimethylchlorosilane) (Bouarab et al. 2004; Choi et al. 2012; Kamenarska et al. 2004; Lion et al. 2006; Wiesemeier et al. 2008). The major disadvantage of GC/GC-MS has been lower sensitivity for less abundant species, and it yields information on the hydrolysis products of lipids, not on the parent compounds, and thus, the identification of lipid classes and the information of fatty acid main location cannot be retrieved.

NMR is a strong nondestructive and nonselective technique to identify a wide variety of lipids without losing chemical information about the analyte environment in biological systems (Gross and Han 2011). It provides unique information about molecular structure and dynamics; however, its sensitivity and resolving power to distinguish individual chemical species is limited and complicated due to considerable number of spin-coupled multiplets. NMR in seaweeds has been mainly employed for structural characterization of purified bioactive lipid compounds and novel oxylipins (Al-Fadhli et al. 2006; Choi et al. 2012; Kousaka et al. 2003; Todd et al. 1993, 1994; Williams et al. 2007) instead of lipidomic studies. It is anticipated that it may develop as a powerful lipidomic tool in the future through the synergistic application of a solution-state and solid-state NMR approaches in seaweeds.

LC (or HPLC) is an analytical tool for separation of different subsets of lipid molecules such as lipid classes, oxylipins, and sterols and is usually coupled with evaporative light scattering detector (ELSD), UV, or MS, recently reviewed by Pati et al. (2016). The lipid analysis is performed using normal phase (NP), reverse phase (RP), or hydrophilic interactions (HILIC). In NPLC and HILIC, lipid molecules are distinguished by their hydrophilic properties and separate them according to their polar head groups. NPLC and HILIC are suitable methods for the separation of lipid classes and different groups of oxylipins (hydroxy, epoxy, oxo, and others). RPLC distinguish lipid molecules by their hydrophobic properties and separates them according to their length and unsaturation. RPLC coupled with MS is the most widely used method for analysis of complex lipids (Al-Fadhli et al. 2006; El-Baroty et al. 2011; Kendel et al. 2015; Kim et al. 2007; Williams et al. 2007) and oxylipins (Barbosa et al. 2015; Bouarab et al. 2004; Choi et al. 2012; Collén et al. 2013; Gaquerel et al. 2007; Küpper et al. 2009; Ritter et al. 2008, 2014,Weinberger et al. 2011) in various seaweeds including the species of *Avrainvillea*, *Chondria*, *Chondrus*, *Cymathere*, *Codium*, *Dilophys*, *Ectocarpus*, *Fucus*, *Gracilaria*, *Laminaria*, *Padina*, *Plocamium*, *Sargassum*, *Solieria*, *Ulva*, and others. Recently, Jacquemoud and Pohnert (2015) developed a protocol for the comprehensive analysis of oxylipins (including extraction, purification, chromatographic, and mass spectrometric procedures) for *G. vermiculophylla* that can be applicable to any other seaweed. They illustrated that a range of oxylipins (HETEs, di-HETEs, prostaglandins, and leukotrienes) can be extracted from seaweeds using methanol followed by enrichment with Waters OASIS HLB SPE cartridge and separation by LC-MS using mobile phase A, 0.1% formic acid (w/v) in 2% acetonitrile in ultra-pure water (v/v), and mobile phase B, 0.1% formic acid (w/v) in acetonitrile on a BEH C18 UPLC column (2.1 mm x 1.7 μm; Waters or a similar C18 column).

Different chromatographic methods coupled with MS have invariably been used for lipid analysis, but it is the development of soft ionization techniques (electrospray ionization mass spectrometry, ESI-MS; atmospheric pressure chemical ionization, APCI; and matrix-assisted laser desorption ionization, MALDI) that has greatly accelerated the field of lipidomics. In APCI, the sample is nebulized and heated so that both solvent and analytes are in the gas phase followed by a corona discharge which ionizes the solvent molecules that subsequently also ionize the analyte molecules. APCI generally yields monocharged ions and is mainly used with small thermally stable nonpolar molecules (<1500 Da) and is mostly applied for glycerolipids, fatty acyls and sterols, and fatty acid esters (Li et al. 2014; Vaz et al. 2015). For MALDI-MS-based lipidomic analysis, the sample is mixed with a matrix that readily forms crystals that aid the ionization process. The fluid mixture of sample and matrix is spotted on a MALDI plate and allowed to dry. To ionize the analyte molecules, a laser is fired at the matrix crystals in the dried-droplet spot, which absorbs the laser energy resulting in desorption and ionization. The ionized matrix molecule then transfers its charge to the analyte, thus ionizing the analyte (Vaz et al. 2015; Pati et al. 2016). ESI-MS, particularly, has emerged as a powerful and indispensable tool for lipidomics, as it allows analysis of thermally labile and nonvolatile compounds and thus can be applied to almost all lipid categories, discussed hereby in detail. (Vaz et al. 2015). In ESI-MS, the lipid sample is nebulized through a highly charged capillary using heated nitrogen gas, producing a fine aerosol that results in evaporation of the solvent and ionization of the molecules after which the ions enter the mass spectrometer. ESI-MS analysis of lipids results in decreased molecular ion decomposition and better reproducibility and has lower detection limits (Gross and Han 2011; Pati et al. 2016; Vaz et al. 2015; Wang et al. 2016). ESI-MS-based lipidomics is classified into two categories depending on whether the lipidomic analysis is being accomplished at a constant or varied lipid concentration:

A. LC-MS-based lipidomics: It is also known as CLASS approach (comprehensive lipidomic analysis by class separation), where the concentration of lipid solution is constantly changing. The lipid molecules are separated by HPLC/LC and analyzed by online MS.
B. Shotgun lipidomics: It is also known as direct infusion in which the concentration of lipid solution delivered to the ion source is constant. In this, the lipid sample is infused directly in the mass spectrometer without prior separation, and lipid species are identified and quantified by specific precursor ion scans (PIS) or neutral loss scans (NLS) reviewed recently by Wang et al. (2016). The lipid solution is delivered by a syringe pump (generally a tightly sealed high quality glass syringe) for direct infusion. Its major disadvantage is clogging of capillary

line and requirement of a relatively high flow rate for ion current stabilization, which makes the consumption of a large quantity of lipid samples and thus the biological material. Moreover, this system is difficult to be automated. These drawbacks of the delivery system have been now overcome with the introduction of robotic nanoflow ion sources such as NanoMate devices (Wang et al. 2016 and references therein). Further, shotgun lipidomics minimizes the difficulties of lipid analysis from alteration in concentration and ion pairing and avoids chromatographic anomalies. A constant interaction between lipid species under constant concentration leads to a constant ratio of ion peak intensities between lipid species of a class and a constant suppression of lipid species both within a class and between lipid classes (Vaz et al. 2015; Wang et al. 2016). A full mass spectrum can be acquired to display the molecular ions of all the species of a lipid class of interest and can be quantified by a direct comparison with their selected internal standards. Individual lipid species are identified according to the characteristic fragment ion(s) yielded from the common head group of a particular lipid class after collision-induced dissociation (Vaz et al. 2015; Wang et al. 2016). If the characteristic product ion is produced as a result of neutral loss of a common fragment, then the lipid species are scanned in NLS mode, while if the characteristic product ion represents the common fragment ion present in the product-ion mass spectra of individual lipid species of the class, then the lipid species are scanned in PIS mode for monitoring this fragment ion. The characteristic molecular ions or MS/MS fragmentation data of each lipid class is summarized in Table 4.1. Shotgun lipidomics is classified into three categories based on different MS approaches for analysis: tandem mass spectrometry-based, high mass accuracy-based, and multidimensional MS-based shotgun lipidomics (Table 4.2), recently reviewed by Wang et al. (2016).

4.4.3 *Lipidomic Data Analysis*

A large amount of data is generated in lipidomic experiments and thus requires preprocessing of data prior to detection and identification of lipid molecules, statistical and biological pathway analysis for reliable outputs, and correct biological interpretation.

4.4.3.1 Preprocessing of Lipidomic Data

Preprocessing involves the application of methods required to generate a peak table for each sample containing all detected peaks and their relative concentrations, with the aim to collect maximum number of true lipid metabolite signals from the data with minimal number of detected artifacts (Han et al. 2012; Vaz et al. 2015;

Table 4.1 Characteristic molecular ion(s) and MS/MS fragmentation data of polar lipid classes in ESI-MS based lipidomic analysis

		Detected ions in MS		Scan mode			
				Neutral loss scan (NLS)		Precursor ion scan (PIS)	
Lipids	References	Positive	Negative	Positive (Da)	Negative (Da)	Positive (m/z)	Negative (m/z)
Monogalactosyldiacylglycerol (MGDG)	da Costa et al. (2015), Kumari et al. (2013a, 2014b, 2015)	**$[M + Na]^+$**, $[M + NH_4]^+$	$[M\text{-}H]^-$	162 (Loss of one hexose residue)		**243** ($C_9H_{16}O_6Na$)	
Digalactosyldiacylglycerol (DGDG)	Maciel et al. (2016), Melo et al. (2015)	**$[M + Na]^+$**, $[M + NH_4]^+$	$[M\text{-}H]^-$	162 (Loss of one hexose residue) 341 (Loss of two hexose residues)		**243** ($C_9H_{16}O_6Na$) 347 $[Hex_{2res} + Na]^+$ 365 $[Hex_2 + Na]^+$ 405 ($C_{15}H_{26}O_{11}Na$)	
Sulfoquinovosyldiacylglycerol (SQDG)		$[M + Na]^+$, $[M + NH_4]^+$	**$[M\text{-}H]^-$**				**225** [Sulfonated glucose – H_2O ($C_6H_9O_7S$)]
1,2-Diacylglyceryl-3-*O*-4′-(*N*,*N*,*N*-trimethyl)-homoserine (DGTS)	da Costa et al. (2015), Roche and Leblond (2010)	**$[M + H]^+$**		**87** ($-CH_2CH_2N^+(CH_3)_3$) 74 ($-CH_2N^+(CH_3)_3$) 59 ($N^+(CH_3)_3$)		**236** ($-C_{10}H_{22}O_5N^+$)	
1,2-Diacylglyceryl-3-*O*-2′-(hydroxymethyl)-(*N*,*N*,*N*-trimethyl)-β-alanine (DGTA)	Roche and Leblond (2010)	**$[M + H]^+$**		**87** ($-CH_2CH_2N^+(CH_3)_3$) 74 ($-CH_2N^+(CH_3)_3$) 59 ($N^+(CH_3)_3$)		**236** ($-C_{10}H_{22}O_5N^+$)	

(continued)

Table 4.2 (continued)

Phosphatidylglycerol (PG)/ Phosphatidic acid (PA)/ Lyso-phosphatidylglycerol (LPG)	da Costa et al. (2015), Kumari et al. (2013a, 2014b, 2015) Maciel et al. (2016) Melo et al. (2015)	$[M + Na]^+$, $[M + NH_4]^+$	**$[M\text{-}H]^-$**		74 (CH_3COOCH_3)		**153** (Glycerol phosphate-H_2O) 171 ($C_3H_7O_2OPO_3H$)
Phosphatidylcholine (PC)/ Lysophosphatidylcholine (LPC)		**$[M + H]^+$**, $[M + Na]^+$				184 [Phosphocholine $HPO_4(CH_2)_2N^+(CH_3)_3$]	
Phosphatidylethanolamine (PE)/ Lysophosphatidylethanolamine (LPE)		**$[M + H]^+$**	$[M\text{-}H]^-$	141 (Phosphoethanolamine) 196 (Glycerol phosphoethanolamine-H_2O)			
Phosphatidylinositol (PI)		$[M + NH_4]^+$	**$[M\text{-}H]^-$**				241 (Cyclic inositol phosphate)
Phosphatidylserine (PS)		$[M\text{-}H]^+$	**$[M\text{-}H]^-$**	185 (Phosphoserine)	87 (Serine)		
Ceramide	Melo et al. (2015)	**$[M + H]^+$**, $[M + Na]^+$, $[M + NH_4]^+$	$[M\text{-}H]^-$			264 (Dehydrated sphingosine)	
Galactosylceramide (GalCer)				162 (Galactose – H_2O) 180 (Galactose)		264 (Dehydrated sphingosine)	
Inositolphosphoceramide (IPC)						241 (Cyclic inositol phosphate) 259 (Inositol monophosphate)	

Note: Characteristic ions most formed in ESI-MS are shown in bold for each lipid class

Table 4.2 Different categories of shotgun lipidomics

I	*Tandem mass spectrometry-based shotgun lipidomics* (Wang et al. 2016)
	It utilizes a class-specific NLS or PIS approach to specifically isolate individual lipid species of a polar class of interest using a triple quadrupole mass spectrometer. Quantification of individual lipid species is done with the help of internal standards (generally two of each lipid class of interest) after comparing the ion intensities of a lipid species with the calibration curve after considering the regression variables of the lipid species
Advantages	
1	All individual species in a particular lipid class can be detected in one MS/MS acquisition directly from a total lipid extract with any commercially available triple quadrupole (i.e., QqQ)-type mass spectrometer
2	It provides global determination of the lipid species of any targeted class at the level of instrumentation sensitivity in a high-throughput fashion
3	It's a simple, efficient, highly sensitive approach and requires less expensive instrumentation
Drawbacks	
1	Fatty acyl substituents of lipid species are not identified as this approach only targets the class-specific head group fragments
2	Some altered ionization conditions cannot be easily recognized during and after the experiments
3	The accurate quantification of the detected lipid species might not be as simple as expected because of the differential fragmentation thermodynamics and kinetics manifest in individual lipid species within each lipid class
II	*High mass accuracy-based shotgun lipidomics* (Wang et al. 2016; Brügger 2014)
	It utilizes advanced hybrid-type mass spectrometers (quadrupole time-of-flight, Q-TOF, or quadrupole OrbiTrap) to sensitively acquire full mass spectra of lipid samples of interest in a survey scan mode and conduct product-ion MS analysis of lipid species to determine all the fragments in an entire mass region of interest. Any interesting PIS and/or NLS can be extracted from the acquired data array of the product-ion mass spectra for identification of a specific lipid class. Quantification is performed from full mass spectra from which ion intensities between individual lipid species of interest and their corresponding internal standard(s) are compared. This approach is also known as data-dependent acquisition shotgun lipidomic approach (Schwudke et al. 2007a). Other variations of this approach such as top-down lipidomics (Schwudke et al. 2007b), bottom-up lipidomics (Schuhmann et al. 2011), and MS(All) method (Almeida et al. 2015) are also popular
Advantages	
1	It provides efficient, broad, and sensitive measurement of lipid species in a high-throughput fashion
2	This approach can be conducted in an untargeted fashion to analyze any lipid species present in a cellular lipidome if the dynamic range of the instrument is permitted
3	It allows one to perform multiple precursor ion and neutral loss scans simultaneously
Drawbacks	
1	Linear dynamic range of quantification depends on the instrument used under experimental conditions in either multi-PIS (NLS) or high mass accuracy strategies
2	Not efficient for analyzing poorly ionized lipids, particularly those in low abundance

(continued)

Table 4.2 (continued)

III	*Multidimensional MS-based shotgun lipidomics* (Wang et al. 2016; Brügger 2014)
	It utilizes the structural characteristics of lipid species to effectively identify individual lipid species including isobaric and isomeric species. The characteristic fragment ions either from the head group or resulted from the neutral loss of the head group are used to identify the lipid class of interest, and PIS or NLS of fatty acyl chains is used to identify the individual molecular species present within the class (Han and Gross 2005; Brügger 2014)
Advantages	
1	It uses mass spectrometer both as a separation tool and an analyzer, thereby significantly minimizing the ion suppression effects
2	It uses multibuilding blocks of individual lipid molecular species to identify their structure and isomers and thus eliminates the presence of any artifactual species present in single PIS- or NLS-based approach
3	It uses the peak contours in multidimensional space, which facilitates refinements in quantitation through two-step quantification approach to extend the linear dynamic range
4	It can be exploited to study the distinctive chemical characteristics of many lipid classes
Drawbacks	
1	It is a relatively low throughput and laborious approach because of the involvement of different procedures (e.g., derivatization) in multiplexed sample preparation
2	It cannot distinguish isomeric species, of which the fragmentation patterns are identical
3	It is not ideal for identification and quantitation of species of an unknown or uncharacterized lipid class since identification of the building blocks of a lipid class has to be predetermined

Wang et al. 2016). This is to be taken into account that there are no framed standard protocols and platforms either for collecting lipidomic data or preprocessing strategies. It basically involves peak detection, peak quantification/normalization, peak grouping, imputation of missing peaks, visualization, isotope correction, and compound identification (Harkewicz and Dennis 2011; Rolim et al. 2015; Vaz et al. 2015; Yetukuri et al. 2008), described briefly in Table 4.3. There are various commercial and open-source software tools as well as a few databases for preprocessing of lipidomic datasets such as LipidBlast, Lipid Data Analyzer, Lipid MS prediction tool, LipidQA, LIMSA, LipidPro, and others enlisted in Table 4.4. In addition, Lipid Profiler (Ejsing et al. 2006) and LipidInspector (Schwudke et al. 2005) are compatible with the lipidomic data acquired using hybrid quadrupole/time-of-flight instruments that can perform multiple precursor ion scans in a single experiment. Fatty acid analysis tool (FAAT) (Leavell and Leary 2006) is useful for the analysis of data coming from Fourier transform mass spectrometry The main functionalities of FAAT include identification of overlapping saturated and unsaturated lipids, assignment of known ions from a user-defined library, and handling of isotopic shifts from stable isotope labeling experiments. The commercial softwares can be efficiently integrated with the machine of the vendor, have advanced graphical interfaces, and are usually well documented. However, the analysis and interpretation of lipidomic data is still challenging, as most of the softwares are usually custom designed for a certain mass spectrometer or data acquisition procedures (Maciel

Table 4.3 Preprocessing steps of a lipidomic data analysis

Preprocessing steps	Description
Peak detection	It involves the detection of each lipid metabolite peak from the raw data for each individual sample. A list of peaks characterized by its mass (m/z) and retention time is obtained. Subsequently, peak quantification is done by determining the peak area through peak integration, and a detection limit (signal-to-noise level) for the intensity is set a priori to eliminate noise peaks in the case of LC-MS-based lipidomic data. For direct infusion data, the intensity of each peak is determined as a (weighted) average over the collection time
Peak quantification/normalization	It is accomplished using internal standards in order to obtain semiquantitative data that can be further statistically analyzed. Multiple internal standards are used to counteract the matrix effects and suppression in response of different lipid classes
Peak grouping	Lipid metabolite peaks are grouped according to their peak positions across groups of samples for their statistical comparison. Peak matching procedures search for peaks across samples within a group of related samples falling within a pre-specified m/z and retention time distance of each other and, subsequently, represent the same metabolite. Nonlinear time alignment approaches are further applied to correct time shifts between samples
Imputation of missing peaks	One or few samples in a group of related samples may sometimes miss one or more metabolite peaks due to sample outliers or experimental/bioinformatic shortcomings. Such missing peaks can be added to the peak table by assuming that they are located at the same position as the identified peaks in the related samples and integrate the measured intensity within these areas. At this stage, a peak group in the list is referred to as a feature, defined by its m/z and RT and intensity. Further analysis of the dataset determines which features correspond to identified metabolites or noise and which remain unidentified
Visualization	The complexity of lipidomic data decrees the use of visualization methods to explore raw and processed data for quality control and validation of results. For every feature, simple graphs of the contributing peaks (overlays of extracted ion chromatograms) or box plots of the intensities are employed to obtain rapid insight into the shape of the peaks and the relative differences between groups of samples
Isotope correction	At low resolution, often peaks from isotope molecules of one lipid overlap with peaks of a lipid with two more hydrogen atoms. Thus, the intensity of the isotope peak needs to be subtracted from the intensity of the second lipid, before all lipids can be properly quantified

(continued)

Table 4.3 (continued)

Preprocessing steps	Description
Compound identification	Lipid metabolite is usually based on queries to internal or external public databases of all known lipid species. Mass spectroscopic resolution and the quality of the chromatographic separation of different classes of lipids are key factors for the accuracy of the identification. At low resolution, there is a risk of ambiguous assignments because of overlapping lipid peaks with small mass differences in their mass, while, at high resolution, the m/z value of a feature is much more accurate and the identification of the corresponding lipid molecule is much more reliable. Several web-based and commercial services are available that can search MS files to identify lipids, but most of them are vendor specific and do not allow user modifications

Note: adapted from Boccard et al. (2010), Harkewicz and Dennis (2011), Rolim et al. (2015), Vaz et al. (2015), and Yetukuri et al. (2008)

et al. 2016; Vaz et al. 2015; Wang et al. 2016). The open-source packages are advantageous and helpful as they provide more flexibility and state-of-the-art methods for data interpretation, and the software can be tailored or extended to the specific needs of the researcher (Maciel et al. 2016; Vaz et al. 2015). Recently, Metabolon and SCIEX together developed the first lipidomic analysis system providing quantitative, specific, and complete information on complex lipids from biological samples, the Lipidyzer™ Platform (https://sciex.com/lipidyzer-platform) that produces accurate data on more than 1000 individual lipid species, and the results can be mapped to pathways at the level of FA metabolism, complex lipid metabolism, and lipid class metabolism. Metabolon's TrueMass™ Complex Lipid Panel technology helps the Lipidyzer Platform to achieve such specificity by using an innovative differential ion mobility spectroscopy-mass spectrometry platform (DMS-MS/MS) that allows only one class of lipid into the MS at a time and thus eliminating cross-class isomers in the analysis.

In addition, there are a series of websites and lipidomic consortia such as Lipid Library (http://lipidlibrary.co.uk), Cyberlipids (http://www.cyberlipid.org), LIPID MAPS (www.lipidmaps.org) and its affiliated sphinGOMAP (http://sphingolab.biology.gatech.edu/), LIPIDAT (www.lipidat.chemystry.ohio-state.edu/home.stm), and Lipidomics Expertise Platform (www.lipidomics-expertise), and similar community-wide efforts in Japan (www.lipidbank.jp) and Europe (www.lipidomics.net) that provide useful information about lipids structure, their functions, and detailed protocols to extract and separate the lipids, lipid standards, and lipidomic expertise.

Table 4.4 Lipid databases for lipidomic data analysis

Database	URL	Details
ChEBI	www.ebi.ac.uk/chebi	A database and ontology for chemical entities of biological interest including lipids
ChemBank	www.chembank.broadinstitute.org	A small-molecule screening and cheminformatics resource database
HMDB	http://www.hmdb.ca	It is a freely available electronic database containing detailed information about small-molecule metabolites found in the human body. It contains 42,003 metabolite entries including both water-soluble and lipid-soluble metabolites as well as metabolites that would be regarded as either abundant (>1 uM) or relatively rare (<1 nM). It supports extensive text, sequence, chemical structure, and relational query searches
KEGG lipids	http://www.genome.jp/kegg-in/get_htext?br08002.kegg	It provides lipid pathway maps as well as associated information such as the name, formula, mass, structure, biochemical reactions, and external links to other public databases
LIPID MAPS	http://www.lipidmaps.org	Database incorporating structure and annotation of lipids. It classifies lipids into eight categories, fatty acyls, glycerolipids, glycerophospholipids, saccharolipids, sphingolipids, sterol lipids, prenol lipids, and polyketides, and contains around 40,360 structures in its LIPID MAPS Structure Database (LMSD), each assigned with a specific LIPID MAPS ID
LipidBank	http://lipidbank.jp	Database for natural lipids including fatty acids, glycerolipids, sphingolipids, sterols and various vitamins. It contains >6000 unique molecular structures and their annotation
LIPIDAT	www.lipidat.chemystry.ohio-state.edu/home.stm	It is a relational database of thermodynamic and associated information on lipid mesophase and crystal polymorphic transitions, containing 15,385 lipid molecular structures
LipidHome	https://www.ebi.ac.uk/metabolights/lipidhome	A database of theoretical lipids optimized for high-throughput mass spectrometry lipidomics

(continued)

Table 4.4 (continued)

Database	URL	Details
LipidomeDB	https://www.k-state.edu/lipid/technology_development/index.html	An online database containing lipid profile data created by Kansas Lipidomics Research Center. It contains extensive metadata derived from biological experiments, as well as much additional information about lipid molecular species and their mass spectral properties. It also provides for continuous annotation of large datasets containing spectral information about identified lipid metabolites
LMSD	www.lipidmaps.org	LIPID MAPS Structure Database containing 300,000 lipid structures
LMSAD	http://lipid.zju.edu.cn/	The Lipid Mass Spectrum Analysis Database (LMSAD) is an object-relational database encompassing MS data, structures, and annotations of biologically relevant lipids. The initial version of this database contains about 80 records, representing mainly sphingolipids
SphingoMap	http://www.sphingomap.org	It is an evolving pathway map for sphingolipid biosynthesis that includes many of the known sphingolipids and glycosphingolipids arranged according to their biosynthetic origin. It uses an annotation suggested by the LIPID MAPS and Functional Glycomics Consortia and displays around 450 compounds

4.4.3.2 Statistical Data Analysis

Lipidomic datasets usually comprise tens to hundreds of identified lipid species, along with a large number of unidentified species, and thus the compound identification is followed by statistical analysis for validation of data, depending on the biological question of interest. The integrated collaboration of lipid profiling and uni-/multivariate statistics in a lipidomic approach helps in discovering potential lipid biomarkers and in-depth understanding of lipid biochemical pathways. The univariate method such as analysis of variance (ANOVA) is often used for the comparison of two or more groups. One drawback of univariate approach is that it may result in too many false-positive findings due to many hypothesis tests being performed (Han et al. 2012; Hendriks et al. 2011; Vaz et al. 2015). Multivariate approaches such as multivariate analysis of variance (MANOVA), permutational multivariate analysis of variance (PERMANOVA), and similarity percentages (SIMPER) are often used for validation of proposed models. The principal component analysis (PCA) and partial-least-squares discriminant analysis (PLS-DA) consider the correlation structure of the data and reduce its dimensionality by constructing so-called latent variables, which are combinations of the original variables, and also facilitating the visualization of the data in two or three dimensions. PLS-DA is mainly used for biomarker discovery and PCA or cluster analysis for obtaining information about the (separation of) groups of samples and/or metabolites (Chen et al. 2016; Hendriks et al. 2011; Kumari et al. 2014b; Melo et al. 2015; Vaz et al. 2015).

4.4.3.3 Bioinformatic Interpretation and Pathway Analysis

The next step is development of bioinformatic and systems biology approaches to link the changes of cellular lipidome to alterations in the biological functions, including the enzymatic activities that are involved in the biosynthesis of the altered lipid classes and molecular species, addressing the biological question of interest. The lipid classes and individual molecular species involved in the biosynthesis of a particular lipid class are clustered, and the known biosynthesis and/or remodeling pathways are utilized to simulate the ion profiles of the lipid class of interest (Haimi et al. 2006; Wenk 2010; Wang et al. 2016 and references therein). A best match between the simulated and determined ion spectra is achieved from simulation based on the known pathways. Numerous parameters involved in the biosynthetic pathways can be derived from the simulation of high-throughput lipidomic datasets that can be used for quantitative interpretation of the pathways involved in adaptive changes in lipid metabolism after any perturbation (Han et al. 2012; Vaz et al. 2015; Wenk 2010; Wang et al. 2016). To facilitate the biological interpretation of lipidomic data, the identified metabolites are integrated and visualized in the context of metabolic pathways obtained from public databases (Haimi et al. 2006; Han et al. 2012; Sreenivasaiah et al. 2012; Vaz et al. 2015; Wenk 2010). Further, a statistical approach toward pathway analysis involving metabolite set enrichment analysis

(MSEA) (Xia and Wishart 2011) based on gene set enrichment analysis (GSEA) can be used in the analysis of gene expression datasets. MSEA is used to investigate the enrichment of predefined groups of related metabolites instead of individual metabolites. MSEA starts with a list of metabolites that have been extracted from the experimental data through statistical approaches such as ANOVA, cluster analysis, PCA, or PLS-DA. Subsequently, overrepresentation of metabolites from a predefined metabolite set is checked by hypergeometric test (Vaz et al. 2015). Similar other metabolite enrichment softwares are available such as MetaGeneAlyse, MetaboAnalyst, MetaMapR, and MPEA (Fukushima and Kusano 2013). The metabolite sets can be constructed according to the desired aim. The construction of correlation networks from metabolic profiles to identify metabolites that are regulated and co-regulated across the conditions can be measured. The alteration of metabolic fluxes in a network can also be pursued for investigating metabolic adaptations at the molecular level (Han et al. 2012; Vaz et al. 2015; Wenk 2010).

However, managing and organizing lipid-related pathways into useful, interactive pathways and networks present a challenge for lipid bioinformatics. Lipidomic data-based pathways and/or network reconstruction analysis mainly relies upon The KEGG, LIPID MAPS consortium, and SphingoMAP databases (Table 4.4). The KEGG PATHWAY database offers information on most of the metabolic pathways including lipid pathways encompassing FA biosynthesis, FA elongation, FA metabolism, steroid biosynthesis, glycerolipid, glycerophospholipid, ether lipid, sphingolipid, AA, LA, ALA metabolism, and biosynthesis of unsaturated FAs. Additionally, KEGG also provides generic pathways (i.e., species-independent pathways) to serve as reference pathways for the reconstruction of context- or organism-specific pathways. The KEGG BRITE (http://www.genome.jp/kegg/brite.html) specifically maintains a collection of hierarchical classifications of lipid species whose reactions and pathways can be viewed. LIPID MAPS biopathway workbench (http://www.biopathwaysworkbench.org/) provides a graphic tool that facilitates to display, edit, and analyze biochemical pathways of lipids. Open-source visualization tools such as BioCyc (www.biocyc.org), CYTOSCAPE (http://cytoscape.org), LipidBANK (http://lipidbank.jp/), MAPPFinder (www.genmapp.org/help_v2/UsingMAPPFinder.htm), MetaCyc (http://metacyc.org/), MarVis-Pathway (http://marvis.gobics.de), PANTHER (www.pantherdb.org), Pathway Editor (www.lipidmaps.org/pathways/pathwayeditor.html), Pathway-Express (http://vortex.cs.wayne.edu/projects.htm), and VANTED (www.vanted.org) enable retrieval and visualization and editing of lipid signaling and metabolic pathways (Table 4.5).

4.5 Application of Lipidomics to the Seaweed Systems Biology

Systems biology is rather a new frontier to explore in seaweed biology with the availability of whole genomes of *Ectocarpus*, *Chondrus*, *Pyropia*, and *Saccharina*, which are promising existing and future models for understanding seaweed

Table 4.5 Software tools for lipidomic data analysis

Software tools	URL	Data type	Details
ALEX	http://mslipidomics.info/contents/	MS/MS	A software-based platform for streamlined data processing, management, and visualization of shotgun lipidomic data acquired using high mass accuracy or high-resolution mass spectrometers
AMDMS-SL	https://pharmacometabolomics.duhs.duke.edu/resources-tools/sanfordburnham-medical-research-institute	MS/MS	This tool has been developed to identify and quantify individual lipid species from the data obtained from the MDMS-SL approach
LIMSA	http://www.helsinki.fi/science/lipids/software.html	MS/MS	It serves as an interface to process data from individual full MS and tandem mass spectra. It searches and integrates peaks in a mass spectrum, matches the peaks with a user-supplied list of expected lipids, corrects for overlap in their isotopic patterns, and quantifies the identified lipid species according to internal standards
LipidBlast	http://fiehnlab.ucdavis.edu/projects/LipidBlast	MS/MS	It is in silico tandem mass spectral library, maintained by Fiehn Lab (University of California-Davis) and freely available for commercial and noncommercial uses. It contains 212, 516 tandem mass spectra for 119, 200 different lipids in 26 lipid classes. It can be successfully applied to analyze MS/MS data from over 40 different types of mass spectrometers
LipidomeDB DCE	https://www.k-state.edu/lipid/analytical_laboratory/analysis_components/lipidomedb_dce/index.html	MS/MS	A tool to process mass spectrometry data acquired after direct infusion of a lipid-containing biological extract containing a cocktail of internal standards into an electrospray source. It has been developed as a joint project of the Kansas Lipidomics Research Center and the K-INBRE Bioinformatics Core Facility
Lipid Data Analyzer (LDA)	http://genome.tugraz.at/lda/lda_description.shtml	LC-MS	A tool for the quantitation of lipids in LC-MS data which is empowered with a 3D algorithm that confines the peak borders in m/z and time direction and the use of theoretical isotopic distribution of an analyte as selection/exclusion criterion. It also provides standardization, a statistical module for results analysis, a batch mode for unattended analysis of several runs, and a 3D viewer for the manual verification

(continued)

Table 4.5 (continued)

Software tools	URL	Data type	Details
Lipid MS prediction tool	www.lipidmaps.org/tools/	LC-MS, MS/MS	This tool performs searches with precursor ion or product-ion peak lists on a variety of lipid classes, where structures are generated by computational methods or are present in the LIPID MAPS Structure Database. It also generates in silico product-ion peak lists for glycosphingolipids
LipidPro	http://www.neurogenetics.biozentrum.uni-wuerzburg.de/verbundprojekte/services/lipidpro/	LC-MS/MS	This tool supports the identification of lipids by interpreting large datasets generated by LC-MS/MS using the advanced data-independent acquisition mode MS(E). It matches the retention time-aligned mass-to-charge ratio data of molecular and fragment ions with a lipid database and generates a report on all identified lipid species
LipidQA	http://lipidqa.dom.wustl.edu/	MS/MS	It is a lipid qualitative/quantitative analysis software platform to identify and quantitate complex lipid molecular species in biological mixtures by data-dependent scanning and fragment-ion database searching
LipidSearch	https://www.thermofisher.com/order/catalog/product/IQLAAEGABSFAPCMBFK	LC-MS, MS/MS	This software tool developed jointly by Prof. Ryo Taguchi and MKI (Tokyo, Japan) is made available by Thermo Fisher Scientific. It enables automatic identification and quantification of cellular lipid species from large amount of mass spectrometric data obtained from both LC-MS and shotgun lipidomic approaches. It contains about 1.5 million lipid ions and their predicted fragmentation patterns. It supports a number of acquisition modes, including PIS, NLS, and product-ion analysis
LipidView™	https://sciex.com/products/software/lipidview-software	MS/MS	A data processing tool for the molecular characterization and quantification of ESI-MS-based lipidomic data. It is based on parent- and fragment-ion masses of lipid species. It streamlines various key steps including automated data processing from template methods, lipid species identification, comprehensive isotope contribution removal, multiple internal standards-based quantification, and visualization and result reporting
LipidXplorer	https://wiki.mpi-cbg.de/lipidx/Main_Page	MS/MS	An open-source software that supports the quantitative characterization of complex lipidomes by interpreting large datasets of shotgun mass spectra

MetaboSearch	http://omicsgeorgetown.edu/metabosearch.html	MS/MS	This tool enables simultaneous m/z search from HMDB, MMCD, Metlin, and LIPID MAPS and integrates the results
MS-Dial	http://prime.psc.riken.jp/MetabolomicsSoftware/MS-DIAL/index.html	MS/MS	This software tool deals with both data-dependent and data-independent MS/MS lipidomic datasets. It also supports compound identification, peak alignment, and principal component analysis on the graphical user interface. The spectrum information is outputted by MassBank, NIST, and Mascot formats, and the organized data matrix (sample vs metabolite) is exported as tab delimited text file
MS-Lamp	http://ms-lamp.igib.res.in/	ESI-MS, MALD-MS	Mass spectrometry-based lipid(ome) analyzer and molecular platform (MS-Lamp) is a stand-alone software capable of aiding and interpreting ESI and/or MALDI-MS spectrometric data of lipids
MZmine2	http://mzmine.sourceforge.net/	LC-MS	A framework for data processing and visualizing of metabolomics data including lipidomic data (not specifically dedicated to lipidomics data)
NIST	http://www.nist.gov/srd/nist1a.cfm	GC, LC-MS, MS/MS	A comprehensive MSRI library covering >200,000 EI spectra for metabolites (including lipids)
SimLipid®	http://premierbiosoft.com/lipid/index.html	LC-MS, LC-MS/MS, MS/MS	It is a high-throughput lipid identification as well as qualitative and quantitative data analysis tool that accepts experimental MS and MS/MS data, enables lipid profiling by searching precursors against the known lipid structures, and elucidates an unknown structure by matching experimental product ions against theoretical fragments of lipids from the database
VaLID	http://neurolipidomics.ca/	MS/MS	It is a web-based software that returns all theoretically possible phospholipids for a given m/z value and MS condition. It links a search engine, a phospholipid DB, and multiple visualization features for identification and dissemination of large-scale lipidomic datasets

biochemical pathways and their interactions with the environment. In seaweeds, extensive targeted and untargeted lipid analysis have been accomplished for profiling of specific lipid classes (e.g., FAs, oxylipins, glycerolipids, and saccharolipids) for different purposes such as chemotaxonomic biomarkers, biotechnological applications, and perturbation to external cues (Illijas et al. 2009; Kumari et al. 2013a, b, 2014b, 2015; Küpper et al. 2009; Rempt et al. 2012; Ritter et al. 2014; Weinberger et al. 2011; Yan et al. 2011). There are only a few lipidomic studies carried out in seaweeds using LC-MS-based CLASS approach or direct infusion shotgun lipidomic approach discussed herein and enlisted in Table 4.6.

4.5.1 Elucidation of Lipidomic Profiling and Identification of Novel Lipids

The first polar lipidome (polar lipid along with its FA composition) of the red seaweed *C. crispus* was elucidated through hydrophilic interaction liquid chromatography-electrospray ionization mass spectrometry (HILIC-ESI-MS) approach (Melo et al. 2015). Saccharolipids comprising of SQDG and DGDG; glycosphingolipids bearing ceramide backbones (galactosylceramides); inositolphosphoceramides; and glycerophospholipids comprising of PC, PG, PA, LPC, LPG, and betaine lipids, as well as some phytyl derivatives, as chlorophylls and pheophytins, were main lipid classes identified with >180 lipid molecular species. Oxylipin-containing saccharolipids, SQDGs (30:1-OH, 32:2-OH, 32:1-OH, 34:4-OH, 34:3-OH, 34:2-OH, 34:1-OH, 36:5-OH, and 36:4-OH) and DGDGs (32:2-OH, 34:2-OH, 34:1-OH, 36:5-OH, 36:4-OH, and 38:3-OH), were detected for the first time indicating the presence of 14:1-OH, 16:1-OH, 18:1-OH, 18:2-OH, and 18:4-OH in chloroplastidic saccharolipid fraction (Melo et al. 2015). Consecutively, da Costa et al. (2015) also reported the polar lipidome of *Codium tomentosum* using CLASS-based lipidomic approach with HILIC-ESI-MS. The polar lipid profile of *C. tomentosum* contained >200 species, including SQDG, sulfoquinovosylmonoacylglycerols (SQMG), MGDG, DGDG, PC, PI, PG, PA, LPC, LPG, and di- and monoacyl betaine lipids. SQMG, some lipid molecular species of monoacyl betaine lipids, were reported for the first time in any green algae. They also detected DGDG containing hydroxy-PUFAs, DGDG 32:3-OH (14:0/18:3-OH), and DGDG 32:2-OH (14:0/18:2) in *C. tomentosum* polar lipidome.

However, earlier reports of using ESI-MS-based targeted lipid profiling also prevail, mainly limited to the isolation and characterization partial lipidome or a few lipid classes. For example, two new monogalactosyldiacylglycerols, (2S)-1-O-(5Z,8Z,11Z,14Z,17Z-eicosapentaenoyl)-2-*O*-(9Z,12Z,15Z-octadecatrienoyl)-3-*O*-ß-D-galactopyranosyl-*sn*-glycerol and (2S)-1-*O*-(9Z,12Z,15Z-octadecatrienoyl) -2-*O*-(6Z,9Z,12Z,15Z-octadecatetraenoyl)-3-*O*ß-D-galactopyranosyl-*sn*-glycerol, were identified from *Sargassum thunbergii* (Mertens ex Roth) Kuntze by using FAB tandem mass spectrometry-based saccharolipid-targeted CLASS approach (Kim et al. 2007). Similarly, Ma et al. (2014) identified ten MGDG molecular species

Table 4.6 A list of seaweeds analyzed by lipidomic approach

Seaweeds	MS approach	Lipids detected	References
Chlorophyta			
Codium tomentosum Stackhouse	*CLASS* HILIC LC-MS^n ESI-LXQ-IT	MGDG, DGDG, SQDG, SQMG, DGTS, MGTS, PG, PC, PI, LPG, LPC, PA	da Costa et al. (2015)
Ulva fasciata Delile	*Targeted CLASS approach* LC-MS^n ESI-QqQ	SQDG, SQMG	El Baz et al. (2013)
Ulva intestinalis Linnaeus	*CLASS* HILIC LC-MS ESI/IT-TOF	SQDG, SQMG	Ragonese et al. (2014)
Ulva lactuca Linnaeus	*Shotgun approach* ESI-Q-TOF-MS^n	MGDG, DGDG, SQDG, DGTS, PG, PC, PS, PI, PE, LPG, LPC, LPE, PA	Kumari et al. (2014a, b)
Ulva rigida C. Agardh	*CLASS* HILIC LC-MS ESI/IT-TOF	SQDG, PC, LPE	Ragonese et al. (2014)
Phaeophyta			
Colpomenia sinuosa (Mertens ex Roth) Derbés & Solier	*CLASS* HILIC LC-MS ESI/IT-TOF	SQMG, PC, PI, LPE	Ragonese et al. (2014)
Cystoseira brachycarpa J. Agardh	*CLASS* HILIC LC-MS ESI/IT-TOF	SQDG, PG, PC	Ragonese et al. (2014)
Dictyota dichotoma (Hudson) J. V. Lamouroux	*CLASS* HILIC LC-MS ESI/IT-TOF	SQDG, SQMG, PC, LPE	Ragonese et al. (2014)
Sargassum thunbergii (Mertens ex Roth) Kuntze	*Targeted CLASS approach* RP-HPLC-FAB MS^n)	MGDGs	Kim et al. (2007)
Sargassum horneri (Turner) C. Agardh	*CLASS* RPLC-MS/MS	MGDGs	Ma et al. (2014)
Taonia atomaria (Woodward) J. Agardh	*Targeted CLASS approach* LC-MS^n ESI-QqQ	SQDG, SQMG	El Baz et al. (2013)
Rhodophyta			
Asparagopsis taxiformis (Delile) Trevisan	*CLASS* HILIC LC-MS ESI/IT-TOF	SQMG	Ragonese et al. (2014)

(continued)

Table 4.6 (continued)

Seaweeds	MS approach	Lipids detected	References
Chondria armata (Kützing) Okamura	*CLASS* Offline TLC-ESI-QTQF-MS^n	MGMG, MGDG, SQMG, PG	Al-Fadhli et al. (2006)
Chondrus crispus Stackhouse	*CLASS* HILIC LC-MS^n ESI-LXQ-IT	DGDG, SQDG, DGTS, PG, PC, PA, LPG, LPC	Melo et al. (2015)
Gracilaria dura (C. Agardh) J. Agardh	*Shotgun approach* ESI-Q-TOF-MS/MS	MGDG, DGDG, SQDG, PG, PC, PS, PI, PE, LPG, LPC, LPE, PA	Kumari et al. (2015)
Palmaria palmata (Linnaeus) F.Weber & D.Mohr	*CLASS* LC-Q-MS^n	MGDG, DGDG, SQDG, PG, PE	Banskota et al. (2014)
Pterocladiella capillacea (S. G. Gmelin) Santelices & Hommersand	*CLASS* HILIC LC-MS ESI/IT-TOF	DGDG, SQDG, SQMG, PG, PC, PI, PS, LPI, LPE	Ragonese et al. (2014)
Pyropia haitanensis (T. J. Chang & B. F. Zheng) N. Kikuchi & M. Miyata	*CLASS* UPLC-Q-TOF-MS	DGDG, lyso-MGDG, lyso-DGDG, SQDG, lyso-SQDG, LPA, LPC, LPE, LPI, LPG, DAG, PIP	Chen et al. (2016)

(mainly containing C14/C16 saturated FAs and C18/C16 unsaturated FA moieties) in *Sargassum horneri* (Turner) C. Agardh using RPLC-MS/MS, with MGDGs containing 18:2 at *sn*-2 position showing reduced triglycerides and FA accumulation in adipocytes. HILIC-ESI/IT-TOF-MS was used to analyze partial lipidomic profiles of a few seaweeds including *Pterocladiella capillacea* (S. G. Gmelin) Santelices & Hommersand containing lipid classes, viz., PG, PC, PI, LPI, PS, LPE, DGDG, SQDG, and SQMG, *Asparagopsis taxiformis* (Delile) Trevisan had a single SGMG, and *Dictyota dichotoma* (Hudson) J. V. Lamouroux mainly contained glycerophospholipids PC, LPE, SQMG, and SQDG (Ragonese et al. 2014). This lipidomic-based advanced approach is envisaged as a promising tool for elucidation of seaweed lipid fingerprints, for understanding their metabolism, dependence of environmental conditions, for and biotechnological development of seaweed edible products and bioactive compounds.

4.5.2 Lipidomic Changes in Seaweed Acclimation Strategies to Abiotic/Biotic Stress

Lipids are an integral part of cellular membrane and, thus, constitute a part of the first line of defense against any abiotic/biotic perturbation/stresses. Lipids play a crucial role in maintaining membrane fluidity in seaweeds by modulating the levels of PUFAs and inducing FA oxidation cascade during oxidative stress, which further leads to induction of defense-related genes/enzymes (Dittami et al. 2011, 2012; Gravot et al. 2010; Kumar et al. 2010a, b, 2011; Kumari et al. 2014b, 2015; Ritter et al. 2014). Recently, the role of lipids and targeted and nontargeted GC-MS-based metabolomic approaches to reveal the role of fatty acids, LC-ESI-MS-based approaches to reveal the role of oxylipins, and other lipid molecules in seaweed defense strategies against salinity, desiccation, metal, diurnal oscillations, wounding, and other stresses has been reviewed (Kumar et al. 2016 and references therein). Thus, hereby, updated information of lipidomic changes in seaweed acclimation strategies to abiotic/biotic stress unraveled by advanced shotgun or CLASS approach is discussed.

ESI-MS-based shotgun lipidomic approach was for the first time used to unravel the physiological roles of polar lipid metabolites including lipid classes, fatty acids, and oxylipins in green seaweed *Ulva lactuca* Linnaeus under nutritional constraints of nitrate and phosphate (Kumari et al. 2014b). They deciphered reactive oxygen species (ROS)-mediated nonenzymatic lipid peroxidation due to nutritional limitation-induced oxidative stress in *U. lactuca*. To combat this nutritional stress, *U. lactuca* thalli undergo lipid remodeling including shift in lipid classes and fatty acids (SFA/UFA), oxylipins (C18- and C20-derived hydroxyoxylipins) to combat nutritional stress. *U. lactuca* accumulated DGDG, SQDG, and DGTS when deprived of either nitrate, phosphate, or both, while supplementation of nutrients, especially nitrate, led to retrieval of lost MGDG. Similarly, glycerophospholipid synthesis was also found to be dependent on phosphate availability, while phosphate deprivation resulted in degradation of PLs to PA or lyso-phospholipids (Kumari et al. 2014b). Successively, Kumari et al. (2015) first time reported polar lipidome of *G. dura* (C. Agardh) J. Agardh and of any seaweed under methyl jasmonate stress using ESI-MS-based shotgun lipidomic approach, highlighting the channeling of fatty acyl chains from MGDG toward the biosynthesis of 13-hydroperoxylinolenic acid, which further directed toward either the jasmonate pathway or other alternative pathways of FA oxidation cascade, analogous to higher plants. These authors showed that MeJA induces a strong dose- and time-dependent oxidative burst, resulting in lipid peroxidation, induction of fatty acid oxidation cascade, hydroxy-oxylipin synthesis, upregulation of 13-LOX pathway, and modulation of lipid acyl chains, along with a shift toward secondary metabolism as a defense strategy to combat the induced oxidative stress. Lipid molecular species belonging to MGDG, PC, PE, and PA were found to be differentially expressed lipid classes in response to MeJA treatment, of which MGDG and PC were the most affected lipid classes due to higher metabolic flux of these classes during lipid metabolism, as they are the primary sites for de novo fatty acid allocation (Ohlrogge and Browse 1995). Also, high levels of PA (40:8, 40:7, 38:5, 38:4, 36:4, and 36:3) and lysolipids, LPC (20:4, 20:3, 18:3, and 18:2), and LPE (20:4) were reported, which were probably generated from PC and PE in the MeJA-treated thalli, indicating higher phospholipase activity and phospholipid turnover in *G. dura*. Also, *G.* dura modulated the lipid acyl chains in such a way that no significant change was observed in the fatty acid profile of the treated thalli as compared with those of the control, except for C16:0, C16:1 (*n*-9), C20:3 (*n*-6), and C20:4 (*n*-6) ($P < 0.05$). This may be a strategy to maintain the membrane fluidity and integrity of membrane to combat oxidative stress (Kumari et al. 2015). Recently, Chen et al. (2016) employed UPLC-Q-TOF-MS and multivariate statistical analysis (PCA and heat maps) to decipher lipidomic changes of *P. haitanensis* in short-term response to high-temperature stress to understand the effect of global warming on *Pyropia* sp. They identified 39 lipid biomarkers belonging to the classes DAG, DGDG, lyso-MGDG, lyso-DGDG, SQDG, lyso-SQDG, lyso-PA, lyso-PC, lyso- PE, lyso-PI, lyso-PG, and PIP differentially regulated in response to high-temperature stress in *Pyropia*. The biomarker-based heat map and box plots showed the decrease in levels of most of these lipid biomarkers (saccharolipids, Lyso-PE, Lyso-PI, Lyso-PG, and PIP) with the application of high-temperature stress (from 20 to 35 °C). This decrease in the level of major photosynthetic membrane lipids due to heat stress was suggested to be a consequence of the impairment of its photosynthetic apparatus. The accumulation of PA in plant cell membranes immediately after exposure to temperature stress concomitant with decrease in the levels of DAG, PE, and PG (Chen et al. 2016) indicated the transfer of the phosphatidyl group to produce PA (a stress biomarker). These lipidomic alterations in *P. haitanensis* are significant for studies regarding photosynthesis rate, signal transduction, and cell membrane stability during acclimation at higher temperature. In another study, UPLC-ESI-Q-TOF-MS analysis-based targeted oxylipin profiling was used to study activated chemical defense induced by increased production of PGE_2 and related toxic lipid compounds in non-native populations of *G. vermiculophylla* (Hammann et al. 2016). These authors reported that wounding of non-native seaweed populations of *G. vermiculophylla* resulted in approximately 390% more production of 15-keto-PGE_2, 90% more PGE_2, 37% more PGA_2, and 96% more 7,8-di-HETE than wounding of native populations. PGE_2, PGA_2, and 7,8-di-HETE are known to deter various biological enemies of *G. vermiculophylla* that cause tissue or cell wounding, and also repelled the mesograzer *Littorina brevicula*, indicating that non-native populations of *G. vermiculophylla* were more defended against herbivory than native populations. This increased capacity for activated chemical defense may be the reason behind their invasive success.

4.6 Integration of Lipidomics with Allied Omics Platforms

Integration of lipidomics with other allied omics (genomics, transcriptomics, proteomics, and metabolomics) platforms offers promising tool to practice systems biology to identify novel genes, enzymes, and lipid metabolic pathways to increase our comprehension of seaweed lipid biochemistry and the role of novel lipid molecules in combating different biotic/abiotic stresses. With the accessibility of whole genome sequences, next-generation mass spectrometers, and advanced bioinformatic tools, lipidomics has assumed a prominent role in systems biology studies through its unique ability to directly identify functional alterations in multiple lipid metabolic and signaling networks. systems biology is an interdisciplinary approach integrating data from different omics disciplines into numerical models with the aim to simulate the physiology of the organism (Kumar et al. 2016). Recently, Loizides-Mangold (2013) proposed a new term "systematic lipidomics" which combines different pathway analyses with MS-based lipidomics. Systematic lipidomics aims to investigate how the perturbance of one pathway influences the lipidome of cells. Pathway targeting in the context of lipid analysis will lead to an increase or decrease in certain lipid metabolites, either directly or indirectly through a cascade of signaling events. There are two approaches in systematic lipidomics: hypothesis-driven approach and data-driven approach (Fig. 4.4). In hypothesis-driven approach, prior knowledge is used to select a pathway hypothesized to have a lipid phenotype. In the data-driven approach, large-scale sets of biological pathways are selected to uncover novel pathway-lipid phenotype relationships and are supported by pathway analysis of genomic data or metabolite pathway enrichment analysis (Loizides-Mangold 2013).

The full genomes and characterization of key genetic pathways driving metabolism of model seaweeds have focused the direction of most of the omics studies toward model seaweeds *Ectocarpus*, *Chondrus*, *Pyropia*, and *Saccharina*. Most of the

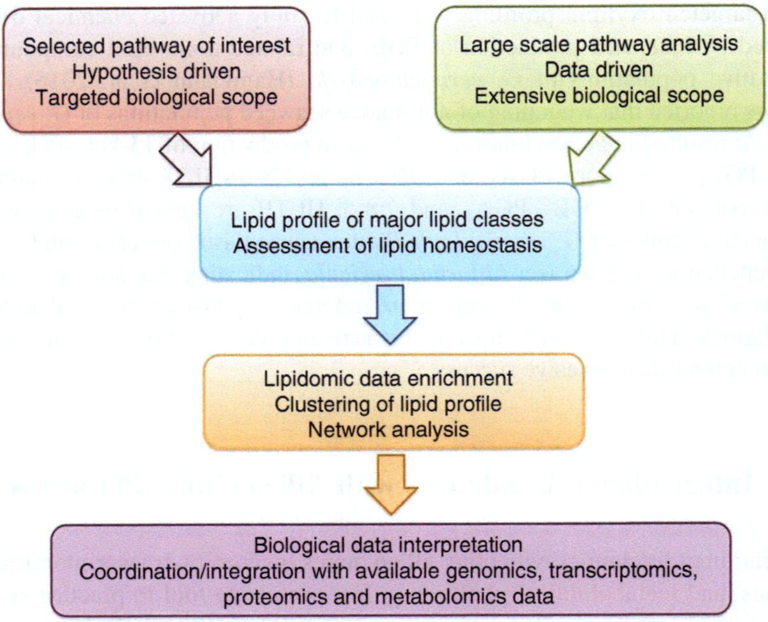

Fig. 4.4 Systematic lipidomic pathway analysis of lipidomic data. Adapted and modified from Loizides-Mangold (2013)[Loizides-Mangold U (2013) On the future of mass-spectrometry-based lipidomics. FEBS J 280:2817–2829]

network-based approaches employed in systems biology mainly include gene to metabolite, gene to protein, transcripts to metabolite, and metabolite to protein network interaction studies (Kumar et al. 2016). The integration of lipidomic data with the gene/protein/transcript is relatively limited since currently available pathway level representation of lipids in databases such as KEGG and SphingoMAP is limited and lacks the level of details required by modern MS-based approaches (Vaz et al. 2015). However, lately developed LIPID MAPS consortium provides all the user-friendly tools for pathway analysis (Tables 4.4 and 4.5), but it is slightly biased toward human/animal lipids and has a few ambiguities in respect to plant lipids in its classification and annotation ontology. LipidomeDB database (https://www.k-state.edu/lipid/technology_development/index.html), curated and maintained by Kansas Lipidomics Research Center, contains extensive metadata derived from biological experiments, as well as additional information about lipid molecular species and their mass spectral properties, and offers provision for continuous annotation of large datasets containing spectral information about identified lipid metabolites (Table 4.4). Its bioinformatic tool LipidomeDB DCE (Zhou et al. 2011) is also very helpful for plant lipid annotation. ARALIP specifically designed on the information available from *Arabidopsis* acyl lipid metabolism (2013) has improved the annotation of *Arabidopsis* genes related to acyl lipid metabolism and is organized by pathways represented by key navigation figures in the "Acyl lipid: pathways" homepage (http://aralip.plantbiology.msu.edu/pathways/pathways); ARALIP is also very helpful for annotation of plant lipids and their pathways representation.

However, only a few studies have been reported in seaweeds correlating gene-metabolite (including lipid molecules) co-expression deciphering their adaptive/acclimation responses to external cues (Dittami et al. 2011, 2012; Gravot et al. 2010; Groisillier et al. 2014; Rousvoal et al. 2011) recently reviewed by Kumar et al. (2016). Tonon et al. (2011) proposed a protocol focusing on integrating heterogeneous knowledge gained on brown algal metabolism in *Ectocarpus*. The resulting abstraction of the system will help in understanding how brown algae cope with changes in abiotic parameters within their unique habitat and in deciphering the mechanisms underlying their acclimation and adaptation, respectively, consequences of the behavior, or the topology of the system resulting from the integrative approach. The transcriptomic and metabolomic analysis of copper stress acclimation in *E. siliculosus* provided insight into the role of lipid and oxylipin metabolic pathways in seaweeds (Ritter et al. 2014). The gene-metabolite co-expression data highlighted the activation of oxylipins and repression of inositol (myoinositol) signaling pathways, together with the regulation of genes encoding for several transcription associated proteins. A significant accumulation of 12-OPDA, PPA1, and PPA2, C20:4 cyclic prostaglandins such as PGA_2 and PGJ_2 with no change in MeJA was observed along with the upregulation of genes belonging to the CYP74 family (an interesting candidate for AOS activity involved in the synthesis of oxylipins). Apart from this, there is hardly any information on correlation network analysis specifically focusing on integration of lipidomic data with either gene or proteome data.

With recent development of lipidomic strategies in seaweeds (da Costa et al. 2015; Kumari et al. 2015; Melo et al. 2015), the futuristic development of lipidome-gene as well as lipidome-proteome co-expression studies to decipher the novel lipid pathways and their regulation under different perturbations can be foreseen. Lipidomics in conjunction with other omic studies could also decipher the seaweed lipid metabolic pathways, which is believed to be identical to higher plants and microalgae, but not studied much. Recently only, it has been found that KAS II, one of the key enzymes involved in FA biosynthesis in higher plants and microalgae, is not found in *Pyropia* sp., and its role is played by its another isoform, KAS I (Chan et al. 2012), as discussed above in Sect. 4.3. Moreover, there are many dissimilarities in the lipid metabolic pathways (specifically in oxylipin biosynthetic and metabolic pathways) in the elucidated genomes of different seaweeds; which may be due to phyla- and/or species-specific variations due to conserved phyla-specific FA compositions in seaweeds (Kumari et al. 2013b). For example, genes for both plastidial and microsomal desaturases are present in the genome of *E. siliculosus*, while *C. crispus* genome does not have any gene for plastidial ACP desaturases. Instead it contains stearoyl desaturase, which produces 18:1 from 18:0 in the endoplasmic reticulum (Cock et al. 2010; Collén et al. 2013). The evolution of conserved lipid biosynthetic pathway and the huge diversity and variability present among different seaweeds can only be explained with integrated systems biology approach of lipidomics, genomics, and proteomics.

Jasmonates are lipid-derived signal molecules belonging to oxylipin class that mediates a plethora of biological processes from stress and defense responses to reproductive development, secondary metabolism, and senescence in plants (Browse 2009; Wasternack et al. 2012). They are generated via the allene oxide synthase (AOS) branch of the lipoxygenase (LOX) pathway of lipid oxidation and exert their effects by orchestrating large-scale reprogramming of gene expression (Kombrink 2012). Methyl jasmonate (MeJA) is one of the most active forms of jasmonic acid in plants produced by methyl ester formation on C1 of JA by JA-specific methyltransferase (Seo et al. 2001). Numerous studies have implicated the roles of MeJA in defense against biotic/abiotic stresses in seaweeds including *C. crispus* (Bouarab et al. 2004; Collén et al. 2013; Gaquerel et al. 2007), *Fucus vesiculosus* Linnaeus (Arnold et al. 2001), *Laminaria digitata* (Hudson) J. V. Lamouroux (Küpper et al. 2009), and *G. dura* (Kumari et al. 2015). However, it was unclear if MeJA was an endogenous compound in these seaweeds until the recent report of MeJA release from the cystocarps of *Grateloupia imbricata* Holmes (Pilar et al. 2016). MeJA was earlier only detected in *Gelidium latifolium* Bornet ex Hauck (Krupina and Dathe 1991), and all other attempts to detect MeJA in seaweeds (*Chondrus*, *Colpomenia*, *Dictyota*, *Ectocarpus*, *Fucus*, *Himanthalia*, *Saccharina*, and *Sargassum*) were not successful (Bouarab et al. 2004; Wiesemeier et al. 2008). Further, adding to the complexity of MeJA pathway in seaweeds, the entire set of enzymes necessary for the biosynthesis of JA from ALA has been identified in *Gracilariopsis* sp. (Hamberg and Gerwick 1993) and *Lithothamnion corallioides* (P. Crouan & H. Crouan) P. Crouan & H. Crouan (Hamberg 1992). The genome sequence of *E. siliculosus* (Cock et al. 2010) also showed the candidate genes for AOS, which catalyzes the initial step of JA biosynthesis and allene oxide cyclase (AOC). On contrary, no candidate genes could be identified for AOS and AOC in the genome of the red alga *C. crispus*, which contained two candidate genes for 12-oxo-phytodienoic acid reductase (12-OPR) and genes involved in ß-oxidation within the JA biosynthetic pathway (Collén et al. 2013). Moreover, no genes for jasmonic acid carboxyl methyl transferase have been identified in the genome of both the seaweeds indicating that an enzyme different from those characterized in land plants may be present in these seaweeds (Cock et al. 2010; Collén et al. 2013). No consensus is present in the scientific community about the occurrence and physiological relevance of MeJA in seaweeds. The combined genomics, transcriptomics, proteomics, and lipidomics study could only unveil the jasmonic acid/methyl jasmonate pathway in seaweeds and can explain whether this pathway is conserved or seaweeds exhibit an entirely different pathway from what is known in higher plants. Additionally, the active forms of jasmonic acid such as methyl jasmonate or jasmonic acid isoleucine conjugate or any other active forms can be identified and their mode of action can be deduced in seaweeds. Therefore, lipid biochemistry is an extremely diverse field with several unresolved questions in seaweeds, and systems biology approaches are the only answer to get an insight into complex seaweed lipid biology and its interactions with other metabolic pathways. It is anticipated that several models of lipid metabolic and signaling networks will be developed for seaweeds in the near future, and systems biology tools and approaches will shape the seaweed research landscape.

4.7 Conclusion

Recent advancements in MS-based approaches particularly the development of (ultra)high-resolution/mass accuracy measurement capabilities and refinement of soft ionization technique have been the driving force behind the successful application and integration of lipidomics in seaweed studies. Seaweed lipidomics has opened up an unprecedented opportunity to explore the lipid diversity, discover new lipid biomarkers, and deepen our understanding of seaweed lipid biochemical pathways that are crucial for adaptation and acclimation of seaweeds in diverse environmental conditions of marine ecosystem. To fully understand these mechanisms, elucidation of seaweed lipidome (by targeted/nontargeted approaches) must be combined with fluxes associated with lipidomic changes. The fluxes may also reveal hitherto uncharacterized lipid biochemical pathways and can be achieved with isotopomer experiments. Lipidomics along with other omics technologies is emerging as an important tool to practice systems biology in seaweeds as well, as lipids comprise a very significant part of the metabolome and play pleiotropic roles in cellular functions. With the increasing availability of genomic, transcriptomic, and proteomic data in seaweeds, context-specific association of lipid molecular species with proteins involved in biosynthesis and metabolism, and the concomitant genes encoding these proteins, several lipid-specific pathways will be deciphered/reconstructed for seaweeds in the future.

However, a future challenge is to develop an integrative workflow that combines different extraction procedures and mass spectrometric approaches to comprehensively assess all the lipid species in a given sample, as it is evident that no single platform is sufficient to fully characterize the lipidome completely. A plethora of different techniques is available, each with their own strengths and weaknesses. It is believed that combining different lipidomic techniques, such as shotgun/LC lipidomics, global profiling using LC-high-resolution MS, and quantitative targeted lipid analyses with LC-QqQ MS and GC-MS, has the potential to yield the most complete lipidomic datasets (Brügger 2014). Also, there is a great need for standardization and consolidation in the field of lipidomics, in terms of sample preparation, data collection methodology, and data management and analysis. Preprocessing is a complicated step in the analysis of lipidomic data, and choices of parameter settings greatly influence the outcome and thus the biological interpretation. More bioinformatic research is required to develop improved algorithms that eventually allow to preprocess data with no or only minimal human intervention. The standardization of procedures for sample preparation and analysis will increase interlaboratory reproducibility and reliability and allow more efficient use of generated lipidomic data. The further development of statistical, bioinformatics, and systems biology methods in seaweeds is another challenge to improve biological interpretation and pathway analysis. It is commendable that seaweed biologists are working in consortium across the world to address the demand of seaweed industry and making new-sophisticated tools in seaweed research available to public domains. Thus, the development of seaweed-dedicated databases encompassing all the omics information can be realized in near future. The systems biology approaches will be more common in seaweeds, and the integration and interpretation of lipidomic data with genome, transcriptome, proteome, and metabolome data will enable new insights in seaweed lipid biomarker discovery and lipid pathways and resolve the remaining questions in seaweed lipid biochemistry.

References

Al Easa HS, Kornprobst J, Rizk AM (1995) Major sterol composition of some algae from Qatar. Phytochemistry 39:373–374

Al-Fadhli A, Wahidulla S, D'Souza L (2006) Glycolipids from the red alga *Chondria armata* (Kütz.) Okamura. Glycobiology 16:902–915

Andreou A, Feussner I (2009) Lipoxygenases-structure and reaction mechanism. Phytochemistry 70:1504–1510

Arnold TM, Targett NM, Tanner CE et al (2001) Evidence for methyl jasmonate-induced phlorotannin production in *Fucus vesiculosus* (Phaeophyceae). J Phycol 37(6):1026–1029

Bano S, Uddin S, Ahmed VU (1990) Marine natural products part XV. An acetylated derivative of a new *N*-acylsphingosine from red alga *Halymenia porphyroides*. Planta Med 56:233–234

Barbosa M, Collado-González M, Andrade PB et al (2015) Nonenzymatic α-linolenic acid derivatives from the sea: macroalgae as novel sources of phytoprostanes. J Agric Food Chem 63:6466–6474

Banskota A, Stefanova R, Sperker S et al (2014) Lipids isolated from the cultivated red alga *Chondrus crispus* inhibit nitric oxide production. J Appl Phycol 26:1565–1571

Barbosa M, Valentão P, Andrade PB (2016) Biologically active oxylipins from enzymatic and nonenzymatic routes in macroalgae. Mar Drugs 14(1):23

Bligh EG, Dyer WJ (1959) A rapid method of total lipid extraction and purification. Can J Biochem Biophysiol 37:911–915

Boccard J, Veuthey JL, Rudaz S (2010) Knowledge discovery in metabolomics: an overview of MS data handling. J Sep Sci 33:290–304

Bouarab K, Adas F, Gaquerel E et al (2004) The innate immunity of a marine red alga involves oxylipins from both the eicosanoid and octadecanoid pathways. Plant Physiol 135:838–1848

Breen EP, Gouin SG, Murphy AF et al (2005) On the mechanism of mitochondrial uncoupling protein 1 function. J Biol Chem 281:2114–2119

Browse J (2009) Jasmonate passes muster: a receptor and targets for the defense hormone. Annu Rev Plant Biol 60:183–205

Brügger B (2014) Lipidomics: analysis of the lipid composition of cells and subcellular organelles by electrospray ionization mass spectrometry. Annu Rev Biochem 83:79–98

Burri L, Hoem N, Banni S et al (2012) Marine omega-3 phospholipids: metabolism and biological activities. Int J Mol Sci 13:15401–15419

Calder PC, Grimble RF (2002) Polyunsaturated fatty acids, inflammation and immunity. J Clin Nutr 56:14–19
Chan CX, Blouin NA, Zhuang Y et al (2012) *Porphyra* (Bangiophyceae) transcriptomes provide insights into red algal development and metabolism. J Phycol 48:1328–1342
Chen J, Li M, Yang R et al (2016) Profiling lipidome changes of *Pyropia haitanensis* in short-term response to high-temperature stress. J Appl Phycol 28:1903–1913
Choi H, Proteau PJ, Byrum T et al (2012) Cymatherelactone and cymatherols A-C, polycyclic oxylipins from the marine brown alga *Cymathere triplicata*. Phytochemistry 73:134–141
Christie WW (1993) Preparation of ester derivates of fatty acids for chromatographic analysis. In: Christie WW (ed) Advances in lipid methodology—II. Oily Press, Dundee, pp 69–111
Cock JM, Coelho SM, Brownlee C et al (2010) The *Ectoarpus* genome sequence: insights into brown algal biology and the evolutionary diversity of the eukaryotes. New Phytol 188:1–4
Collén J, Porcel B, Carré W et al (2013) Genome structure and metabolic features in the red seaweed *Chondrus crispus* shed light on evolution of the Archaeplastida. Proc Natl Acad Sci U S A 110:5247–5252
da Costa E, Melo T, Moreira ASP et al (2015) Decoding bioactive polar lipid profile of the macroalgae *Codium tomentosum* from a sustainable IMTA system using a lipidomic approach. Algal Res 12:388–397
Dembitsky VM (1996) Betaine ether-linked glycerolipids: chemistry and biology. Prog Lipid Res 35:1–51
Dembitsky VM, Rezanka T, Rozentsvet OA (1993) Lipid composition of three macrophytes from the Caspian Sea. Phytochemistry 33:1015–1019
Dembitsky VM, Rozentsvet OA (1990) Phospholipid composition of some marine red algae. Phytochemistry 29:3149–3152
Dembitsky VM, Rozentsvet OA (1996) Distribution of polar lipids in some marine, brackish and freshwater green macrophytes. Phytochemistry 41:483–488
Dembitsky VM, Rozentsvet OA, Pechenkina EE (1990) Glycolipids, phospholipids and fatty acids of brown algae species. Phytochemistry 29:3417–3421
Dennis EA (2009) Lipidomics joins the omics evolution. Proc Natl Acad Sci U S A 106:2089–2090
Dittami SM, Gravot A, Goulitquer S et al (2012) Towards deciphering dynamic changes and evolutionary mechanisms involved in the adaptation to low salinities in *Ectocarpus* (brown algae). Plant J 71:366–377
Dittami SM, Gravot A, Renault D (2011) Integrative analysis of metabolite and transcript abundance during the short-term response to saline and oxidative stress in the brown alga *Ectocarpus siliculosus*. Plant Cell Environ 34:629–642
Eichenberger W, Araki S, Muller DG (1993) Betaine lipids and phospholipids in brown algae. Phytochemistry 34:1323–1333
Ejsing CSE, Duchoslav J, Sampaio K et al (2006) Automated identification and quantification of glycerophospholipid molecular species by multiple precursor ion scanning. Anal Chem 78:6202–6214
El-Baroty GS, El-Baz FK, Abd-Elmoein A et al (2011) Evaluation of glycolipids of some Egyptian marine algae as a source of bioactive substances. Int Res J Pharmacy 2:165–174
El-Shoubaky GA, Moustafa AMY, Salem EAE (2008) Comparative phytochemical investigation of beneficial essential fatty acids on a variety of marine seaweeds algae. Res J Phytochem 2:18–26
El Baz FK, El Baroty GS, Abd El Baky HH et al (2013) Structural characterization and biological activity of sulfolipids from selected marine algae. Grasas Aceites 64:561–571
Fahy E, Subramaniam S, Brown HA et al (2005) A comprehensive classification system for lipids. J Lipid Res 46:839–861
Fahy E, Subramaniam S, Murphy RC et al (2009) Update of the LIPID MAPS comprehensive classification system for lipids. J Lipid Res 50:S9–S14
Fahy E, Cotter D, Sud M et al (2011) Lipid classification, structures and tools. Biochim Biophys Acta 1811:637–647

Folch J, Lees M, Sloane-Stanley GH (1957) A simple method for the isolation and purification of total lipids from animal tissues. J Biol Chem 226:497–509

Food and Agriculture Organization of the United Nations (2014) Fisheries and Aquaculture Information and Statistics Services [WWW document]. http://www.fao.org/figis/. Accessed 26 July 2014

Fukushima A, Kusano M (2013) Recent progress in the development of metabolome databases for plant systems biology. Front Plant Sci 4:73

Galloway AWE, Britton-Simmons KH, Duggins DO et al (2012) Fatty acid signatures differentiate marine macrophytes at ordinal and family ranks. J Phycol 48:956–965

Gaquerel E, Hervé C, Labrière C et al (2007) Evidence for oxylipin synthesis and induction of a new polyunsaturated fatty acid hydroxylase activity in *Chondrus crispus* in response to methyl jasmonate. Biochim Biophys Acta 1771:565–575

Gerwick WH, Bernart MW, Moghaddam MF et al (1990) Icosanoids from the Rhodophyta: a new metabolism in the algae. Hydrobiologia 204(205):621–628

Gerwick WH, Proteau PJ, Nagle DG et al (1993) Biologically active oxylipins from seaweeds. Hydrobiologia 260(261):653–665

Graeve M, Kattner G, Wiencke C et al (2002) Fatty acid composition of Arctic and Antarctic macroalgae: indicator of phylogenetic and trophic relationships. Mar Ecol Prog Ser 231:67–74

Gravot A, Dittami SM, Rousvoal S et al (2010) Diurnal oscillations of metabolite abundances and gene analysis provide new insights into central metabolic processes of the brown alga *Ectocarpus siliculosus*. New Phytol 188:98–110

Groisillier Z, Shao G, Michel S et al (2014) Mannitol metabolism in brown algae involves a new phosphatase family. J Exp Bot 65:559–570

Gross RW, Han X (2011) Lipidomics at the interface of structure and function in systems biology. Chem Biol 18:284–291

Haimi P, Uphoff A, Hermansson M, Somerharju P (2006) Software tools for analysis of mass spectrometric lipidome data. Anal Chem 78:8324–8331

Hamberg M (1992) Metabolism of 6,9,12-octadecatrienoic acid in the red alga *Lithothamnion coralliodes*: mechanism of formation of a conjugated tetraene fatty acid. Biochem Biophys Res Commun 188:1220–1227

Hamberg M, Gerwick WH (1993) Biosynthesis of vicinal dihydroxy fatty acids in the red alga *Gracilariopsis lemaneiformis*: identification of a sodium-dependent 12- lipoxygenase and a hydroperoxide isomerase. Arch Biochem Biophys 305:115–122

Hammann M, Rempt M, Pohnert G et al (2016) Increased potential for wound activated production of prostaglandin E_2 and related toxic compounds in non-native populations of *Gracilaria vermiculophylla*. Harmful Algae 51:81–88

Han X, Jiang X (2009) A review of lipidomic technologies applicable to sphingolipidomics and their relevant applications. Eur J Lipid Sci Technol 111:39–52

Han X, Gross RW (2003) Global analysis of cellular lipidomes directly from crude extracts of biological samples by electrospray ionization mass spectrometry: a bridge to lipidomics. J Lipid Res 44:1071–1079

Han X, Yang K, Gross RW (2012) Multi-dimensional mass spectrometry-based shotgun lipidomics and novel strategies for lipidomic analyses. Mass Spectrom Rev 31:134–178

Harkewicz R, Dennis EA (2011) Applications of mass spectrometry to lipids and membranes. Annu Rev Biochem 80:301–325

Harwood JL, Guschina IA (2009) The versatility of algae and their lipid metabolism. Biochimie 91:679–684

Hendriks MMWB, Eeuwijk FA, Jellema RH et al (2011) Data-processing strategies for metabolomics studies. Trends Anal Chem 30:1685–1698

Hofmann M, Eichenberger W (1997) Lipid and fatty acid composition of the marine brown alga *Dictyopteris membranacea*. Plant Cell Physiol 38:1046–1052

Hofmann M, Eichenberger W (1998) Radiolabelling studies on the lipid metabolism in the marine brown alga *Dictyopteris membranacea*. Plant Cell Physiol 39:508–515

Illijas IM, Indy JR, Yasui H et al (2009) Lipid and fatty acid composition of a little known and rarely collected *Exophyllum wentii* Weber- van Bosse from Bali island, Indonesia J Oleo Sci 58:103–110

Jacquemoud D, Pohnert G (2015) Extraction and analysis of oxylipins from macroalgae illustrated on the example *Gracilaria vermiculophylla*. Methods Mol Biol 1308:159–172

Kamenarska Z, Stefanov K, Dimitrova-Konaklieva S et al (2004) Chemical composition and biological activity of the brackish water green alga Cladophora Rivularis (L.) Hoek. Bot Mar 47:215–221

Kanamoto H, Takemura M, Ohyama K (2011) Identification of a cyclooxygenases gene from the red alga *Gracilaria vermiculophylla* and bioconversion of arachidonic acid to $PGF_{2\alpha}$ in engineered *Escherichia coli*. Appl Microbiol Biotechnol 91:1121–1129

Kato M, Miho S, Adachi K et al (1996) Distribution of betaine lipids in marine algae. Phytochemistry 42:1341–1345

Kendel M, Wielgosz-Collin G, Bertrand S et al (2015) Lipid composition, fatty acids and sterols in the seaweeds *Ulva armoricana*, and *Solieria chordalis* from Brittany (France): an analysis from nutritional, chemotaxonomic, and antiproliferative activity perspectives. Mar Drugs 13:5606–5628

Khotimchenko SV (2002) Distribution of glyceroglycolipids in marine algae and grasses. Chem Nat Compd 38:223–229

Khotimchenko SV (2003) The fatty acid composition of glycolipids of marine macrophytes. Russ J Mar Biol 29:126–128

Khotimchenko SV, Klochkova NG, Vaśkovsky VE (1990) Polar lipids of marine macrophytic algae as chemotaxonomic markers. Biochem Syst Eco 18:93–101

Khotimchenko SV, Kulikova IV (1999) Lipids of two species of brown algae of the genus *Laminaria*. Chem Nat Compd 35:17–20

Khotimchenko SV, Kulikova IV, Vaśkovski VE (2000) Distribution of ceramidephosphoinositol in red seaweeds. Russ J Mar Biol 26:286–288

Khotimchenko SV, Vaśkovsky VE (2004) An inositol-containing sphingolipid from the red alga *Gracilaria verrucosa*. Russ J Bioorg Chem 30:168–171

Khotimchenko SV, Vaśkovsky VE, Titlyanova TV (2002) Fatty acids from the Pacific coast of North California. Bot Mar 45:17–22

Khozin-goldberg I (2016) Lipid metabolism in microalgae. In: Borowitzka MA, Beardall J, Raven JA (eds) The physiology of microalgae, vol 6. Springer, Cham, pp 413–484

Khozin-goldberg I, Zvi C (2011) Unraveling algal lipid metabolism: recent advances in gene identification. Biochimie 93:91–100

Kim S, Ta QV (2011) Potential beneficial effects of marine algal sterols on human health. In: Kim S (ed) Advances in food and nutrition research, marine medicinal foods: implications and applications, micro and macroalgae, vol 64. Elsevier, USQQQ, pp 191–198

Kim YH, Kim EH, Lee C et al (2007) Two new monogalactosyl diacylglycerols from brown alga *Sargassum thunbergii*. Lipids 42:395–399

Kim M, Dubacq TJ, Thomas J et al (1996) Seasonal variations in triacylglycerols and fatty acids in Fucus serratus. Phytochemistry 43:49–55

Kishimoto K, Urade R, Ogawa T et al (2001) Nondestructive quantification of neutral lipids by thin-layer chromatography and laser-fluorescent scanning: suitable methods for "lipidome" analysis. Biochem Biophys Res Commun 281:657–662

Kombrink E (2012) Chemical and genetic exploration of jasmonate biosynthesis and signaling paths. Planta 236:1351–1366

Kousaka K, Ogi N, Akazawa Y et al (2003) Novel oxylipin metabolites from the brown alga *Eisenia bicyclis*. J Nat Prod 66:1318–1323

Krupina MV, Dathe W (1991) Occurrence of jasmonic acid in the red alga *Gelidium latifolium*. Z Naturforsch 46:1127–1129

Kulikova IV, Khotimchenko SV (2000) Lipids of different thallus regions of the brown alga *Sargassum miyabei* from the sea of Japan. Russ J Mar Biol 26:54–57

Kumar M, Kumari P, Gupta V et al (2010a) Biochemical responses of red alga *Gracilaria corticata* (Gracilariales, Rhodophyta) to salinity induced oxidative stress. J Exp Mar Biol Ecol 391:27–34

Kumar M, Kumari P, Gupta V et al (2010b) Oxidative stress induced biomarkers in marine macroalgae *Ulva lactuca* (Ulvales, Chlorophyta) under cadmium stress. Biometals 23:315–325

Kumar M, Gupta V, Trivedi N et al (2011) Desiccation induced oxidative stress and its biochemical responses in intertidal red alga *Gracilaria corticata* (Gracilariales, Rhodophyta). Environ Exp Bot 72:194–201

Kumar M, Kuzhiumparambil U, Pernice M et al (2016) Metabolomics: an emerging frontier of systems biology in marine macrophytes. Algal Res 16:76–92

Kumari P, Kumar M, Gupta V et al (2010) Tropical marine macroalgae as potential sources of nutritionally important PUFAs. Food Chem 120:749–757

Kumari P, Reddy CRK, Jha B (2011) Comparative evaluation and selection of a method for lipid and fatty acid extraction from macroalgae. Anal Biochem 415:134–144

Kumari P, Kumar M, Reddy CRK et al (2013a) Algal lipids, fatty acids and sterols. In: Domínguez H (ed) Functional ingredients from algae for food and nutraceuticals. Woodhead Publishing Ltd, Cambridge, UK, pp 87–134

Kumari P, Bijo AJ, Mantri VA et al (2013b) Fatty acid profiling of tropical marine macroalgae: an analysis from chemotaxonomic and nutritional perspectives. Phytochemistry 86:44–56

Kumari P, Reddy CRK, Jha B (2014a) Quantification of select endogenous hydroxy-oxylipins from tropical marine macroalgae. Mar Biotechnol 16:74–87

Kumari P, Reddy CRK, Jha B (2014b) Nitrate and phosphate regimes induced lipidomic and biochemical alterations in the intertidal macroalga *Ulva lactuca* (Ulvophyceae, Chlorophyta). Plant Cell Physiol 55:52–63

Kumari P, Reddy CRK, Jha B (2015) Methyl jasmonate-induced lipidomic and biochemical alterations in the intertidal macroalga *Gracilaria dura* (Gracilariaceae, Rhodophyta). Plant Cell Physiol 56:1877–1889

Künzler K, Eichenberger W (1997) Betaine lipids and zwitterionic phospholipids in plants and fungi. Phytochemistry 46:883–892

Küpper FC, Gaquerel E, Boneberg E et al (2006) Early events in the perception of lipopolysaccharides in the brown alga *Laminaria digitata* include an oxidative burst and activation of fatty acid oxidation cascades. J Exp Bot 57:1991–1999

Küpper FC, Gaquerel E, Cosse A et al (2009) Free fatty acids and methyl jasmonate trigger defense reactions in *Laminaria digitata*. Plant Cell Physiol 50:789–800

Leavell MD, Leary JA (2006) Fatty acid analysis tool (FAAT): an FT-ICR MS lipid analysis algorithm. Anal Chem 78:5497–5503

Li X, Fan X, Han L et al (2002) Fatty acids of some algae from the Bohai Sea. Phytochemistry 59:157–161

Li M, Yang L, Bai Y et al (2014) Analytical methods in lipidomics and their applications. Anal Chem 86:161–175

Lion U, Wiesemeier T, Weinberger F et al (2006) Phospholipases and galactolipases trigger oxylipin-mediated wound activated defence in the red alga *Gracilaria chilensis* against epiphytes. Chem Bio Chem 7:457–462

Liu QY, Reith ME (1994) Isolation of a gametophyte-specific cDNA encoding a lipoxygenase from the red alga *Porphyra purpurea*. Mol Mar Biol Biotechnol 3:206–209

Lo JM, Wang WL, Chiang YM et al (2001) Ceramides from the Taiwan red alga *Ceratodictyon spongiosum* and symbiotic sponge *Sigmadocia symbiotica*. J Chinese Chem Soc (Taipei, Taiwan) 48:821–826

Loizides-Mangold U (2013) On the future of mass-spectrometry-based lipidomics. FEBS J 280:2817–2829

Ma AC, Chen Z, Wang T et al (2014) Isolation of the molecular species of monogalactosyldiacylglycerols from brown edible seaweed *Sargassum horneri* and their inhibitory effects on triglyceride accumulation in 3T3-L1 adipocytes. J Agric Food Chem 62:11157–11162

Maciel E, Leal MC, Lillebo AI et al (2016) Biopropspecting of marine macrophytes using MS-based lipidomics as a new approach. Mar Drugs 14:49–76

Makewicz A, Gribi C, Eichenberger W (1997) Lipids of *Ectocarpus fasciculatus* (Phaeophyceae). Incorporation of [l-^{14}C] Oleate and the role of TAG and MGDG in lipid metabolism. Plant Cell Physiol 38:952–960

Matyash V, Liebisch G, Kurzchalia TV et al (2008) Lipid extraction by methyl-tert-butyl ether for high-throughput lipidomics. J Lipid Res 49:1137–1146

Melo T, Alves E, Azevedo V et al (2015) Lipidomics as a new approach for the bioprospecting of marine macroalgae-unraveling the polar lipid and fatty acid composition of *Chondrus crispus*. Algal Res 8:181–191

Michalik L, Auwerx J, Berger JP et al (2006) Peroxisome proliferator-activated receptors. International Union of Pharmacology, LXI. Pharmacol Rev 58:726–741

Mizusawa N, Wada H (2012) The role of lipids in photosystem II. Biochim Biophys Acta 1817:194–208

Muller DG, Eichenberger W (1994) Betaine lipid content and species delimitation in *Ectocarpus*, *Feldmannia* and *Hincksia* (Ectocarpales, Phaeophyceae). Eur J Phycol 29:219–225

Nagle DG, Gerwick WH (1990) Constanolactones a and B, novel cyclopropyl hydroxyl eicosanoids from the temperate red alga *Constantinea simplex*. Tetrahed Lett 31:2995–2998

Nakamura Y, Sasaki N, Kobayashi M et al (2013) The first symbiont-free genome sequence of marine red alga, Susabi-nori (*Pyropia yezoensis*). PLoS One 8(3):e57122

Nishitsuji K, Arimoto A, Iwai K et al (2016) A draft genome of the brown alga, *Cladosiphon okamuranus*, S-strain: a platform for future studies of 'mozuku' biolgy. DNA Res 23:561–570

Nylund GM, Weinberger F, Rempt M et al (2011) Metabolomic assessment of induced and activated chemical defence in the invasive red alga *Gracilaria vermiculophylla*. PLoS One 6(12):e29359

Ohlrogge J, Browse J (1995) Lipid biosynthesis. Plant Cell 7:957–970

Pati S, Nie B, Arnold RD et al (2016) Extraction, chromatographic and mass spectrometric methods for lipid analysis. Biomed Chromatogr 30:695–709

Pilar G, Olegario B, Rafael RR (2016) Occurrence of jasmonates during cystocarp development in the red alga *Grateloupia imbricata*. J Phycol 52(6):1085–1093

Proteau PJ, Gerwick WH (1993) Divinyl ethers and hydroxy fatty acids from three species of *Laminaria* (brown algae). Lipids 28:783–787

Ragonese C, Tedone L, Beccaria M et al (2014) Characterisation of lipid fraction of marine macroalgae by means of chromatography techniques coupled to mass spectrometry. Food Chem 145:932–940

Rempt M, Weinberger F, Grosser K et al (2012) Conserved and species-specific oxylipin pathways in the wound-activated chemical defense of the noninvasive red alga *Gracilaria chilensis* and the invasive *Gracilaria vermiculophylla*. Beilstein J Org Chem 8:283–289

Ritter A, Goulitquer S, Salaün J et al (2008) Copper stress induces biosynthesis of octadecanoid and eicosanoid oxygenated derivatives in the brown algal kelp *Laminaria digitata*. New Phytol 180:809–821

Ritter A, Dittami SM, Goulitquer S et al (2014) Transcriptomic and metabolomic analysis of copper stress acclimation in *Ectocarpus siliculosus* highlights signaling and tolerance mechanisms in brown algae. BMC Plant Biol 14:116

Roche SA, Leblond JD (2010) Betaine lipids in chlorarachniophytes. Phycol Res 58 (4):298–305

Rolim AEH, Henrique-Araújo R, Ferraz EG et al (2015) Lipidomics in the study of lipid metabolism: current perspectives in the omic sciences. Gene 554:131–139

Rousvoal S, Groisillier A, Dittami SM et al (2011) Mannitol-1-phosphate dehydrogenase activity in *Ectocarpus siliculosus*, a key role for mannitol synthesis in brown algae. Planta 233:261–273

Rozentsvet OA, Dembitsky VM, Zuhikova VS (1995) Lipids from macrophytes of the middle Volga. Phytochemistry 38:1209–1213

Sanina NM, Goncharova SN, Kostetsky EY (2004) Fatty acid composition of individual polar lipid classes from marine macrophytes. Phytochemistry 65:721–730

Sanina NM, Goncharova SN, Kostetsky EY (2008) Seasonal changes of fatty acid composition and thermotropic behavior of polar lipids from marine macrophytes. Phytochemistry 69:1517–1527

Schreiner M (2006) Optimization of solvent extraction and direct transmethylation methods for the analysis of egg yolk. Lipids Int J Food Prop 9:573–581

Schwudke DJ, Oegema L, Burton E et al (2005) Lipid profiling by multiple precursor and neutral loss scanning driven by the data-dependent acquisition. Anal Chem 78:585–595

Schwudke D, Hannich JT, Surendranath V et al (2007a) Top-down lipidomic screens by multivariate analysis of high-resolution survey mass spectra. Anal Chem 79: 4083–4093

Schwudke D, Liebisch G, Herzog R et al (2007b) Shotgun lipidomics by tandem mass spectrometry under data-dependent acquisition control. Methods Enzymol 433:175–91

Seo HS, Song JT, Cheong JJ et al (2001) Jasmonic acid carboxyl methyltransferase: a key enzyme for jasmonate-regulated plant responses. Proc Natl Acad Sci U S A 98:4788–4793

Sreenivasaiah PK, Rani S, Cayetano J et al (2012) IPAVS: integrated pathway resources, analysis and visualization system. Nucleic Acids Res 40:D803–D808

Todd JS, Proteau PJ, Gerwick WH (1993) Egregiachlorides A-C: new chlorinated oxylipins from the marine brown alga *Egregia menziesii.* Tetrahedron Lett 34:7689–7692

Todd JS, Proteau PJ, Gerwick WH (1994) The absolute configuration of ecklonialactones a, B, and E, novel oxylipins from brown algae of the genera *Ecklonia* and *Egregia.* J Nat Prod 57:171–174

Tonon T, Eveillard D, Prigent S et al (2011) Towards systems biology in brown algae to explore acclimation and adaptation to the shore environment. OMICS 12:883–892

Tremolieres A, Siegenthaler PA (1998) Role of acyl lipids in the function of photosynthetic membranes in higher plants. In: Siegenthaler PA, Murata N (eds) Lipids in photosynthesis: structure, function and genetics. Kluwer, Dordrecht, pp 145–173

Van Meer G (2005) Cellular lipidomics. The EMBO J 24:3159–3165

Vaśkovsky VE, Khotimchenko SV, Xia B et al (1996) Polar lipids and fatty acids of some marine macrophytes from the Yellow Sea. Phytochemistry 42:1347–1356

Vaz FM, Pras-Raves M, Bootsma AH et al (2015) Principles and practice of lipidomics. J Inherit Metab Dis 38:41–52

Vegiopoulos A, Muller-Decker K, Strzoda D et al (2010) Cyclooxygenase-2 controls energy homeostasis in mice by de novo recruitment of brown adipocytes. Science 328:1158–1161

Wang M, Wang C, Han RH et al (2016) Novel advances in shotgun lipidomics for biology and medicine. Prog Lipid Res 61:83–108

Wasternack C, Forner S, Strnad M et al (2012) Jasmonates in flower and seed development. Biochimie 95(1):79–85

Weinberger F, Lion U, Delage L et al (2011) Up-regulation of lipoxygenase, phospholipase, and oxylipin-production in the induced chemical defense of the red alga *Gracilaria chilensis* against epiphytes. J Chem Ecol 37:677–686

Wenk MR (2010) Lipidomics: new tools and applications. Cell 143:888–895

Wiesemeier T, Jahn K, Pohnert G (2008) No evidence for the induction of brown algal chemical defense by the phytohormones jasmonic acid and methyl jasmonate. J Chem Ecol 34:1523–1531

Williams DE, Sturgeon CM, Roberge M et al (2007) Nigricanosides a and B, antimitotic glycolipids isolated from the green alga *Avrainvillea nigricans* collected in Dominica. J Am Chem Soc 129:5822–5823

Xia J, Wishart DS (2011) Web-based inference of biological patterns, functions and pathways from metabolomic data using Metabo analyst. Nat Protoc 6:743–760

Yan X, Chen D, Xu J et al (2011) Profiles of photosynthetic glycerolipids in three strains of *Skeletonema* determined by UPLC-Q-TOF-MS. J Appl Phycol 23:271–282

Ye N, Zhang X, Zhao F (2015) Saccharina genomes provide novel insight into kelp biology. Nat Commun 6:1–11

Yetukuri L, Ekroos K, Vidal-Puig A et al (2008) Informatics and computational strategies for the study of lipids. Mol Bio Syst 4:121–127

Zhou Z, MArepally SR, Nune DS et al (2011) Lipidome DB data calculation environment: online processing of direct-infusion mass spectral data for lipid profiles. Lipids 46:879–884

Chapter 5
Volatiles in the Aquatic Marine Ecosystem: Ethylene and Related Plant Hormones and Sporulation in Red Seaweeds

P. Garcia-Jimenez and R.R. Robaina

Abstract Reports on the production of volatile compounds in algae have been focused on what they produce rather than on their functions. In this scenario, a myriad of fatty acid derivatives, nitrogen-containing compounds, organic halogen compounds, sulphur compounds and compounds derived from transferase activity have been described. Recently, a broad range of these volatile compounds has also been identified under a physiological complex pattern in algae, inferring that seaweeds must somehow integrate signals not only to reply to environmental status but also as a response to their growth and development capacity. Evidence comes from algal mats, which suffer sudden increments in the number of reproductive structures and correspondingly abrupt decreases in biomass. Specifically, the emission of airborne substances, such as ethylene and dimethylsulphide (DMS), has provided valuable information on their participation in algal physiology. DMS is affected by environmental factors such as light and salinity. In addition, the time course of DMS and ethylene release has revealed that the synthesis of ethylene, via the alternative route of DMSP lyase, is not a priority in the red seaweeds. Furthermore, ethylene has a significant effect on the formation of reproductive structures and the sporulation of the red seaweeds *Grateloupia imbricata* and *Pterocladiella capillacea*. Our data suggest that the presence of putative receptors and the response of ethylene could be influenced by the length of exposure to this volatile.

Unfortunately, despite the importance of all these facts in the formation of reproductive structures and spore germination, there is to date scant information at a molecular level, and little is known about the role of genes on these processes. In this chapter, we aim to compile the current knowledge of volatile compounds in algae. We will discuss the relation between the emission of volatiles and algal physiology and mainly focus on the involvement of volatiles in the reproduction of red seaweeds. As far as possible, we will try to unveil the molecular mechanisms of the perception of volatiles.

P. Garcia-Jimenez (✉) • R.R. Robaina
Department Biology, Facultad de Ciencias del Mar, Universidad de Las Palmas de Gran Canaria, Las Palmas de Gran Canaria, Canary Islands 35017, Spain
e-mail: pilar.garcia@ulpgc.es

© Springer International Publishing AG 2017

M. Kumar, P. Ralph (eds.), *Systems Biology of Marine Ecosystems*,
DOI 10.1007/978-3-319-62094-7_5

Keywords Ethylene • Jasmonates • Red algae • Sporogenesis • Seaweed reproduction • Volatiles

Contents

5.1 Introduction ... 100
5.2 Volatile Compounds Are All Around: How Do Algae Smell? ... 101
5.3 Complexity of Factors Drives Volatile Emission by Algae ... 102
5.4 Profiling the Physiological Responses of Algae to Volatiles ... 104
5.5 Ethylene and Methyl Jasmonate as Priming Volatile Growth Regulators that Affect Reproduction in Seaweeds ... 106
5.6 Genetic 'Dissection' of Volatile Emission in the Reproduction of Macroalgae ... 109
5.7 Conclusions ... 111
References ... 112

5.1 Introduction

Plants produce and emit large amounts of volatile organic compounds. The influence of scent is deemed to play several roles at the physiological and ecological level as defence compounds, as stabilisers and as protection for the plasma membrane against higher temperature and alterations of nearby flowering plants (Qualley and Dudareva 2010). One of the main roles of airborne compounds is also their involvement as signals to other organisms, and in this, volatiles facilitate interactions with the environment.

Volatile organic compounds in marine algae can also act as chemical communicators, where algae sense gaseous signals. Moreover, algae emit volatiles that vary both quantitatively and qualitatively depending on time and space. In recent years, a broad range of these volatiles have been identified in a physiological complex pattern, inferring that algae must also somehow integrate signals not only to reply to environmental status but also as a response to their growth and development capacity. Evidence comes from macroalga mats, which suffer sudden increments in the number of reproductive structures and correspondingly abrupt decreases in biomass (Garcia-Jimenez personal communication). To date, of the most important proceeding in algae, reproduction events are not still well understood and remain disputed. Reproduction is usually stimulated by environmental factors such as light, temperature and availability of nutrients that influence the transition from the vegetative phase to the reproductive phase (Cole and Sheath 1990). Furthermore, as described below, volatile production has also seen to be affected by (a)biotic factors. Thus, the effects of several interacting environmental factors, and the fact that one of these factors can also vary in one of the multiple aspects, such as light with its intensity, quality, the length of the day and the number of photons, make it difficult to offer a comprehensive interpretation of the role of volatiles in reproductive processes.

Moreover, responses to volatile growth regulators are reported to show great inconsistency in algae. Assuming uniform volatile sampling, the key to understanding this reasoning is the cellular competence of algae with respect to plant hormones, as Baweja et al. (2009) indicated. Competence is defined as the status in which a cell must be able to perceive, transduce and respond to a signal (Osborne and McManus 2005). There was a confused but apparently accepted idea that any part of any alga was competent with respect to plant growth regulators, but this is not so (Baweja et al. 2009). All in all, our knowledge of the role of volatiles in algal reproduction is so little and so disperse but truly incredible to discover.

5.2 Volatile Compounds Are All Around: How Do Algae Smell?

Very briefly, algae smell of short-chained aromatic volatiles, the most common components as benzaldehyde, which contribute to the flavour of processed algae and algae-based products, something of great importance in the industry, since their identification would determine the quality and freshness of a product as their organoleptic features are related to processed foods (Yamamoto et al. 2014). Other well-characterised volatile organic compounds have been identified, such as alkanes, alkenes, alcohols, terpenes, carbonyls, esters, halogen, compounds containing sulphur and acids (Loreto and Schnitzler 2010; Klaschka and Kolossa-Gehring 2007; Holopainen 2004; Bleecker 2000).

Halogenated compounds are the most extensively studied volatiles. The production of halomethanes from macroalgae has been reported to vary by up to three orders of magnitude depending on the geographical location (Schall et al. 1994; Manley et al. 1992) and by anthropogenic effect (Goodwin et al. 1997).

Specifically, chlorinated compounds have been encountered in algae, in spite of being relatively rare in living organisms. Their interest lies in their antibacterial properties, such as the green alga *Chara globularis*, reportedly emitting against *Staphylococcus aureus* (Bankova et al. 2001).

Micro- and macroalgae also contribute to a large extent to the production of bromine and polybromomethanes in marine environments (Paul and Pohnert 2011; Carpenter et al. 2003), and their presence in the atmosphere may directly affect human health, cloud condensation nuclei formation and climate (Pirjola et al. 2000). Moreover, bromine is known to be the halogen often found in marine-derived compounds (Abrahamsson et al. 1995). It is estimated that around 70% of the world's bromoform is produced by macroalgae (Carpenter and Liss 2000).

Furthermore, seaweeds also produce other haloforms, whose emission rates depend on light intensity and pH changes. For example, halomethane production has been reported in the green algae *Enteromorpha* and *Ulva*, in the giant kelp *Macrocystis* (Goodwin et al. 1997) and in the brown seaweeds *Ascophyllum*, *Fucus* and *Laminaria* and in the red seaweeds *Gelidium*, *Asparagopsis*, *Eucheuma* and *Mastocarpus* (Garcia-Jimenez et al. 2013; Mtolera et al. 1996 and references herein).

An intense bouquet from planktonic microalgae has also been associated with fatty acid transformation, mainly halogenated fatty acids and fatty acid-derived metabolites. Otherwise, fatty acids render long-chain aldehydes (LCA) by means of enzyme activities. More specifically, volatile compounds from long-chain aldehydes have been quantified and detected simultaneously with nitrogen compounds in the seagrass *Zostera marina* (Kawasaki et al. 1998) and in the red seaweed *Gelidium arbuscula* (Garcia-Jimenez et al. 2013). The presence of LCA-forming activities, both in seagrasses and seaweeds, has been hypothesised as part of the evolution from green algae to land plants (Kawasaki et al. 1998).

Consequently, studies of volatile organic compounds released by algae have focused more on the identities of all of the compounds produced than on their metabolic origins or potential functions. Moreover, the scarcity of standardised data obscures the exact role of these volatile compounds. Volatile screening varies enormously with respect to the amounts released, time monitoring and trapping methods used, making comparison difficult and generating confusion and producing speculative results.

Different approaches have been taken to clarify how far different experimental factors and biosynthesis pathways can influence the production of these volatile compounds. All of the myriad volatiles identified derive from different precursors, including amino acids, fatty acids and carotenoids, whose biosynthetic pathways are shared with the non-volatile secondary metabolites. Hence, several authors have established a classification of volatiles released depending on their functional groups, as a benchmark to improve the comprehension of the role of these compounds, in order to further infer the putative functions of these volatiles and their enzyme pathways. In the red seaweed *Gelidium arbuscula*, for instance, six volatile compound groups were generated—compounds consisting of methyl alkyl, amines, lipid oxidation derivatives, halides, sulphur and ethylene—in order to classify volatiles released under different experimental conditions and to further study their roles (Garcia-Jimenez et al. 2013).

5.3 Complexity of Factors Drives Volatile Emission by Algae

A set of factors affects the production of the myriad of volatiles reported. Light availability, desiccation, tissue age, kind of algae, wounded algae and grazing can influence production and rates of release of volatiles , moreover, with defined periods of temporal evolution (Leedham Elvidge et al. 2015). For instance, algae with large surface fronds, as *Laminaria digitata* from intertidal area, have been reported to release nearly 514–742 ng total volatiles detected g^{-1} dw after 4–6 h of desiccation, compared with other algae as *Fucus* and *Enteromorpha*, which retain water for longer periods, with rates of 110–170 ng total volatiles detected g^{-1} dw (Bravo-Linares et al. 2010). Moreover, in the brown alga *Laurencia dendroidea*, the accumulation and transport of secondary metabolites were associated with the presence of vesicle transport into the cells (Reis et al. 2013).

Additionally, when in vitro production of volatiles was analysed under different light and salinity conditions, the profiles of volatile organic compounds generated by the red seaweed *Gelidium arbuscula* differed with the light and salinity treatments tested (Garcia-Jimenez et al. 2013).

After mimicking the desiccation that *G. arbuscula* thalli experienced during low tides, the ethylene and the composition of gases emitted were also affected by treatments of exogenous ethylene (Garcia-Jimenez et al. 2013).

In broad-spectrum white light, most of the compounds generated by *G. arbuscula* were methyl alkyl compounds, amines and derivatives of lipid oxidation, while in darkness only methyl alkyl and amine compounds predominated. Despite the apparent similarity in the volatile emission profiles, white light affected the level of amines, constituting 50% of relative area counts liberated, in a different manner from those liberated in darkness (35% of relative area counts; Garcia-Jimenez et al. 2013). Comparing with the red light, we observed that not only did the profile of volatiles differ, the relative area counts of compounds derived from lipid oxidation also increased in red light (50% of relative area counts) compared to white light (20% of relative area counts; Garcia-Jimenez et al. 2013).

Natural emissions of DMS, the most abundant sulphur-containing compounds (Andreae et al. 1995), have also been determined for two green algae, *Ulva intestinalis* (Plettner et al. 2005) and *Enteromorpha clathrata* (Steinke et al. 1996), and for the red alga *Polysiphonia hendryi* (Van Alstyne and Houser 2003). By all means, DMS seems to play an important ecological role insofar as it contributes to the albedo (van Rijssel and Gieskes 2002). Meanwhile, dipropyl disulphide in the brown seaweed *Dictyopteris prolifera* has been differentially determined in protoplasts compared with thalli of this alga (Fujimura et al. 1994).

In *Gelidium arbuscula*, levels of dimethyl sulphide (DMS) were lower following thalli exposure to darkness and red light compared to those obtained under broad-spectrum white light. The in vivo emission of DMS in broad-spectrum white light was estimated as 3.6 ± 0.2 nmol h^{-1} g^{-1} fw (Garcia-Jimenez et al. 2013). Moreover, DMS was also affected by salinity in *G. arbuscula*. DMS breaks down dimethylsulphoniopropionate (DMSP) by the action of the enzyme DMSP lyase. Consequently, the more DMSP is broken down, the more DMS is liberated. In algae, DMSP is accumulated as an osmoticum to foster acclimation to high salinity (Otte et al. 2004; Stefels 2000; Dacey et al. 1987). DMS in the red seaweed *G. arbuscula* has been determined as 2.4 ± 0.02 nmol DMS h^{-1} g^{-1} fw at 60 psu, while when salinity fell to 36 psu, DMS liberation was 20.6 ± 0.015 nmol DMS h^{-1} g^{-1} fw (Garcia-Jimenez et al. 2013).

We were also able to determine endogenous jasmonates levels in four different seaweeds. Beyond the fact that production of jasmonates relies on the corresponding metabolism of seaweeds (availability of substrates, biosynthetic enzymes and emission routes), brown algae emitted a high concentration of jasmonates compared with the green and red algae (Table 5.1). No comparative information on the presence of methyl jasmonate between seaweeds has been reported previously, although our results infer that methyl jasmonate could be involved in several tasks. Without discerning the exact role of this compound among algal genera, the increment in volatile concentration in brown algae could be explained by the existence of an environmental control (algae located in the upper-intertidal area, where desiccation and herbivory are intense and could produce more volatiles; Hay 1996) and by the cellular structure (volatile compounds may be concentrated in the outer meristem; Tugwell and Branch 1989).

Table 5.1 Rates of jasmonates (mM jasmonates mg^{-1} fw h^{-1}) release in four macroalgae from intertidal coast of Gran Canaria (Canary Islands)

	Rates of jasmonates released (mM mg^{-1} fw h^{-1})	*n*
Green alga		
Enteromorpha sp.	1.05 ± 0.12	24
Red algae (infertile thalli)		
Grateloupia imbricata	0.96 ± 0.14	37
Gelidium arbuscula	1.05 ± 0.11	43
Brown alga		
Fucus sp.	1.34 ± 0.13[a]	21

n, number of samples analysed
[a]Significant differences ($p < 0.01$)

5.4 Profiling the Physiological Responses of Algae to Volatiles

It seems surprising that in spite of the massive production of volatiles released by algae, this has not been clearly related to any physiological events. In general terms, the constitutive emissions of volatiles, as opposed to those induced as responses to external attacks, could be due to specific changes during developmental stages, although the origin of volatiles remains to be properly discerned. Marine brown algae produce several C_{11} hydrocarbons, some of which are recognised as sexual pheromones that attract gametes (Müller 1989). Moreover, the emission of volatiles has been involved in algal successions and in blooms of microalgae (Takamo et al. 2003). The question of whether the emissions lead to an adaptive advantage or, on the contrary, it is merely to avoid toxic metabolites remains ambiguous. Hence, the understanding of the synthesis routes of the volatile compounds and how far physiological factors influence volatile production are issues that must be resolved in order to characterise algal response.

Most of the biosynthetic pathways for the production of the volatile compounds rely on S-adenosylmethionine (SAM), either as a substrate for key enzyme reactions (as in the synthesis of ethylene and polyamines) or as a source of a methyl group (as in the synthesis of jasmonates, salicylates or brassinosteroids) (Roje 2006; Yang and Hoffmann 1984). The presence of transferase enzyme activity and the production of halide forms would reflect its metabolic involvement as a means of eliminating halogenated metabolites.

Apart from transferases, other strategies to trap halogens in organic molecules have been reported through SAM-dependent enzymes and halogenation enzymes (La Barre et al. 2010). In particular, the presence of the most abundant haloperoxidases deals with a particular physiological event in algae. The oxidation of halogens, through H_2O_2 as substrate, would correspond with the alteration of cell strength, favouring the substrate adhesion of algal spores and propagules (Bitton et al. 2006, 2007). Specifically, brown seaweeds show a high oxidised bromine compound content, which results in compounds involved in adhesion of Laminariales (Salgado et al. 2009).

Other volatiles, such as ethylene, derived from the oxidation of 1-aminocyclopr opane-1-carboxylic acid (ACC) from SAM (Vanden Driessche et al. 1997), have also been determined. Ethylene affected the cap formation and development rates in the unicellular alga *Acetabularia* in a manner that depended on the timing of ethylene application (Vanden Driessche et al. 1998). Ethylene production was also noted through the decomposition of ethephon, a compound that spontaneously decomposes to generate ethylene. In the green seaweed *Enteromorpha intestinalis*, the addition of ethephon reduces chlorophyll levels compared to thalli without ethephon (Plettner et al. 2005).

Some authors have suggested an alternative to the conventional ethylene biosynthesis pathway through ACC for ethylene in marine organisms. Dimethylsulphoniopropionate (DMSP) lyase (EC 4.4.1.3) would act on DMSP to produce volatile DMS and acrylate, and acrylate would be converted to ethylene by decarboxylation (Plettner et al. 2005; Niki et al. 2000; Yoch et al. 1997). At a physiological level, DMS could scavenge reactive oxygen intermediates from cells, while DMSP would maintain salinity acclimatisation of algae (Sunda et al. 2002).

Furthermore, the time course of ethylene and DMS production in vitro in the red seaweed *G. arbuscula* suggests that ethylene is mainly produced through the ACC synthase and the ACC oxidase pathway (Garcia-Jimenez et al. 2013; Garcia-Jimenez and Robaina 2012). As these authors reported, two considerations can be highlighted. Initially, the time course of ethylene and DMS production revealed that DMS levels remained unaltered, while ethylene release increased. Therefore, it is understood that ethylene is not a result of the final transformation of DMSP. Secondly, the determination of enzyme activities involved in both pathways showed that ACC synthase and ACC oxidase-specific activities are predominant compared to DMSP lyase in this red alga (Fig. 5.1).

Moreover, the jasmonic acid and its relative compounds have been determined in the unicellular green algae *Euglena* and *Chlorella* (Ueda et al. 1991a, b) and in the red alga *Gelidium* (Krupina and Dathe 1991). The content of methyl jasmonate varies during the cell cycle of the Chlorophyta *Scenedesmus acutus* (Christov et al. 1996), in response to temperature stress in *Scenedesmus incrassulatus* (Christov et al. 2001) and for anti-herbivory activity in the brown alga *Fucus vesiculosus* (Arnold et al. 2001). In *Laminaria digitata*, production of methyl jasmonate seems to be involved in protection and against its brown endophyte (Küpper et al. 2009). In the red seaweed *Chondrus crispus*, methyl jasmonate activates the oxidative metabolism of polyunsaturated fatty acids, leading to the production of hydroperoxides and oxygenated fatty acids (Gaquerel 2005; Bouarab et al. 2004), inducing resistance activities involved in defence reactions (Gaquerel et al. 2009).

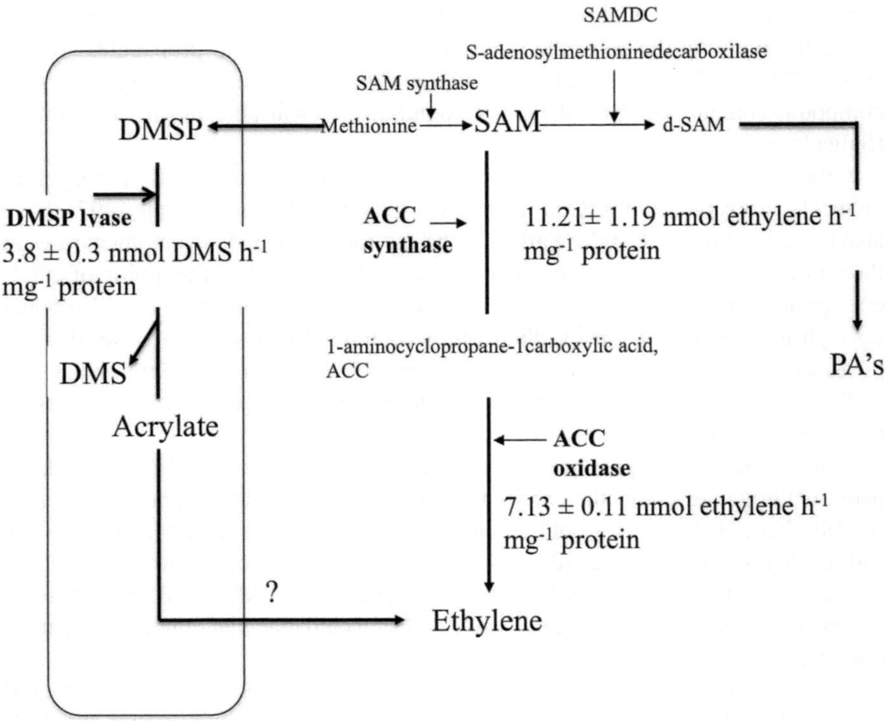

Fig. 5.1 The ethylene biosynthetic pathway and its connection with DMSP lyase and ACC synthetase and oxidase routes. *DMSP lyase* dimethylsulphoniopropionate lyase, *ACC* 1-aminocyclopropane-1-carboxylic acid

5.5 Ethylene and Methyl Jasmonate as Priming Volatile Growth Regulators that Affect Reproduction in Seaweeds

Excitingly, red seaweeds also produce and perceive ethylene and methyl jasmonate with major physiological implications, such as reproduction. Depending on the ethylene biosynthetic route, the enzyme activities of ACC synthase and ACC oxidase are reported, and ethylene production has been correlated with these enzyme activities (Garcia-Jimenez and Robaina 2012). Ethylene affects the in vitro development of tetrasporangial branches in the red seaweed *Pterocladiella capillacea* (Fig. 5.2a) and the incremental number of cystocarps in *G. imbricata* (Garcia-Jimenez and Robaina 2012), while the application of an inhibitor of the perception of ethylene, such as silver thiosulphate (STS), eliminates this ethylene response.

The increment in the number of tetrasporangial branches in *P. capillacea* is also dependent on the duration of ethylene exposure, increasing nearly 200-fold in thalli treated with ethylene for 30 min compared to untreated thalli. In vitro, the increase in ethylene production after an initial exposure to exogenous ethylene suggests an

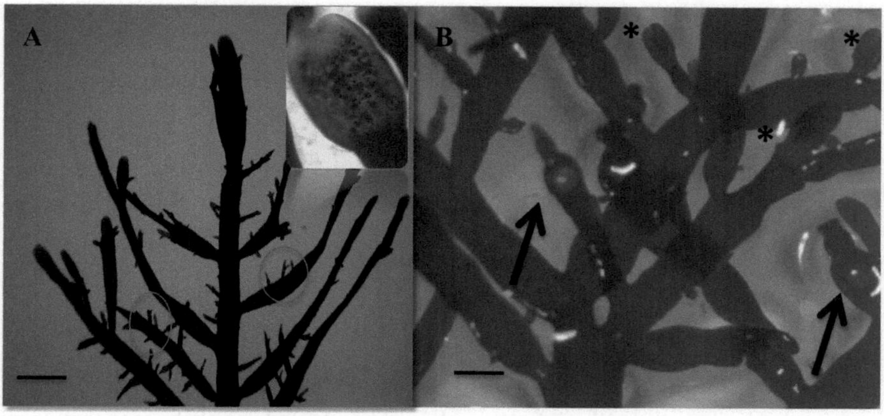

Fig. 5.2 Gelidiaceae. (**a**) Development of tetrasporangial branches in thalli of *Pterocladiella capillacea* treated with ethylene. (**b**) Mixed reproductive phases in the same individual of *Gelidium arbuscula* treated with methyl jasmonate. *Arrows*, carposporangial branches; *, tetrasporangial branches. Bar scale is 0.75 mm

initial stimulation of ethylene receptors and their saturation after prolonged exposure (Garcia-Jimenez et al. 2013; Garcia-Jimenez and Robaina 2012). Recently, the formation of spermatia and zygotospores in the gametophytes of *Pyropia yezoensis* after the exogenous treatment with the ethylene precursor ACC was reported (Uji et al. 2016).

The term jasmonates is used to describe lipid derivatives synthesised via the octadecanoid pathway and is mainly represented by JA, JA derivatives and its ester methyl jasmonate (Wasternack 2007; Baldwin et al. 2006; Liechti et al. 2006; Liechti and Farmer 2006; Matsui 2006). In particular, methyl jasmonate is derived from linolenic acid, via lipoxygenase, in which the synthesis of methyl jasmonate activates the oxidative metabolism of polyunsaturated fatty acids, generating reactive oxygen species (in the form of O_2, H_2O_2 or OH–) and oxidised derivatives of polyunsaturated fatty acids (Weinberger et al. 2011; Miller et al. 2010). The redox-active compound 1-phenyl-3-pyrazolidinone (phenidone) inhibits lipoxygenase activity by reducing the active form to an inactive form (Bruinsma et al. 2010).

As in higher plants where the presence of jasmonic acid and its derivative has been found to regulate various physiological processes of plant development, such as in the maturation and germination of pollen grains, and it may also have a role in anther dehiscence (Liechti and Farmer 2006; Devoto and Turner 2003), a burst of endogenous jasmonates was seen in cystocarps containing thalli compared with infertile thalli of the red seaweed *Grateloupia imbricata* (Garcia-Jimenez et al. 2016).

Likewise, in the course of in vitro maturation of the cystocarps of *G. imbricata*, the rates of jasmonates release after an exogenous methyl jasmonate treatment were significantly higher in thalli containing mature cystocarps than in those ones with immature cystocarps (Table 5.2). We have reported that these higher levels of

Table 5.2 Occurrence of methyl jasmonate in thalli of *Grateloupia imbricata* naturally collected and during in vitro development of cystocarps

	Infertile thalli (non-visible cystocarps)	Fertile thalli (visible cystocarps)
	(mM jasmonates mg^{-1} fw h^{-1})	
Naturally collected thalli	0.95 ± 0.12	1.27 ± 0.20[a]
	Thalli with immature cystocarps	Thalli with mature cystocarps
	(mM jasmonates mg^{-1} fw h^{-1})	
In vitro cystocarp development	0.76 ± 0.10	1.32 ± 0.18[a]

[a]Significant difference ($p < 0.01$)

jasmonates in mature thalli might be explained by physiological changes that occur when the thalli transform from the vegetative to reproductive state. Moreover, when phenidone was added as an inhibitor of jasmonate synthesis, not only did emission reduce, the number of mature cystocarps also diminished. Hence, the amounts of methyl jasmonate seem to be closely related to the maturation and production of cystocarps in the red macroalga *Grateloupia imbricata*. The involvement of methyl jasmonate in the reproductive state of thalli would indicate a possible function for it during the maturation process, acting as a growth regulator (Garcia-Jimenez et al. 2016).

Exogenous application of methyl jasmonate also reduced the cystocarp maturation period in the red alga *G. imbricata*. Although a reduction in the developmental period for reproductive structures has been previously reported in thalli from other red seaweeds that were treated with growth regulators like spermine (Sacramento et al. 2004) and ethylene (Garcia-Jimenez and Robaina 2012), the shortening effect of the maturation period in presence of methyl jasmonate also triggered *G. imbricata* cystocarps to open, carpospores to be released and the presence of dehiscent cavities 48 h after exposure, all of which are indicative of cystocarp dehiscence (Fig. 5.3a, b, Garcia-Jimenez et al. 2016).

What was striking is that we have also observed that when thalli of *Gelidium arbuscula* were treated with exogenous methyl jasmonate, it provoked an alteration of the reproductive cycle, incrementing firstly the number of tetrasporangial branches and later the presence of a mixed reproductive phase, in which the carposporangia and tetrasporangia coexisted in the same individual, without visible spores at 72 h after treatment (Fig. 5.2b). Contrary to what one would think, the mixed reproductive phase in algae has been previously described in Ceramiales and the red alga *Gracilaria tikvahiae* (Kim and Lee 1989 and references herein), but the difference is that in *G. arbuscula*, it occurs in the diploid phase (i.e. tetrasporophyte thalli), although we can only infer that some type of sex-determining mechanism(s) induced by methyl jasmonate could be involved in the development of *G. arbuscula*.

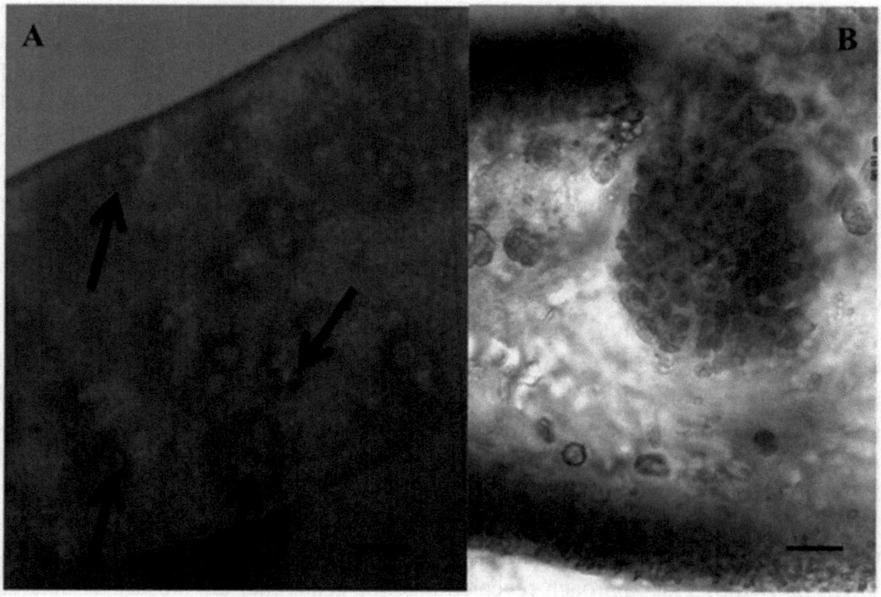

Fig. 5.3 Role of methyl jasmonate in the red seaweed *Grateloupia imbricata*. (**a**) Longitudinal section of thalli of *G. imbricata* showing dehiscent cystocarps (*arrows*). Scale bar 0.3 mm. (**b**) Longitudinal section of thalli containing a mature cystocarp releasing spores. Scale bar 0.2 mm

These issues undoubtedly open up an important field to study the role of methyl jasmonate and ethylene in reproduction and to depict as yet unexplored underlying molecular and genetic mechanism(s) of volatile emissions in red seaweeds.

5.6 Genetic 'Dissection' of Volatile Emission in the Reproduction of Macroalgae

The study of whole genome has identified genes related with aspects of life cycles and stress tolerance mechanisms of several marine algae. Albeit genomic information is still scarce, it has allowed assigning genes and gene families to particular events, confirming gene functions experimentally and providing insights into the biology of these organisms (Cock et al. 2010; Collén et al. 2007; Barbier et al. 2005).

Despite the importance of volatile growth regulators in the success of forming reproductive structures and spore germination in seaweeds, not many have been identified at a molecular level to date, and little is known about the role of genes in these processes. Pursuant to our description, we understand that a feature of reproduction in algae should be the activation of multiple responses involving gene interactions and cross talk with many molecular pathways. This is where the main complexities have emerged at a molecular level. Firstly, if the response is related to chemical stimuli, as volatiles are, the sensing model will comprise a molecule—a chemical ligand—which will bind to a specific receptor. Secondly, the presence of a type of receptor does not mean that its nucleotide sequences are uniform in all organisms and/or development stages, and moreover, these sequences can likely be identified. This situation is recurrent in seaweeds since it is not feasible to compare gene sequences from red algae with those from green algae and plants or unicellular versus multicellular algae either. Furthermore, the existence of multiple non-annotated protein sequences in algae and the presence of nonhomologous genes in other organisms hinder the sequence comparison.

With this scenario, and even more so given the scarcity of genes in public database and genes related to reproduction events, a major gene repertoire for the red seaweed *Grateloupia imbricata* was developed with the aim of uncovering potential genomic mechanisms related to different physiological events (GenBank sequences record with BioProject record PRJNA309128 and BioSample record SAMN04420758).

In particular, the focus of the genetic research in this seaweed has turned to identifying genes related to important traits of development such as carposporogenesis in *G. imbricata*. This approach revealed a set of genes that are closely associated with the synthesis of ethylene and methyl jasmonate. Furthermore, *Grateloupia* transcriptome revealed a number of sequences encoding proteins that target ethylene synthesis and different kinases which could involve response to ethylene signalling and that the perception of stimuli, such as gaseous hormones, could activate different membrane receptors and signalling molecules. We have proposed the ethylene receptor structure through a detailed analysis of the receptor sequences deposited in different databases and comparison with our transcriptome. Basically, we have hypothesised that the *Grateloupia* receptor could be made up of clusters of proteins for ethylene perception and for signal transduction, which is formed, for the most part, by the PAS and histidine kinase domains (Fig. 5.4). Although we are far from sure of all the elements in the receptor structure of *G. imbricata*, the data open up an important framework to study seaweed receptors.

Jasmonates are synthesised by complex and intriguing molecular machinery (Sasaki et al. 2001). In the *Grateloupia* transcriptome, we have reported an extensive network of methyl jasmonate-responsive genes: from genes encoding lipoxygenase and SAM methyl transferases, which catalyse the methyl transfer, to methyl jasmonate-responsive genes such as tyrosine aminotransferase, glutathione

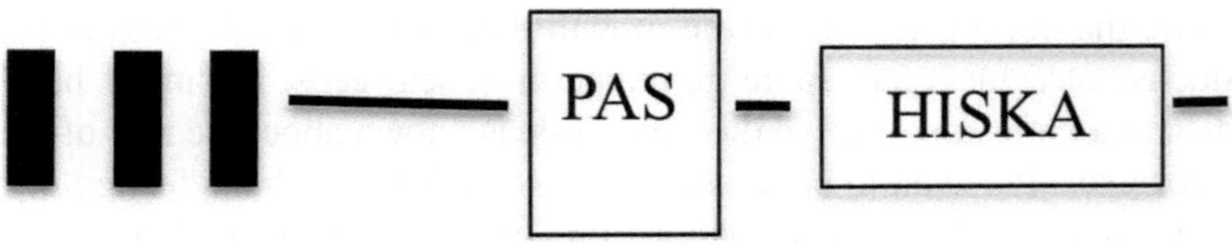

Fig. 5.4 Depicted model of an ethylene receptor proposed for *Grateloupia imbricata*

S-transferase and cytochrome P450, which are not directly related to jasmonate biosynthesis. The response to methyl jasmonate, mediated in *G. imbricata* by enzymes, including 1-deoxy-D-xylulose 5-phosphate synthase, 1-deoxy-D-xylulose 5-phosphate-reductosisomerase, farnesyl diphosphate synthase and geranyl diphosphate synthase was also determined. The genes encoding these enzymes form part of the mevalonate-phenylpropanoid- and tyrosine-derived pathways.

Unlike a genome comparison, the choice of a candidate gene to represent the manifestation of a trait (i.e. cystocarp maturation, sporulation) could help to discern the role of volatiles in reproduction. In particular the ornithine decarboxylase (ODC) gene, which encodes the synthesis of the main protein (ODC, EC. 4.1.1.17), in charge of polyamine biosynthesis has been a target gene for the study of reproduction in red seaweeds (Garcia-Jimenez and Robaina 2015).

When infertile *G. imbricata* thalli were treated with ethylene or methyl jasmonate as elicitors of cystocarp development, different *ODC* expression patterns were observed. Depending on the type of elicitors of cystocarp development, the *ODC* gene expression was differentially correlated with the elicitation period or with the disclosure period (first visible developing cystocarps; Montero-Fernandez et al. 2016). Moreover, transcription factor-related motifs and regulatory sequences related to methyl jasmonate and ethylene have been identified in the 5′ upstream region of ODC. Their functional involvement with this gene remains to be determined.

Obviously, the issue is still more complex owing to chemical diversity of volatile compounds and to genes involved in biosynthesis and signalling pathways of different volatiles. Despite high-throughput data from seaweed which are enabling the study of functional genomics with approaches about how gene regulation works, the volatile receptors and signalling patterns of these volatile compounds in seaweed, why some of the signalling patterns emerge instead of others and how we could control them remain to be known.

We are aware that the understanding of genes represents the first step towards the comprehension of seaweed system biology. However, this is not that easy. For instance, seaweeds are subjected to extreme environmental conditions in which different gene pathways act, mainly some aiming to favour the release of spores in low tide periods and others related to stress situations which are coincident with the desiccation. In this sense to reach a complete awareness of system biology, an exhaustive picture of gene network and interactions would be necessary.

We are far from this knowledge in seaweeds, but once the gene network is identified, the bridge between gene function and traits relevant to alga reproduction will undoubtedly be of utmost interest to related events.

5.7 Conclusions

Overall, the information summarised here emphasises the myriad of volatiles emitted by seaweeds, which are classified according to their functional chemical groups. Whereas a lot of basic information on the type of algal volatile emissions have accumulated in recent years, we still lack understanding of how the emission rates affect physiological events. Specifically, we underscore the importance of volatiles in reproductive responses. Though there is encouraging evidence on the extent to which volatiles affect carposporogenesis and tetrasporogenesis in several red seaweeds, how the volatile signal is perceived and transmitted remains unclear.

Recent transcriptomic analysis of the red alga *Grateloupia imbricata* has managed to isolate the biosynthetic genes involved in volatile synthesis. A putative model of the ethylene receptor has been hypothesised, and a candidate gene for reproduction, such as the *ODC* gene, has been proposed. *ODC* expression has been altered by different volatile compounds, namely, ethylene and methyl jasmonate, which act as elicitors of carposporogenesis. Though non-functional roles have been assigned, transcription factors on the 5′ upstream region of ODC, related to these elicitors, have been also identified. We are aware that more information and research are needed, but amazing perspectives have been opened up.

Acknowledgements The authors would like to thank Olegario Brito-Romano for his skilful technical assistance. This work has been carried out at the Biology Department of the Facultad de Ciencias del Mar, Universidad de Las Palmas de Gran Canaria.

References

Abrahamsson K, Ekdahl A, Collén J, Pedersén M (1995) Marine algae. A source of trichloroethylene and perchloroethylene. Limnol Oceanogr 40(7):1321–1326

Andreae MO, Elbert W, de Mora SJ (1995) Biogenic sulfur emissions and aerosols over the tropical South Atlantic: 3. Atmospheric dimethylsulfide, aerosols and cloud condensation nuclei. J Geophys Res 100:11335–11356

Arnold TM, Targett NM, Tanner CE, Hatch WI, Ferrari KE (2001) Evidence for metil jasmonate-induced phlorotannin production in *Fucus vesiculosus* (Phaeophyceae). J Phycol 37:1025–1029

Baldwin IT, Halitschke R, Paschold A, von Dahl CC, Preston CA (2006) Volatile signaling in plant–plant interactions: ‘talking trees’ in the genomics era. Science 311:812–815

Bankova V, Stepanov K, Dimitrova-Konaklieva ST, Keremedchieva G, Frette X, Nikolova C, Kujimgiev A, Popov S (2001) Secondary metabolites and lipids in *Chara globularis* Thuill. Hydrobiologia 457:199–203

Barbier G, Oesterhelt C, Larson MD, Halgren RG, Wilkerson C, Garavito RM, Benning C, Weber APM (2005) Comparative genomics of two closely related unicellular thermo-acidophilic red algae, *Galdieria sulphuraria* and *Cyanidioschyzon merolae*, reveals the molecular basis of the metabolic flexibility of *Galdieria sulphuraria* and significant differences in carbohydrate metabolism of both algae. Plant Physiol 137:460–474

Baweja P, Sahoo D, Garcia-Jimenez P, Robaina RR (2009) Seaweed tissue culture as applied to biotechnology: problems, achievements and prospects. Phycol Res 57:45–58

Bitton R, Ben-Yehuda M, Davidovich M, Balazs Y, Potin P, Delage L, Colin C, Bianco-Peled H (2006) Structure of algal-born phenolic polymeric adhesives. Macromol Biosci 7:1280–1289

Bitton R, Berglin M, Elwing H, Colin C, Delage L, Potin P, Bianco-Peled H (2007) The influence of halide-mediated oxidation on algae-born adhesives. Macromol Biosci 7:1280–1289

Bleecker AB (2000) ETHYLENE: a gaseous signal molecule in plants. Annu Rev Cell Dev Biol 16:1–18

Bouarab K, Adas F, Gaquerel E, Kloareg B, Salaün JP, Potin P (2004) The innate immunity of a marine red alga involves oxylipins from both the eicosanoid and octadecanoid pathways. Plant Physiol 135:1838–1848

Bravo-Linares CM, Mudge SM, Loyola-Sepulveda RH (2010) Production of volatile organic compounds (vocs) by temperate macroalgae. The use of solid phase microextraction (spme) coupled to GC-MS as method of analysis. J Chil Chem Soc 55(2):227–232

Bruinsma M, van Broekhoven S, Poelman EH, Posthumus MA, Müller MJ, van Loon JJA, Dicke M (2010) Inhibition of lipoxygenase affects induction of both direct and indirect plant defences against herbivorous insects. Oecologia 162:393–404

Carpenter LJ, Liss PS (2000) On temperate sources of bromoform and other reactive organic bromines gases. J Geophys Res Atmos 105:20539–205547

Carpenter LJ, Liss PS, Penkett SA (2003) Marine organohalogens in the atmosphere over the Atlantic and Southern Oceans. J Geophys Res 108(D9):4256. doi:10.1029/2002JD002769

Christov C, Fournadzieva S, Bozhcova M, Cherkezov N, Pilarski P, Zafirova T (1996) Abscisic acid and jasmonates content during the ontogenesis of *Scenedesmus acutus*. In: Ordog V, Szigeti J, Pulz O (eds) Proceedings of a conference on progress in plant sciences from plant breeding to growth regulation. Pannon University of Agricultural Sciences, Mosonmagyarovár, pp 155–162

Christov C, Pounevacc I, Bozhkova M, Toncheva T, Fournadzieva S, Zafirova T (2001) Influence of temperature and methyl jasmonate on *Scenedesmus incrassulatus*. Biol Plant 44(3):367–371

Cock JM, Sterck L, Rouze P, Scornet D, Allen AE et al (2010) The *Ectocarpus* genome and the independent evolution of multicellularity in brown algae. Nature 465:617–621. doi:10.1038/nature09016

Cole KM, Sheath RG (1990) Biology of the red algae. Cambridge University Press, Cambridge, New York, Port Chester, Melbourne, Sydney

Collén J, Guisle-Marsollier I, Léger JL, Boyen C (2007) Response of the transcriptome of the intertidal red seaweed *Chondrus crispus* to controlled and natural stresses. New Phytol 176:45–55

Dacey JWH, King GM, Wakeham SG (1987) Factors controlling emission of dimethylsulfide from salt marshes. Nature 330:643–645

Devoto A, Turner JG (2003) Regulation of jasmonate-mediated plant responses in *Arabidopsis*. Ann Bot 92:329–337

Fujimura T, Kawai T, Kajiwara T, Iwhida Y (1994) Volatile components in protoplasts isolated from the marine brown alga *Dictyopteris prolifera* (Dictyotales). Papers Plant Tissue Culture Letters 11(1):34–39

Gaquerel E (2005) Caractérisation et rôle des oxylipines dans la signalisation lipidique cellulaire chez le gamétophyte de *Chondrus crispus*. PhD thesis, Université de Bretagne Occidentale, Brest

Gaquerel E, Weinhold A, Baldwin IT (2009) Molecular interactions between the specialist herbivore *Manduca sexta* (Lepidoptera, Sphigidae) and its natural host *Nicotiana attenuata*. VIII. An unbiased GCxGC-ToFMS analysis of the plant's elicited volatile emissions. Plant Physiol 149(3):1408–1423

Garcia-Jimenez P, Robaina RR (2012) Effects of ethylene on tetrasporogenesis in *Pterocladiella capillacea* (Rhodophyta). J Phycol 48:710–715

Garcia-Jimenez P, Robaina RR (2015) On reproduction in red algae: further research needed at the molecular level. Front Plant Sci 6:1–6

Garcia-Jimenez P, Brito-Romano O, Robaina RR (2013) Production of volatiles by the red seaweed *Gelidium arbuscula* (Rhodophyta): emission of ethylene and dimethyl sulphide. J Phycol 49:661–669

Garcia-Jimenez P, Brito-Romano O, Robaina RR (2016) Occurrence of jasmonates during cystocarp development in the red alga *Grateloupia imbricata*. J Phycol 52:1085–1093

Goodwin KD, North WJ, Lidstrom ME (1997) Production of bromoform and dibromomethane by Giant Kelp: factors affecting release and comparison to anthropogenic bromine sources. Limnol Oceanogr 42:1725–1734

Hay ME (1996) Marine chemical ecology: what's known and what's next? J Exp Mar Biol Ecol 200:103–134

Holopainen JK (2004) Multiple functions of inducible plant volatiles. Trends Plant Sci 9:529–533

Kawasaki W, Matsui K, Akakabe Y, Itai N, Kajiwara T (1998) Volatiles from *Zostera marina*. Phytochemistry 47(1):27–29

Kim GH, Lee IK (1989) Mixed phases reproduction of *Platythamnion yezoense* Inagaki in culture. Korean J Phycol 4(2):111–119

Klaschka U, Kolossa-Gehring M (2007) Fragrances in the environment: pleasant odours for nature? Environ Sci Pollut Res 14:44–52

Krupina MV, Dathe W (1991) Occurrence of jasmonic acid in the red alga *Gelidium latifolium*. Z Naturforsch 46:1127–1129

Küpper FC, Gaquerel E, Cosse A, Adas F, Peters AF, Müller DG, Kloareg B, Salaün JP, Potin P (2009) Free fatty acids and methyl jasmonate trigger defense reactions in *Laminaria digitata*. Plant Cell Physiol 50(4):789–800

La Barre S, Potin P, Leblanc C, Delage L (2010) The halogenated metabolism of brown algae (Phaeophyta), its biological importance and its environmental significance. Mar Drugs 8(4):988–1010. doi:10.3390/md8040988

Leedham Elvidge EC, Phang SM, Sturges WT, Malin G (2015) The effect of desiccation on the emission of volatile bromocarbons from two common temperate macroalgae. Biogeosciences 12:387–398

Liechti R, Farmer EE (2006) Jasmonate biochemical pathway. Sci STKE 322:1–3

Liechti R, Gfeller A, Farmer EE (2006) Jasmonate signaling pathway. Sci STKE 14. doi:10.1126/stke.3222006cm2

Loreto F, Schnitzler JP (2010) Abiotic stress and induced BVOCS. Trends Plant Sci 15:154–166

Manley SL, Goodwin K, North WJ (1992) Laboratory production of bromoform, methylene bromide, and methyl iodide by macroalgae and distribution in nearshore southern California waters. Limnol Oceanogr 37:1652–1659

Matsui K (2006) Green leaf volatiles: hydroperoxide lyase pathway of oxylipin metabolism. Curr Opin Plant Biol 9:274–280

Miller G, Suzuki N, Ciftci-Yilmaz S, Mittler R (2010) Reactive oxygen species homeostasis and signaling during drought and salinity stresses. Plant Cell Environ 33:453–467

Montero-Fernández M, Robaina RR, Garcia-Jimenez P (2016) In silico characterization of DNA motifs associated with the differential expression of the ornithine decarboxylase gene during in vitro cystocarp development in the red seaweed Grateloupia imbricata. J Plant Physiol 195:31–38

Mtolera MSP, Collén J, Pedersén M, Ekdahl A, Abrahamsson K, Semesi AK (1996) Stress-induced production of volatile halogenated organic compounds in *Eucheuma denticulatum* (Rhodophyta) caused by elevated pH and high light intensities. Eur J Phycol 32(1):89–95

Müller DG (1989) The role of pheromones in sexual reproduction of brown algae. In: Coleman AW, Goff LJ, Stein-Taylor JR (eds) Algae as experimental systems. Alna r Liss, Inc, New York, pp 201–213

Niki T, Kunugi M, Otsuki A (2000) DMSP-lyase activity in five marine phytoplankton species: its potential importance in DMS production. Mar Biol 136:759–764

Osborne DJ, McManus MT (2005) Hormones, signals and target cells in plant development. Cambridge University Press, Cambridge

Otte ML, Wilson G, Morris JT, Moran BM (2004) Dimethylsulphoniopropionate (DMSP) and related compounds in higher plants. J Exp Bot 55:1919–1925

Paul C, Pohnert G (2011) Production and role of volatile halogenated compounds from marine algae. Nat Prod Rep 28:186–195

Pirjola L, O'Dowd CD, Brooks IM, Kulmala M (2000) Can new particle formation occur in the clean marine boundary layer? J Geophys Res 105:26531–26546

Plettner I, Steinke M, Malin G (2005) Ethene (ethylene) production in the marine macroalga *Ulva* (*Enteromorpha*) *intestinales* L. (Chlorophyta, Ulvophyceae): effect of light stress and co-production with dimethylsulphide. Plant Cell Environ 28:1136–1145

Qualley A, Dudareva N (2010) Plant volatiles. eLS. doi:10.1002/9780470015902.a0000910

Reis VM, Oliveira LS, Passos RMF, Viana NB, Mermelstein C (2013) Traffic of secondary metabolites to cell surface in the red alga *Laurencia dendroidea* depends on a two-step transport by the cytoskeleton. PLoS One 8(5):e63929. doi:10.1371/journal.pone.0063929

Roje S (2006) S-Adenosyl-L-methionine: beyond the universal methyl group donor. Phytochemistry 67:1686–1698

Sacramento AT, Garcia-Jimenez P, Alcazar R, Tiburcio AF, Robaina RR (2004) Influence of polyamines on the sporulation of *Grateloupia* (Halymeniaceae, Rhodophyta). J Phycol 40:887–894

Salgado LT, Cinelli LP, Viana NB, Tomazetto de Carvalho R, Mourão P, Teixeira VL, Farina M, Amado Filho GM (2009) Vanadium bromoperoxidase catalyzes the formation of high-molecular-weight complexes between brown algal phenolic substances and alginates. J Phycol 45:193–202

Sasaki Y, Asamizu E, Shibata D, Nakamura Y, Kaneko T, Awai K, Amagai M, Kuwata C, Tsugane T, Masuda T, Shimada H, Takamiya K, Ohta H, Tabata S (2001) Monitoring of methyl jasmonate-responsive genes in *Arabidopsis* by cDNA macroarray: self-activation of jasmonic acid biosynthesis and crosstalk with other phytohormone signaling pathways. DNA Res 8:153–161

Schall C, Laturnus F, Heumann KG (1994) Biogenic volatile organoiodine and organobromine compounds released from polar macroalgae. Chemosphere 28:1315–1324

Stefels J (2000) Physiological aspects of the production and conversion of DMSP in marine algae and higher plants. J Sea Res 43:183–197

Steinke M, Daniel C, Kirst GO (1996) DMSP lyase in marine macro- and microalgae: intraspecific differences in cleavage activity. In: Kiene RP, Visscher PT, Keller M, Kirst GO (eds) Biological and environmental chemistry of DMSP and related sulfonium compounds. Plenum, New York, pp 317–324

Sunda W, Kieber DJ, Kiene RP, Huntsman S (2002) An antioxidant function for DMSP and DMS in marine algae. Nature 418:317–320

Takamo K, Igarashi S, Mikai H, Hino S (2003) Causation of reversal simultaneity for diatom biomass and density of *Phormidium tenue* during the warm season in eutrophic Lake Barato, Japan. Limnology 4:73–78

Tugwell S, Branch GM (1989) Differential polyphenolic distribution among tissues in the kelps *Ecklonia maxima, Laminaria pallida* and *Macrocystis angustifolia* in relation to plant-defense theory. J Exp Mar Biol Ecol 129:219–230. doi:10.1016/0022-0981(89) 90104-4

Ueda J, Miyamoto K, Sato T, Momotani Y (1991a) Identification of jasmonic acid from *Euglena gracilis* Z as a plant growth regulator. Agric Biol Chem 55(1):275–276

Ueda J, Miyamoto K, Aoki M, Hirata T, Sato T, Momotani Y (1991b) Identification of jasmonic acid in *Chlorella* and *Spirulina*. Bull Univ Osaka Prefect Ser B Agric Biol 43:103–108

Uji T, Matsuda R, Takechi K, Takano H, Mizuta H, Takio S (2016) Ethylene regulation of sexual reproduction in the marine red alga *Pyropia yezoensis* (Rhodophyta). J Appl Phycol 28:3501. doi:10.1007/s10811-016-0904-6

Van Alstyne KL, Houser LT (2003) Dimethylsulfide release during macroinvertebrate grazing and its role as an activated chemical defense. Mar Ecol Prog Ser 250:175–181

van Rijssel M, Gieskes WC (2002) Temperature, light, and the dimethylsulfoniopropionate (DMSP) content of *Emiliania huxleyi* (Prymnesiophyceae). J Sea Res 48:17–27

Vanden Driessche T, De Vries PG, Guisset JL (1997) Differentiation, growth and morphogenesis: *Acetabularia* as a model system. New Phytol 135:1–20

Vanden Driessche T, Kevers C, Collet M, Gaspar T (1998) *Acetabularia mediterranea* and ethylene: production in relation with development, circadian rhythms in emission, and response to external application. J Plant Physiol 133:635–639

Wasternack C (2007) Jasmonates: an update on biosynthesis, signal transduction and action in plant stress response, growth and development. Ann Bot 100:681–697

Weinberger F, Lion U, Delage L, Kloareg B, Potin P, Beltrán J, Flores V, Faugeron S, Correa J, Pohnert G (2011) Up-regulation of lipoxygenase, phospholipase, and oxylipin-production in the induced chemical defense of the red alga *Gracilaria chilensis* against epiphytes. J Chem Ecol 37:677–686

Yamamoto M, Baldermann S, Yoshikawa K, Fujita A, Mase N, Watanabe N (2014) Determination of volatile compounds in four commercial samples of Japanese green algae using solid phase microextraction gas chromatography mass spectrometry. Sci World J 2014:289780. doi:10.1155/2014/289780

Yang SF, Hoffmann NE (1984) Ethylene biosynthesis and its regulation in higher plants. Annu Rev Plant Physiol 35:155–189

Yoch DC, Ansede JH, Rabinowitz S (1997) Evidence for intracellular and extracellular dimethylsulfoniopropionate (DMSP) lyases and DMSP uptake sites in two species of marine bacteria. Appl Environ Microbiol 63:3182–3188

Part II

Chapter 6
Abiotic Stress of Seagrasses: Recent Advances in Transcriptomics, Genomics, and Systems Biology

E.E. Malandrakis, T. Danis, A. Iona, and A. Exadactylos

Abstract Seagrasses are a unique taxonomic rank of Plantae, totally submerged to marine environment. They evolved from terrestrial plants so this progression provoked alterations in genome, providing adaptation abilities to aquatic environment. The development of high-throughput technologies, such as omics, bridges the gap between genome and phenotype, shedding some light on molecular mechanisms that regulate seagrass tolerance to abiotic stress.

Keywords Seagrasses • Abiotic stress • Transcriptomics • Genomics • Systems biology

Contents

6.1 Introduction 120
6.2 Systems Biology 120
  6.2.1 Genomics 121
  6.2.2 Transcriptomics 123
  6.2.3 Metabolomics 124
6.3 Abiotic Stress of Seagrasses 125
  6.3.1 Acidification 125
  6.3.2 Light Stress 126
  6.3.3 Temperature 127
  6.3.4 Salinity 127
6.4 Perspectives 128
References 129

E.E. Malandrakis • T. Danis • A. Exadactylos (✉)
Department of Ichthyology and Aquatic Environment, School of Agricultural Sciences, University of Thessaly, Volos 38446, Greece
e-mail: exadact@uth.gr

A. Iona
Hellenic National Oceanographic Data Centre-HCMR, Elliniko, PC 166 04, Agios Kosmas, Attiki, Greece

© Springer International Publishing AG 2017

M. Kumar, P. Ralph (eds.), *Systems Biology of Marine Ecosystems*,
DOI 10.1007/978-3-319-62094-7_6

6.1 Introduction

Seagrasses are flowering monocotyledonous plants which grow in the marine environment and belong all in the order Alismatales. Seagrasses constitute a valuable part of the aquatic ecosystems, by providing food and shelter mainly in nursery grounds. Their cosmopolitan expansion renders them as a very important part of the global marine biodiversity scaffold. Seagrass habitats have been rated as those of the most economically important part of the earth biome (Costanza et al. 1998). Furthermore, because of their sensitivity to water quality, seagrasses can be utilized as bioindicators of the ecosystem health. They prevent erosion by absorbing the forces of the waves and currents, which impact shallower waters. Seagrasses also act as carbon traps for a decent amount of the total carbon fixed in the oceans. Their high capacity to store organic carbon is explained as a result of their high primary production and their capacity to filter particles from the water column. Organic carbon is stably stored for several meters in the seabed (Fourqurean et al. 2012), unlike terrestrial forests that decline as a result of deforestation and increased land used by humans. Large-scale research on seagrasses will contribute toward biodiversity monitoring, conservation, and management policies in the near future.

6.2 Systems Biology

Biological systems are quite complex, and they cannot become comprehensible by focusing in individual parts. We can draw a parallel between systems biology and a satellite that views earth, with rivers, mountains, oceans, and continents, instead of studying a specific river or a mountain.

Systems biology is defined as the science of computational modeling of complex biological systems using modern holistic approaches. It studies the dynamic interactions in a biological object, in order to understand all the intrinsic complex processes among living organisms, by combining experimental designs with mathematical models. Systems biology is commonly an interdisciplinary approach of computational science, molecular biology, and biochemistry, investigating and predicting the network components and their systems interactions, with new high-throughput techniques, by merging all these data into dynamical simulation models. It combines traditional biological research strategies with theoretical disciplines including physics, engineering, computer science, and mathematics (Green 2017). Thus, the development of systems biology is a multidimensional process with different theoretical approaches.

Seagrasses can be confronted as complex biological systems of their molecular ingredients such as metabolites, RNA, DNA, proteins, ions, and their interactions within (Cramer et al. 2011). The rapid development of high-throughput technologies, also known as "omics" (genomics, proteomics, metabolomics, transcriptomics, ionomics), conventionalizes the application of systems biology in seagrasses (Nicholson and Wilson 2003). The integration of constantly developing computer algorithms with experimental data will potentially produce mathematical models of such biological systems (Hiroaki 2002). As long as this biological information is enriched, the gap between the mathematical models and the observed data will converge. Along these lines, it becomes feasible to integrate molecular functions, their interactions, the molecular pathways, the networks they participate in, and the dynamics among networks. This kind of data requires high-quality datasets, available computational methods, and high-level computational capacity. Finally, all these pieces of information need vastly available databases for comparative analysis of such biological networks.

6.2.1 Genomics

Seagrasses have evolved from terrestrial plants and exhibit a spectacular repertoire of adaptation mechanisms in order to survive submerged in the marine environment. Seagrasses settled in the sea about 100 million years ago (Hemminga and Duarte 2000) and belong in 5 monocotyledon families, containing 12 genera. This evolutionary pathway has provoked alterations in genes and genomes that reflect the ability of seagrasses to develop and grow in aquatic environments. For example, the photosynthetic process is differentiated according to depth gradient, and epiphytes usually cover a large part of the photosynthetically active leaves of the plants (Dalla Via et al. 1998). Wissler et al. (2011) identified in seagrasses 51 genes as positively selected, belonging in a few metabolic pathways, photosynthesis, and translation machinery (ribosomes), which diverged after the split of the common ancestor from terrestrial monocotyledonous plants. Transcriptomic analysis in *Zostera muelleri* suggested that genes involved in ethylene metabolism and signaling functions are missing and that there is a gene loss in processes related with ethylene biosynthesis and signaling in the species in question (Golicz et al. 2015).

The introduction of state-of-the-art genomic technologies in seagrass studies motivated groundbreaking advances, resulting in novel scientific fields. Genomic inventions deliver fast, inexpensive, and accurate genome information, having tremendous advantages over conventional methods. The development of such new technologies applied to molecular biology and genomics, such as next-generation sequencing methods and high-throughput genotyping, allows the rapid increase of availability of genomic resources in seagrasses. These resources will provide valuable insights to the research community and will help in the determination of the genetic factors involved in several biological aspects of seagrass species. Bioinformatics analyses are being made easier as the quantity of the available software is increased and the annotations provided by databases are continuously improving, especially for terrestrial plants. Available depositories are enriched in a daily basis, and an extensive usage of databases such as UniProt, Nr, and RefSeq is made in order to annotate genes from seagrasses. The rapidly increasing literature on the molecular biology of seagrasses is mainly supported by gene annotations of terrestrial plants such as *Arabidopsis thaliana*, *Oryza sativa*, *Vitis vinifera*, etc. This is quite reasonable, taking into account the fact that the first complete plant genome to be determined was that of the plant *Arabidopsis thaliana*, which is available since 2000 (*Arabidopsis* Genome 2000), while a well-annotated genome for seagrass species is not available yet in public databases. In that sense, the assignment of sequenced unigenes in the adequate biochemical pathways could be somewhat tricky. Genomic information on seagrass species is continuously enriched, and the first genome sequenced was that of *Zostera marina* (Olsen et al. 2016), which produced important outcomes for seagrass genomics, although genome sizes vary among seagrasses (Koce et al. 2003), with that of *Posidonia oceanica* being substantially larger than that of *Zostera marina*. This is to date attributed to the number and length of chromosomes, transposable elements, repetitive regions, etc. Therefore, genomic information on different seagrass species will help out the scientific community unravel the evolutionary forces exerted on the seagrasses.

Seagrasses colonized the marine environment and formed a widespread and highly important specific habitat. During this transition, gene gains and losses shaped their ability to survive and boom in the sea. Gene gains facilitated processes such as ion regulation, nutrient uptake, and O_2/CO_2 exchange. In other words, these operational modes render seagrasses to survive in elevated salinities. Their cell walls contain low-methylated pectins and sulfated galactans (Aquino et al. 2005), features interestingly shared with macroalgae. On the other hand, gene families implicated in stomatal development, ethylene and terpenoid biosynthesis, ultraviolet protection, and far-red sensing have been all over omitted (Kong et al. 2014; Golicz et al. 2015; Lee et al. 2016; Olsen et al. 2016).

Life on the seabed is associated with anatomical adaptations such as lack of stomata. Their surfaces are covered by a thin cuticle, across which the gas and nutrient transfer takes place (Hemminga and Duarte 2000; Kuo and Den Hartog 2007). Seagrasses retained the ability to exchange gases with their environment with "algal-like" cell walls. In *Arabidopsis EPF1/2* genes act as regulators of stomatal development by interacting with the leucine-rich repeat-containing receptor-like protein TMM (Nadeau and Sack 2002; Hunt and Gray 2009). The *SPEECHLESS* (*SPCH*) gene, which regulates the expression of both *TMM* and *EPF1/2*, and the *SCREAM2 (SCRM2)*, which contribute toward meristemoid formation, have been omitted in *Zostera marina*. The entire pathway that differentiates meristemoid mother cells (MMC) to guard cells, which are produced to form stomata, has been lost (Olsen et al. 2016).

Another molecular pathway lost in *Zostera marina* is that of ethylene biosynthesis and signaling (Olsen et al. 2016). Ethylene is an important phytohormone in terrestrial plants, since it has been demonstrated that it enhances maturation and modulates cell division in *Arabidopsis* sp. roots (Ortega-Martinez et al. 2007). The genome of *Zostera marina* became the cornerstone among the large-scale studies trying to elucidate seagrass evolution and transition to the sea.

6.2.2 Transcriptomics

Transcriptomics for differential global gene expression studies depend upon the reconstruction of the transcriptome of a given species. The transcriptome reconstruction mainly falls into two categories, depending on the usage of a reference genome: the genome-guided transcriptome assembly and the de novo transcriptome assembly. For organisms without reference genomes, only the second option is feasible, but for organisms with known reference genomes, both options are available. D'Esposito et al. (2016) summarized that RNA-seq for transcriptome analysis of non-model organisms is very efficient and cost-effective. Even so, the genome-guided approach is highly preferable, although de novo assembly can extend information on already existing genomes.

In Unix-based environment, a plethora of de novo transcriptome assemblers has been developed, with the most common being Trinity (Haas et al. 2013), Trans-ABySS (Robertson et al. 2010), SOAPdenovo-Trans (Xie et al. 2014), and Velvet/Oases (Schulz et al. 2012). Beyond open-source software, which is extensively used in research, commercial solutions have been developed accordingly. Differential gene expression is quantified, in a holistic point of view, via counting "raw" reads that map uniquely to each contig (transcript). In order to discover genes differentially expressed between two or more groups, expression values must be normalized. The normalization procedure is related with the appropriate statistical strategy adopted (Dillies et al. 2013).

In many cases the understanding of the information acquired by microarrays or RNA-seq data is far beyond a list of differentially expressed transcripts. Thus, results from such experiments remain essentially unexplored. Toward this end, data can be further explored with gene networks, through data reduction and clustering. Gene ontology information can be useful in order to prioritize specific genes in gene networks. Comparative analysis of transcription among different conditions was primarily utilized through EST analysis (Reusch et al. 2008). Comparative EST sequencing shed light to the pleiotropic effect of stress. A substantial proportion of the sequences in a dataset remain unannotated (e.g., Kong et al. 2014). A database available to download ESTs from *Posidonia oceanica*, *Zostera marina*, and *Nanozostera noltii* is Dr. Zombo from the Institute for Evolution and Biodiversity at the University of Münster (Wissler et al. 2009). Additionally, the complete transcriptome of *Posidonia oceanica* and the first gene catalogues for this plant (D'Esposito et al. 2016) are a reliable tool to begin an assay in order to restrict human-driven environmental changes on this Mediterranean endemic species. Dattolo et al. (2013) studied on the in situ acclimation of *P. oceanica* to different depths and revealed networks and pathways involved in response to depth gradients. Seagrasses must cope with different levels of light irradiance at these different depths. Furthermore, Dattolo et al. (2014) have also shown that light-associated gene expression is connected with seagrass depth distribution; RuBisCO subunits in *P. oceanica* were negatively regulated. Interestingly, Dattolo et al. (2013) observed increased reactive oxygen species (ROS) in *P. oceanica* in low light conditions, which is possibly a result of an energy-costly process of the immune system. Procaccini et al. (2017) demonstrate that transcription rate in shallow plants is higher than those in deep waters. This is supported by upregulation of chlorophyll a/b-binding protein and RuBisCO-activated proteins. In shallow plants the translation of RuBisCO protein is in maximal abundance before dawn, so that by sunrise the plants are ready to take maximum advantage of light for photosynthesis. Thus, RuBisCO levels decreased to a minimum at sunset. On the other hand, deepwater plants exhibit low RuBisCO levels at sunrise due to the upregulation of their corresponding transcripts (ribulose-bisphosphate carboxylase small chain 5B) which cause later during the day sunlight increased RuBisCO levels (Procaccini et al. 2017). In terms of cellular energetic metabolism, Procaccini et al. (2017) observed at sunrise the same high levels of glyceraldehyde 3-phosphate dehydrogenase (GADPH) among shallow and deepwater plants, thus supporting the fact that respiratory responses were the highest during the full light hours of the day. It has been shown that respiration rate is also similar for other seagrasses like *Z. marina* regardless of depth distribution, even though in shallow water plants the photosynthetic rate was higher (Dennison and Alberte 1986).

6.2.3 Metabolomics

Metabolomics is the study of the intermediates and the products of metabolism with high-throughput techniques. In plant metabolomics, primary and secondary metabolites are usually referred; the first is directly implicated in development, growth, and reproduction, and the second does not affect survivability of organisms, but may affect homeostasis in a long-term impairment.

Nuclear magnetic resonance (NMR) and mass spectrometry (MS) techniques are among the most recent technologies available to perform throughput metabolite profiling and are currently commonly used, for instance, on documenting interactions of marine plants with their environment. The most common techniques to separate molecules before identification are gas chromatography (GC) and liquid chromatography (LC), even if other techniques have been developed. For example, reverse-phase HPLC was utilized to measure phenolic substances in marine seagrasses (Arnold et al. 2012).

Environmental disturbances can alter the metabolomic profile of seagrasses. Cellular mechanisms act toward sustaining homeostasis and tolerating stressful periods. *Zostera marina* drafts the carbon nitrogen metabolism by the alanine, GABA, and 2-oxoglutarate shunt as anoxia tolerance mechanisms (Hasler-Sheetal et al. 2015). A recent application of metabolomics is the identification of bioactive compounds from *Syringodium isoetifolium* and *Cymodocea serrulata* which could act as antifouling agents. These substances were characterized as lipidic metabolites such as of high molecular weight fatty acids and its esters (Iyapparaj et al. 2014).

6.3 Abiotic Stress of Seagrasses

In terrestrial plants the sensing of stress induces signaling pathways that include kinase cascades, ROS production, ion channel activation, and production of hormones ethylene (ET), abscisic acid (ABA), and jasmonic acid (JA) (Delker et al. 2006). These pathways activate genes responsible for stress allostasis. Marine plants have been adapted to cope with stress accordingly. Seagrasses have evolved mechanisms to survive in different environments, through modifying their physiological functions. They are able to cope with conditions of light limitation, activate mechanisms, and change light spectrum through metabolic adjustments such as downregulation of RuBisCO and enhanced proteolysis. Noteworthy for marine and also freshwater plants is the oxygen transport, due to its low solubility. Oxygen is transported to whole plant tissues, when photosynthesis releases oxygen into a special tissue which is presented in all seagrasses, aerenchyma. In most seagrass meadows, an ancient symbiosis takes place between seagrasses, bivalves, and their sulfide-oxidizing gill bacteria which decay organic matter and reduce sulfide stress on plant tissues (Papenbrock 2012). *Zostera marina* plants from a southern European population (Italy) managed to overcome a simulated heat wave by returning gene expression profile in contrast to plants acquired from Denmark (Franssen et al. 2011). In the other hand, Winters et al. (2011) concluded that both populations suffered heat wave equally, although the photochemical activity fully recovered in the southern population. Noteworthy is the fact that the northern population demonstrated reduced ability to recover its photo-physiological functions. Seagrasses, due to their cosmopolitan extent, can help understand the adaptation to wide range of temperatures, salinity tolerance, ocean acidification, and light gradients.

6.3.1 Acidification

Rising carbon dioxide levels in the atmosphere are acidifying the world's oceans and threatening the survival of seagrasses among other marine organisms. Over the past years, there has been much focus on studying the potential impacts of ocean acidification (Boyd 2011). Underwater hydrothermal vents can serve as natural laboratories for ocean acidification, due to the higher CO_2 concentrations.

Lauritano et al. (2015) studied the gene expression of 35 stress-related genes in *Posidonia oceanica* exposed to volcanic vents in southern Italy. Among them, notable was the induction of the antioxidant enzymes peroxiredoxin Q (*PRXQ*) and glutathione peroxidase (*GPX*). Both are reactive oxygen species (ROS) scavenging enzymes, which require the donation of electrons from thioredoxin and glutathione, respectively. Nevertheless, responses to elevated CO_2 are affected by nutrient availability, mostly nitrogen, because carbon and nitrogen metabolisms affect each other (Touchette and Burkholder 2007). Ow et al. (2016) observed that respiration in *Halodule uninervis* with high levels of CO_2 was reduced by nitrate enrichment. On the other hand, in *Thalassia hemprichii* net primary production and growth are not affected by nitrate enrichment. However, *Zostera marina* seems resistant to CO_2 enrichment, whereas ocean acidification will increase productivity of seagrass meadows in specific coastal areas (Palacios and Zimmerman 2007). Jiang et al. (2010) studied the effects of CO_2 enrichment on photosynthesis, growth, and biochemical composition of *T. hemprichii*. Their results indicate that *T. hemprichii* may respond positively to CO_2-induced ocean acidification due to its enhanced relative electron transport rate (RETR), minimum saturating irradiance (E(k)), and enhanced carbon/nitrogen ratio.

However, seagrass habitats can provide true shelter to associated communities of invertebrate taxa in highly acidified environments (Garrard et al. 2014). Arnold et al. (2012) found decreased phenolic substances in near undersea CO_2 vents, which suggest that ocean acidification may alter coastal carbon fluxes by affecting rates of seagrass decomposition, grazing, and disease diffusion.

6.3.2 Light Stress

The depth of the euphotic zone, where photosynthesis occurs, depends largely on the concentration of suspended materials in the water column. Thus, with more dissolved materials, such as in coastal waters, the depth of the euphotic zone will be shallow, perhaps only a few meters deep. Seagrasses are distributed in depth gradient; thus, they are adapted to survive in various light intensities. Seagrasses cope with shifted light spectrum, compared to their terrestrial counterparts, where ultraviolet, red, and far-red light gradually do not penetrate. Olsen et al. (2016) observed that genes associated with UV sensing and response and red/far-red receptors have been eliminated through adaptive gene loss. Genes of UV light resistance were evolutionary lost in *Zostera muelleri* (Lee et al. 2016). Dattolo et al. (2014) revealed that light-associated gene expression profile is connected with depth distribution. Furthermore, the photosynthetic light-harvesting complex B (LHCB) genes are more abundant in *Z. marina* probably by enhancing photosynthetic performance at lower irradiances (Olsen et al. 2016). Phytochrome PHYC is absent in seagrasses probably due to their marine lifestyle, because the role of this receptor is a red light detection (Franklin et al.2003). Greco et al. (2013) observed photoreceptors for blue and red in *P. oceanica* suggesting that the importance of these genes lie within the water column. Brassinosteroids provide resistance to low light stress in plants (Saini et al. 2015), and these hormones were also detected in seagrasses (Olsen et al. 2016). On the other hand, photosystems I and II have been expanded (gene gain) for increased performance in attenuated light. Light attenuation due to both natural and anthropogenically driven processes leads to reduced photosynthesis. The submarine light deterioration is one among the most serious threats experienced by the seagrass meadows across the globe (Waycott et al. 2009). Within this context a plethora of bioindicators has been developed to monitor such environmental pressures (reviewed in McMahon et al. 2013).

6.3.3 Temperature

The somewhat hidden thermal tolerance of seagrasses allows them to inhabit a wide range of habitats in the marine environment. Out of tolerance temperature variations or abrupt temperature shifts would probably introduce stress. Depending on the severity of the stressor, it may lead to irreparable disturbance of homeostasis and death. Long-term exposure may result in habitat degradation and loss.

Differential gene expression is an excellent approach to analyze cellular disturbance at the level of transcription. Franssen et al. (2011) used this technique to study different thermal adaptations as a response to heat stress. In *Zostera marina* subjected to heat stress, the consistently upregulated genes included pectinesterases, proteins involved in the synthesis of ribosomal chloroplast proteins, and proteins involved in protein folding, which contain immunophilins. Pectinesterases are enzymes implicated in cell wall modification. In *Arabidopsis*, 66 ORFs have been annotated as pectinesterases, some of which are ubiquitously expressed and others are expressed during specific developmental stages.

Several temperature-responsive pathways include photosynthesis-related (photosystems I and II) and carbon fixation through photosynthesis such as RuBisCO (Reusch et al. 2008). Indeed, Campbell et al. (2006) suggested that photosynthetic condition of seagrasses is likely to suffer irreparable effects in seawater temperatures as high as 40–45°C.

6.3.4 Salinity

High salinity causes both hyperionic and hyperosmotic stress effects; the results of these potentially cause death. Most commonly, stress is caused by high ionic content in the medium. There are growing concerns on increased salinities in marine estuarine areas worldwide. Although increased salinity can disturb carbon and O_2 balance in cells, *Thalassia testudinum*, *Halodule wrightii*, and *Ruppia maritima* seem to actively tolerate a wide range of salinities (35–50 psu) (Koch et al. 2007).

Salinity stress may cause ion channel gating, which leads to signal transduction pathway activation. In response to salinity fluctuation, many plants accumulate organic solutes to achieve osmotic adjustment. Sucrose and proline appear to be the principle organic osmoprotectants in *Zostera marina* (Ye and Zhao 2003) and *Ruppia maritima* (Murphy et al. 2003). Khalafallah et al. (2013) suggested the same mechanism of tolerance to increased salinity for the seagrass *Halodule uninervis*. Apart from sucrose and proline, free amino acids were significantly increased in the vegetative parts of *Halodule uninervis*. The recorded data showed that with increasing salinity for a few days, soluble sugars, free amino acids, and proline were significantly increased in the vegetative parts of *Halodule uninervis*, while long-term effect of salinities (60 and 65 psu) resulted in a highly significant reduction in these solutes. A recent study by Piro et al. (2015) pertains to the response of *Cymodocea nodosa* to hypersalinity conditions (43 psu) over 30 days. They observed through protein expression a decline in the expression of leaf proteins' level in stressed plants. Additionally, a downregulation of structural PSI, PSII proteins, and RuBisCO was illustrated. However, key enzymes involved in glycolysis showed higher accumulation levels suggesting a change in carbon metabolism in stressed plants (Piro et al. 2015). Moreover, the overexpression of cytochrome b559, which is a necessary receptor for PSII, is an indication that stressed plants in hypersaline conditions possess reparative mechanisms (Piro et al. 2015). Their findings are in concordance with previous studies (Muramatsu et al.2002; Kong et al. 2014) indicating that seagrasses in order to tolerate hypersaline conditions decrease photosynthetic conditions and increase osmoregulation mechanisms.

6.4 Perspectives

The amount of the genomic information on seagrasses is increasing exponentially, and the information on seagrass evolution and adaptation is contributing to improve our understanding on how different species are expected to respond to abiotic stress and adapt to the changing environment. New advances in systems biology holistic approaches for gene discovery and functional genomics will give a deeper insight to the molecular mechanisms that regulate the seagrass responses to tolerance acquisition to abiotic stress (Exadactylos 2015). In the Mediterranean Sea, for example, *Cymodocea nodosa*, a euryhaline and eurythermal species, will adapt the global change, while *Posidonia oceanica*, a stenothermal-stenohaline species, will decline. The molecular approach of their adaptive tolerance mechanisms will help us to better understand their evolutionary trends, as well as to identify stress-related genes and use the seagrasses as bioindicators of the ecosystem health.

There is a plethora of techniques that can be used to understand gene functions. Functional genomics through gene manipulation have not yet been adopted in seagrasses in order to fully comprehend gene functions and networks. Among them, gene knockdown is the foremost used technique in not only aquatic plants.

The CRISPR-Cas9 system, a flexible and robust technique for genome editing to analyze the molecular basis of abiotic stress response, is considered nowadays as the future solution for efficient and precise gene modifications which could in a longer perspective lead to creation of abiotic stress-tolerant crop plants (Hsu et al. 2014; Kumar and Jain 2015; Bortesi and Fischer 2015; Younis et al. 2014; Jain 2015). RNA interference (RNAi) is another gene regulatory approach in functional genomics. Recent studies have hinted possible roles of RNAi-related processes in plant stress adaptation (IAASA, Pocket K No. 34 RNAi for Crop Improvement, www.isaaa.org/resources/publications/pocketk/34/). Last but not least, gene fusions to reporter genes such as green fluorescent protein (GFP) can be used in large-scale experiments through the recombination-based strategy to illustrate the seagrass stress degree due to light attenuation.

For the appropriate use of seagrasses as marine bioindicators for the assessment of the environmental status in support of the Marine Strategy Framework Directive (MSFD) implementation, it is important to distinguish between anthropogenic-induced impacts and climatic long-term environmental changes in order to improve management and conservation activities.

References

Aquino RS, Landeira-Fernandez AM, Valente AP, Andrade LR, Mourao PAS (2005) Occurrence of sulfated galactans in marine angiosperms: evolutionary implications. Glycobiology 15:11–20

Arabidopsis Genome I (2000) Analysis of the genome sequence of the flowering plant *Arabidopsis thaliana*. Nature 408:796–815

Arnold T, Mealey C, Leahey H, Miller AW, Hall-Spencer JM, Milazzo M, Maers K (2012) Ocean acidification and the loss of phenolic substances in marine plants. PLoS One 7:e35107

Bortesi L, Fischer R (2015) The CRISPR/Cas9 system for plant genome editing and beyond. Biotechnol Adv 33:41–52

Boyd PW (2011) Beyond ocean acidification. Nat Geosci 4:273–274

Campbell SJ, Mckenzie LJ, Kerville SP (2006) Photosynthetic responses of seven tropical seagrasses to elevated seawater temperature. J Exp Mar Biol Ecol 330:455–468

Costanza R, D'Arge R, De Groot R, Farber S, Grasso M, Hannon B, Limburg K, Naeem S, O'Neill RV, Paruelo J, Raskin RG, Sutton P, Van den Belt M (1998) The value of the world's ecosystem services and natural capital (Reprinted from Nature, vol 387, pg 253, 1997). Ecol Econ 25:3–15

Cramer G, Urano K, Delrot S, Pezzotti M, Shinozaki K (2011) Effects of abiotic stress on plants: a systems biology perspective. BMC Plant Biol 11:163

Dalla Via J, Sturmbauer C, Schonweger G, Sotz E, Mathekowitsch S, Stifter M, Reinhard R (1998) Light gradients and meadow structure in *Posidonia oceanica*: ecomorphological and functional correlates. Mar Ecol Prog Ser 163:267–278

Dattolo E, Gu J, Bayer PE, Mazzuca S, Scrra IA, Spadafora A, Bernado L, Natali L, Cavallini A, Procaccini G (2013) Acclimation to different depths by the marine angiosperm *Posidonia oceanica*: transcriptomic and proteomic profiles. Front Plant Sci 195:1–15

Dattolo E, Ruocco M, Brunet C, Lorenti M, Lauritano C, D'Esposito D, De Luca P, Sanges R, Mazzuca S, Procaccini G (2014) Response of the seagrass *Posidonia oceanica* to different light environments: insights from a combined molecular and photo-physiological study. Mar Environ Res 101:225–236

Dennison WC, Alberte RS (1986) Photoadaptation and growth of *Zostera marina* L. (eelgrass) transplants along a depth gradient. J Exp Mar Biol Ecol 98:265–282

D'Esposito D, Orru L, Dattolo E, Bernardo L, Lamontara A, Orsini L, Serra IA, Mazzuca S, Procaccini G (2016) Annotation of "de novo" transcriptome in the seagrass *Posidonia oceanica* and SSR markers development. Sci Data Nat 3:160115

Kuo J, Den Hartog C (2007) Seagrasses morphology, anatomy, and ultrastructure. In: AWD L et al (eds) Seagrasses: biology, ecology and conservation. Springer, Dordrecht

Delker C, Stenzel I, Hause B, Miersch O, Feussner I, Wasternack C (2006) Jasmonate biosynthesis in *Arabidopsis thaliana*—enzymes, products, regulation. Plant Biol 8:297–306

Dillies MA, Rau A, Aubert J, Hennequet-Antier C, Jeanmougin M, Servant N, Keime C, Marot G, Castel D, Estelle J, Guernec G, Jagla B, Jouneau L, Laloe D, Le Gall C, Schaeffer B, Le Crom S, Guedj M, Jaffrezic F, French Statomique C (2013) A comprehensive evaluation of normalization methods for Illumina high-throughput RNA sequencing data analysis. Brief Bioinform 14:671–683

Exadactylos A (2015) Molecular approach of seagrasses response related to tolerance acquisition to abiotic stress. In: Çalişkan M, Oz GC, Kavakli IH, Ozcan B (eds) Molecular approaches to genetic diversity. InTech-Open Access, Rijeka, pp 75–91

Fourqurean JW, Duarte CM, Kennedy H, Marba N, Holmer M, Mateo MA, Apostolaki ET, Kendrick GA, Krause-Jensen D, Mcglathery KJ, Serrano O (2012) Seagrass ecosystems as a globally significant carbon stock. Nat Geosci 5:505–509

Franssen SU, Gu J, Bergmann N, Winters G, Klostermeier UC, Rosenstiel P, Bornberg-Bauer E, Reusch TBH (2011) Transcriptomic resilience to global warming in the seagrass *Zostera marina*, a marine foundation species. Proc Natl Acad Sci U S A 108:19276–19281

Franklin L, Osmond CB, Larkum AD (2003) Photoinhibition, UV-B and algal photosynthesis. In: Larkum AD, Douglas S, Raven J (eds) Photosynthesis in algae, vol 14. Springer, Dordrecht, pp 351–384

Garrard SL, Gambi MC, Scipione MB, Patti FP, Lorenti M, Zupo V, Paterson DM, Buia MC (2014) Indirect effects may buffer negative responses of seagrass invertebrate communities to ocean acidification. J Exp Mar Biol Ecol 461:31–38

Golicz AA, Schliep M, Lee HT, Larkum AW, Dolferus R, Batley J, Chan CK, Sablok G, Ralph PJ, Edwards D (2015) Genome-wide survey of the seagrass *Zostera muelleri* suggests modification of the ethylene signalling network. J Exp Bot 66:1489–1498

Greco M, Chiappetta A, Bruno L, Bitonti MB (2013) Effects of light deficiency on genome methylation in *Posidonia oceanica*. Mar Ecol Prog Ser 473:103–114

Green S (2017) Philosophy of systems biology. Perspectives from scientists and philosophers. Springer, Cham. doi:10.1007/978-3-319-47000-9

Haas BJ, Papanicolaou A, Yassour M, Grabherr M, Blood PD, Bowden J, Couger MB, Eccles D, Li B, Lieber M, Macmanes MD, Ott M, Orvis J, Pochet N, Strozzi F, Weeks N, Westerman R, William T, Dewey CN, Henschel R, Leduc RD, Friedman N, Regev A (2013) De novo transcript sequence reconstruction from RNA-seq using the trinity platform for reference generation and analysis. Nat Protoc 8:1494–1512

Hasler-Sheetal H, Fragner L, Holmer M, Weckwerth W (2015) Diurnal effects of anoxia on the metabolome of the seagrass *Zostera marina*. Metabolomics 11:1208–1218

Hemminga MA, Duarte CM (2000) Ecology. Cambridge University Press, Cambridge

Hiroaki K (2002) Computational systems biology. Nature 450:206–210

Hunt L, Gray JE (2009) The signaling peptide EPF2 controls asymmetric cell divisions during stomatal development. Curr Biol 19:864–869

Hsu P, Lander E, Zhang F (2014) Development and applications of CRISPR-Cas9 for genome engineering. Cell 157:1262–1278

Iyapparaj P, Revathi P, Ramasubburayan R, Prakash S, Palavesam A, Immanuel G, Anantharaman P, Sautreau A, Hellio C (2014) Antifouling and toxic properties of the bioactive metabolites from the seagrasses *Syringodium isoetifolium* and *Cymodocea serrulata*. Ecotoxicol Environ Saf 103:54–60

Jain M (2015) Function genomics of abiotic stress tolerance in plants: a CRISPR approach. Front Plant Sci 6:375

Jiang ZJ, Huang XP, Zhang JP (2010) Effects of CO_2 enrichment on photosynthesis, growth, and biochemical composition of seagrass *Thalassia hemprichii* (Ehrenb.) Aschers. PubMed 10:904–913

Khalafallah AA, Geneid YA, Shaetaey SA, Shaaban B (2013) Responses of the seagrass *Halodule uninervis* (Forssk.) Aschers. to hypersaline conditions. Egypt J Aquat Res 39:167–176

Koce JD, Vilhar B, Bohanec B, Dermastia M (2003) Genome size of Adriatic seagrasses. Aquat Bot 77:17–25

Koch MS, Schopmeyer SA, Kyhn-Hansen C, Madden CJ, Peters JS (2007) Tropical seagrass species tolerance to hypersalinity stress. Aquat Bot 86:14–24

Kong FN, Li H, Sun PP, Zhou Y, Mao YX (2014) De novo assembly and characterization of the transcriptome of seagrass *Zostera marina* using Illumina paired-end sequencing. PLoS One 9:e112245

Kumar V, Jain M (2015) The CRISPR-Cas system for plant genome editing: advances and opportunities. J Exp Bot 66:47–57

Lauritano C, Ruocco M, Dattolo E, Buia MC, Silva J, Santos R, Olive I, Costa MM, Procaccini G (2015) Response of key stress-related genes of the seagrass *Posidonia oceanica* in the vicinity of submarine volcanic vents. Biogeosciences 12:4185–4194

Mcmahon K, Collier C, Lavery PS (2013) Identifying robust bioindicators of light stress in seagrasses: a meta-analysis. Ecol Indic 30:7–15

Murphy LR, Kinsey ST, Durako MJ (2003) Physiological effects of short-term salinity changes on *Ruppia maritima*. Aquat Bot 75:293–309

Muramatsu Y, Harada A, Ohwaki Y, Kasahara Y, Takagi S, Fukuhara T (2002) Salt-tolerant ATPase activity in the plasma membrane of the marine angiosperm Zostera marina L. Plant Cell Physiol 43(10):1137–1145

Nadeau JA, Sack FD (2002) Control of stomatal distribution on the *Arabidopsis* leaf surface. Science 296:1697–1700

Nicholson JK, Wilson ID (2003) Understanding 'global' systems biology: metabolomics and the continuum of metabolism. Nat Rev 2:668–676

Olsen JL, Rouze P, Verhelst B, Lin YC, Bayer T, Collen J, Dattolo E, De Paoli E, Dittami S, Maumus F, Michel G, Kersting A, Lauritano C, Lohaus R, Topel M, Tonon T, Vanneste K, Amirebrahimi M, Brakel J, Bostrom C, Chovatia M, Grimwood J, Jenkins JW, Jueterbock A, Mraz A, Stam WT, Tice H, Bornberg-Bauer E, Green PJ, Pearson GA, Procaccini G, Duarte CM, Schmutz J, Reusch TB, Van de Peer Y (2016) The genome of the seagrass *Zostera marina* reveals angiosperm adaptation to the sea. Nature 530:331–335

Ortega-Martinez O, Pernas M, Carol RJ, Dolan L (2007) Ethylene modulates stem cell division in the *Arabidopsis thaliana* root. Science 317:507–510

Ow YX, Vogel N, Holtum JAM, Flores F, Uthicke S (2016) Nitrate fertilization does not enhance CO_2 responses in two tropical seagrasses species. PMC 6:23093

Papenbrock J (2012) Highlights in seagrasses' phylogeny, physiology, and metabolism: what makes them special? ISRN Bot:1–15

Palacios SL, Zimmerman RC (2007) Response of eelgrass Zostera marina to CO2 enrichment: possible impacts of climate change and potential for remediation of coastal habitats. Mar Ecol Prog Ser 344:1–13

Piro A, Marin-Guirao L, Serra IA, Spadafora A, Sandoval-Gip JM, Bernardeau-Esteller J, Fernandez JM, Mazzuca S (2015) The modulation of leaf metabolism plays a role in salt tolerance of *Cymodocea nodosa* exposed to hypersaline stress in mesocosms. Front Plant Sci 6:464

Procaccini G, Ruocco M, Marin-Guiraro L, Dattolo E, Brunet C, D'Esposito D, Lauritano C, Mazzuca S, Serra IA, Bernardo L, Piro A, Beer S, Bjork M, Gullstrom M, Buapet P, Rasmusson LM, Felisberto P, Gobert S, Runcie JW, Silva J, Olive I, Costa MM, Barrote I, Santos R (2017) Depth-specific fluctuations of gene expression and protein abundance modulate the photophysiology in the seagrass *Posidonia oceanica*. Sci Rep Nat 4:42890

Reusch TBH, Veron AS, Preuss C, Weiner J, Wissler L, Beck A, Klages S, Kube M, Reinhardt R, Bornberg-Bauer E (2008) Comparative analysis of expressed sequence tag (EST) libraries in the seagrass *Zostera marina* subjected to temperature stress. Mar Biotechnol 10:297–309

Robertson G, Schein J, Chiu R, Corbett R, Field M, Jackman SD, Mungall K, Lee S, Okada HM, Qian JQ, Griffith M, Raymond A, Thiessen N, Cezard T, Butterfield YS, Newsome R, Chan SK, She R, Varhol R, Kamoh B, Prabhu AL, Tam A, Zhao YJ, Moore RA, Hirst M, Marra MA, Jones SJM, Hoodless PA, Birol I (2010) De novo assembly and analysis of RNA-seq data. Nat Methods 7:909–U62

Saini S, Sharma I, Pati PK (2015) Versatile roles of brassinosteroid in plants in the context of its homoeostasis, signaling and crosstalks. Front Plant Sci 6:950

Schulz MH, Zerbino DR, Vingron M, Birney E (2012) Oases: robust de novo RNA-seq assembly across the dynamic range of expression levels. Bioinformatics 28:1086–91092

Touchette BW, Burkholder JM (2007) Carbon and nitrogen metabolism in the seagrass, Zostera marina L.: environmental control of enzymes involved in carbon allocation and nitrogen assimilation. J Exp Mar Biol Ecol 350:216–233

Lee HT, Golicz AA, Bayer PE, Jiao Y, Tang H, Paterson AH, Sablok G, Krishnaraj RR, Kenneth Chan C-K, Batley J, Kendrick GA, AWD L, Ralph PJ, Edwards D (2016) The genome of a southern hemisphere seagrass species (*Zostera muelleri*). Plant Physiol 172:272–283

Younis A, Siddique M, Kim C, Lim K (2014) RNA interference (RNAi) induced gene silencing: a promising approach of hi-tech plant breeding. Int J Biol Sci 10(10):1150–1158

Waycott M, Duarte CM, Carruthers TJ, Orth RJ, Dennison WC, Olyarnik S, Calladine A, Fourqurean JW, Heck KL Jr, Hughes AR, Kendrick GA, Kenworthy WJ, Short FT, Williams SL (2009) Accelerating loss of seagrasses across the globe threatens coastal ecosystems. Proc Natl Acad Sci U S A 106:12377–12381

Winters G, Nelle P, Fricke B, Rauch G, Reusch TBH (2011) Effects of a simulated heat wave on photophysiology and gene expression of high- and low-latitude populations of *Zostera marina*. Mar Ecol Prog Ser 435:83–95

Wissler L, Codoner FM, Gu J, Reusch TB, Olsen JL, Procaccini G, Bornberg-Bauer E (2011) Back to the sea twice: identifying candidate plant genes for molecular evolution to marine life. BMC Evol Biol 11:8

Wissler L, Dattolo E, Moore AD, Reusch TB, Olsen JL, Migliaccio M, Bornberg-Bauer E, Procaccini G (2009) Dr. Zompo: an online data repository for *Zostera marina* and *Posidonia oceanica* ESTs. Database (Oxford) 2009:bap009

Xie YL, Wu GX, Tang JB, Luo RB, Patterson J, Liu SL, Huang WH, He GZ, Gu SC, Li SK, Zhou X, Lam TW, Li YR, Xu X, Wong GKS, Wang J (2014) SOAPdenovo-trans: de novo transcriptome assembly with short RNA-Seq reads. Bioinformatics 30:1660–1666

Ye CJ, Zhao KF (2003) Osmotically active compounds and their localization in the marine halophyte eelgrass. Biol Plant 46:137–140

Chapter 7
Photobiology of Seagrasses: A Systems Biology Perspective

Pimchanok Buapet

Abstract In addition to their well-recognized ecosystem services, seagrass meadows have recently gained considerable attention for their significant role in the marine carbon cycle. Accurate estimation of the carbon budgets of seagrass meadows does require a systems understanding of photosynthetic processes particularly in the dynamic environments of their coastal habitats and inevitably in the global change scenarios. Also, the frequent fluctuations in physicochemical factors in seagrass habitats can directly affect their photosynthetic activity and its interplay with other connecting metabolic pathways. With high-throughput experiments being increasingly common, combining the "omics" approaches can help to elucidate complex photosynthetic responses in such dynamic environment of seagrass habitats. Here, seagrass photosynthesis basics with an emphasis on its unique adaptation to submerged conditions, as well as research progress in molecular physiology are summarized. The knowledge gaps are discussed as well as further prospects for integrating photophysiology with systems biology.

Keywords Seagrass • Photosynthesis • Stress physiology • Omics

Contents

7.1 General Overview of the Photosynthetic Processes in Seagrass 134
7.1.1 Light-Harvesting and Light Reactions 134
7.1.2 Carbon Reactions 135
7.1.3 Photorespiration and Alternative Electron Flows 136
7.2 Photosynthesis in a Dynamic Environment: Research Advances in Seagrasses 138
7.2.1 Light: Photoacclimation and Photoprotection 138
7.2.2 Salinity Stress and Its Impact on Photosynthesis 141
7.2.3 Heat Stress and Its Impact on Photosynthesis 142
7.2.4 Effect of CO_2 Enrichment on Seagrass Primary Production 143
7.2.5 Phytotoxins 144

P. Buapet (✉)
Department of Biology, Faculty of Science, Prince of Songkla University, Hat Yai, Songkhla, Thailand, 90112

Coastal Oceanography and Climate Change Research Center, Prince of Songkla University, Hat Yai, Songkhla, Thailand, 90112
e-mail: pimchanok.b@psu.ac.th

© Springer International Publishing AG 2017

M. Kumar, P. Ralph (eds.), *Systems Biology of Marine Ecosystems*,
DOI 10.1007/978-3-319-62094-7_7

7.3 Knowledge Gaps 146
7.3.1 The Function of Photoreceptors 147
7.3.2 Mechanistic Understanding of Photosynthetic Carbon Fixation 147
7.3.3 Linkage Between Photosynthesis and Other Metabolic Pathways 148
7.4 Further Prospects for Integrating Seagrass Photophysiology with Systems Biology 151
7.5 Conclusion 153
References 154

7.1 General Overview of the Photosynthetic Processes in Seagrass

The successful colonization of seagrass, the only group of marine angiosperms, likely reflects their ability to cope with environmental fluctuations in the marine systems. Photosynthesis is the main target of physiological regulation in that it plays the central role in energy metabolisms that determine production and growth of plants. Although angiosperms share essentially the same physiology and biochemistry, seagrass exhibits adaptations to marine life which affect their photosynthetic activity.

7.1.1 Light-Harvesting and Light Reactions

Photosynthetic reactions begin with the absorption of light energy followed by electron transport within the thylakoid membrane, ATP synthesis and production of reducing equivalent (NADPH) to be further used to assimilate CO_2 (Taiz et al. 2015). At least five major protein complexes, located in the inner thylakoid membrane, are involved in these processes: photosystem II (PSII), cytochrome b6f complex, photosystem I (PSI), ATP synthase, and ferredoxin-NADP $^+$-oxidoreductase complex. Both photosystems (PSII and PSI) form supercomplexes with light-harvesting complexes LHCII and LHCI, respectively. These light-harvesting complexes contain antenna molecules (chlorophyll a/b and carotenoids) and binding proteins working in concert to harvest light energy. Several proteins associated with PSII and PSI supercomplexes have been identified and characterized (Wicke et al. 2011).

Aquatic environment dramatically affects light quantity and quality. While the intensity of light is reduced with depths, wavelength composition of downwelling light is also altered as red, and far-red wavelengths are largely attenuated by the water column, whereas blue light is highly scattered in the coastal water (Zimmerman 2003; Ragni and Ribera d'Alcala 2004). It is, thus, expected that light-harvesting and the photosynthetic machinery of seagrasses are adjusted according to such light attenuation. It appears that seagrasses have similar regions of maximum absorption (blue 400–500 nm and red 600–700 nm; Drake et al. 2003; Mvungi et al. 2012) and the same composition of light-harvesting pigments for photosynthesis as other higher plants, i.e., chlorophyll a, chlorophyll b, and a variety of xanthophylls and carotenoids (Casazza and Mazzella 2002). However, the siphonaxanthin-like pigment has been reported in the seagrasses *Posidonia oceanica* and *Halophila stipulacea*, and its concentration increases with depth (Casazza and Mazzella 2002). This group of pigment is specifically found in the light-harvesting complex of green alga and has been proposed to play a role in harvesting light in the green region (510–550 nm) and transfer absorbed energy to chlorophylls (Wang et al. 2013). From such finding, it is suggested that seagrass might have undergone light-harvesting adaptation to the underwater light environment.

Recent research provides further evidence of photosynthetic adaptation associated with light-harvesting and the photosynthetic machinery of seagrass. Wissler et al. (2011) identified candidate genes associated with the molecular evolution of seagrass. This work revealed that gene class encoding photosynthetic proteins have been significantly enriched in *Zostera marina* and *Posidonia oceanica*, whereas a number of genes encoding photosynthetic antenna proteins and involved in photosynthetic electron transport such as light-harvesting complex 5 (LHCB5), ferredoxin 3 (ATFD3), PSI subunit L (PSAL), PSII subunit O-2, oxygen-evolving binding (PSBO2), PSII subunit R (PSBR), and chlorophyll-binding protein (LHCA3) have been identified as positively selected genes, indicating that many constituents of the photosynthetic pathway have acquired sequence changes upon adaptation to the sea (Wissler et al. 2011). Olsen et al. (2016) examined the genome of the seagrass *Zostera marina* and showed that PSI and PSII of *Z. marina* remain similar to those of other terrestrial and aquatic plants, but chlorophyll a/b-binding proteins of light-harvesting complex II subtype Lhcb of *Z. marina* become more diversified. There are experimental indications from the studies in higher plants and algae that these PSII-associated chlorophyll a/b-binding proteins are the main targets of regulation under changing light condition (Kouřil et al. 2013).

7.1.2 Carbon Reactions

Similar to many other aquatic plants, seagrass lacks stomata and dissolved CO_2 is absorbed directly into the epidermal cells. Correspondingly, the sequenced genome of the seagrass *Zostera marina* has revealed a complete loss of the genes associated with stomatal differentiation (Olsen et al. 2016). Seagrasses are able to utilize HCO_3^- in addition to CO_2 as an exogenous carbon source. Such ability gives seagrasses competitive advantage in a marine environment where HCO_3^- is the dominant form of dissolved inorganic carbon and diffusion rate of CO_2 is slow. The thick stagnant layer adjacent to the leaf surface or boundary layer plays a part in impeding the CO_2 acquisition of seagrass from the water body as CO_2 can only be transferred via molecular diffusion in such layer (Beer 1989). The most common carbon-concentrating system of seagrasses is associated with extracellular carbonic anhydrase. It has been proposed that this enzyme catalyzes the formation of CO_2 from

HCO_3^- at the boundary layer. The newly formed CO_2 then diffuses into the cell (Moroney et al. 1985; Mercado et al. 1997, 1998; Larsson and Axelsson 1999; Beer and Rehnberg 1997; Invers et al. 1999; Borum et al. 2016; Ow et al. 2016). An additional system proposed in seagrasses involves the formation of CO_2 from HCO_3^- in the diffusion boundary layer by lowering pH and shifting DIC equilibrium via protons extrusion (Hellblom et al. 2001; Uku et al. 2005). The efficiency and limitation of such carbon acquisition mechanisms differ among species and the environmental conditions in which seagrasses grow (Hellblom et al. 2001; Uku et al. 2005). Recent work has also shown that seagrass in the intertidal areas can utilize carbon from both dissolved inorganic carbon pools and atmospheric CO_2 depending on the tidal levels (Park et al. 2016). More negative δ13C values were observed in leaf tissues of seagrasses from intertidal areas which suggests that these plants utilize atmospheric CO_2 directly during low tide exposure (Park et al. 2016).

Identifying seagrasses as C3 or C4 plants remains challenging. Although seagrasses are generally regarded as C3 plants, previous studies have shown contradictory results, and certain species appeared to have C3–C4 intermediate characteristics (see review by Touchette and Burkholder 2000). The C4 Kranz anatomy identified by chloroplast-rich bundle sheath cells is not observed in seagrasses although several aquatic species of Hydrocharitaceae and certain species of Alismataceae are able to operate Kranz-less single-cell C4 photosynthesis (Bowes et al. 2002). Results from pulse-chase ^{14}C experiments have suggested that the photosynthetic carbon fixation of several seagrass species such as *Halodule wrightii*, *H. uninervis*, *Syringodium filiforme*, *S. isoetifolium*, *Thalassia testudinum*, *T. hemprichii*, *Thalassodendron ciliatum*, *Halophila spinulosa*, and *H. stipulacea* follows the C3 pathway with phosphoglycerate being the first major stable organic compound and malate and aspartate being a small fraction of the labeled substances (Andrews and Abel 1979; Benedict et al. 1980; Beer and Waisel 1979; Beer et al. 1980; Beer and Wetzel 1982).

7.1.3 Photorespiration and Alternative Electron Flows

Ribulose-1,5-bisphosphate carboxylase oxygenase (Rubisco) is able to catalyze both carboxylation and oxygenation reactions. Such dual function of Rubisco is common to all photosynthetic organisms. The oxygenation of ribulose-1,5-bisphosphate (RuBP) generates the toxic by-product 2-phosphoglycolate (2PG) which is subsequently metabolized into the Calvin cycle intermediate, phosphoglycerate (PGA), via a series of reactions called photorespiration (Bauwe et al. 2010). Photorespiration takes place in three cell components: chloroplast, peroxisome, and mitochondria (Bauwe et al. 2010). This process also causes the loss of carbon and nitrogen as CO_2 and NH_4 and consumes ATP and reducing equivalents. It has been estimated that in C3 plants, photorespiration can lead to up to 30% decrease in primary production under present-day atmospheric CO_2:O_2. As a consequence, it has been viewed as a wasteful process hindering photosynthetic productivity in C3 plants (Zhu et al. 2004). Nevertheless, photorespiration has been proven to be crucial for photosynthetic organisms as mutants lacking key photorespiratory enzymes often exhibit poor performance, stress symptoms, and lethality under present-day atmospheric condition (Bauwe et al. 2010). Moreover, a number of studies suggest that photorespiration acts as safety valve adjusting redox homoeostasis under stress conditions (Voss et al. 2013) or a source of intermediate metabolites for other metabolic pathways of the plant (Novitskaya et al. 2002).

Previous studies have reported photorespiratory activities in certain species based on evidence from ^{14}C pulse-chase experiments showing O_2-enhanced ^{14}C incorporation into photorespiratory intermediates (Burris et al. 1976; Andrews and Abel 1979), inhibited photosynthetic rates in the presence of high O_2 (Downton et al. 1976), O_2-stimulated respiration in the light (Hough 1976), activities of glycolate metabolizing enzymes (Tolbert 1976), and enhanced photosynthetic rates under low O_2 condition (Buapet et al. 2013a). Nevertheless, it has been suggested that HCO_3^- utilization systems in seagrass might be able to efficiently maintain high CO_2 availability around Rubisco fixation site thus suppressing photorespiration (Beer 1989). As a result, the role of photorespiration, as well as other alternative electron sinks, has long been ignored. A few recent studies, however, have shown that photorespiration in the temperate seagrasses, *Z. marina* and *Ruppia maritima*, is enhanced by carbon limitation (Buapet et al. 2013a). Since carbon availability occasionally becomes limiting in stagnant shallow coastal waters (Buapet et al. 2013b), the impact of photorespiration on seagrass primary production might be more prominent in the natural setting.

Photosynthetic organisms have flexible electron transports. Linear electron flow is the principal pathway in which the electrons are transported from PSII to PSI and eventually used to generate NADPH via ferredoxin-$NADP^+$-oxidoreductase. Also, plants have other alternative electron flow pathways, e.g., pseudocyclic electron flow or the Mehler reaction and cyclic electron flow around PSI (Allen 2003). In pseudocyclic electron flow or the Mehler reaction, molecular oxygen acts as an electron acceptor instead of $NADP^+$, forming superoxide which is subsequently detoxified into water (Heber 2002). Two pathways, PGR5- and NDH-dependent pathways, have been identified to be involved in cyclic electron flow around PSI (Johnson 2011). These alternative electron flow and linear electron flow pathways appear to operate simultaneously. The rates, however, depend largely on the plant growing conditions (Allen 2003). Studies have shown conflicting conclusions regarding contribution and physiological role of each alternative electron flow pathway (Cornic and Briantais 1991; Biehler and Fock 1996; Lovelock and Winter 1996; Badger et al. 2000; Foyer and Noctor 2000; Proctor and Smirnoff 2011; Driever and Baker 2011; Kramer and Evans 2011). It is proposed that these pathways serve as photoprotection by generating trans-thylakoid pH gradient, thus inducing energy dissipation by non-photochemical quenching (Johnson 2011; Johnson et al. 2014), while helping adjusting ATP/NAPDH production (Peng et al. 2009; Kramer and Evans 2011). Recent studies have reported suboptimal growth rates in cyclic electron flow mutants, while plants lacking the Mehler reaction were unaffected, suggesting a more relevant contribution of cyclic electron flow when compared to the Mehler reaction (Lakshmanan et al. 2015). Regarding seagrass, the Mehler reaction has been investigated in the seagrass *Z. marina*, and its contribution to the electron flow appears to be minor when compared to photorespiration (Buapet and Björk 2016). Although alternative electron flows have been mentioned in several seagrass studies, the works focusing on their contribution and physiological role remain scarce.

7.2 Photosynthesis in a Dynamic Environment: Research Advances in Seagrasses

Seagrass meadows, particularly in the shallow water, are subjected to dynamic environmental conditions. Daily and seasonal fluctuations of various physicochemical factors such as light, temperature, and salinity have been reported in many seagrass habitats across the bioregions. In the global change scenarios, such fluctuations can be exacerbated as extreme weather events become more frequent. Additionally, interactions between these environmental factors and a future increase in dissolved CO_2 might impose a profound effect on seagrass metabolisms. To successfully maintain positive carbon balance and growth, seagrass must be able to adjust their photosynthetic systems to these short- and long-term changes. Here, recent research addressing effects of abiotic stress factors on photophysiology of seagrasses are summarized.

7.2.1 Light: Photoacclimation and Photoprotection

Seasonal and daily fluctuation in light quantity and quality are common in seagrass habitats (Kenworthy and Haunert 1991; Gallegos and Kenworthy 1996; Kahn et al. 2013). Furthermore, anthropogenic activities causing sedimentation and eutrophication can lead to a shift in light regimes (Schmidt et al. 2012; Yaakub et al. 2014). The effects of varying light conditions on the photosynthetic activity of seagrasses have been extensively studied. Early efforts have been invested in determining the light requirement for seagrass to maintain its carbon balance (Dennison and Alberte 1982; Pérez and Romero 1992; Abal et al. 1994; Zimmerman et al. 1995; Kenworthy and Fonseca 1996), while more recent works have focused on photoacclimation mechanisms to both low light and high light in relation to depth distributions (Olivé et al. 2013; Dattolo et al. 2013, 2014; Park et al. 2016). Pulse amplitude modulation (PAM) fluorometry has been the most frequently used techniques for the investigation of photophysiology of seagrasses. The most common PAM parameters used to describe photosynthetic activity are the effective photochemical efficiency of PSII (Φ_{PSII}) which provides an estimation of the proportion of the light absorbed by PSII that is utilized in photochemistry and the maximum photochemical efficiencies of PSII (F_v/F_m) which indicates stress associated with photoinhibition (Maxwell and Johnson 2000). Nevertheless, fast chlorophyll a fluorescence transient (OJIP) analysis (see review in Stirbet and Govindjee 2011) and investigations at molecular levels have increased significantly during the recent years.

Large non-photosynthetic tissue in seagrasses resulted in high respiratory oxygen demand, hence the high minimum light requirement to maintain the whole-plant positive carbon balance. Furthermore, seagrasses depend largely on the transfer of photosynthetically derived oxygen to the rhizosphere to prevent sulfide intrusion (Brodersen et al. 2015a, b, discussed further in the section *Phytotoxins*). Consequently, seagrasses are regarded as highly sensitive to a decrease in light intensity (reviewed in Ralph et al. 2007). For this reason, much attention has been focused on determining the effect of light limitation on seagrass photosynthesis (reviewed in Ralph et al. 2007). In general, plant responses to low light condition include an increase in number of chloroplasts, chlorophyll contents, and other light-harvesting pigments per unit leaf area (Murchie and Horton 1997) as well as an increase in chlorophyll b/a ratios which have been shown to indicate a larger PSII light-harvesting antenna (Bailey et al. 2001). Modulation of photosynthetic apparatus under a low light condition such as state transition (Mullineaux and Emlyn-Jones 2005), an increase in PSI reaction center components (Bailey et al. 2001) and an alteration in the activation state of Rubisco (Salvucci and Ogren 1996) have been observed. Seagrasses grown under low light condition exhibit a common shade plant response measured as the photosynthesis-irradiance characteristics such as an increase in a slope of the light-limiting range of the photosynthesis-irradiance curve (α) and lower light saturation point (Ek) when compared to those of seagrasses under high light (Silva et al. 2013; Howarth and Durako 2013; Dattolo et al. 2014; Park et al. 2016). Enhancement of light-harvesting capacity is achieved by means of increasing chlorophyll a, chlorophyll b, and carotenoid contents and antenna size (reviewed in Ralph et al. 2007; Silva et al. 2013; Howarth and Durako 2013; Dattolo et al. 2014). Chlorophyll b/a ratio (a proxy of PSII antenna size) was found to be higher in seagrasses growing under lower light intensity (Dennison and Alberte 1982; Longstaff and Dennison 1999; Lamote, and Dunton 2006; Dattolo et al. 2014). Accumulating evidence suggests a regulatory role of PSI in shade adaptation in seagrasses. *Thalassia testudinum* grown in deeper site exhibited larger PSI antenna, estimated as total chlorophylls associated with isolated PSI, when compared to that of the shallow site (Major and Dunton 2002). It was proposed that an increase in PSI antenna size might help to reduce the light requirement to reach maximum photosynthesis, thus optimizing the photosynthetic efficiency under low light intensity. In the extreme case of *H. stipulacea* with exceptionally wide range of depth distribution (10–50 m), the protein levels of PSI iron-sulfur center (PsaC) relative to PSII protein D1 (PsbA) as well as the functionality of PSI (estimated from the peak heights of fluorescence emission at 717 nm at low-temperature using excitation wavelength at 435 nm (Sharon et al. 2011) increase in the deep water population. A similar result was observed in *Arabidopsis thaliana* (Bailey et al. 2001), and it was proposed to be a result of an increase in ATP demand in relation to NADPH which can be supplied by cyclic electron transports around PSI (Bailey et al. 2001; Sharon et al. 2011). In contrast, opposing results have been reported in recent studies in the seagrass *P. oceanica* and *Z. muelleri* (Dattolo et al. 2014;

Schliep et al. 2015). In these work, seagrasses under low light condition exhibited downregulation of PSI gene expression (PSI reaction center subunit V in Dattolo et al. 2014 and PS I reaction center subunit IV B in Schliep et al. 2015). Although the physiological and molecular mechanisms of these differentially expressed genes and proteins remain unclear, these cited research provide a basis for further investigation of seagrass molecular adaptation to limiting light.

On the contrary, intertidal and shallow subtidal species might experience periodic or prolonged high irradiance during low tides (Ralph and Burchett 1995; Hanelt and Figueroa 2012). At the physiological level, fast chlorophyll fluorescence induction curves revealed a closure of PSII reaction centers, which likely contributed to the decline in effective quantum efficiency under higher irradiance (York et al. 2013). To prevent photodamage due to high light, plants use different photoprotection strategies. Lowering light perception by reducing light-harvesting antenna size measured as chlorophyll b/a ratio was reported in *P. oceanica* (Dattolo et al. 2014). At the molecular level, Dattolo et al. (2013, 2014) highlighted the possible roles of chlorophyll a/b-binding proteins (CABs) and Rubisco activase in photoprotection and photoacclimation of the seagrass *P. oceanica*. In these works, seagrasses growing in deep (lower light intensity) and shallow (higher light intensity) sites exhibited distinct transcriptional and proteomic profiles regarding CABs (Dattolo et al. 2014): transcript level of chlorophyll a/b-binding proteins (chlorophyll a/b-binding protein 1 (Pooc_B_c293), 21 (Pooc_B_c132), 131 (Pooc_B_c386)) was found to be higher in the seagrass growing in shallow site. CABs are thought to involve in light-harvesting regulation under different light conditions, and certain members of CAB family have been proposed to play a photoprotective role via state transition (Pietrzykowska et al. 2014) and excess light energy dissipation (Li et al. 2000). Dattolo et al. (2013) also reported higher protein level of Rubisco activase in the seagrass growing in shallow site. Rubisco activase regulates photosynthesis by modulating Rubisco activation under various conditions, and its activity has been shown to increase with light intensity (Lan et al. 1992; Portis 2003). Additionally, xanthophyll cycle has been proposed to serve a role in excess energy dissipation in various seagrass species such as *Z. capricorni* (Flanigan and Critchley 1996), *Z. marina* (Ralph et al. 2002), *P. sinuosa* (Collier et al. 2008), *T. testudinum* (Howarth and Durako 2013), *P. oceanica*, and *C. nodosa* (Marín-Guirao et al. 2013a; Dattolo et al. 2014). At the molecular level, *P. oceanica* in shallow areas show an upregulation of the xanthophyll cycle-related gene, zeaxanthin de-epoxidase, as well as an increase in violaxanthin and a consequential increase in the sum of xanthophyll cycle pigment pool compared to the deep meadow. Additionally, a few works have proposed that anthocyanin also serve a role in photoprotection in seagrasses. Anthocyanin accumulation in leaves has been widely reported in various seagrass species in the intertidal to the shallow subtidal area across the tropical bioregions (Novak and Short 2010; Ragavan et al. 2013; Kaewsrikhaw and Prathep 2014). Novak and Short (2011a, b, 2012) conducted extensive studies on the role of anthocyanin in *T. testudinum* (from Florida Keys, USA). From these observations, it was concluded that anthocyanin accumulation provides photoprotection for *T. testudinum* by absorbing both ultraviolet (absorption maxima 280 nm) and visible wavelengths (absorption maxima 500–550 nm Novak and Short 2011a, b) and reflecting across a wide region (600–640 nm Gausman 1982). Such absorbed light energy is not transferred to the photosynthetic pigments, thus reducing excitation energy reaching the photosystems. Production of anthocyanin is also stimulated upon the exposure to UV-B, indicating that UV-B plays a critical role in seagrass light signal transduction pathways (Novak and Short 2011b). On the other hand, no clear indication of the photoprotective function of anthocyanin was observed in the seagrass *H. ovalis* (Southern Thailand, Kaewsrikhaw and Prathep 2014). In this work, *H. ovalis* exhibited higher anthocyanin content in the rainy season when irradiance was less strong when compared to that in the dry season.

7.2.2 Salinity Stress and Its Impact on Photosynthesis

Photosynthetic responses to salinity have been quantified in the seagrass species such as *P. oceanica*, *C. nodosa*, *T. testudinum*, and *Z. japonica* (Shafer et al. 2011; Marín-Guirao et al. 2011; Marín-Guirao et al. 2013b; Sandoval-Gil et al. 2012; Howarth and Durako 2013; Sandoval-Gil et al. 2014; Piro et al. 2015a). Nevertheless, the mechanisms which salinity induces photosynthetic damage are not yet fully understood. Chlorophyll fluorescence investigations have shown a reduction in Φ_{PSII} and F_v/F_m (Howarth and Durako 2013) in hypersaline treatment. Interestingly, photoprotection appears to contribute to salinity tolerance as demonstrated in the seagrass *P. oceanica* (Marín-Guirao et al. 2013a, b). In this work, no difference in maximum photochemical efficiencies was detected in under hypersaline stress whereas non-photochemical quenching (NPQ) was significantly enhanced. This suggests that under such unfavorable condition when carbon assimilation is compromised, excess energy dissipation via NPQ plays a significant role in protecting the photosynthetic apparatus of this seagrass (Marín-Guirao et al. 2013a, b). It has been suggested that a decrease in chlorophyll density observed under hypersaline stress could also be a result of a downregulation of light energy harvesting in order to minimize oxidative stress (Sandoval-Gil et al. 2012).

While photosynthetic rates, measured as oxygen evolution rates, decrease under salinity stress, many studies have reported an increase in respiratory rates in seagrasses (Shafer et al. 2011; Marín-Guirao et al. 2011, 2013b; Sandoval-Gil et al. 2012). For example, dark respiration rates in salt-stressed plants were almost double of control plants (Marín-Guirao et al. 2013b). This shift in carbon metabolisms affects plant carbon balance and might subsequently reduce their growth rates and lead to mortality. Such an increase in respiratory demands might be a result of more energy needed to accommodate an increase in the activity of osmoregulatory processes. Piro et al. (2015a) provide evidence of a shift in carbon metabolisms in *C. nodosa* under salinity stress from proteomics data. Here, it was shown that the structural proteins and enzymes associated with photosynthesis (PSII, PSI, Rubisco) became less abundant, whereas the respiratory enzymes (cytosolic glyceraldehyde-3-phosphate dehydrogenase, enolase2, and triose phosphate isomerase) became more abundant under hypersaline stress. This work also proposed a role in repair processes of the photosynthetic machinery for cytochrome b559 alpha subunit of the PSII initial complex (Piro et al. 2015a).

7.2.3 Heat Stress and Its Impact on Photosynthesis

Shallow water seagrasses encounter daily and seasonal fluctuations in temperature. The ongoing increase in mean global ocean temperatures, as well as more frequent extreme heat events, brought about by global warming, can further affect seagrass metabolisms. It is well established that high temperature increases respiration/photosynthesis ratios, thus negatively affecting the plant carbon balance (Marsh et al. 1986; Greve et al. 2003; Lee et al. 2007; Olsen et al. 2012; Rasmusson 2015; Marín-Guirao et al. 2016) leading to reduced growth rates and biomass (Massa et al. 2009; Olsen et al. 2012). In general, heat-tolerant plants can effectively maintain their balance in carbon metabolisms between respiratory rates and photosynthetic rates under high temperature (Marín-Guirao et al. 2016).

Exposure to heat shock in the seagrass *Z. noltii* has led to a downregulation of certain CABs, indicating a control of energy balance or apoptosis (Massa et al. 2011). A large volume of recent works has focused on the effects of increasing temperature on the photosynthetic function of seagrasses from different temperature regimes. Transcriptomic profiling of the seagrass *Zostera marina* populations from northern and southern Europe (Franssen et al. 2011) has shown that heat stress induced an upregulation of genes associated with heat shock proteins in both populations. However, the seagrass from the southern population (adapted to higher temperature) recovered their transcriptomic profiles after heat stress, while the seagrass from the northern population did not recover and exhibited an upregulation of genes associated with protein degradation. Similarly, Winters et al. (2011) compare the effects of a simulated heat wave on the photophysiology and gene expressions of *Z. marina* from the Adriatic Sea and North and Baltic Seas. All the populations showed a similar response to an increase in temperature, e.g., F_v/F_m, and electron transport rates declined except the southern population. A similar concept was investigated in the two populations of *P. oceanica* from different depths (Marín-Guirao et al. 2016). The photosynthetic function of the plants from the deep population was more sensitive to heat as shown by a larger decrease in ϕPSII, an increase in basal chlorophyll a fluorescence, and a downregulation of the genes associated with electron transport chain and carbon fixation. Additionally, the genes encoding for the PSII core proteins D1 and D2 were downregulated under heat stress, resulting in a halt of PSII repair, whereas the expressions of these genes remained unchanged in shallow population. The ability to activate photoprotective mechanisms (observed as an increase in NPQ) and enhance electron flow and carbon assimilation under heat stress was suggested to contribute to higher heat tolerance in the shallow population.

Interestingly, an increase in the expressions of the genes encoding Rubisco small subunit was observed in parallel with an increase in the expressions of the genes encoding Rubisco activase. The positive effect of high temperature (27–33°C) has also been demonstrated in a tropical seagrass, *H. uninervis*, in which photosynthesis and growth increased with increased temperature while respiration remained constant (Collier et al. 2011). These works indicate that responses to heat stress vary considerably depending on the plant history.

7.2.4 *Effect of CO_2 Enrichment on Seagrass Primary Production*

Considering the projected increase in dissolved CO_2 in the global change scenarios, seagrass is expected to benefit from an increase in carbon availability. Therefore, the effects of CO_2 enrichment on photosynthesis have been intensively investigated in seagrasses for the past few years (Alexandre et al. 2012; Campbell and Fourqurean 2013a,b; Martínez-Crego et al. 2014; Ow et al. 2015, 2016; Cox et al. 2016; Borum et al. 2016). Additionally, the more ecological relevant experimental setting using in situ mesocosms has become increasingly common (Campbell and Fourqurean 2013a,b; Cox et al. 2016). These studies, however, focused on comparing photosynthetic and growth rates between control and CO_2-enriched plants, while only a few works have addressed the mechanistic adaptation of seagrass photobiology to high CO_2.

The positive effects of increased CO_2 have been reported in various seagrass species such as *Z. noltii*, *C. serrulata*, *H. uninervis*, *T. hemprichii*, and *Amphibolis antarctica* (Alexandre et al. 2012; Burnell et al. 2014; Ow et al. 2015, 2016). In these seagrasses, growth rates, net primary productivity, the maximum photosynthetic rates and the photosynthetic efficiency of CO_2-enriched plants were higher than controls. Additionally, synergistic effects between an increase in dissolved CO_2 and irradiance have been reported by Ow et al. (2016). Therefore seagrass responses to an increase in CO_2 might vary depending on the light environments. It was suggested that the seagrass *C. serrulata* grown under low light, being short of ATP production, is more dependent on the passive CO_2 uptake and thus benefits more from an increase in dissolved CO_2 (Ow et al. 2016). This is confirmed by a greater increase in the maximum photosynthetic rates (P_{max}) and photosynthetic efficiency (α) as a response to high CO_2 in *C. serrulata* under low light. Similarly, other studies have suggested that species with low HCO_3^- utilizing efficiency would respond more positively to an increase in dissolved CO_2 more than the efficient HCO_3^- users (Campbell and Fourqurean 2013b; Borum et al. 2016). Borum et al. (2016) examined the HCO_3^- utilization efficiency of nine seagrass species and their subsequent responses to CO_2 enrichment. Here *A. antarctica* which appeared to be among the least efficient HCO_3^- users exhibited the strongest response to an increase in CO_2.

However, such response was observed only at saturating CO_2 concentration (274 μM) which is much higher than the predicted level in 2100 (~24 μM). It is also possible that in high CO_2 condition, the costly HCO_3^- utilization systems will be downregulated; thus the energy demand for carbon acquisition will be reduced (Burnell et al. 2014). Although a positive growth response was observed in CO_2-enriched *A. antarctica*, there was no clear indication of a downregulation in HCO_3^- utilization system (Burnell et al. 2014). On the contrary, a number of studies found no significant or limited effect of increased CO_2 on seagrass photosynthesis and primary production (Campbell and Fourqurean 2013a; Martínez-Crego et al. 2014; Cox et al. 2016; Borum et al. 2016).

7.2.5 Phytotoxins

Due to its proximity to shores, seagrass meadows are highly exposed to contamination via industrial runoff, waste discharges, and leachates (Lin et al. 2016a). Previous studies have shown that exposure to heavy metals (copper, cadmium, lead, and zinc), herbicides, and petrochemical causes a decrease in photosynthetic activity indicated by a decline in ϕPSII and pigment concentration in the seagrass *Z. muelleri* (previously *Z. capricorni*) (Haynes et al. 2000; Macinnis-Ng and Ralph 2002, 2004a, b), *H. ovalis* (Ralph and Burchett 1998; Wilkinson et al. 2015a; Wilkinson et al. 2015b), and *C. nodosa* (Llagostera et al. 2016). Nevertheless, the cellular mechanisms for toxicity effects on photosynthesis in this group of plants remain to be elucidated. Excessive heavy metals generally disrupt photosynthetic activity, causing an imbalance in metabolisms. However, the mechanism of action of each metal differs depending on its target site in the photosynthetic pathway. In higher plants and bacteria, PSII seems to be the most sensitive site (Barón et al. 1995; Dewez et al. 2005). Targets for certain metals such as copper and zinc also include oxygen-evolving complex, quinone B, pheophytin (Mohanty et al. 1989; Yruela et al. 1996), and PSI ferredoxin-NADP$^+$ oxidoreductase (Rijstenbil et al. 1994; Shioi et al. 1978; Van Assche and Clijsters 1990). Inhibition of these components does not only decrease the photosynthetic electron transport, but it may as well cause damage to the photosystems by over-excitation and overproduction of reactive oxygen species (ROS) (Valko et al. 2006; Sahi and Sharma 2005). It has been reported that heavy metals (cadmium and mercury) induce responses associated with oxidative stress such as a decrease in glutathione pool (GSH), an increase of lipid peroxidation, and induction of an enzyme glutathione transferase (GST) in *P. oceanica* (Hamoutène et al. 1996; Ranvier et al. 2000; Ferrat et al. 2002a, b, c; Ferrat et al. 2003). Recent work has also highlighted a role of antioxidative systems in heavy metal responses (Lin et al. 2016b). In this work, *Z. japonica* exposed to copper, lead, and cadmium exhibited higher activity of antioxidant enzymes, catalase (CAT), superoxide dismutase (SOD), and glutathione peroxidase (GPX), whereas more severe lipid peroxidation was found in high heavy metal concentrations (Lin et al. 2016b). In addition to antioxidative systems, the heavy metal-binding ligands, phytochelatins (PCs) and metallothioneins (MTs), have been proposed to play a role in metal homoeostasis in plants (Cobbett and Goldsbrough 2002). Accumulation and synthesis of PCs following heavy metal exposure (cadmium and lead) have been observed in the seagrasses *T. testudinum* (Alvarez-Legorreta et al. 2008) and *T. hemprichii* (Tupan et al. 2014). Transcript-encoding metallothioneins (including metallothionein-like protein) and putative type II metallothioneins have been identified in *Z. marina* and *P. oceanica*, respectively (Giordani et al. 2000; Cozza et al. 2006; Kong et al. 2014; Olsen et al. 2016). In *P. oceanica*, an increase in transcript level of putative type II metallothionein after exposure to copper and cadmium has been reported (Giordani et al. 2000; Cozza et al. 2006). These results point to the importance of ROS scavenging and heavy metal detoxification in seagrass metal tolerance. Uptake and accumulation of pesticides, herbicides, and trace metals into the seagrass biomass and the possibility of using seagrasses as a bioindicator of such contaminants have been explored (Ferrat et al. 2003; Govers et al. 2014; Lin et al. 2016a; Bonanno and Di Martino 2016). Much more research effort is still needed to establish the relationship between duration and dose of exposure and rapid physiological responses associated with photosynthetic mechanisms, antioxidative systems, and heavy metal homoeostasis which tend to precede those that manifest at the whole plant or higher levels.

In sediments of seagrass habitats, microbial reduction of sulfate in anoxic condition generates sulphide which is a potent phytotoxin (Lamers et al. 2013). Generally, seagrass releases oxygen to the rhizosphere via aerenchyma which then oxidizes sulfide to nontoxic form (Hasler-Sheetal and Holmer 2015; Brodersen et al. 2015a,b, 2016). A recent study also suggested that once sulfide enters plant cells, it is detoxified into thiols and further metabolized in sulfur metabolic pathways (Hasler-Sheetal and Holmer 2015). Nevertheless, following an event of organic matter load, excessive organic matter breakdown by microbes can lead to an increase in concentrations of sulfide in porewater (Pérez et al. 2007; Govers et al. 2014). Additionally, limiting light brought about by sedimentation, overgrowth of epiphytes, or algal blooms might lower seagrass tolerance to sulfide by decreasing photosynthetic rates, thereby reducing photosynthetically derived oxygen available for transport to belowground tissue (Goodman et al. 1995; Brodersen et al. 2015b). Sulfide exposure was found to decrease maximum photosynthetic rates (calculated as O_2 evolution), and chlorophyll a concentration and increase light compensation point of *Z. marina* (Goodman et al. 1995; Holmer and Bondgaard 2001). When exposed to high sulfide concentrations (>100 mM), the photosynthetic activity of *Z. marina* was fully inhibited after 6 days (Holmer and Bondgaard 2001). Sulfide has been proposed to act as an inactivator of metalloenzymes such as oxygen-evolving complex of PSII (Goodman et al. 1995; Armstrong et al. 1996; Fürtig et al. 1996) and inhibitor of cytochrome c oxidase in mitochondrial respiratory electron transport chain (Goodman et al. 1995; Holmer and Bondgaard 2001). Nevertheless, the mechanisms of sulfide toxicity and plant strategies to cope with sulfide intrusion at the molecular level remain to be further investigated.

7.3 Knowledge Gaps

7.3.1 *The Function of Photoreceptors*

Plants perceive changes in the light environment using a variety of photoreceptors, e.g., red and far-red light-absorbing phytochromes and UV-A/blue light-absorbing cryptochromes. The perception of light signals primes plants photoadaptation, enabling avoidance of stress imposed by limiting and high light. Additionally, it also plays a significant role in plant growth and development. These photoreceptors are also found in some bacteria, fungi, and algae (Duanmu et al. 2014; Rensing et al. 2016). Phytochrome is the best-characterized class of photoreceptors. Multiple gene duplication events and subfunctionalization have given rise to various phytochromes with different functions in bryophytes, pteridophytes, and higher plants (Rensing et al. 2016). Such diversity of phytochromes in these groups of plants might allow them to cope with fluctuation of light conditions with greater plasticity (Li et al. 2011). Light signals have found to trigger a profound alteration in transcriptomes of macroalgae and higher plants (Facella et al. 2008; Deng et al. 2012). Additionally, it has been shown in various plant species that light can induce significant changes in expressions of the genes involved in photosynthetic light reactions, photorespiratory pathway, and photosynthetic carbon reactions (Casal and Yanovsky 2005; Jung et al. 2008; Monnier et al. 2010; Ono et al. 2010; Lehmann et al. 2011). Interestingly, some of these responsive genes are also transcription factors (Tepperman et al. 2001; Zhang et al. 2008). The regulatory mechanisms of these transcription factors in a putative transcriptional cascade of photoreceptor-mediated responses are, however, not yet fully understood (Casal and Yanovsky 2005; Duanmu et al. 2014).

There are many studies on photoreceptor-mediated physiological responses in terrestrial plants. On the contrary, knowledge on how seagrasses sense the light environment is still lacking although significant daily and seasonal fluctuations in light quantity and quality have been observed in seagrass habitats (Kenworthy and Haunert 1991; Gallegos and Kenworthy 1996; Kahn et al. 2013). Genes associated with photoreceptors have been identified in the seagrass *P. oceanica* including phytochrome A (PoPHYA), phytochrome B (PoPHYB), phytochrome C (PoPHYC), cryptochrome 1 (CRY1), and cryptochrome 2 (CRY2) (Greco et al. 2013; Dattolo et al. 2014). Greco et al. (2013) reported changes in methylation status of the gene PoPHYB as well as chlorophyll a/b-binding proteins and phosphoenolpyruvate carboxylase (PoPPC4) in response to a light limitation in *P. oceanica*. From these findings, it is suggested that PHYB methylation might play a regulatory role in transcriptional cascade underlying photoacclimatory responses of this seagrass. Dattolo et al. (2014b) have shown that the expression of photosynthetic genes follows circadian light cycle, indicating a possible regulatory role of the photoreceptors. Recently Dattolo et al. (2017) have conducted reciprocal light experiments and revealed a downregulation of PHYA and CRY1 in *P. oceanica* upon exposure to higher light intensity and an upregulation of the same genes upon exposure to lower light intensity (Dattolo et al. 2017). As for *Z. marina*, the photoreceptors including

PHYA, PHYB, CRY1, phototropin 1 (PHO1), and phototropin 2-like have been identified (Kong et al. 2014; Olsen et al. 2016). Phytochromes C and UVR8 transcripts are, however, absent in *Z. marina* (Olsen et al. 2016).

Although existing studies (Greco et al. 2013; Dattolo et al. 2014; Kong et al. 2014; Olsen et al. 2016) suggest that seagrass physiological responses may be regulated by photoreceptors. In aquatic condition, light quality is altered with depth. As red and far-red wavelengths are largely attenuated by the water column (Ragni and Ribera d'Alcala 2004), the possible functions of phytochromes in such condition remain unclear and how they are involved in light signaling may be different from the terrestrial systems. Further research should investigate how light is sensed by seagrasses at different light environments and the signaling pathways associated with light-dependent essential processes such as photosynthesis as it might reveal one of the key adaptations of this group of plants to the sea.

7.3.2 Mechanistic Understanding of Photosynthetic Carbon Fixation

It is generally taken for granted that seagrass possesses the same photosynthetic pathway as terrestrial higher plants. Nevertheless, the key photosynthetic processes of seagrasses are still not completely understood. Recent genome assemblies of the two seagrasses *Z. marina* and *Z. muelleri* have pointed out that seagrass adaptation to the aquatic environment has shaped their physiological features so that they in many ways differ significantly from those of the terrestrial higher plants. While chlorophyll fluorescence technique allows extensive investigation of the light-dependent reactions of photosynthesis, little is known about the mode of photosynthetic carbon fixation of seagrasses. In general, seagrasses are regarded as C3 plants (discussed in the previous section); however, certain seagrass species exhibit C4 characteristics, while others were identified as C3–C4 intermediates (Benedict and Scott 1976; Andrews and Abel 1979; Beer and Waisel 1979; Waghmode and Joshi 1983; Bowes and Salvucci 1989). It is difficult to compare these results since different measurements and parameters have been used such as a photosynthetic quantum efficiency, $\delta^{13}C$ values, first stable photosynthetic products and the activities of photosynthetic enzymes. Although there are indications that C4 photosynthesis occurs in certain seagrass species, and that it might be inducible under certain environmental conditions, no further investigation has been conducted on this aspect. On the contrary, much more information has been accumulated in the terrestrial plant systems regarding mode of carbon assimilation. Differential gene expressions between C3 and C4 species have been revealed, whereas the responsive genes in the facultative C4 plant and the key genes attributing to C3–C4 intermediate characteristics in certain plant species have been identified (Rao et al. 2008; Bräutigam et al. 2011; Külahoglu et al. 2014; Schulze et al. 2016).

In addition to unclear C3 or C4 nature of seagrass photosynthesis, a basic understanding of Rubisco kinetics in seagrasses is poor. In many terrestrial plant species, Rubisco is activated by light via an enzyme rubisco activase (Zhang and Portis 1999). A decline in photosynthetic activity under moderate heat stress in these plants has been attributed to either a decreasing capacity of photosynthetic electron transport or a reduction in carbon fixation capacity as a result of an inactivation of Rubisco activase which is heat-sensitive (Salvucci and Crafts-Brandner 2004; Salvucci et al. 2001; Salvucci and Crafts-Brandner 2004; Portis et al. 2008). It has also been suggested that the heat tolerance of photosynthesis is controlled by the thermal properties of Rubisco activase (Salvucci and Crafts-Brandner 2004). Little is known about such role of Rubisco activase in seagrasses. Until now, it has been addressed only in the studies of the seagrass, *P. oceanica*, in relation to photoacclimation (Dattolo et al. 2013, 2014) and heat stress (Marín-Guirao et al. 2016). Global warming and the rise in atmospheric CO_2 will likely increase the temperature of seawater in the coming decades, particularly in the shallow areas. Identifying the major limiting process at elevated temperature is essential when predicting the photosynthetic response of seagrasses. In terrestrial plants system, it has been demonstrated that when taking into consideration the regulation of Rubisco activase, kinetic properties of Rubisco can effectively predict the temperature response of photosynthesis (Crafts-Brandner and Salvucci 2000).

The most common carbon-concentrating mechanism (CCM) proposed to operate in seagrasses is associated with extracellular carbonic anhydrase (discussed in the previous section). This conclusion, however, is mainly drawn from inhibitors experiments. Identification and localization of extracellular carbonic anhydrase, as well as analysis of molecular responses induced by low CO_2 availability, could provide further evidence for CCM in this group of plants. For example, advanced understanding of CCM induction and regulation in algae was obtained from comparative transcriptome, proteome, and metabolome analyses when algae were transferred from high to low carbon availability. As a result, genes, proteins, and key metabolites that may be involved with the CCM including carbonic anhydrase have been identified (Yamano and Fukuzawa 2009; Renberg et al. 2010; Baba et al. 2011; Ramanan et al. 2012; Winck et al. 2013).

7.3.3 *Linkage Between Photosynthesis and Other Metabolic Pathways*

In the past decade, much more effort has been made toward understanding the photobiology of seagrasses at physiological to molecular levels (as discussed earlier). However, knowledge on the interactions between photosynthesis and other metabolic pathways across the whole system is still lacking. It is important to have a better understanding of complex interplay within the plant primary metabolism networks because such interactions might affect the responses to stresses at a whole-plant level.

7.3.3.1 Alternative Electron Flows

It is challenging to fully understand the interconnectivity within the photosynthetic network itself. Photosynthetic electron transport is highly flexible, and many alternative electron flow pathways have been identified. Since the role of these pathways has long been understudied in seagrass systems, little is known about how they operate and their relative contribution to seagrass primary production under changing environments. Although the capacity of alternative electron flows is generally low in optimal conditions, these processes likely play a substantial role in redox regulation under stress conditions (Ort and Baker 2002; Peltier and Cournac 2002).

An obvious link between photosynthetic carbon assimilation and alternative electron flows is that both processes utilize photosynthetically derived electrons. Relative contributions of each electron sink vary depending on the relative amounts of CO_2 and O_2 surrounding the active site of Rubisco which is influenced by the future scenarios of warm and high CO_2 conditions (Bloom 2015). CO_2 and temperature impose antagonistic effects on carbon fixation and photorespiration. As increased CO_2 availability favors carbon fixation, an increase in temperature promotes photorespiratory activity. The solubility of CO_2 decreases more than the solubility of O_2 with rising temperature, resulting in a lower dissolved CO_2:O_2 ratio. Also, the oxygenase activity of Rubisco is stimulated by increasing temperature to a greater extent than the carboxylase activity. Photosynthetic carbon assimilation, thus the production of carbohydrate, becomes less efficient under higher temperature (Ehleringer et al. 1997). In this condition, altered gene expression in photorespiratory pathways might be expected since plants have to increase the capacity to metabolize phosphoglycolate generated from Rubisco oxygenation of RuBP (Hodges et al. 2016; Timm et al. 2016). It has been suggested that suppressed photorespiration by CO_2-enrichment contributed, in part, to an increase in effective photochemical efficiency observed in the seagrasses *Z. noltii* (Alexandre et al. 2012). There is evidence of photorespiration and the Mehler reaction operating in the seagrass *Z. marina* (Buapet et al. 2013a; Buapet and Björk 2016). However, knowledge of rates and regulations of these alternative electron flow in relation to photosynthesis as well as in response to the future high CO_2 and temperature is still missing. Given that these processes might be upregulated under stress conditions, it is of high importance to consider them in order to get accurate estimates of seagrass primary production.

In addition to being diverted paths from carbon assimilation, alternative electron flows might play photoprotective roles under unfavorable conditions. It has been proposed that photorespiration, the Mehler reaction, chlororespiration, and cyclic electron flow may serve as safety valves when light energy is in excess, thus mitigating photoinhibition (see review by Osmond and Grace 1995; Wingler et al. 2000; Voss et al. 2013). The common function of these processes is maintaining the redox balance of the photosynthetic electron transport chain, while the mode of action is, however, slightly different. The proposed photoprotective roles of alternative electron flows have been supported by many studies in terrestrial systems. For example, an increase in the expression of photorespiratory genes, as well as proteins associated with photorespiration, has been observed when plants are subjected to energy imbalance (Lepistö et al. 2009; Abogadallah 2011). Tobacco overexpressing chloroplastic glutamine synthetase (GS2) exhibits higher tolerance to strong light compared to those with a lower level of GS2 (Kozaki and Takeba 1996). Similarly, enhanced cyclic electron flow around PSI, chlororespiration, and the Mehler reaction were observed in various plant species under mild stress conditions (Heber 2002; Makino et al. 2002; Rumeau et al. 2007; Huang et al. 2012). These alternative electron flows coordinate in their protective roles, and the contribution of each process under certain stress condition differs among plant species (Heber 2002; Makino et al. 2002; Rumeau et al. 2007).

7.3.3.2 Photosynthesis, Respiration, and Nitrogen Metabolisms

Photosynthesis and respiration are primary pathways of carbon and energy metabolism in plants. The balance between photosynthetic production and respiratory breakdown of carbohydrates determines plant carbon gain and potentially growth. These two processes are interrelated via an exchange of metabolites such as ATP, reducing equivalents and other carbon-containing intermediates. Photosynthesis and respiration respond to changes in environmental factors in a different manner and thus result in an alteration of photosynthesis to cellular respiration ratio. The effects of increased temperature and salinity on photosynthesis and respiration have been investigated in seagrasses, and a clear shift in plant carbon metabolisms has been observed (as discussed in the previous section). Cellular respiration in the seagrass *Z. marina* is inhibited in the light, whereas the respiratory rates, as well as expressions of respiratory genes, show diel variations (Rasmusson and Björk 2014; Rasmusson 2015). Although it has been proposed that such variations could be a result of an interplay between photosynthesis, photorespiration, and cellular respiration, the mechanisms involved are not completely understood. In terrestrial plants, high CO_2 increases the respiratory breakdown of carbohydrates (Ainsworth et al. 2006; Li et al. 2013). Such an increase in respiratory activity was found in conjunction with an accumulation of carbohydrates as well as an increase in transcripts associated with glycolysis, tricarboxylic acid cycle, and mitochondrial electron transport (Rogers et al. 2004; Gillespie et al. 2012). The opposing utilization and release of O_2 and CO_2 between photosynthesis, respiration, and photorespiration make it challenging to assess the rates of these processes in the light using classical gas exchange method. Additionally, lacunae present in all seagrasses and the potential internal recycling of both O_2 and CO_2 interfere with gas exchange measurement and manipulations of available O_2 and CO_2.

The interaction between carbon and nitrogen metabolisms in the context of increasing CO_2 has been widely investigated in the terrestrial model plants (Yong et al. 2000; Bloom et al. 2002; Cousins and Blooms 2004; Takatani et al. 2014). Suppression of nitrate assimilation by a downregulation of nitrate and nitrite reductase activities was demonstrated in wheat exposed to elevated CO_2 (Bloom et al. 2002), whereas *Arabidopsis* mutant with low nitrate reductase activity exhibited nitrogen deficiency symptoms under high CO_2, i.e., decreased nitrogen content and reduced shoot/root ratio (Takatani et al. 2014). Both carbon and nitrogen assimilation compete for ATP, reducing equivalents and ferredoxin supplied by the light-dependent reactions of photosynthesis. Enhanced photosynthetic carbon assimilation induced by high CO_2 under light-limited condition decreases the amount of ATP and reducing powers supplied to nitrogen assimilation and eventually leads to nitrogen limitation (Bloom 2015). Additionally, a reduction in photosynthetic activity under long-term exposure to high CO_2 is more pronounced under nitrogen limitation (Stitt and Krapp 1999; Sun et al. 2002; Sanz-sàez et al. 2010). Despite a growing number of studies on the effect of increased CO_2 on seagrass primary production, simultaneous impact on nitrogen metabolisms has been scarcely studied. Alexandre et al. (2012) investigated the effect of predicted future CO_2 level on carbon and nitrogen metabolisms in *Z. noltii* in the mesocosm experiment. Here an increase in the photosynthetic activity was observed, while nitrate uptake and nitrogen content in the leaves of CO_2-enriched plants were found to be significantly lower than controls. Thus no positive effect of CO_2 enrichment on the growth rates of *Z. noltii* in these experiments is likely to be a result of nitrogen limitation. This suggests that nitrogen might become limiting for *Z. noltii* in the future CO_2 level. Similar declines in the leaf nitrogen content under high CO_2 have been reported in *T. hemprichii* (Jiang et al. 2010) and *T. testudinum* (Campbell and Fourqurean 2013a). This shift of limiting factor under high CO_2 might be the underlying reason for a limited or no positive effect of CO_2 enrichment observed in many studies conducted on seagrasses (as discussed in the previous section). On the contrary, CO_2 enrichment did not affect nitrate uptake and nitrate reductase activity of *H. uninervis*, whereas it increased nitrate assimilation in *T. hemprichii* grown under high nitrate level (Ow et al. 2016). Such interaction might as well affect downstream pathways such as ammonium assimilation which also requires ATP and NADPH. A Recent study by Pernice et al. (2016) has revealed a fast increase in expression of glutamine synthetase (GS), a key enzyme for ammonium assimilation, following exposure to a pulse of ammonium supply. This indicated that GS could be used as a molecular marker of nitrogen assimilation and should be included in further investigations of seagrass metabolisms. In-depth physiological studies are needed in order to elucidate the connection between these two vital processes: photosynthesis and nitrogen assimilation. This subject is of great ecological relevance as inorganic nitrogen concentrations are commonly low in seagrass-dominated habitats (Lee and Dunton 1999; McGlathery et al. 2001).

7.4 Further Prospects for Integrating Seagrass Photophysiology with Systems Biology

Seagrasses are rapidly declining worldwide. Therefore, there is a critical need to characterize the impacts of environmental changes on seagrasses, particularly those associated with the future global change scenarios. Better mechanistic understanding of photosynthesis is necessary to assess how environmental constraints affect seagrass primary production and its carbon budget. Although photosynthesis plays a significant role as a central carbon and energy metabolisms process, it is clearly important to investigate not only photobiology but also the connected metabolic pathways in order to predict seagrass responses to environmental changes effectively. With high-throughput experiments being increasingly common, combining the "omics" approaches can help us to gain a systemic understanding of photosynthesis within plant complex metabolic networks. Although being in its early stage, many seagrass scientists have pioneered research in this field (Reusch et al. 2008; Franssen et al. 2011; Winters et al. 2011; Dattolo et al. 2013, 2014; Mazzuca et al. 2013; Piro et al. 2015a, b; Kumar et al. 2016a). It is only recently that the first genomic data has become available from *Z. marina*, a common seagrass species of the northern hemisphere (Olsen et al. 2016) followed by its relative from the southern hemisphere, *Z. muelleri* (Lee et al. 2016). The availability of full genome sequences for these species means that we can now adopt a systems approach to investigate the metabolic networks of these seagrasses and how environmental changes modulate their metabolisms. Detailed reviews of the utilization of omics techniques and molecular profiling in seagrass studies in the past decade and concepts and applications of metabolomics in the marine macrophytes studies have been recently presented by Davey et al. (2016) and Kumar et al. (2016b), respectively. Here, omics approaches are briefly described. Further, examples of omics investigations in model photosynthetic organisms which can be adopted to address the knowledge gaps related to seagrass photobiology are discussed.

Transcriptomics and proteomics analyses have been proven to provide a good systemic overview in seagrass metabolisms. The number of transcriptomic and proteomic studies carried out in seagrass is growing (Franssen et al. 2011; Winters et al. 2011; Dattolo et al. 2013, 2014; Mazzuca et al. 2013; Piro et al. 2015a, b; Kumar et al. 2016a). In these works, stressors were applied to seagrasses to unravel the underlying metabolic pathways which are involved in stress responses or tolerance. Comparative transcriptomics and proteomics focus on differential gene and protein expression levels of the entire transcriptome and proteome induced by environmental changes (Franssen et al. 2011; Winters et al. 2011; Dattolo et al. 2013, 2014; Mazzuca et al. 2013; Piro et al. 2015a, b; Kumar et al. 2016a). Transcripts and proteins differentially expressed due to the environmental stimuli such as high temperature, salinity, and light gradient have been identified, as well as the metabolic pathway they are part of. Metabolomics is another new frontier in plant systems biology. It greatly complements transcriptome and proteome analysis as the difference in quality and quantity of metabolites readily reflect cellular biochemical processes. Metabolome analysis provides comprehensive and dynamic views of plant primary and secondary metabolites under changing environments. Although remained unexplored in seagrasses, it holds great potential for seagrass research (see extended review by Kumar et al. 2016b).

To better understand the photosynthetic mechanisms and interconnectivity between photosynthesis and other metabolic pathways of seagrasses, it is crucial to conduct a parallel analysis of transcript, protein, and metabolic profiles. Combined omics data have provided valuable insights into mechanisms of carbon acquisition and accumulation in the green alga, *Chlamydomonas reinhardtii* (reviewed in Winck et al. 2013 and Wang et al. 2015). Time-series multi-"omics" analyses have identified carbonic anhydrase as the main component of carbon-concentrating mechanisms (CCMs) in *C. reinhardtii* and elucidated regulatory processes of CCM as well as interplay between photosynthetic carbon metabolism and other metabolic pathways in response to carbon limitation (Wienkoop et al. 2010; Baba et al. 2011; Winck et al. 2013; Wang et al. 2015). Additionally, advanced understanding of carbon metabolism and its coordination with other metabolic pathways such as photorespiration and nitrogen metabolism have been established in many model photosynthetic organisms such as *Arabidopsis*, rice, tomato, and maize using multi-"omics" analyses (reviewed in Fukushima and Kusano 2014; Sato and Yanagisawa 2014). For example, it was shown that responses of plant metabolism to increasing CO_2 depend largely on the nitrogen conditions. While similar approach can be adopted to seagrass studies, the availability of full genome sequences for certain seagrass species makes it possible to develop a genome-scale metabolic model to investigate fluxes between photosynthesis and other metabolic processes in these seagrasses. It has been shown that simulation of plant metabolisms using genome-scale model can successfully predict fluxes between fundamental metabolic pathways such as glycolysis and the TCA cycle in *Arabidopsis* (Poolman et al. 2009; Williams et al. 2010) and cooperation between mesophyll cells and bundle sheath cells during C4 photosynthesis (De Oliveira Dal'Molin et al. 2010).

Advanced insights into the function and regulation of the photosynthesis have come from functional genomic studies in which photosynthetic processes have been dissected by the molecular tools (Dent et al. 2001, 2005). While omics data provides an indication of regulatory roles of certain genes or proteins, the future challenge for seagrass research is the development of mutagenesis experiment which is a powerful approach to study the function of the genes of interest. Furthermore, proteomics techniques can be applied to investigate protein-protein interaction, posttranslational modifications, and organization of proteins in multi-protein complexes which remain unexplored in seagrasses. Additionally, chloroplast-targeted transcriptomes and proteomes can provide insights on chloroplast structures and function (Eckardt 2012; Petersen et al. 2013; Chang et al. 2015). These studies have led to better understanding of chloroplast protein components and photosynthetic responses to various stressors, as well as interconnectivity between chloroplast-encoded and nuclear-encoded proteins. Purification of intact chloroplast for quantitative proteomic study has recently been developed in *P. oceanica* (Piro et al. 2015a, b), and future organelle-specific analysis will most likely improve our knowledge on seagrass photosynthetic mechanisms.

7.5 Conclusion

System biology is an area of research that needs development and represents the key to a significantly improved understanding of the regulation of photosynthesis and its interaction with other metabolic pathways. The integration of omics approaches will help to unravel plant dynamic and complex responses to changing environments. Such knowledge is extremely crucial for predicting impacts of global climate change on seagrasses.

References

Abal EG, Loneragan N, Bowen P, Perry C, Udy J, Dennison WC (1994) Biological and morphological responses of the seagrass *Zostera capricorni* Aschers. to light intensity. J Exp Mar Biol Ecol 178:113–129

Abogadallah GM (2011) Differential regulation of photorespiratory gene expression by moderate and severe salt and drought stress in relation to oxidative stress. Plant Sci 180:540–547

Ainsworth EA, Rogers A, Vodkin LO, Walter A, Schurr U (2006) The effects of elevated CO_2 concentration on soybean gene expression. An analysis of growing and mature leaves. Plant Physiol 142:135–147

Alexandre A, Silva J, Buapet P, Björk M, Santos R (2012) Effects of CO_2 enrichment on photosynthesis, growth, and nitrogen metabolism of the seagrass *Zostera noltii*. Ecol Evol 2:2620–2630

Allen JE (2003) Cyclic, pseudocyclic and noncyclic photophosphorylation: new links in the chain. Trends Plant Sci 8:15–19

Alvarez-Legorreta T, Mendoza-Cozat D, Moreno-Sanchez R, Gold-Bouchot G (2008) Thiol peptides induction in the seagrass *Thalassia testudinum* (Banks ex König) in response to cadmium exposure. Aquat Toxicol 86:12–19

Andrews TJ, Abel KM (1979) Photosynthetic carbon metabolism in seagrass: C labeling evidence for the C3 pathway. Plant Physiol 63:650–656

Armstrong J, Afreen-Zobayed F, Armstrong W (1996) Phragmites die-back: sulphide- and acetic acid-induced bud and root death, lignifications, and blockages within aeration and vascular systems. New Phytol 134:601–614

Baba M, Suzuki I, Shiraiwa Y (2011) Proteomic analysis of high-CO_2-inducible extracellular proteins in the unicellular green alga, *Chlamydomonas reinhardtii*. Plant Cell Physiol 52:1302–1314

Badger MR, Von Caemmerer S, Ruuska S, Nakano H (2000) Electron flow to oxygen in higher plants and algae: rates and control of direct photoreduction (Mehler reaction) and rubisco oxygenase. Philos Trans R Soc Lond B Biol Sci 355:1433–1445

Bailey S, Walters RG, Jansson S, Horton P (2001) Acclimation of *Arabidopsis thaliana* to the light environment: the existence of separate low light and high light responses. Planta 213:794–801

Barón M, Arellano JB, Gorgé JL (1995) Copper and photosystem II: a controversial relationship. Physiol Plant 94:174–180

Bauwe H, Hagemann M, Fernie AR (2010) Photorespiration: players, partners and origin. Trends Plant Sci 15:330–336

Beer S (1989) Photosynthesis and photorespiration in marine angiosperms. Aquat Bot 34:153–166

Beer S, Eshel A, Waisel Y (1980) Carbon metabolism in seagrasses activities of carbon fixing enzymes in relation to internal salt concentrations. J Exp Bot 31:1027–1033

Beer S, Rehnberg J (1997) The acquisition of inorganic carbon by the seagrass *Zostera marina*. Aquat Bot 56:277–283

Beer S, Waisel Y (1979) Some photosynthetic carbon fixation properties of seagrasses. Aquat Bot 7:129–138

Beer S, Wetzel RG (1982) Photosynthetic carbon fixation pathways in *Zostera marina* and three Florida seagrasses. Aquat Bot 13:141–146

Benedict CR, Scott JR (1976) Photosynthetic carbon metabolism of a marine grass. Plant Physiol 57:876–880

Benedict CR, Wong WWL, Wong IHH (1980) Fractionation of the stable isotopes of inorganic carbon by seagrasses. Plant Physiol 65:512–517

Biehler K, Fock H (1996) Evidence for the contribution of the Mehler-peroxidase reaction in dissipating excess electrons in drought-stressed wheat. Plant Physiol 112:265–272

Bloom AJ (2015) Photorespiration and nitrate assimilation: a major intersection between plant carbon and nitrogen. Photosynth Res 123:117–128

Bloom AJ, Smart DR, Nguyen DT, Searles PS (2002) Nitrogen assimilation and growth of wheat under elevated carbon dioxide. Proc Natl Acad Sci U S A 99:1730–1735

Bonanno G, Di Martino V (2016) Seagrass *Cymodocea nodosa* as a trace element biomonitor: bioaccumulation patterns and biomonitoring uses. J Geochem Explor 169:43–49

Borum J, Pedersen O, Kotula L, Fraser MW, Statton J, Colmer TD, Kendrick GA (2016) Photosynthetic response to globally increasing CO_2 of co-occurring temperate seagrass species. Plant Cell Environ 39:1240–1250

Bowes G, Salvucci ME (1989) Plasticity in the photosynthetic carbon metabolism of submersed aquatic macrophytes. Aquat Bot 34:233–266

Bowes G, Rao SK, Estavillo GM, Reiskind JB (2002) C4 mechanisms in aquatic angiosperms: comparisons with terrestrial C4 systems. Funct Plant Biol 29:379–392

Bräutigam A, Kajala K, Wullenweber J (2011) An mRNA blueprint for C4 photosynthesis derived from comparative transcriptomics of closely related C3 and C4 species. Plant Physiol 155:142–156

Brodersen KE, Nielsen DA, Ralph PJ, Kühl M (2015a) Oxic microshield and local pH enhancement protects *Zostera muelleri* from sediment derived hydrogen sulphide. New Phytol 205:1264–1276

Brodersen KE, Lichtenberg M, Paz L-C, Kühl M (2015b) Epiphyte-cover on seagrass *(Zostera marina* L.) leaves impedes plant performance and radial O_2 loss from the below-ground tissue. Front Mar Sci 2:58

Brodersen KE, Koren K, Lichtenberg M, Kühl M (2016) Nanoparticle-based measurements of pH and O_2 dynamics in the rhizosphere of *Zostera marina* L.: effects of temperature elevation and light-dark transitions. Plant Cell Environ 39: 1619–1630

Buapet P, Björk M (2016) The role of O_2 as an electron acceptor alternative to CO_2 in photosynthesis of the common marine angiosperm *Zostera marina* L. Photosynth Res 129:59–69

Buapet P, Rasmusseon LM, Gullström M, Björk M (2013a) Photorespiration and carbon limitation determine productivity in temperate seagrasses. PLoS One 8(12):e83804

Buapet P, Gullström M, Björk M (2013b) Photosynthetic activity of seagrasses and macroalgae in temperate shallow waters can alter seawater pH and total inorganic carbon content at the scale of a coastal embayment. Mar Freshw Res 64:1040–1048

Burnell OW, Connell SD, Irving AD, Watling JR, Russell BD (2014) Contemporary reliance on bicarbonate acquisition predicts increased growth of seagrass Amphibolis antarctica in a high-CO_2 world. Conserv Physiol 2:1–11

Burris JE, Holm-Hansen O, Black CC (1976) Glycine and serine production in marine plants as a measure of photorespiration. Aust J Plant Physiol 3:87–92

Campbell JE, Fourqurean JW (2013a) Effects of in situ CO_2 enrichment on the structural and chemical characteristics of the seagrass *Thalassia testudinum*. Mar Biol 160:1465–1475

Campbell JE, Fourqurean JW (2013b) Mechanisms of bicarbonate use influence the photosynthetic carbon dioxide sensitivity of tropical seagrasses. Limnol Oceanogr 58:839–848

Casal JJ, Yanovsky MJ (2005) Regulation of gene expression by light. Int J Dev Biol 49:501–511

Casazza G, Mazzella L (2002) Photosynthetic pigment composition of marine angiosperms: preliminary characterization of Mediterranean seagrasses. Bull Mar Sci 71:1171–1181

Chang L, Guo A, Jin X, Yang Q, Wang D, Sun Y, Huang Q, Wang L, Peng C, Wang X (2015) The beta subunit of glyceraldehyde 3-phosphate dehydrogenase is an important factor for maintaining photosynthesis and plant development under salt stress- based on an integrative analysis of the structural, physiological and proteomic changes in chloroplasts in *Thellungiella halophila*. Plant Sci 236:223–238

Cobbett C, Goldsbrough P (2002) Phytochelatins and metallothioneins: roles in heavy metal detoxification and homeostasis. Annu Rev Plant Biol 53:159–182

Collier CJ, Lavery PS, Ralph PJ, Masini RJ (2008) Physiological characteristics of the seagrass *Posidonia sinuosa* along a depth-related gradient of light availability. Mar Ecol Prog Ser 353:65–79

Collier CJ, Uthicke S, Waycott M (2011) Thermal tolerance of two seagrass species at contrasting light levels: implications for future distribution in the Great Barrier Reef. Limnol Oceanogr 56:2200–2210

Cornic G, Briantais JM (1991) Partitioning of photosynthetic electron flow between CO_2 and O_2 reduction in a C3 leaf (*Phaseolus vulgaris* L.) at different CO_2 concentrations and during drought stress. Planta 183:178–184

Cousins AB, Bloom AJ (2004) Oxygen consumption during leaf nitrate assimilation in a C3 and C4 plant: the role of mitochondrial respiration. Plant Cell Environ 27:1537–1545

Cox TE, Gazeau F, Alliouane S, Hendriks IE, Mahacek P, Le Fur A, Gattuso J-P (2016) Effects of in situ CO_2 enrichment on structural characteristics, photosynthesis, and growth of the Mediterranean seagrass *Posidonia oceanica*. Biogeosciences 13:2179–2194

Cozza R, Pangaro T, Maestrini P, Giordani T, Natali L, Cavallini A (2006) Isolation of putative type 2 metallothionein encoding sequences and spatial expression pattern in the seagrass *Posidonia oceanica*. Aquat Bot 85:317–323

Crafts-Brandner SJ, Salvucci ME (2000) Rubisco activase constrains the photosynthetic potential of leaves at high temperature and CO_2. Proc Natl Acad Sci U S A 97:13430–13435

Dattolo E, Gu J, Bayer PE, Mazzuca S, Serra IA, Spadafora A, Bernardo L, Natali L, Cavallini A, Procaccini G (2013) Acclimation to different depths by the marine angiosperm *Posidonia oceanica*: transcriptomic and proteomic profiles. Front Plant Sci 4:1–15

Dattolo E, Marín-Guirao L, Ruiz JM, Procaccini G (2017) Long-term acclimation to reciprocal light conditions suggests depth-related selection in the marine foundation species *Posidonia oceanica*. Ecol Evol 7:1148–1164

Dattolo E, Ruocco M, Brunet C, Lorenti M, Lauritano C, D'Esposito D, De Luca P, Sanges R, Mazzuca S, Procaccini G (2014) Response of the seagrass *Posidonia oceanica* to different light environments: insights from a combined molecular and photo-physicological study. Mar Environ Res 101:225–236

Davey PA, Pernice M, Sablok G, Larkum A, Lee HT, Golicz A, Edwards D, Dolferus R, Ralph P (2016) The emergence of molecular profiling and omics techniques in seagrass biology; furthering our understanding of seagrasses. Funct Integr Genomics 16:465–480

De Oliveira Dal'Molin CG, Quek LE, Palfreyman RW, Brumbley SM, Nielsen LK (2010) AraGEM, a genome-scale reconstruction of the primary metabolic network in *Arabidopsis*. Plant Physiol 152:579–589

Deng Y, Yao J, Wang X, Guo H, Duan D (2012) Transcriptome sequencing and comparative analysis of *Saccharina japonica* (Laminariales, Phaeophyceae) under blue light induction. PLoS One 7(6):e39704

Dennison WC, Alberte RS (1982) Photosynthetic response of *Zostera marina* L. (eelgrass) to *in situ* manipulations of light intensity. Oecologia 55:137–144

Dent RM, Haglund CM, Chin BL, Kobayashi MC, Niyogi KK (2005) Functional genomics of eukaryotic photosynthesis using insertional mutagenesis of *Chlamydomonas reinhardtii*. Plant Physiol 137:545–556

Dent RM, Han M, Niyogi KK (2001) Functional genomics of plant photosynthesis in the fast lane using *Chlamydomonas reinhardtii*. Trends Plant Sci 6:364–371

Dewez D, Geoffroy L, Vernet G, Popovic R (2005) Determination of photosynthetic and enzymatic biomarkers sensitivity used to evaluate toxic effects of copper and fludioxonil in alga *Scenedesmus obliquus*. Aquat Toxicol 74:150–159

Downton WJS, Bishop DG, Larkum AWD, Osmond CB (1976) Oxygen inhibition of photosynthetic oxygen evolution in marine plants. Aust J Plant Physiol 3:73–79

Drake LA, Dobbs FC, Zimmerman RC (2003) Effects of epiphyte load on optical properties and photosynthetic potential of the seagrasses *Thalassia testudinum* Banks ex König and *Zostera marina* L. Limnol Oceanogr 48:456–463

Driever SM, Baker NR (2011) The water-water cycle in leaves is not a major alternative electron sink for dissipation of excess excitation energy when CO_2 assimilation is restricted. Plant Cell Environ 34:837–846

Duanmu D, Bachy C, Sudek S, Wong CH, Jiménez V, Rockwell NC, Martin SS, Ngan CY, Reistetter EM, van Baren MJ, Price DC, Wei CL, Reyes-Prieto A, Lagarias JC, Worden A (2014) Marine algae and land plants share conserved phytochrome signaling systems. Proc Natl Acad Sci U S A 111:15827–15832

Eckardt NA (2012) Mapping the barley chloroplast transcriptome. Plant Cell 24:3

Ehleringer JR, Cerling TE, Helliker BR (1997) C4 photosynthesis, atmospheric CO_2, and climate. Oecologia 112:285–299

Facella P, Lopez L, Carbone F, Galbraith DW, Giuliano G, Perrotta G (2008) Diurnal and circadian rhythms in the tomato transcriptome and their modulation by cryptochrome photoreceptors. PLoS One 3(7):e2798

Ferrat L, Bingert A, Roméo M, Gnassia-Barelli M, Pergent-Martini C (2002a) Mercury uptake and enzymatic response of *Posidonia oceanica* after an exposure to organic and inorganic forms. Environ Toxicol Chem 21:2365–2371

Ferrat L, Roméo M, Gnassia-Barelli M, Pergent-Martini C (2002b) Effects of mercury on antioxidant mechanisms in the marine phanerogam *Posidonia oceanica*. Dis Aquat Org 50:157–160

Ferrat L, Pergent-Martini C, Romeo M, Fernandez C (2002c) Is glutathione transferase (GST) activity in *Posidonia oceanica* a stress response to mercury exposure? Bull Mar Sci 71:1183–1190

Ferrat L, Gnassia-Barelli M, Pergent-Martini C, Roméo M (2003) Mercury and non-protein thiol compounds in the seagrass *Posidonia oceanica*. Comp Biochem Physiol C 134:147–155

Flanigan YS, Critchley C (1996) Light response of D1 turnover and photosystem II efficiency in the seagrass *Zostera capricorni*. Planta 198:319–323

Foyer CH, Noctor G (2000) Oxygen processing in photosynthesis: regulation and signalling. New Phytol 146:359–388

Franssen SU, Gu J, Bergmann N, Winters G, Klostermeier UC, Rosenstiel P, Bornberg-Bauer E, Reusch TBH (2011) Transcriptomic resilience to global warming in the seagrass *Zostera marina*, a marine foundation species. Proc Natl Acad Sci U S A 108:19276–19281

Fukushima A, Kusano M (2014) A network perspective on nitrogen metabolism from model to crop plants using integrated 'omics' approaches. J Exp Bot 65:5619–5630

Fürtig K, Rüegsegger A, Brunold C, Brändle R (1996) Sulphide utilization and injuries in hypoxic roots and rhizomes of common reed (*Phragmites australis*). Folia Geobot Phytotax 31:143–151

Gallegos CL, Kenworthy JW (1996) Seagrass depth limits in the Indian River Lagoon (Florida): application of an optical water quality model. Estuar Coast Mar Sci 42:267–288

Gillespie KM, Xu F, Richter KT, McGrath JM, Markelz RC, Ort DR (2012) Greater antioxidant and respiratory metabolism in field-grown soybean exposed to elevated O_3 under both ambient and elevated CO_2. Plant Cell Environ 35:169–184

Giordani T, Natali L, Maserti BE, Taddei S, Cavallini A (2000) Characterization and expression of DNA sequences encoding putative type-II Metallothioneins in the seagrass *Posidonia oceanica*. Plant Physiol 123:1571–1582

Goodman JL, Moore KA, Dennison WC (1995) Photosynthetic responses of eelgrass (*Zostera marina*) to light and sediment sulfide in a shallow barrier island lagoon. Aquat Bot 50:37–47

Govers LL, Lamers LPM, Bouma TJ, Eygensteyn J, de Brouwer JHF, Hendriks AJ, Huijberse CM, van Katwijk MM (2014) Seagrasses as indicators for coastal trace metal pollution: a global meta-analysis serving as a benchmark, and a Caribbean case study. Environ Pollut 195:210–217

Greco M, Chiappetta A, Bruno L, Bitonti MB (2013) Effects of light deficiency on genome methylation in *Posidonia oceanica*. Mar Ecol Prog Ser 473:103–114

Greve TM, Borum J, Pedersen O (2003) Meristematic oxygen variability in eelgrass (*Zostera marina*). Limnol Oceanograp 48:210–216

Gausman HW (1982) Visible light reflectance, transmittance, and absorptance of differently pigmented cotton leaves. Remote Sens Environ 13:233–238

Hamoutène D, Roméo M, Gnassia M, Lafaurie M (1996) Cadmium effects on oxidative metabolism in a marine seagrass: *Posidonia oceanica*. Bull Environ Contam Toxicol 56:327–334

Hanelt D, Figueroa FL (2012) Physiological and photomorphogenic effects of light on marine macrophytes. In: Wiencke C et al (eds) Seaweed biology. Springer, Berlin Heidelberg, pp 3–23

Hasler-Sheetal H, Holmer M (2015) Sulfide intrusion and detoxification in the seagrass *Zostera marina*. PLoS One 10(6):e0129136

Haynes D, Ralph P, Prange J, Dennison B (2000) Impact of the herbicide diuron on photosynthesis in three species of tropical seagrass. Mar Pollut Bull 41:288–293

Heber U (2002) Irrungen, Wirrungen? The Mehler reaction in relation to cyclic electron transport in C3 plants. Photosynth Res 73:223–231

Hellblom F, Beer S, Björk M, Axelsson L (2001) A buffer sensitive inorganic carbon utilization system in *Zostera marina*. Aquat Bot 69:55–62

Hodges M, Dellero Y, Keech O, Betti M, Raghavendra AS, Sage R, Zhu X, Allen DK, Weber APM (2016) Perspectives for a better understanding of the metabolic integration of photorespiration within a complex plant primary metabolism network. J Exp Bot 67:3015–3026

Holmer M, Bondgaard EJ (2001) Photosynthetic and growth response of eelgrass to low oxygen and high sulfide concentrations during hypoxic events. Aquat Bot 70:29–38

Hough RA (1976) Light and dark respiration and release of organic carbon in marine macrophytes of the Great Barrier Reef region, Australia. Aust J Plant Physiol 3:63–68

Howarth JF, Durako MJ (2013) Diurnal variation in chlorophyll fluorescence of *Thalassia testudinum* seedlings in response to controlled salinity and light conditions. Mar Biol 160:591–605

Huang W, Yang S-J, Zhang S-B, Zhang J-L, Cao K-F (2012) Cyclic electron flow plays an important role in photoprotection for the resurrection plant *Paraboea rufescens* under drought stress. Planta 235:819–828

Invers O, Pérez M, Romero J (1999) Bicarbonate utilization in seagrasses: role of carbonic anhydrase in *Posidonia oceanica* and *Cymodocea nodosa* photosynthesis. J Exp Mar Biol Ecol 235:125–133

Jiang ZJ, Huang X-P, Zhang J-P (2010) Effects of CO_2 enrichment on photosynthesis, growth, and biochemical composition of seagrass *Thalassia hemprichii* (Ehrenb.) Aschers. J Integr Plant Biol 52:904–913

Johnson GN (2011) Physiology of PSI cyclic electron transport in higher plants. Biochim Biophys Acta 1807:384–389

Johnson X, Steinbeck Dent RM, Takahashi H, Richaud P, Ozawa S, Houille-Vernes L, Petroutsos D, Rappaport F, Grossman AR, Niyogi KK, Hippler M, Alric J (2014) Proton gradient regulation 5-mediated cyclic electron flow under ATP- or redox-limited conditions: a study of Delta ATpase pgr5 and Delta rbcL pgr5 mutants in the green alga *Chlamydomonas reinhardtii*. Plant Physiol 165:438–452

Jung K-H, Lee J, Dardick C, Seo Y-S, Cao P, Canlas P, Phetsom J, Xu X, Ouyang S, An K, Cho Y-J, Lee G-C, Lee Y, An G, Ronald PC (2008) Identification and functional analysis of light-responsive unique genes and gene family members in rice. PLoS Genet 4(8):e1000164

Kaewsrikhaw R, Prathep A (2014) The effect of habitats, densities and seasons on morphology, anatomy and pigment content of the seagrass *Halophila ovalis* (R.Br.) Hook.f. at Haad Chao Mai National Park, Southern Thailand. Aquat Bot 116:69–75

Kahn AE, Beal JL, Durako MJ (2013) Diurnal and tidal variability in the photobiology of the seagrass *Halophila johnsonii* in a riverine versus marine habitat. Estuar Coast Mar 36:430–443

Kenworthy WJ, Fonseca MS (1996) Light requirements of seagrasses *Halodule wrightii* and *Syringodium filiforme* derived from the relationship between diffuse light attenuation and maximum depth distribution. Estuaries 19:740–750

Kenworthy WJ, Haunert DE (1991) The light requirements of seagrasses. In: Proceedings of a workshop to examine the capability of water quality criteria, standards and monitoring programs to protect seagrasses. NOAA Tech Mem NMFS-SEFC-287, NOAA

Kong F, Li H, Sun P, Zhou Y, Mao Y (2014) De novo assembly and characterization of the transcriptome of seagrass *Zostera marina* using Illumina paired-end sequencing. PLoS One 9(11):e112245

Kouřil R, Wientjes E, Bultema JB, Croce R, Boekema EJ (2013) High-light vs. Low light: effect of light acclimation on photosystem II composition and organization in *Arabidopsis thaliana*. Biochim Biophys Acta 1827:411–419

Kozaki A, Takeba G (1996) Photorespiration protects C3 plants from photooxidation. Nature 384:557–560

Kramer DM, Evans JR (2011) The importance of energy balance in improving photosynthetic productivity. Plant Physiol 155:70–78

Külahoglu C, Denton AK, Sommer M (2014) Comparative transcriptome atlases reveal altered gene expression modules between two Cleomaceae C3 and C4 plant species. Plant Cell 26:3243–3260

Kumar M, Kuzhiumparambila U, Pernice M, Jiang Z, Ralph PJ (2016b) Metabolomics: an emerging frontier of systems biology in marine macrophytes. Algal Res 16:76–92

Kumar M, Padula MP, Davey P, Pernice M, Jiang Z, Sablok G, Contreras-Porcia L, Ralph PJ (2016a) Proteome analysis reveals extensive light stress-response reprogramming in the seagrass *Zostera muelleri* (Alismatales, Zosteraceae) metabolism. Front Plant Sci 7:2023

Lakshmanan M, Lim S-H, Mohanty B, Kim JK, Ha S-H, Lee D-Y (2015) Unraveling the light-specific metabolic and regulatory signatures of rice through combined in silico modeling and multiomics analysis. Plant Physiol 169:3002–3020

Lamers LPM, Govers LL, Janssen ICJM, Geurts JJM, Van der Welle MEW, Van Katwijk MM (2013) Sulfide as a soil phytotoxin: a review. Front Plant Sci 4:1–14

Lamote M, Dunton KH (2006) Effects of drift macroalgae and light attenuation on chlorophyll fluorescence and sediment sulfides in the seagrass *Thalassia testudinum*. J Exp Mar Biol Ecol 334:174–186

Lan Y, Woodrow IE, Mott KA (1992) Light-dependent changes in ribulose bisphosphate carboxylase activase activity in leaves. Plant Physiol 99:304–309

Larsson C, Axelsson L (1999) Bicarbonate uptake and utilization in marine macroalgae. Eur J Phycol 34:79–86

Lee H, Golicz AA, Bayer PE, Jiao Y, Tang H, Paterson AH, Edwards D (2016) The genome of a southern hemisphere seagrass species (*Zostera muelleri*). Plant Physiol 172:272–283

Lee KS, Dunton KH (1999) Inorganic nitrogen acquisition in the seagrass *Thalassia testudinum*: development of a whole-plant nitrogen budget. Limnol Oceanogr 44:1204–1215

Lee K-S, Park SR, Kim YK (2007) Effects of irradiance, temperature and nutrients on growth dynamics of seagrasses: a review. J Exp Mar Biol Ecol 350:144–175

Lehmann P, Nöthen J, Schmidt von Braun S, Bohnsack MT, Mirus O, Schleiff E (2011) Transitions of gene expression induced by short-term blue light. Plant Biol 13:349–361

Lepistö A, Kangasjärvi S, Luomala EM, Brader G, Sipari N, Keränen M, Keinänen M, Rintamäki E (2009) Chloroplast NADPH-thioredoxin reductase interacts with photoperiodic development in Arabidopsis. Plant Physiol 149:1261–1276

Li J, Li G, Wang H, Deng ZW (2011) Phytochrome signaling mechanisms. Arabidopsis Book 9:e0148

Li X, Zhang G, Sun B, Zhang S, Zhang Y, Liao Y, Zhou Y, Xia X, Shi K, Yu J (2013) Stimulated leaf dark respiration in tomato in an elevated carbon dioxide atmosphere. Sci Rep 3:3433

Li X-P, Björkman O, Shih C, Grossman AR, Rosenquist M, Jansson S, Niyogi KK (2000) A pigment-binding protein essential for regulation of photosynthetic light harvesting. Nature 403:391–395

Lin H, Sun T, Xue S, Jiang X (2016a) Heavy metal spatial variation, bioaccumulation, and risk assessment of *Zostera japonica* habitat in the Yellow River Estuary, China. Sci Total Environ 541:435–443

Lin H, Sun T, Zhou Y, Zhang X (2016b) Anti-oxidative feedback and biomarkers in the intertidal seagrass *Zostera japonica* induced by exposure to copper, lead and cadmium. Mar Pollut Bul 109:325–333

Llagostera I, Cervantes D, Sanmartí N, Romero J, Pérez M (2016) Effects of copper exposure on photosynthesis and growth of the seagrass *Cymodocea nodosa*: an experimental assessment. Bull Environ Contam Toxicol 97:374–379

Longstaff BJ, Dennison WC (1999) Seagrass survival during pulsed turbidity events: the effects of light deprivation on the seagrasses *Halodule pinifolia* and *Halophila ovalis*. Aquat Bot 65:105–121

Lovelock CE, Winter K (1996) Oxygen-dependent electron transport and protection from photoinhibition in leaves of tropical tree species. Planta 198:580–587

Macinnis-Ng CMO, Ralph PJ (2002) Towards a more ecologically relevant assessment of the impact of heavy metals on the photosynthesis of the seagrass, *Zostera capricorni*. Mar Pollut Bull 45:100–106

Macinnis-Ng CMO, Ralph PJ (2004a) In situ impact of multiple pulses of metal and herbicide on the seagrass *Zostera capricorni*. Aquat Toxicol 67:227–237

Macinnis-Ng CMO, Ralph PJ (2004b) Variations in sensitivity to copper and zinc among three isolated populations of the seagrass, *Zostera capricorni*. J Exp Mar Biol Ecol 302:63–83

Major KM, Dunton KH (2002) Variations in light-harvesting characteristics of the seagrass Thalassia testudinum: evidence for photoacclimation. J Exp Mar Biol Ecol 275:173–189

Makino A, Miyake C, Yokota A (2002) Physiological functions of the water–water cycle (Mehler reaction) and the cyclic electron flow around PSI in rice leaves. Plant Cell Physiol 43:1017–1026

Marín-Guirao L, Ruiz JM, Dattolo E, Garcia-Munoz R, Procaccini G (2016) Physiological and molecular evidence of differential short-term heat tolerance in Mediterranean seagrasses. Sci Rep 6:28615

Marín-Guirao L, Ruiz JM, Sandoval-Gil JM, Bernardeau-Esteller J, Stinco CM, Meléndez-Martínez AJ (2013a) Xanthophyll cycle-related photoprotective mechanism in the Mediterranean seagrasses *Posidonia oceanica* and *Cymodocea nodosa* under normal and stressful hypersaline conditions. Aquatic Bot 109:14–24

Marín-Guirao L, Sandoval-Gil JM, Bernardeau-Esteller J, Ruiz JM, Sánchez-Lizaso JL (2013b) Responses of the Mediterranean seagrass *Posidonia ocean*ica to hypersaline stress duration and recovery. Mar Environ Res 84:60–75

Marín-Guirao L, Sandoval-Gil JM, Ruiz JM, Sánchez-Lizaso JL (2011) Photosynthesis, growth and survival of the Mediterranean seagrass *Posidonia oceanica* in response to simulated salinity increases in a laboratory mesocosm system. Estuar Coast Shelf Sci 92:286–296

Marsh JA, Dennison WC, Alberte RS (1986) Effect of temperature on photosynthesis and respiration in eelgrass (*Zostera marina* L.) J Exp Mar Biol Ecol 101:257–267

Martínez-Crego B, Olivé I, Santos R (2014) CO_2 and nutrient-driven changes across multiple levels of organization in *Zostera noltii* ecosystems. Biogeosciences 11:7237–7249

Massa SI, Arnaud-Haond S, Pearson GA, Serrao EA (2009) Temperature tolerance and survival of intertidal populations of the seagrass *Zostera n*oltii (Hornemann) in Southern Europe (Ria Formosa, Portugal). Hydrobiologia 619:195–201

Massa SI, Pearson GA, Aires T, Kube M, Olsen JL, Reinhardt R (2011) Expressed sequence tags from heat-shocked seagrass *Zostera noltii* (Hornemann) from its southern distribution range. Mar Genomics 4:181–188

Maxwell K, Johnson GN (2000) Chlorophyll fluorescence-a practical guide. J Exp Bot 51:659–668

Mazzuca S, Björk M, Beer S, Felisberto P, Gobert S, Procaccini G, Runcie J et al (2013) Establishing research strategies, methodologies and technologies to link genomics and proteomics to seagrass productivity, community metabolism and ecosystem carbon fluxes. Front Plant Sci 4:38

Mcglathery KJ, Berg P, Marino R (2001) Using porewater profiles to assess nutrient availability in seagrass-vegetated carbonate sediments. Biogeochemistry 56:239–263

Mercado JM, Gordillo FJL, Figueroa FL, Niell FX (1998) External carbonic anhydrase and affinity for inorganic carbon in intertidal macroalgae. J Exp Mar Biol Ecol 221:209–220

Mercado JM, Niell FX, Figueroa FL (1997) Regulation of mechanism for HCO_3^- use by the inorganic carbon level in *Porphyra leucosticta* thus in Le Jolis (Rhodophyta). Planta 201:319–325

Mohanty N, Vass I, Demeter S (1989) Copper toxicity affects photosystem II electron transport at the secondary quinone acceptor, QB. Plant Physiol 90:175–179

Monnier A, Liverani S, Bouvet R, Jesson B, Smith J, Mosser J, Corellou F, Bouget F-Y (2010) Orchestrated transcription of biological processes in the marine picoeukaryote Ostreococcus exposed to light/dark cycles. BMC Genomics 11:192

Moroney JV, Husic HD, Tolbert NE (1985) Effect of carbonic anhydrase inhibitors on inorganic carbon accumulation by *Chlamydomonas reinhardtii*. Plant Physiol 79:177–183
Mullineaux CW, Emlyn-Jones D (2005) State transitions: an example of acclimation to low-light stress. J Exp Bot 56:389–393
Murchie EH, Horton P (1997) Acclimation of photosynthesis to irradiance and spectral quality in British plant species: chlorophyll content, photosynthetic capacity and habitat preference. Plant Cell Environ 20:438–448
Mvungi EF, Lyimo TJ, Björk M (2012) When *Zostera marina* is intermixed with *Ulva*, its photosynthesis is reduced by increased pH and lower light, but not by changes in light quality. Aquat Bot 102:44–49
Novak AB, Short FT (2010) Leaf reddening in seagrasses. Bot Mar 53:93–97
Novak AB, Short FT (2011a) Leaf reddening in the seagrass *Thalassia testudinum* in relation to anthocyanin, seagrass physiology and morphology, and plant protection. Mar Biol 158:1403–1416
Novak AB, Short FT (2011b) UV-B induced leaf reddening and supports photosynthesis in the seagrass *Thalassia testudinum*. J Exp Mar Biol Ecol 409:136–142
Novak AB, Short FT (2012) Permanent and transient leaf reddening in the seagrass *Thalassia testudinum*. Bull Mar Sci 88:305–315
Novitskaya L, Trevanlon SJ, Driscoll S, Foyer CH, Noctor G (2002) How does photorespiration modulate leaf amino acid contents? A dual approach through modelling and metabolite analysis. Plant Cell Environ 258:821–835
Olivé I, Vergara JJ, Pérez-Lloréns JL (2013) Photosynthetic and morphological photoacclimation of the seagrass *Cymodocea nodosa* to season, depth and leaf position. Mar Biol 160:285–297
Olsen JL, Rouzé P, Verhelst B, Lin Y-C, Bayer T, Collén J et al (2016) The genome of the seagrass *Zostera marina* reveals angiosperm adaptation to the sea. Nature 530:331–335
Olsen Y, Sanchez-Camacho M, Marbá N, Duarte CM (2012) Mediterranean seagrass growth and demography responses to experimental warming. Estuar Coast 35:2505–1213
Ono N, Ishida K, Yamashino T, Nakanishi H, Satoe S, Tabata S, Mizuno T (2010) Genomewide characterization of the light-responsive and clock-controlled output pathways in Lotus japonicas with special emphasis of its uniqueness. Plant Cell Physiol 51:1800–1814
Ort DR, Baker NR (2002) A photoprotective role for O_2 as an alternative electron sink in photosynthesis? Curr Opin Plant Biol 5:193–198
Osmond CB, Grace SC (1995) Perspectives on photoinhibition and photorespiration in the field: quintessential inefficiencies of the light and dark reactions of photosynthesis? J Exp Bot 46:1351–1362
Ow YX, Collier CJ, Uthicke S (2015) Responses of three tropical seagrass species to CO_2 enrichment. Mar Biol 162:1005–1017
Ow YX, Uthicke S, Collier CJ (2016) Light levels affect carbon utilisation in tropical seagrass under ocean acidification. PLoS One 11(3):e0150352
Park SR, Kim S, Kim YK, Kang CK, Lee KS (2016) Photoacclimatory responses of *Zostera marina* in the intertidal and subtidal zones. PLoS One 11(5):e0156214
Peltier G, Cournac L (2002) Chlororespiration. Annu Rev Plant Biol 53:523–550
Peng L, Fukao Y, Fujiwara M, Takami T, Shikanai T (2009) Efficient operation of NAD(P)H dehydrogenase requires supercomplex formation with photosystem I via minor LHCI in *Arabidopsis*. Plant Cell 21:3623–3640
Pérez M, Invers O, Ruiz JM, Frederiksen MS, Holmer M (2007) Physiological responses of the seagrass *Posidonia oceanica* to elevated organic matter content in sediments: an experimental assessment. J Exp Mar Biol Ecol 344:49–160
Pérez M, Romero J (1992) Photosynthetic response to light and temperature of the seagrass *Cymodocea nodosa* and the prediction of its seasonality. Aquat Bot 43:51–62
Pernice M, Sinutok S, Sablok G, Commault A, Schliep M, Macreadie P, Rasheed M, Ralph P (2016) Molecular physiology reveals ammonium uptake and related gene expression in the seagrass *Zostera muelleri*. Mar Environ Res 122:126–134

Petersen J, Rogowska-Wrzesinska A, Jensen OL (2013) Functional proteomics of barley and barley chloroplasts-strategies, methods and perspectives. Front Plant Sci 4:52

Pietrzykowska M, Suorsa M, Semchonok DA, Tikkanen M, Boekema EJ, Aro E-M, Jansson S (2014) The light-harvesting chlorophyll a/b binding proteins Lhcb1 and Lhcb2 play complementary roles during state transitions in *Arabidopsis*. Plant Cell 26:3646–3660

Piro A, Marín-Guirao L, Serra IA, Spadafora A, Sandoval-Gil JM, Bernardeau-Esteller J, Mazzuca S (2015a) The modulation of leaf metabolism plays a role in salt tolerance of *Cymodocea nodosa* exposed to hypersaline stress in mesocosms. Front Plant Sci 6:464

Piro A, Serra IL, Spadafora A, Cardilio M, Bianco L, Perrotta G, Santos R, Mazzuca S (2015b) Purification of intact chloroplasts from marine plant *Posidonia oceanica* suitable for organelle proteomics. Proteomics 15:4159–4174

Poolman MG, Miguet L, Sweetlove LJ, Fell DA (2009) A genome-scale metabolic model of *Arabidopsis* and some of its properties. Plant Physiol 151:1570–1581

Portis AR (2003) Rubisco activase: Rubisco's catalytic chaperone. Photosynth Res 75:11–27

Portis AR, Li CS, Wang DF, Salvucci ME (2008) Regulation of Rubisco activase and its interaction with Rubisco. J Exp Bot 59:1597–1604

Proctor MCF, Smirnoff N (2011) Ecophysiology of photosynthesis in bryophytes: major roles for oxygen photoreduction and non-photochemical quenching? Physiol Plant 141:130–140

Ragavan P, Saxena A, Mohan PM, Coomar T, Ragavan A (2013) Leaf reddening in seagrasses of Andaman and Nicobar Islands. Trop Ecol 54:269–273

Ragni M, Ribera d'Alcala M (2004) Light as an information carrier underwater. J Plankton Res 26:433–443

Ralph PJ, Burchett MD (1995) Photosynthetic responses of the seagrass *Halophila ovalis* (R Br.) Hook. f. to high irradiance stress, using chlorophyll a fluorescence. Aquat Bot 51:55–66

Ralph PJ, Durako MJ, Enríquez S, Collier CJ, Doblin MA (2007) Impact of light limitation on seagrasses. J Exp Mar Biol Ecol 350:176–193

Ralph PJ, Polk SM, Moore KA, Orth RJ, Smith WO (2002) Operation of the xanthophyll cycle in the seagrass *Zostera marina* in response to variable irradiance. J Exp Mar Biol Ecol 271:189–207

Ralph PJ, Burchett MD (1998) Photosynthetic response of *Halophila ovalis* to heavy metal stress. Environ Pollut 103:91–101

Ramanan VK, Shen L, Moore JH, Saykin AJ (2012) Pathway analysis of genomic data: concepts, methods, and prospects for future development. Trends Genet 28:323–332

Ranvier S, Gnassia-Barelli M, Pergent G, Capiomont A, Roméo M (2000) The effect of mercury on glutathione S transferase in the marine phanerogam *Posidonia oceanica*. Bot Mar 43:161–168

Rao SK, Reiskind JB, Bowes G (2008) Kinetic analyses of recombinant isoforms of phosphorenolpyruvate carboxylase from *Hydrilla verticillata* leaves and the impact of substituting a C4-signature serine. Plant Sci 174:475–483

Rasmusson LM (2015) Seagrass respiration an assessment of oxygen consumption patterns of temperate marine macrophytes. Doctoral thesis, Stockholm University

Rasmusson LM, Björk M (2014) Determining light suppression of mitochondrial respiration for three temperate marine macrophytes using the Kok method. Bot Mar 57:483–486

Renberg L, Johansson AI, Shutova T, Stenlund H, Aksmann A, Raven JA, Gardeström P, Moritz T, Samuelsson G (2010) A metabolomic approach to study major metabolite changes during acclimation to limiting CO_2 in *Chlamydomonas reinhardtii*. Plant Physiol 154:187–196

Rensing SA, Sheerin DJ, Hiltbrunner A (2016) Phytochromes: more than meets the eye. Trends Plant Sci 21:543–546

Reusch TH, Veron A, Preuss C, Weiner J, Wissler L, Beck A, Klages S, Kube M, Reinhardt R, Bornberg-Bauer E (2008) Comparative analysis of expressed sequence tag (EST) librariesin the seagrass *Zostera marina* subjected to temperature stress. Mar Biotechnol 10:297–309

Rijstenbil JW, Derksen JWM, Gerringa LJA, Poortvliet TCW, Sandee A, Van Den Berg M, Van Drie J, Wijnholds JA (1994) Oxidative stress induced by copper: defense and damage in the marine planktonic diatom *Ditylum brightwellii* (Grunow) West, grown in continuous cultures with high and low zinc levels. Mar Biol 119:583–590

Rogers A, Allen DJ, Davey PA, Morgan PB, Ainsworth EA, Bernacchi CJ et al (2004) Leaf photosynthesis and carbohydrate dynamics of soybeans grown throughout their life-cycle under Free-Air Carbon dioxide Enrichment. Plant Cell Environ 27:449–458

Rumeau D, Peltier G, Cournac L (2007) Chlororespiration and cyclic electron flow around PSI during photosynthesis and plant stress response. Plant Cell Environ 30:1041–1051

Sahi SV, Sharma NC (2005) Phytoremediation of lead. In: Shtangeeva I (ed) Trace and ultratrace elements in plants and soils, Series advances in ecological researches. Witpress, Southampton, Boston, pp 209–222

Salvucci ME, Crafts-Brandner SJ (2004) Mechanism for deactivation of Rubisco under moderate heat stress. Physiol Plant 122:513–519

Salvucci ME, Ogren WL (1996) The mechanism of Rubisco activase: insights from studies of the properties and structure of the enzyme. Photosynth Res 47:1–11

Salvucci ME, Osteryoung KW, Crafts-Brandner SJ, Vierling E (2001) Exceptional sensitivity of Rubisco activase to thermal denaturation in vitro and in vivo. Plant Physiol 127:1053–1064

Sandoval-Gil JM, Marín-Guirao L, Ruiz JM (2012) The effect of salinity increase on the photosynthesis, growth and survival of the Mediterranean seagrass *Cymodocea nodosa*. Estuar Coast Shelf Sci 115:260–271

Sandoval-Gil JM, Ruiz JM, Marín-Guirao L, Bernardeau-Esteller J, Sánchez-Lizaso JL (2014) Ecophysiological plasticity of shallow and deep populations of the Mediterranean seagrasses *Posidonia oceanica* and *Cymodocea nodosa* in response to hypersaline stress. Mar Environ Res 95:39–61

Sanz-sàez A, Erice G, Aranjuelo I, Nogués S, Irigoyen JJ, Sánchezdíaz M (2010) Photosynthetic down-regulation under elevated CO_2 exposure can be prevented by nitrogen supply in nodulated alfalfa. J Plant Physiol 167:1558–1565

Sato S, Yanagisawa S (2014) Characterization of metabolic states of *Arabidopsis thaliana* under diverse carbon and nitrogen nutrient conditions via targeted metabolomic analysis. Plant Cell Physiol 55:306–319

Schliep M, Pernice M, Sinutok S, Bryant CV, York PH, Rasheed MA, Ralph PJ (2015) Evaluation of reference genes for RT-qPCR studies in the seagrass *Zostera muelleri* exposed to light limitation. Sci Rep 5:17051

Schmidt AL, Wysmyk JKC, Craig SE, Lotze HK (2012) Regional-scale effects of eutrophication on ecosystem structure and services of seagrass beds. Limnol Oceanogr 57:1389–1402

Schulze S, Westhoff P, Gowik U (2016) Glycine decarboxylase in C3, C4 and C3–C4 intermediate species. Curr Opin Plant Biol 31:29–35

Shafer DJ, Kaldy JE, Sherman TD, Marko KM (2011) Effects of salinity on photosynthesis and respiration of the seagrass *Zostera japonica*: a comparison of two established populations in North America. Aquat Bot 95:214–220

Sharon Y, Levitan O, Spungin D, Berman-Frank I, Beer S (2011) Photoacclimation of the seagrass *Halophila stipulacea* to the dim irradiance at its 48–meter depth limit. Limnol Oceanogr 56:357–362

Shioi Y, Tamai H, Sasa T (1978) Effects of copper on photosynthetic electron transport systems in spinach chloroplasts. Plant Cell Physiol 19:203–209

Silva J, Barrote I, Costa MM, Albano S, Santos R (2013) Physiological responses of *Zostera marina* and *Cymodocea nodosa* to light-limitation stress. PLoS One 8(11):e81058

Stirbet A, Govindjee (2011) On the relation between the Kautsky effect (chlorophyll a fluorescence induction) and Photosystem II: Basics and applications of the OJIP fluorescence transient. J Photochem Photobiol B 104:236–257

Stitt M, Krapp A (1999) The interaction between elevated carbon dioxide and nitrogen nutrition: the physiological and molecular background. Plant Cell Environ 22:583–621

Sun J, Gibson KM, Kiirats O, Okita TW, Edwards GE (2002) Interactions of nitrate and CO_2 enrichment on growth, carbohydrates, and Rubisco in *Arabidopsis* starch mutants: significance of starch and hexose. Plant Physiol 130:1573–1583

Taiz L, Zeiger E, Moller I, Murphy A (2015) Plant physiology and development, 6th edn. Sinauer Associates Inc, Sunderland, MA

Takatani N, Ito T, Kiba T, Mori M, Miyamoto T, Maeda S-I (2014) Effects of high CO_2 on growth and metabolism of *Arabidopsis* seedlings during growth with a constantly limited supply of nitrogen. Plant Cell Physiol 55:281–292

Tepperman JM, Zhu T, Chang HS, Wang X, Quail PH (2001) Multiple transcription-factor genes are early targets of phytochrome a signalling. Proc Natl Acad Sci U S A 98:9437–9442

Timm S, Florian A, Fernie AR, Bauwe H (2016) The regulatory interplay between photorespiration and photosynthesis. J Exp Bot 67:2923–2929

Tolbert NE (1976) Glycolate oxidase and glycolate dehydrogenase in marine algae and plants. Aust J Plant Physiol 3:129–132

Touchette BW, Burkholder JM (2000) Overview of the physiological ecology of carbon metabolism in seagrasses. J Exp Mar Biol Ecol 250:169–205

Tupan CI, Herawati EY, Arfiati D (2014) Detection of phytochelatin and glutathione in seagrass *Thalassia hemprichii* as a detoxification mechanism due to lead heavy metal exposure. Aquat Sci Technol 2:67–78

Uku J, Beer S, Björk M (2005) Buffer sensitivity of photosynthetic carbon utilisation in eight tropical seagrasses. Mar Biol 147:1085–1090

Valko M, Rhodes CJ, Moncol J, Izakovich J, Masur M (2006) Free radicals, metals and antioxidants in oxidative stress-induced cancer. Chem Biol Interact 160:1–40

Van Assche F, Clijsters H (1990) Effects of metals on enzyme activity in plants. Plant Cell Environ 13:195–206

Voss I, Sunil B, Scheibe R, Raghavendra AS (2013) Emerging concept for the role of photo respiration as an important part of abiotic stress response. Plant Biol 15:713–722

Waghmode AP, Joshi GV (1983) Significance of phosphoglycollate phosphatase and 3-phosph oglycerate phosphatase in photosynthetic carbon assimilation in some marine plants. Photosynthetica 17:193–197

Wang W, Qin X, Sang M, Chen D, Wang K, Lin R, Lu C, Shen J, Kuang T (2013) Spectral and functional studies on siphonaxanthin-type light-harvesting complex of photosystem II from *Bryopsis corticulans*. Photosynth Res 117:267–279

Wang Y, Stessman DJ, Spalding MH (2015) The CO_2 concentrating mechanism and photosynthetic carbon assimilation in limiting CO_2: how *Chlamydomonas* works against the gradient. Plant J 82:429–448

Wicke S, Schneeweiss GM, de Pamphilis CW, Müller KF, Quandt D (2011) The evolution of the plastid chromosome in land plants: gene content, gene order, gene function. Plant Mol Biol 76:273–297

Wienkoop S, Weiß J, May P, Kempa S, Irgang S, Recuenco-Munoz L, Pietzke M, Schwemmer T, Rupprecht J, Egelhofer V, Weckwerth W (2010) Targeted proteomics for *Chlamydomonas reinhardtii* combined with rapid subcellular protein fractionation, metabolomics and metabolic flux analyses. Mol BioSyst 6:1018–1031

Wilkinson AD, Collier CJ, Flores F, Mercurio P, O'Brien J, Ralph PJ, Negri AP (2015a) A miniature bioassay for testing the acute phytotoxicity of photosystem II herbicides on seagrass. PLoS One 10(2):e0117541

Wilkinson AD, Collier CJ, Flores F, Negri AP (2015b) Acute and additive toxicity of ten photosystem–II herbicides to seagrass. Sci Rep 5:17443

Williams TCR, Poolman MG, Howden AJM, Schwarzlander M, Fell DA, Ratcliffe RG, Sweetlove LJ (2010) A genome-scale metabolic model accurately predicts fluxes in central carbon metabolism under stress conditions. Plant Physiol 154:311–323

Winck FV, Melo DOP, Fernando A, Barrios G (2013) Carbon acquisition and accumulation in microalgae *Chlamydomonas*: insights from "omics" approaches. J Proteome 94:207–218

Wingler A, Lea PJ, Quick WP, Leegood RC (2000) Photorespiration: metabolic pathways and their role in stress protection. Philos Trans R Soc Lond Ser B Biol Sci 355:1517–1529

Winters G, Nelle P, Fricke B, Rauch G, Reusch TBH (2011) Effects of a simulated heat wave on photophysiology and gene expression of high- and low-latitude populations of *Zostera marina*. Mar Ecol Prog Ser 435:83–95

Wissler L, Codoñer FM, Gu J, Reusch TBH, Olsen JL, Procaccini G, Bornberg-Bauer E (2011) Back to the sea twice: identifying candidate plant genes for molecular evolution to marine life. BMC Evol Biol 11:8

Yaakub SM, Chen E, Bouma TJ, Erftemeijer PLA, Todd PA (2014) Chronic light reduction reduces overall resilience to additional shading stress in the seagrass *Halophila ovalis*. Mar Pollut Bull 83:467–474

Yamano T, Fukuzawa H (2009) Carbon-concentrating mechanism in a green alga, *Chlamydomonas reinhardtii*, revealed by transcriptome analyses. J Basic Microbiol 49:42–51

Yong JWH, Wong SC, Letham DS, Hocart CH, Farquhar GD (2000) Effects of elevated CO_2 and nitrogen nutrition on cytokinins in the xylem sap and leaves of cotton. Plant Physiol 124:767–779

York PH, Gruber RK, Hill R, Ralph PJ, Booth DJ (2013) Physiological and morphological responses of the temperate seagrass *Zostera muelleri* to multiple stressors: investigating the interactive effects of light and temperature. PLoS One 8(10):e76377

Yruela I, Pueyo JJ, Alonso PJ, Picorel R (1996) Photoinhibition of photosystem II from higher plants: effect of copper inhibition. J Biol Chem 271:27408–27415

Zhang N, Portis AR (1999) Mechanism of light regulation of Rubisco: a specific role for the larger Rubisco activase isoform involving reductive activation by thioredoxin-f. Proc Natl Acad Sci 96:9438–9443

Zhang XN, Wu Y, Tobias JW, Brunk BP, Deitzer GF, Liu D (2008) HFR1 is crucial for transcriptome regulation in the cryptochrome 1-mediated early response to blue light in *Arabidopsis thaliana*. PLoS One 3(10):e3563

Zhu X-G, Portis AR, Long SP (2004) Would transformation of C3 crop plants with foreign Rubisco increase productivity? A computational analysis extrapolating from kinetic properties to canopy photosynthesis. Plant Cell Environ 278:155–165

Zimmerman RC (2003) A biooptical model of irradiance distribution and photosynthesis in seagrass canopies. Limnol Oceanogr 48:568–585

Zimmerman RC, Reguzzoni JL, Alberte RS (1995) Eelgrass (*Zostera marin*a L.) transplants in San Francisco Bay: role of light availability on metabolism, growth and survival. Aquat Bot 51:67–86

Chapter 8
Systems Biology and the Seagrass Paradox: Adaptation, Acclimation, and Survival of Marine Angiosperms in a Changing Ocean Climate

Richard C. Zimmerman

Abstract Predicting adaptive fitness to any environment requires mechanistic understanding of environmental influence on metabolic networks that control energy assimilation, growth, and reproduction. Although the potential impacts of environment on gene products are myriad, important phenotypic responses are often regulated by a few key points in metabolic networks where externally supplied resources or physiological reaction substrates limit reaction kinetics. Environmental resources commonly limiting seagrass productivity, survival, and growth include light and CO_2 availability that control carbon assimilation and sucrose formation. Phosphate availability can also be important in oligotrophic tropical environments, particularly in the presence of carbonate sediments. Temperature and macronutrient oversupply (eutrophication) can act as confounding stressors, particularly in temperate environments. Photoacclimation can be regulated by electron transport pathways residing in the chloroplast stroma, but stress responses are often manifest by the expression of generalized stress response proteins, both of which appear to be affected by temperature and CO_2 availability. A systems approach is employed to explore (1) the responses of seagrasses to the combined impacts of environmental limiting factors that control fundamental physiological processes leading to whole-plant performance; (2) sediment diagenetic processes that facilitate nutrient remineralization, carbon sequestration, and toxin neutralization; (3) interactions with other organisms induced by trophic cascades; and (4) impacts of human-induced climate change that affect system dynamics at numerous points in the network.

Keywords Seagrass • Photosynthesis • Temperature • Light • CO_2 • Climate • Metabolism

R.C. Zimmerman (✉)
Department of Ocean, Earth and Atmospheric Sciences, Old Dominion University, Norfolk, VA 23529, USA
e-mail: rzimmerm@odu.edu

© Springer International Publishing AG 2017

M. Kumar, P. Ralph (eds.), *Systems Biology of Marine Ecosystems*,
DOI 10.1007/978-3-319-62094-7_8

Contents

8.1 A Few Words About the Systems Approach 168
8.2 Seagrasses as Biological Systems 169
8.2.1 Key Anatomical Features and Functions 169
8.2.2 Sensitivity to Environmental Change Results from High Light Requirements 171
8.2.3 A Surprising Mechanism Responsible for High Light Requirements 172
8.2.4 Integrating Transcriptomic Information into Predictive Seagrass Models 175
8.3 Seagrasses as Ecosystem Engineers 177
8.3.1 Seagrass-Sediment Interactions 177
8.3.2 Interactions with Other Organisms 179
8.3.3 Interactions with Humans 180
8.4 Implications of the Seagrass Paradox for the Future of Seagrass Systems 182
References 183

8.1 A Few Words About the Systems Approach

The complexity of life has always fascinated mankind, but the development of a systems perspective in the biological sciences has been a gradual one. During the twentieth century, the science of biology evolved from descriptive natural history to the development of mechanistic insights derived from painstaking reductionist experimentation increasingly informed by evolutionary theory. However, integrating that knowledge across levels of organization from molecule through individuals to ecosystems has been hindered by important gaps in our understanding of the genome, its physiological regulation, and organismal responses to environmental forcing. The dawn of the twenty-first century has witnessed significant breakthroughs in data collection, processing, and interpretation at all levels or organization such that it is now possible to translate the complex network diagrams of metabolic processes, food webs, and even global biogeochemical cycles into predictive mathematical models. In addition to providing a quantitative understanding, the resulting system-wide perspective provides pathways for exploring emergent properties not readily apparent in the individual pieces of reductionist information used to construct the networks. The ability to influence emergent properties by manipulating specific components of complex systems helps provide rational approaches for identifying critical steps that may be responsive to external forcing.

Seagrasses (marine angiosperms) provide critical ecosystem services in shallow coastal seas through the world and are presently at risk from human impacts derived from coastal development, regional eutrophication, resource extraction, and global climate change (Orth et al. 2006). Growing interest into all scientific aspects of seagrasses during the past 30 years has provided a remarkable archive of information (e.g., Larkum et al. 2006) from which to develop a systems-level appreciation of these remarkable plants from molecules to ecosystems. This chapter will explore physiological, or bottom-up, aspects of seagrasses that must adapt to local environments in order to transform solar energy and dissolved nutrients into biochemical products required to sustain individual plants. We will also explore seagrasses as a component of complex coastal food webs and place particular emphasis on the importance of trophic interactions in maintaining the viability of seagrass ecosystems. Finally, we will examine seagrasses as biogeochemical agents that stabilize coastal sediments; sequester CO_2 through direct burial of organic carbon in coastal ocean sediments, recently termed “blue carbon” (McLeod et al. 2011); and trap remineralized CO_2 by generating alkalinity through carbonate dissolution and hydrogen sulfide precipitation in sedimentary pore waters.

8.2 Seagrasses as Biological Systems

Seagrasses represent an ecological assemblage of about 66 species of angiosperms distributed across four taxonomic families within the superorder Alismatiflorae (Monocotyledonae, den Hartog and Kuo 2006). The somewhat obscure evolutionary origins appear to involve independent derivations from freshwater and estuarine ancestors sometime in the Lower Cretaceous (Les et al. 1997). Despite this polyphyletic origin, seagrasses share many common properties with each other that facilitate an aquatic life history. These traits include (1) the full submergence of all plant structures; (2) tolerance of saline water; (3) a secure anchoring system consisting of metabolically functional roots and rhizomes capable of assimilating nutrients and tolerating anoxic, permanently flooded sediments; and (4) hydrophilous pollination mechanisms (Larkum et al. 2006; Jackson et al. 2009).

8.2.1 Key Anatomical Features and Functions

The seagrass leaf has been heavily modified from the terrestrial archetype to accommodate a submerged aquatic existence. Chloroplasts are concentrated in the epidermis rather than the mesophyll (Kuo and den Hartog 2006) (Fig 8.1a). The cuticle is extremely thin and water permeable to permit gas, solute, and water exchange with the epidermis. There are no stomata. However, the thin water-permeable cuticle renders seagrasses highly vulnerable to desiccation upon exposure to air (Leuschner et al. 1998; Björk et al. 1999). These leaf modifications are more reminiscent of macrophytic algae than the terrestrial vascular plants from which seagrasses are derived and provide a strong example of convergent evolution by the seagrasses and seaweeds to the solution of common problems. The photosynthetic apparatus also shows a high capacity for photoacclimation to different light environments, which protects it from photoinhibition in the bright, blue light of clear tropical waters and facilitates light harvesting in the dim, green waters of many temperate environments (Ralph et al. 2002; Cummings and Zimmerman 2003).

Gas-filled lacunae (i.e., aerenchyma) run continuously from the leaves through the rhizomes to the roots. They provide buoyancy to maintain the leaves in a vertical orientation that minimizes self-shading and optimizes photosynthesis of the entire canopy (Zimmerman 2006). The lacunae transport life-sustaining oxygen to roots

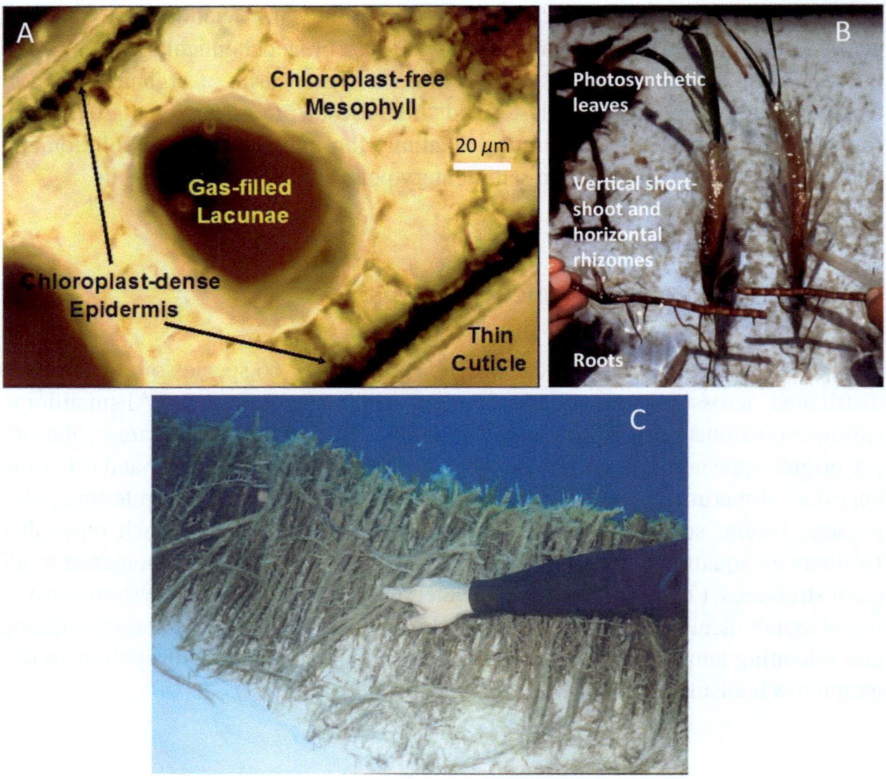

Fig. 8.1 **A.** Photomicrograph of a cross section of a turtlegrass (*Thalassia testudinum* Banks ex. König) leaf showing two layers of chloroplast-dense epidermis, a chloroplast-free mesophyll, and air-filled lacunae running parallel to the central axis of the leaf. Image reproduced from Zimmerman (2006). **B.** Short shoots of turtlegrass (*Thalassia testudinum*, Banks ex Koenig) consisting of green photosynthetic leaves, and vertical rhizomes with old leaf scars connected by horizontal rhisomes containing roots. **C.** Subterranean turtlegrass rizomes emerging from an eroded bank of carbonate sediment on the Great Bahama Bank reveal comsiderable potential for blue carbon burial bu this species

and rhizomes located in permanently flooded anoxic sediments. The roots and rhizomes (Fig 8.1b) provide a solid anchor in the unconsolidated sediments (Fig 8.1c), and the roots assimilate inorganic nutrients from a pore-water reservoir that is generally unavailable to rootless macrophytic algae and phytoplankton (Zimmerman et al. 1987). The roots and rhizomes have low rates of biomass-specific respiration, relative to the leaves, that minimizes carbon demand by these non-photosynthetic tissues (Zimmerman et al. 1989; Zimmerman and Alberte 1996). However, unlike wetland angiosperms that continuously transport air from emergent leaves to belowground tissues rooted in flooded soils, the submerged nature of seagrass leaves restricts belowground transport of oxygen to daylight periods when photosynthesis loads the lacunae with oxygen. As a result, seagrass roots and rhizomes must tolerate daily periods of anoxia that can approach 12 h or more. As in other angiosperms, root/rhizome anoxia interrupts the translocation of sucrose from leaves and induces alcoholic fermentation in these belowground tissues (Pregnall et al. 1984; Smith et al. 1984; Zimmerman and Alberte 1996). When deprived of oxygen, most facultative anaerobes (including most angiosperms) exhibit a Pasteur effect that increases the rate of glycolysis to maintain ATP production necessary for metabolic function. The resulting ability to tolerate anoxia is linked to the supply of glucose, and anoxic death results when tissues become depleted of sugar reserves and cannot maintain ATP and NADPH production (Bailey-Serres and Voesenek 2008). Seagrass roots, however, exhibit a low-oxygen quiescence strategy (LOQS), in which the onset of anoxia reduces the rate of metabolic glucose consumption and ATP production. The reduced nighttime carbon demand of anaerobic tissues extends the period of anoxia tolerated by eelgrass roots and further reduces carbon demand by belowground tissues (Smith et al. 1988; Zimmerman et al. 1996). The LOQS appears to be highly conserved among vascular plants to maintain cellular homeostasis under low-O_2 stress and enables the synthesis of proteins involved in metabolite transport, protection from reactive oxygen species (ROS), and chaperone activity (Voesenek and Bailey-Serres 2015).

Finally, all species of seagrasses (except *Enhalus* spp.) are adapted for completely submerged pollination. The filiform pollen grains of *Zostera marina* tumble as they encounter turbulent velocity gradients induced by the anatomy of flowers and inflorescences, increasing the probability of attachment to the stigmatic surfaces (Ackerman 2006). In *Thalassia testudinum*, invertebrates may also facilitate pollen transfer from male flowers to female stigma (van Tussenbroek et al. 2016).

8.2.2 Sensitivity to Environmental Change Results from High Light Requirements

Although seagrasses possess a remarkable assemblage of successful adaptations to a submerged aquatic life history, their light requirements for survival are 10- to 20-fold higher than many marine autotrophs, which renders them vulnerable to a variety of habitat disturbances, often of anthropogenic origin (Duarte 1991; Short and Wyllie-Echeverria 1996). More than 40 large-scale seagrass declines involving at least 24 different species were reported worldwide during the twentieth century, and the losses have continued to the point where seagrasses may be nearing a crisis with respect to global sustainability (Orth et al. 2006; Short et al. 2011). Eutrophication, which promotes the growth of light-absorbing phytoplankton and opportunistic macroalgae, is a critical factor in many of these disturbance events, including the extensive and persistent losses throughout the Chesapeake Bay, USA (Orth and Moore 1983). However, even when algal competitors are brought under control, seagrass recovery can be hampered by the destabilization and resuspension of marine sediments that further reduces light availability to the benthos (Lawson et al. 2007; Carr et al. 2010; McPherson et al. 2011).

Seagrasses are also vulnerable to elevated summertime water temperature, despite considerable physiological tolerance to short-term temperature elevation. At least part of the vulnerability to extreme summer temperature results from the differential responses of photosynthesis and respiration. Eelgrass near the Chesapeake Bay, for example, are commonly exposed to seasonal temperature excursions ranging from wintertime lows near 5°C to summertime highs exceeding 25°C. Prolonged summer temperatures can approach 30°C during particularly warm years, causing eelgrass meadows to die back throughout the southern Chesapeake Bay (Orth and Moore 1983; Moore and Jarvis 2008). This thermal intolerance has been linked to the differential effects of temperature on photosynthesis and respiration (Evans et al. 1986; Zimmerman et al. 1989). In essence, the Q_{10} response for eelgrass respiration is about 2.5, but only 1.98 for photosynthesis. Thus, temperature causes respiration rates to increase more dramatically than photosynthesis rates, making eelgrass vulnerable to negative carbon balance above 25°C.

8.2.3 A Surprising Mechanism Responsible for High Light Requirements

Much of the paradoxical vulnerability of this otherwise highly adapted group of marine angiosperms to light limitation and thermal stress can be attributed to their relatively poor ability to extract dissolved inorganic carbon (DIC) from seawater to support photosynthesis. Seawater is an alkaline medium, and most of the DIC is present in the form of bicarbonate ($HCO_3^- \cong 2$ mM). However, CO_2 is the exclusive substrate for the carboxylation reaction of Rubisco, and the concentration of aqueous CO_2 [$CO_{2(aq)}$] is sufficiently low (10 to 20 μM) that passive diffusion of CO_2 cannot satisfy the photosynthetic demand for inorganic carbon in most cases (Raven 2010). Consequently, many marine autotrophs possess carbon-concentrating mechanisms for extracting CO_2 from the more abundant pool of HCO_3^- in seawater (Falkowski and Raven 2007; Raven and Beardall 2014).

Although seagrass leaves are capable of extracellular dehydration and direct uptake of HCO_3^- (Al-Moghrabi et al. 1996; Beer and Rehnberg 1997), light-saturated photosynthesis of many species is CO_2 limited. The instantaneous positive response to increased $CO_{2(aq)}$ demonstrates a constitutive ability for light harvesting, electron transport, and carbon fixation at rates that vastly exceed the C-limited photosynthetic capacity of seagrass leaves in the present-day ocean (Durako 1993; Zimmerman et al. 1995; Beer and Koch 1996; Invers et al. 2001). Higher rates of photosynthesis resulting from elevated $CO_{2(aq)}$ can reduce the light requirements for daily carbon balance (Zimmerman et al. 1997) and promote the accumulation of carbon reserves that increases vegetative shoot proliferation and flowering shoot differentiation (Palacios and Zimmerman 2007). Coupling these experimentally derived physiological responses with a geometrically explicit formulation of radiative transfer through submerged plant canopies led to the development of *GrassLight* (Fig 8.2), a bio-optical model that permits quantitative exploration of the impacts of

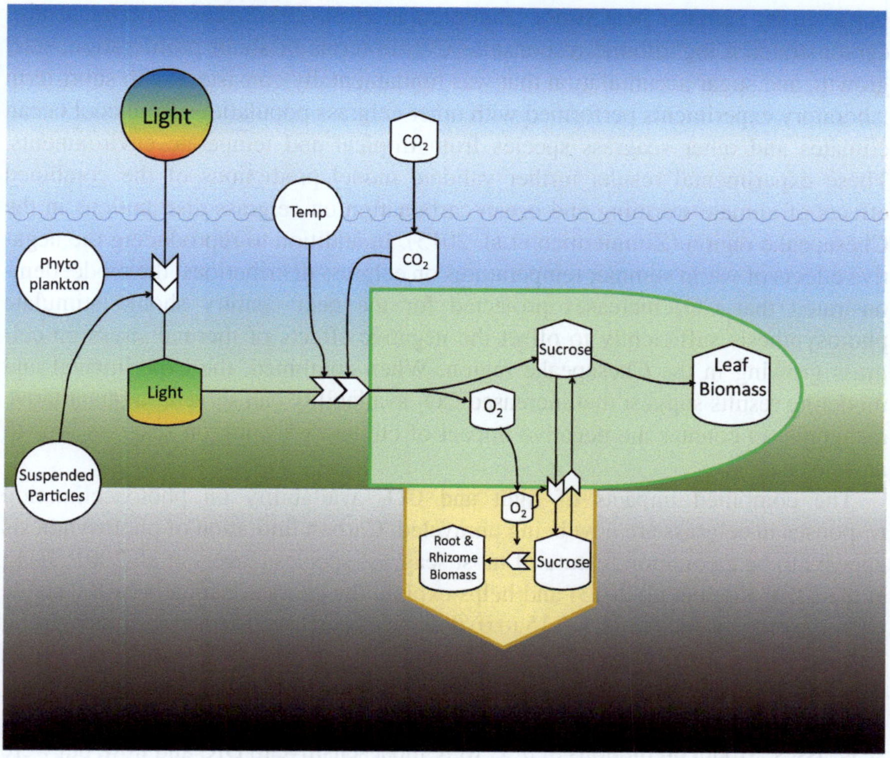

Fig. 8.2 Schematic diagram illustrating the flow of energy and sucrose to support growth and respiration, as mediated by temperature and CO_2 availability in the bio-optical model *GrassLight* using the energy circuit language of Odum (1983). Open circles represent donor-controlled processes external to the model. Tank symbols represent depletable resources capable of interacting with other model components. *Block arrows* represent interactions between components (work gates) that can be either positive or negative, depending on the interaction. Detailed mathematical relations are described in Zimmerman (2003) and Zimmerman et al. (2015)

environmental forcing (e.g., light, temperature, flow, DIC, epiphytes, etc.) on photosynthesis, metabolic carbon balance, shoot density, and depth distribution (Zimmerman 2003; Zimmerman et al. 2015). The model predicted that doubling the atmospheric CO_2 concentration would yield a 35% increase in the density and spatial distribution of eelgrass in Elkhorn Slough, California, USA, a turbid estuary where eelgrass distributions are currently restricted to very shallow (<2 m) depths (Zimmerman et al. 1994; Zimmerman and Caffrey 2002; Zimmerman 2006).

A high-CO_2 world may also reduce seagrass vulnerability to summertime thermal stress, a situation that has increasingly limited the survival of eelgrass near the southern limit of their distribution along the east coast of North America (Moore and Jarvis 2008; Moore et al. 2012). However, prolonged experimental stimulation of eelgrass photosynthesis via CO_2 can enhance the summertime survival, growth, and proliferation of perennial eelgrass from the Chesapeake region that is regularly impacted by summer heat stress (Zimmerman et al. 2017). The experiment also demonstrated a logarithmic response to CO_2 in terms of shoot proliferation, size, growth, and sugar accumulation that was fundamentally consistent with short-term laboratory experiments performed with other eelgrass populations from cool ocean climates and other seagrass species from tropical and temperate environments. These experimental results further validate model predictions of the combined effects of climate warming and ocean carbonation on eelgrass distributions in the Chesapeake region (Zimmerman et al. 2015). In addition to reproducing the negative effects of warm summer temperatures on eelgrass distributions, the model demonstrated that CO_2 increases projected for the next century should stimulate photosynthesis sufficiently to offset the negative effects of thermal stress on eelgrass growing in the Chesapeake region. When combined, the experimental and modeling results suggest that increased CO_2 availability can serve as a quantitative antagonist to counter the negative impact of climate warming on seagrass growth and survival.

The combined impacts of light and CO_2 availability on photoacclimation responses in eelgrass are now being unraveled. Carbon limitation of photosynthesis appears to be a common feature of most seagrass species (Invers et al. 2001; Jiang et al. 2010; Koch et al. 2013) and helps explain the characteristically heavy stable carbon isotope values ($\delta^{13}C = -15$ to 0) that have been used to trace seagrass carbon through marine food webs (Hemminga and Mateo 1996). McPherson et al. (2015) developed a mathematical model parameterized by laboratory experiments in order to predict the impacts of substrate availability on carbon isotope fractionation in seagrasses. Model predictions of δ^{13}C were most sensitive to DIC and flow, but were less sensitive to DIC source [$CO_{2(aq)}$ vs. HCO_3^-], indicating that carbon limitation of Rubisco was the primary driver of seagrass δ^{13}C. Accurate model predictions of specific δ^{13}C values reported for a variety of seagrass taxa from different environments provided a systems-level understanding of the environmental control of carbon isotope composition of seagrasses. The mathematical relationships embodied in this model will become increasingly important for predicting the response of these ecosystem engineers to local processes that affect light availability and flow, as well as global impacts of climate warming and ocean acidification.

A long-term experiment conducted with *Zostera marina* L. (eelgrass) grown in controlled outdoor aquaria revealed predictive increases in absolute growth, plant size, and flowering that could be traced to increased rates of light-saturated photosynthetic and sucrose formation under CO_2 enrichment (Celebi 2016; Zimmerman et al. 2017). In contrast to these increases in whole-plant performance characteristics, photosynthetic and photoprotective pigment content in eelgrass leaves was downregulated, suggesting an important role for Rubisco in balancing redox state in the chloroplast, which regulates expression of light-harvesting complexes (Backhausen and Scheibe 1999; Pfannschmidt 2003; Hanke et al. 2009; Hüner et al. 2012). These regulatory mechanisms are generally thought to be controlled by the redox state of Q_A (a plastoquinone) in the thylakoid membrane that depends on the continuity of electron transport under various limiting conditions (Pfannschmidt 2003; Pfannschmidt and Yang 2012). Thus, in addition to capturing energy for metabolism, the photosynthetic machinery performs important sensory functions which further explain the interdependent regulation of pigment composition and optical properties of eelgrass leaves by CO_2, temperature, and light.

8.2.4 Integrating Transcriptomic Information into Predictive Seagrass Models

As the above discussion indicates, mathematical modeling of ecophysiological processes can provide deep insights into complex systems. The application of systems analysis to subcellular processes has been hindered by available technology and, particularly in the case of seagrasses, the lack of a properly annotated genome on which to base transcriptomic information. The study of seagrass biology, in particular, remains in the nascent stages of linking ecophysiology with genomic responses (Procaccini et al. 2012). However, the explosion in "omics" technology during the last two decades has opened a wealth of opportunities to exploit this information in the development of genome-scale metabolic models that offer a "top-down" perspective on cellular function (Kim et al. 2012). The complete genome of *Zostera marina* L. was recently published, revealing unique insights into the genomic losses and gains involved in adapting to the marine environment (Olsen et al. 2016). Transcriptomic analysis of *Posidonia oceanica* responses to different light environments revealed the upregulation of genes involved in photoacclimation and photoprotection by plants growing in shallow vs. deep water, suggesting shallow-growing plants were experiencing stressful light conditions (Dattolo et al. 2014). Such genome-scale (top-down) approaches are becoming extremely useful for identifying phenomenological correlations that can be mapped onto virtual subcellular mechanisms that can serve as a scaffold for future experimental investigations (Bruggeman and Westerhoff 2007).

In contrast, bottom-up approaches rely on explicitly defined mathematical relationships (e.g., photosynthesis vs. irradiance, Michaelis-Menten enzyme kinetics, Arrhenius equation for temperature, etc.) that can be integrated to predict system behavior (e.g., growth) in response to external drivers (e.g., light, temperature, CO_2, and nutrient availability). As such, they rely heavily on experimentally determined rate constants for accurate parameterization. The bio-optical seagrass model *GrassLight* (Zimmerman 2003; Zimmerman et al. 2015) is an example of such a bottom-up model. It currently possesses significant skill in predicting the light-limited distribution of eelgrass in the Chesapeake region under present-day conditions and is consistent with the observations of dense, productive *Posidonia oceanica* meadows growing in the warm, shallow, and volcanically acidified waters surrounding the Castello Aragonese on the island of Ischia, Bay of Naples, Italy (Hall-Spencer et al. 2008). As with eelgrass, the productivity and geographic distribution of this species is typically inhibited by high summertime water temperature (Zupo et al. 1997; Celebi et al. 2006).

Although the *GrassLight* model is capable of predicting population-level responses to thermal stress and potential impacts of elevated CO_2 on whole-plant energy balance and depth distribution in the coming century using static physiological properties (Zimmerman et al. 2015), it does not presently allow for genetically controlled acclimation responses that introduce temporal variations into these properties. Flux balance analysis offers a potential mathematical approach for bridging this "top-down" vs. "bottom-up" contrast by constraining the flow of metabolites through metabolic networks defined by genomic analyses (Orth et al. 2010). The networks, constrained by steady-state flux stoichiometry between subcellular compartments, can be driven by ecological conditions (e.g., temperature, light, CO_2, and external nutrient availability) to predict non-steady-state rates of growth and chemical composition resulting from the metabolic network (Levering et al. 2016). A conceptual model coupling ecological and transcriptomic functions is illustrated in Fig 8.3. The new model combines

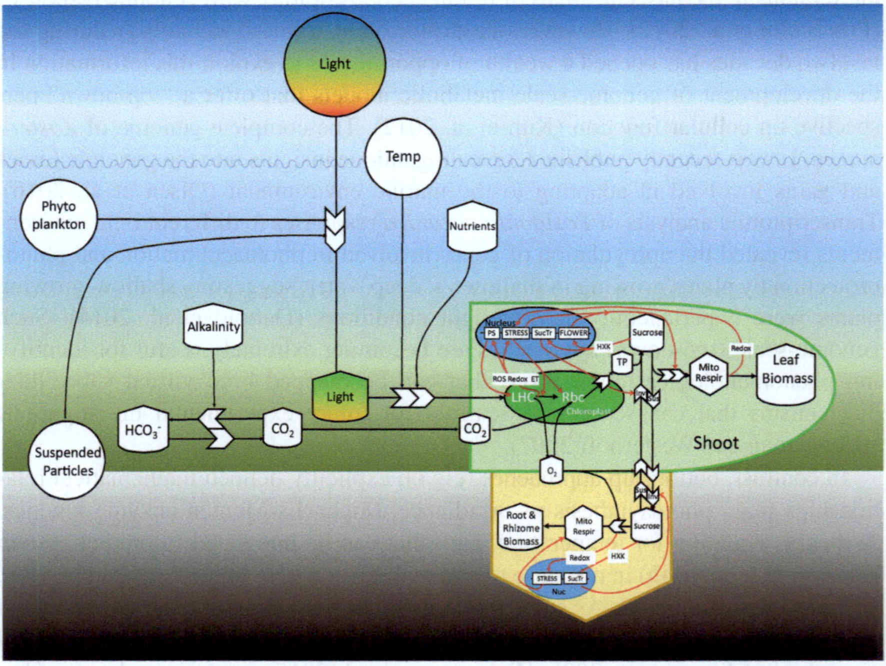

Fig. 8.3 Schematic diagram illustrating the flow of energy and sucrose to support growth and respiration (*black lines*), signal transduction pathways that potentially regulate gene expression and putative expression targets (*red lines*) to be integrated into the model. Effect of light and CO_2 on plant metabolism is mediated primarily by photosynthetic processes in the chloroplasts. Temperature implicitly affects all biochemical processes but is not shown in this diagram, for simplicity. Similarly, stress responses exert broad effects, and feedbacks to all metabolic pathways are implied but not illustrated for simplicity. *Chpl* chloroplast, *LHC* light-harvesting complex, *Rbc* Rubisco, *TP* triose phosphates, *INV* cell wall and cytoplasmic invertases, *SUS* sucrose synthase, *HXK* hexokinases, *Nuc* nucleus, *PS* photosynthesis genes, *STRESS* stress-related genes (HSPs, metallothionein, etc.), *FLOWER* genes responsible for flowering, *SucTr* sucrose transport genes, *ET* electron transport, *ROS* reactive oxygen species

the energy-based metabolic processes (black arrows) already incorporated into *GrassLight* with feedback controls on physiological potential mediated by differential gene expression (red arrows). Genetically controlled responses to environmental forcing are identified by key pathways that control, e.g., the function of chloroplast LHCs, carbon fixation by Rubisco, the allocation of sucrose to growth and respiration, as well as translocation between shoots and root/rhizomes. This integrated approach to environmental forcing through differential transcriptomics that control physiology leading to whole-plant performance will provide a powerful tool for future field investigations of seagrass responses to climate change.

8.3 Seagrasses as Ecosystem Engineers

In addition to representing complex biological systems in their own right, seagrasses interact with a larger ecosystem of sedimentary, surficial, and planktonic microbes, epiphytic algae, several trophic levels of grazers and predators (Duffy 2006), and even pathogens (Lamb et al. 2017). As will be discussed below, seagrasses are more than just another "brick in the wall"—they are true engineers of ecosystems that would not exist in their absence, as evidenced by the disappearance of associated services in the wake of local extirpations (Short and Wyllie-Echeverria 1996; Lamb et al. 2017). This appreciation of seagrass connectivity led to the early adoption of the ecosystems approach to the study of seagrass interactions with their environment and other trophic levels (McRoy and Hellfrich 1977; Phillips and McRoy 1980).

8.3.1 Seagrass-Sediment Interactions

In addition to their importance as a source of organic carbon for higher trophic levels and burial, seagrasses are significant biogeochemical agents in their own right (Fig 8.4). By transporting as much as 6% of the photosynthetically produced O_2 below ground to support aerobic respiration of roots and rhizomes, seagrasses stimulate oxidative remineralization of organic carbon buried in the sediment (Smith et al. 1984; Bodensteiner 2006; Hu and Burdige 2008), potentially limiting their role as blue carbon sinks. However, the corrosive environment generated by the release of respiratory CO_2 stimulates the dissolution of sedimentary carbonates, thereby increasing alkalinity and trapping the remineralized C as HCO_3^- that does not re-equilibrate with the atmosphere upon release from the sediment (Burdige and Zimmerman 2003; Burdige et al. 2008; Burdige et al. 2010). Sulfate reduction in sediments also generates alkalinity, but the low iron content of most carbonate sediments limits pyrite formation, and virtually all of the sulfide is oxidized back to sulfate upon contact with O_2 in the overlying bottom water, consuming alkalinity in the process.

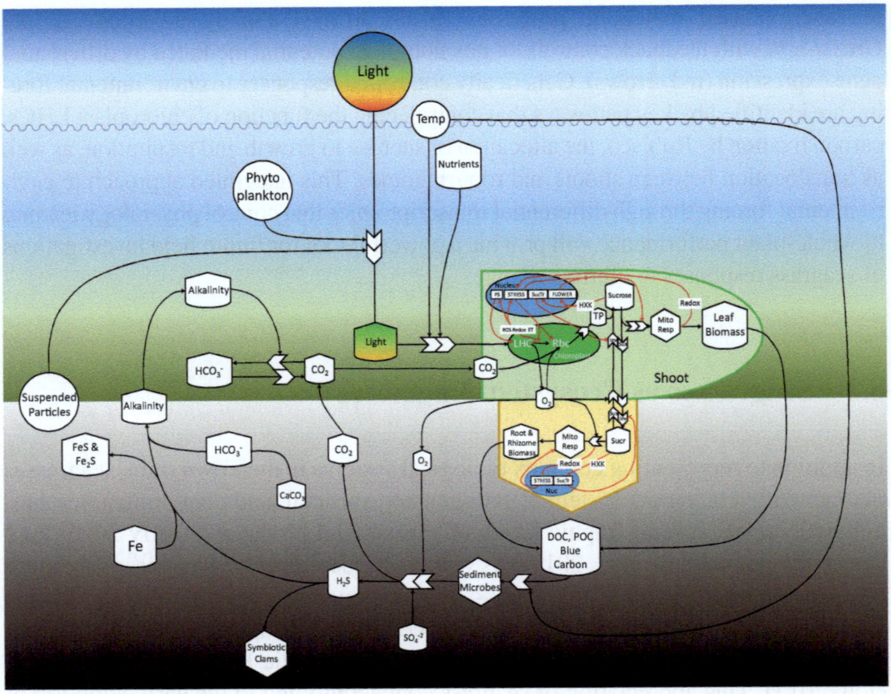

Fig. 8.4 Schematic diagram illustrating seagrass-sediment interactions that lead to the burial and remineralization of organic carbon, the generation of alkalinity, and CO_2 sequestration through carbonate dissolution and sulfide precipitation. Open circles represent donor-controlled processes external to the model. Tank symbols represent depletable resources capable of interacting with other model components. Block arrows represent interactions between components (work gates) that can be either positive or negative, depending on the interaction

Temperate seagrass meadows, in contrast, grow in siliciclastic sediments that generally possess little carbonate for acid titration. However, they are often sufficiently rich in iron that a significant fraction of the microbially generated sulfide is precipitated as pyrite (FeS_2). The *net* burial of these sulfide minerals, which is common in most estuarine and coastal sediments (Goldhaber 2003), will result in a net flux of alkalinity from these sediments to the overlying water (Berner et al. 1970), and this flux may play an important role in the capacity of the surface ocean to absorb and neutralize atmospheric CO_2 (Hu and Cai 2011; Faber et al. 2012; Faber et al. 2014). Sulfide precipitation also protects seagrasses from the toxic effects of high dissolved sulfide concentrations (Pedersen et al. 2004; Borum et al. 2005).

Although the quantitative impact of climate change on these processes is uncertain, enhanced productivity and metabolic activity of seagrasses and carbon burial associated with rising CO_2 and temperature should increase sulfide production and burial in iron-rich sediments and alkalinity flux to the overlying waters. Ongoing work will increase our quantitative understanding of these processes, leading to better estimates of carbon sequestration potential of seagrass systems in the context of a changing climate.

8.3.2 *Interactions with Other Organisms*

Seagrasses interact with a diverse array of epiphytic microorganisms, macroalgae, and metazoans (Fig 8.5). The leaf epiphyte complex consists of (1) all organisms that grow attached to or crawl over the leaf surface, (2) the associated extracellular matrix deposited on the leaves by these organisms, and (3) mineral and organic particles embedded in the organic matrix. This complex provides a significant fraction of the overall productivity of seagrass ecosystems (e.g., Penhale 1977; Mazzella and Alberte 1986; Klumpp et al. 1992), as well as refuge and food for an assemblage of invertebrates and fish (e.g., Orth and Montfrans 1984; van Montrfrans et al. 1984; Neckles et al. 1994; Short et al. 2001). A modest epiphyte layer may benefit seagrasses by preventing damage from ultraviolet radiation (Trocine et al. 1981) or repelling potential leaf grazers (Karez et al. 2000). When epiphytes accumulate to high densities from eutrophication and/or removal of grazers, they may also have negative effects on their seagrass hosts—creating physical barriers to light

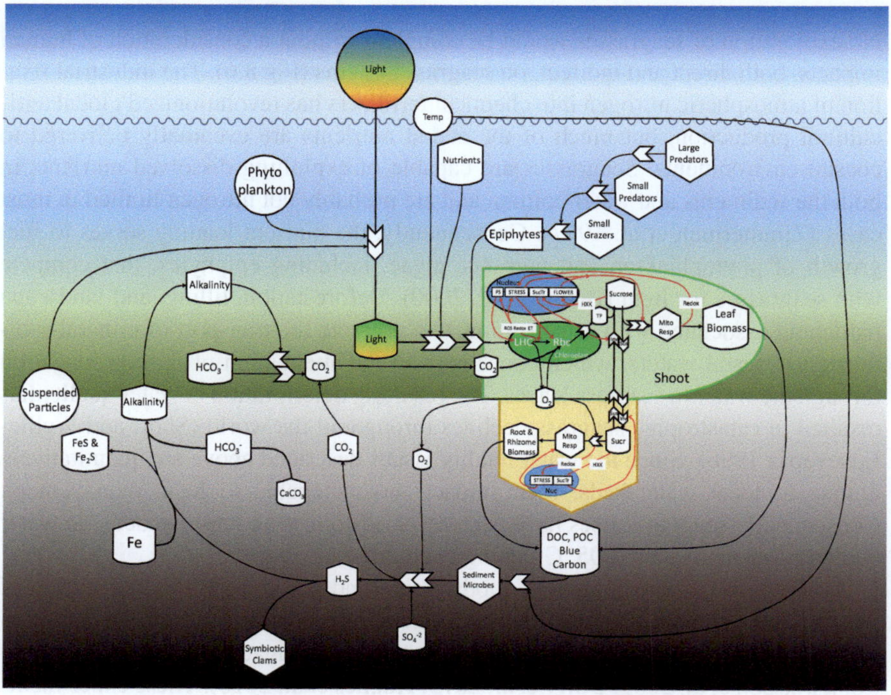

Fig. 8.5 Schematic diagram illustrating interaction between seagrasses, leaf epiphytes, and higher trophic levels that control their abundances. Open circles represent donor-controlled processes external to the model. Tank symbols represent depletable resources capable of interacting with other model components. Block arrows represent interactions between components (work gates) that can be either positive or negative, depending on the interaction

absorption (Bulthuis and Woelkerling 1983) (Losee and Wtezel 1983; Dalla Via et al. 1988; Cebrián et al. 1999; Brush and Nixon 2002), nutrient uptake, gas exchange, or a combination of these factors (Sand-Jensen 1977; van Montrfrans et al. 1984; Sand-Jensen et al. 1985).

Some animals (e.g., sea urchins, turtles, dugongs, manatees, and migratory waterfowl) will graze directly on seagrass leaves. However, most seagrass biomass is probably consumed via detrital pathways involving microbial degradation of drifting wrack and organic carbon buried in the sediment, although there are few quantitative estimates of the magnitude of the detrital flux (Heck et al. 2008). Remineralization of the buried organic carbon can fuel secondary chemosynthesis that supports invertebrate-microbial chemosynthetic symbioses that consume sedimentary sulfide (Burdige 2006), a potential seagrass toxin (Borum et al. 2005; Holmer and Nielsen 2007), and plays an important role in the economically important spiny lobster fishery of the Bahamas (Higgs et al. 2016).

8.3.3 Interactions with Humans

No exploration of seagrasses would be complete without a consideration of human impacts, both direct and indirect, on seagrass systems (Fig 8.6). The industrial fixation of atmospheric nitrogen into chemical fertilizers has revolutionized global agricultural production, but much of the added nutrients are eventually delivered to coastal environments. Seagrasses are capable of exploiting dissolved nutrients in both the sediments and water column and are probably not nitrogen limited in most cases (Zimmerman et al. 1987). Consequently, the nutrient loading serves to fuel growth of phytoplankton and nuisance algae, including epiphytes, that compete with seagrasses for light (Batiuk et al. 2000). Deforestation, tilling, and landscape hardening associated with urbanization also promote the transport of mineral sediments to coastal waters, which contributes to the attenuation of light propagation through the water to seagrass leaves. The resulting deterioration in water quality has resulted in catastrophic seagrass declines throughout the world (Short and Wyllie-Echeverria 1996). Since the water quality issues described above are quantitatively unique to local watersheds, their impacts are amenable to local management. Consequently seagrass losses have been reversed following improvements in water quality (Tomasko et al. 2005; Greening et al. 2011).

Anthropogenic perturbation of biodiversity, and particularly the decimation of top predators, can have comparable and sometimes greater impacts on seagrass survival than eutrophication alone, by releasing epiphytic algae from mesograzer control (Eriksson et al. 2009; Duffy et al. 2015; Hughes et al. 2016). These experiments (and others cited therein) show that biodiversity is a strong predictor of fundamental ecosystem processes controlling seagrass biomass that may be as important as bottom-up processes (e.g., light, nutrients, CO_2, and temperature) in regulating seagrass productivity.

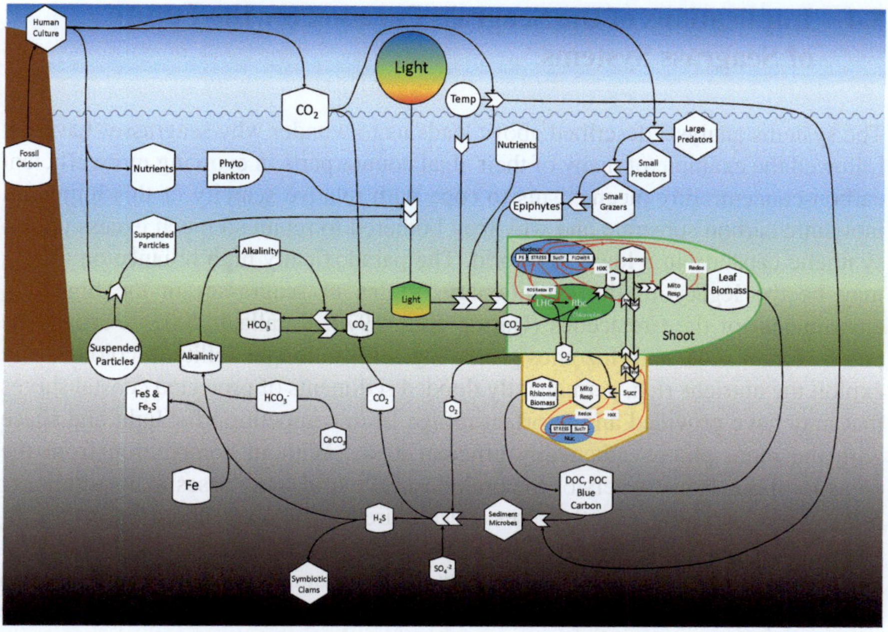

Fig. 8.6 Schematic diagram illustrating the cumulative human impacts, including fossil fuel combustion that generates CO_2, sediment loading, eutrophication, and exploitation of top predators that control leaf epiphytes. Open circles represent donor-controlled processes external to the model. Tank symbols represent depletable resources capable of interacting with other model components. Block arrows represent interactions between components (work gates) that can be either positive or negative, depending on the interaction

In addition to the local processes described above, human activity is altering fundamental biogeochemical processes through the combustion of fossil fuels that increase the concentrations of CO_2 and other greenhouse gases in the atmosphere. Increasing temperature will have direct effects on seagrass respiration and photosynthesis that determine metabolic carbon balance (Evans et al. 1986; Zimmerman et al. 1989), and climate warming is already having an impact on seagrass survival (Short and Neckles 1999; Moore and Jarvis 2008; Moore et al. 2012). However, as described above, increasing availability of CO_2 derived from anthropogenic fossil fuel combustion may stimulate seagrass photosynthesis sufficiently to compensate for at least some of the temperature increase (Palacios and Zimmerman 2007; Jiang et al. 2010; Campbell and Fourqurean 2013; Zimmerman et al. 2017).

The impacts of a warmer climate on seagrass systems will depend on thermal tolerance and temperature optima for photosynthesis, respiration, and growth of individual species. A warmer climate is likely to accelerate microbial activity that controls important processes including the redox state of the sedimentary pore water, blue carbon storage, sulfide production, and carbonate dissolution.

8.4 Implications of the Seagrass Paradox for the Future of Seagrass Systems

The systems analysis described above leads us to wonder why seagrasses have not followed the example of many of their algal counterparts in evolving more efficient carbon-concentrating mechanisms to cope with relative scarcity of this important inorganic carbon substrate and why they bothered to retain so much excess photosynthetic capacity in a low-CO_2 world. The paradoxically high reliance on $CO_{2(aq)}$ may be the legacy of the evolutionary origins of seagrasses in the high-CO_2 environment of the Cretaceous (Hartog et al. 1979; Retallack 2001). Further, the possession of roots and rhizomes gives seagrasses a unique ability to colonize and exploit the nutrient-rich, permanently flooded sediments of protected coastal shores that may have provided an important refuge from competition for light and space with the more photosynthetically efficient algae. Without competition from the algae in oligotrophic waters of the preindustrial world, seagrasses may simply not have experienced the selective pressure required to increase photosynthetic efficiency, or to reallocate resources invested in unnecessary photosynthetic capacity, in these shallow, high light environments. Unfortunately, anthropogenic alteration of the coastal ocean is probably occurring too rapidly to facilitate adaptive change in this group of organisms that has persisted for 100 million years.

Despite their inefficiency in extracting CO_2 for photosynthesis in seawater, seagrasses possess marvelous physiological mechanisms to protect the leaves from photoinhibition in high light environments. Carbon limitation imposes a relatively low threshold for irradiance saturation of photosynthesis that should make seagrasses extremely vulnerable to photoinhibition. Yet, paradoxically, they thrive best in high light environments. Acclimation processes that may protect seagrass leaves involve dynamic xanthophyll cycling capable of modulating non-photochemical quenching on time scales of minutes (Ralph et al. 2002) to the adjustment of light-harvesting pigment concentrations over relatively long time scales involving weeks to months (Cummings and Zimmerman 2003; Celebi 2016). Carbon limitation of photosynthesis combined with high O_2 concentration in the leaf epidermis also appears to promote the dissimilatory photorespiration reaction of Rubisco (Buapet et al. 2013; Celebi 2016). The photorespiratory pathway is often considered a wasteful relic of this enzyme's evolution in a high-CO_2 world because it lowers the efficiency of photosynthesis by oxidizing carbon substrates from the Calvin cycle (Xue et al. 2010), reducing net photosynthesis by 25% or more. In doing so, however, photorespiration may provide a CO_2-responsive "clutch" that facilitates the turnover of Rubisco and maintains the flow of electrons generated by photosystem II, thereby preventing photoinhibition in even the highest light environments.

It is ironic to consider that increasing the atmospheric concentration of CO_2, perhaps the most significant global impact of anthropogenic climate change, may offset some of the insults visited on seagrass populations by anthropogenic eutrophication and climate warming. Our ability to develop a predictive understanding of the future trajectory of seagrass-based ecosystems to climate change will require a greater mechanistic understanding of the physiological and genomic responses of these remarkable submerged vascular plants to the simultaneous impacts of ocean acidification, climate warming, and eutrophication that are altering ecosystem function across the globe. In the case of seagrass systems, CO_2 appears to be a master variable that can exert significant impacts on light requirements, temperature tolerance, and even reproductive success. However, even among those species exhibiting initial positive responses to increased [CO_2], long-term effects can be difficult to predict (Arp 1991; Woodward 2002). In terrestrial systems, higher rates of carbon fixation can increase nutrient demand and/or dilute the nutritional quality of plant biomass, producing a cascade of effects on rates of grazing and decomposition (Field and Mooney 1986; Cotrufo et al. 2002). Further, ocean acidification resulting from increased atmospheric CO_2 may produce significant impacts on organisms, particularly calcareous species that presently serve to keep leaf epiphytes and algal competitors in check (Hall-Spencer et al. 2008).

Clearly, seagrasses possess a number of characteristics that are consistent with the paradigm of adaptation, yet CO_2 limitation of photosynthesis demonstrates that they are less than perfectly adapted to a submerged aquatic existence. Like the panda's thumb (Gould 1978), these less-than-perfect features may help reveal the constraint of evolutionary history on the optimization of systems-level performance in this remarkable group of aquatic angiosperms that we are just beginning to understand.

Acknowledgments Thanks to Manoj Kumar and Peter Ralph for inviting this contribution to this book. Financial support was provided by the US National Science Foundation (Awards OCE-1061823 and OCE-1635403).

References

Ackerman J (2006) Sexual reproduction of seagrasses: pollination in the marine context. In: Larkum A, Orth R, Duarte C (eds) Seagrasses: biology, ecology and conservation. Springer, Dordrecht

Al-Moghrabi S, Goiran C, Allemand D, Speziale N, Jaubert J (1996) Inorganic carbon uptake for photosynthesis by the symbiotic coral-dinoflagellate association II. Mechanism for bicarbonate uptake. J Exp Mar Biol Ecol 199:227–248

Arp W (1991) Effects of source-sink relations on photosynthetic acclimation to elevated CO_2. Plant Cell Environ 14:869–876

Backhausen J, Scheibe R (1999) Adaptation of tobacco plants to elevated CO_2: influence of leaf age on changes in physiology, redox states and NADP-malate dehydrogenase activity. J Exp Bot 50:665–675

Bailey-Serres J, Voesenek LACJ (2008) Flooding stress: acclimations and genetic diversity. Annu Rev Plant Biol 59:313–339

Batiuk R, Bergstrom P, Kemp M, Koch E, Murray L, Stevenson J, Bartleson R, Carter V, Rybicki N, Landwehr J, Gallegos C, Karrh L, Naylor M, Wilcox D, Moore K, Ailstock S, Teichberg M (2000) Chesapeake Bay submerged aquatic vegetation water quality and habitat-based requirements and restoration targets: a second synthesis. Chesapeake Bay Program Office, Annapolis, MD

Beer S, Koch E (1996) Photosynthesis of marine macroalgae and seagrasses in globally changing CO_2 environments. Mar Ecol Prog Ser 141:199–204
Beer S, Rehnberg J (1997) The acquisition of inorganic carbon by the seagrass *Zostera marina*. Aquat Bot 56:277–283
Berner R, Scott M, Thomlinson C (1970) Carbonate alkalinity in the pore waters of anoxic marine sediments. Limnol Oceanogr 15:544–549
Björk M, Uku J, Weil A, Beer S (1999) Photosynthetic tolerances to desiccation of tropical intertidal seagrasses. Mar Ecol Prog Ser 191:121–126
Bodensteiner L (2006) The impact of light availability on benthic oxygen release by the seagrasses *Thalassia testudinum* (Banks ex König) and *Zostera marina* (L.). MS thesis, San Jose State University, San Jose
Borum J, Pedersen O, Greve TM, Frankovich TA, Zieman JC, Fourqurean JW, Madden CJ (2005) The potential role of plant oxygen and sulphide dynamics in die-off events of the tropical seagrass, *Thalassia testudinum*. J Ecol 93:148–158
Bruggeman FJ, Westerhoff HV (2007) The nature of systems biology. Trends Microbiol 15:45–50
Brush M, Nixon S (2002) Direct measurements of light attenuation by epiphytes on eelgrass *Zostera marina*. Mar Ecol Prog Ser 238:73–79
Buapet P, Rasmusson LM, Gullström M, Björk M (2013) Photorespiration and carbon limitation determine productivity in temperate seagrasses. PLoS One 8:e83804
Bulthuis D, Woelkerling W (1983) Biomass accumulation and shading effects of epiphytes on leaves of the seagrass, *Heterozostera tasmanica*, in Victoria, Australia. Aquat Bot 16:137–148
Burdige D, Hu X, Zimmerman R (2010) The widespread occurrence of coupled carbonate dissolution/reprecipitation in surface sediments on the Bahamas Bank. Am J Sci I310:492–521
Burdige D, Zimmerman R (2003) Impact of seagrass density on carbonate dissolution in Bahamian sediments. Limnol Oceanogr 47:1751–1763
Burdige D, Zimmerman R, Hu X (2008) Rates of carbonate dissolution in permeable sediments estimated from pore-water profiles: the role of sea grasses. Limnol Oceanogr 53:549–565
Burdige DJ (2006) Geochemistry of marine sediments. Princeton University Press, Princeton, NJ
Campbell J, Fourqurean J (2013) Effects of in situ CO_2 enrichment on the structural and chemical characteristics of the seagrass *Thalassia testudinum*. Mar Biol 160:1465–1475
Carr J, D'Odorico P, McGlathery K, Wiberg P (2010) Stability and bistability of seagrass ecosystems in shallow coastal lagoons: role of feedbacks with sediment resuspension and light attenuation. J Geophys Res 115:G03011
Cebrián J, Enríquez S, Fortes M, Agawin N, Vermaat J, Duarte C (1999) Epiphyte accrual on *Posidonia oceanica* (L.) Delile leaves: implications for light absorption. Bot Mar 42:123–128
Celebi B (2016) Potential impacts of climate change on photochemistry of *Zostera marina* L. PhD, Old Dominion University, Norfolk
Celebi B, Cemal Gucu A, Ok M, Sakinan S, Alkoglu E (2006) Hydrographic indications to understand the absence of *Posidonia oceanica* in the Levant Sea (Eastern Mediterranean). In: Gambi M, Borg J, Buia M, Di Carlo G, Pergent-Martini C, Pergent G, Procaccini G (eds) Proceedings of the mediterranean seagrass workshop. Biologia Marina Mediterranea
Cotrufo M, Ineson P, Scott Y (2002) Elevated CO_2 reduces the nitrogen concentration of plant tissues. Glob Chang Biol 4:43–54
Cummings M, Zimmerman R (2003) Light harvesting and the package effect in *Thalassia testudinum* Koenig and *Zostera marina* L.: optical constraints on photoacclimation. Aquat Bot 75:261–274
Dalla Via J, Sturmbauer C, Schönweger C, Sötz E, Mathekowitsch M, Stifter M, Reiger R (1988) Light gradients and meadow structure in *Posidonia oceanica*: ecomorphological and functional correlates. Mar Ecol Prog Ser 163:267–278
Dattolo E, Ruocco M, Brunet C, Lorenti M, Lauritano C, D'Esposito D, De Luca P, Sanges R, Mazzuca S, Procaccini G (2014) Response of the seagrass *Posidonia oceanica* to different light environments: insights from a combined molecular and photo-physiological study. Mar Environ Res 101:225–236

den Hartog C, Kuo J (2006) Taxonomy and biogeography of seagrasses. In: Larkum A, Orth R, Duarte C (eds) Seagrasses: biology, ecology and conservation. Springer, Dordrecht

Duarte C (1991) Seagrass depth limits. Aquat Bot 40:363–377

Duffy J (2006) Biodiversity and the functioning of seagrass ecosystems. Mar Ecol Prog Ser 311:233–250

Duffy JE, Reynolds PL, Boström C, Coyer JA, Cusson M, Donadi S, Douglass JG, Eklöf JS, Engelen AH, Eriksson BK, Fredriksen S, Gamfeldt L, Gustafsson C, Hoarau G, Hori M, Hovel K, Iken K, Lefcheck JS, Moksnes P-O, Nakaoka M, O'Connor MI, Olsen JL, Richardson JP, Ruesink JL, Sotka EE, Thormar J, Whalen MA, Stachowicz JJ (2015) Biodiversity mediates top–down control in eelgrass ecosystems: a global comparative-experimental approach. Ecol Lett 18(7):696–705

Durako M (1993) Photosynthetic utilization of $CO_{2(aq)}$ and HCO_3^- in *Thalassia testudinum* (Hydrocharitacae). Mar Biol 115:373–380

Eriksson BK, Ljunggren L, Sandström A, Johansson G, Mattila J, Rubach A, Råberg S, Snickars M (2009) Declines in predatory fish promote bloom-forming macroalgae. Ecol Appl 19:1975–1988

Evans AS, Webb KL, Penhale PA (1986) Photosynthetic temperature acclimation in two coexisting seagrasses, *Zostera marina* L. and *Ruppia maritima* L. Aquat Bot 24:185–197

Faber P, Evrard V, Woodland R, Cartwright I, Cook P (2014) Pore-water exchange driven by tidal pumping causes alkalinity export in two intertidal inlets. Limnol Oceanogr 59:1749–1763

Faber P, Kessler A, Bull J, McKelvie I, Meysman F, Cook P (2012) The role of alkalinity generation in controlling the fluxes of CO_2 during exposure and inundation on tidal flats. Biogeosciences 9:4087–4097

Falkowski P, Raven J (2007) Aquatic photosynthesis. Princeton, Princeton, NJ

Field C, Mooney H (1986) The photosynthesis-nitrogen relationship in wild plants. In: Givnish T (ed) On the economy of plant form and function. Cambridge University Press, Cambridge

Goldhaber M (2003) Sulfur-rich sediments. In: Mackenzie F (ed) Treatise on geochemistry. Elsevier, Amsterdam

Gould S (1978) The panda's peculiar thumb. Nat Hist 87:20–30

Greening HS, Cross L, Sherwood E (2011) A multiscale approach to seagrass recovery in Tampa Bay, Florida. Ecol Restor 29:82–93

Hall-Spencer J, Rodolfo-Metalpa R, Martin S, Ransome E, Fine M, Turner S, Rowley S, Tedesco D, Buia M-C (2008) Volcanic carbon dioxide vents show ecosystem effects of ocean acidification. Nature 454:96–99

Hanke G, Holtgrefe S, König N, Strodtkötter I, Voss I, Scheibe R (2009) Use of transgenic plants to uncover strategies for maintenance of redox homeostasis during photosynthesis. In: Jean-Pierre J (ed) Advances in botanical research, vol 52. Elsevier, Amsterdam, pp 207–251

Hartog CD, Loenhoud PJV, Roelofs JGM, Van de Sande JCPM (1979) Chromosome number from three seagrasses from the Netherlands Antilles. Aquat Bot 7:267–271

Heck KL, Carmthers TJB, Duarte CM, Hughes AR, Kendrick G, Orth RJ, Williams SW (2008) Trophic transfers from seagrass meadows subsidize diverse marine and terrestrial consumers. Ecosystems 11:1198–1210

Hemminga M, Mateo M (1996) Stable carbon isotopes in seagrasses: variability in ratios and use in ecological studies. Mar Ecol Prog Ser 140:285–298

Higgs ND, Newton J, Attrill MJ (2016) Caribbean spiny lobster fishery is underpinned by trophic subsidies from chemosynthetic primary production. Curr Biol 26(24):3393–3398

Holmer M, Nielsen R (2007) Effects of filamentous algal mats on sulfide invasion in eelgrass (*Zostera marina*). J Exp Mar Biol Ecol 353:245–252

Hu X, Burdige D (2008) Shallow marine carbonate dissolution and early diagenesis: implications from an incubation study. J Mar Res 66:489–527

Hu X, Cai W (2011) An assessment of ocean margin anaerobic processes on oceanic alkalinity budget. Global Biogeochem Cycles 25(3):GB3003

Hughes BB, Hammerstrom KK, Grant NE, Hoshijima U, Eby R, Wasson K (2016) Trophic cascades on the edge: fostering seagrass resilience via a novel pathway. Oecologia 182:231–241

Hüner N, Dahal K, Hollis L, Bode R, Rosso D, Krol M, Ivanov A (2012) Chloroplast redox imbalance governs phenotypic plasticity: the "grand design of photosynthesis" revisited. Front Plant Sci 3:1–12

Invers O, Zimmerman R, Alberte R, Perez M, Romero J (2001) Inorganic carbon sources for seagrass photosynthesis: an experimental evaluation for bicarbonate use in temperate species. J Exp Mar Biol Ecol 265:203–217

Jackson M, Ishizawa K, Ito O (2009) Evolution and mechanisms of plant tolerance to flooding stress. Ann Bot 103:137–142

Jiang Z, Huang X, Zhang J (2010) Effects of CO_2 enrichment on photosynthesis, growth and biochemical composition of seagrass *Thalassia hemprichii* (Ehrenb.) Aschers. J Integr Plant Biol 52:904–913

Karez R, Englebert S, Sommer U (2000) 'Co-consumption' and 'protective coating': two new proposed effects of epiphytes on their macroalgal hosts in mesograzer-epiphyte-host interactions. Mar Ecol Prog Ser 205:85–93

Kim TY, Sohn SB, Kim YB, Kim WJ, Lee SY (2012) Recent advances in reconstruction and applications of genome-scale metabolic models. Curr Opin Biotechnol 23:617–623

Klumpp D, Salita-Espinosa J, Fortes M (1992) The role of epiphytic periphyton and macroinvertebrate grazers in the trophic flux of a tropical seagrass community. Aquat Bot 43:327–349

Koch M, Bowes G, Ross C, Zhang X-H (2013) Climate change and ocean acidification effects on seagrasses and marine macroalgae. Glob Chang Biol 19:103–132

Kuo J, den Hartog C (2006) Seagrass morphology, anatomy and ultrastructure. In: Larkum A, Orth R, Duarte C (eds) Seagrasses: biology, ecology and conservation. Springer, Dordrecht

Lamb JB, van de Water JAJM, Bourne DG, Altier C, Hein MY, Fiorenza EA, Abu N, Jompa J, Harvell CD (2017) Seagrass ecosystems reduce exposure to bacterial pathogens of humans, fishes, and invertebrates. Science 355:731–733

Larkum A, Orth R, Duarte C (eds) (2006) Seagrasses: biology, ecology and conservation. Springer, Dordrecht

Lawson S, Wiberg P, McGlathery K, Fugate D (2007) Wind-driven sediment suspension controls light availability in a shallow coastal lagoon. Estuar Coasts 30:101–112

Les D, Cleland M, Waycott M (1997) Phylogenetic studies in Alismatidae, II: evolution of marine angiosperms (seagrasses) and hydrophily. Syst Bot 22:443–463

Leuschner C, Landwehr S, Mehlig U (1998) Limitation of carbon assimilation of intertidal *Zostera noltii* and *Z. marina* by desiccation at low tide. Aquat Bot 62:171–176

Levering J, Broddrick J, Dupont C, Peers G, Beeri K, Mayers J, Gallina A, Allen A, Palsson B, Zengler K (2016) Genome-scale model reveals metabolic basis of biomass partitioning in a marine diatom. PLoS One 11:e0155038

Losee R, Wetzel R (1983) Selective light attenuation by the periphyton complex. In: Wetzel R (ed) Periphyton of freshwater ecosystems. Dr. W. Junk, Dordrecht

Mazzella L, Alberte RS (1986) Light adaptation and the role of autotrophic epiphytes in primary production of the temperate seagrass, *Zostera marina* L. J Exp Mar Biol Ecol 100:165–180

McLeod E, Chmura G, Bouillon S, Salm R, Björk M, Duarte C, Loevlock C, Schlesinger W, Silliman B (2011) A blueprint for blue carbon: toward an improved understanding of the role of vegetated coastal habitats in sequestering CO_2. Front Ecol Environ 9:552–560

McPherson M, Hill V, Zimmerman R, Dierssen H (2011) The optical properties of Greater Florida Bay: implications for seagrass abundance. Estuar Coasts 34:1150–1160

McPherson ML, Zimmerman RC, Hill VJ (2015) Predicting carbon isotope discrimination in eelgrass (*Zostera marina* L.) from the environmental parameters—light, flow, and [DIC]. Limnol Oceanogr 60:1875–1889

McRoy C, Hellfrich C (eds) (1977) Seagrass ecosystems: a scientific perspective. Marcel Dekker, New York

Moore K, Jarvis J (2008) Environmental factors affecting summertime eelgrass diebacks in the lower Chesapeake Bay: implications for long-term persistence. J Coast Res 55:135–147

Moore K, Shields E, Parrish D, Orth R (2012) Eelgrass survival in two contrasting systems: role of turbidity and summer water temperatures. Mar Ecol Prog Ser 448:247–258

Neckles H, Koepfler E, Haas L, Wetzel R, Orth R (1994) Dynamics of epiphytic photoautotrophs and heterotrophs in *Zostera marina* (eelgrass) microcosms: responses to nutrient enrichment and grazing. Estuaries 17:597–605

Odum H (1983) Systems ecology. An introduction. Wiley, New York

Olsen JL, Rouzé P, Verhelst B, Lin Y-C, Bayer T, Collen J, Dattolo E, De Paoli E, Dittami S, Maumus F, Michel G, Kersting A, Lauritano C, Lohaus R, Töpel M, Tonon T, Vanneste K, Amirebrahimi M, Brakel J, Boström C, Chovatia M, Grimwood J, Jenkins JW, Jueterbock A, Mraz A, Stam WT, Tice H, Bornberg-Bauer E, Green PJ, Pearson GA, Procaccini G, Duarte CM, Schmutz J, Reusch TBH, Van de Peer Y (2016) The genome of the seagrass *Zostera marina* reveals angiosperm adaptation to the sea. Nature 530:331–335

Orth JD, Thiele I, Palsson BO (2010) What is flux balance analysis? Nat Biotechnol 28:245–248

Orth R, Carruthers T, Dennison W, Duarte C, Fourqurean J, Heck K, Hughes A, Kendrick G, Kenworthy W, Olyarnik S, Short F, Waycott M, Williams S (2006) A global crisis for seagrass ecosystems. Bioscience 56:987–996

Orth R, Moore K (1983) Chesapeake Bay: an unprecedented decline in submerged aquatic vegetation. Science 222:51–53

Orth RJ, Montfrans JV (1984) Epiphyte-seagrass relationships with an emphasis on the role of micrograzing: a review. Aquat Bot 18:43–69

Palacios S, Zimmerman R (2007) Response of eelgrass (*Zostera marina* L.) to CO_2 enrichment: possible impacts of climate change and potential for remediation of coastal habitats. Mar Ecol Prog Ser 344:1–13

Pedersen O, Binzer T, Borum J (2004) Sulphide intrusion in eelgrass (*Zostera marina* L.) Plant Cell Environ 27:595–602

Penhale PA (1977) Macrophyte-epiphyte biomass and productivity in an eelgrass (*Zostera marina* L.) community. J Exp Mar Biol Ecol 26:211–224

Pfannschmidt T (2003) Chloroplast redox signals: how photosynthesis controls its own genes. Trends Plant Sci 8:33–41

Pfannschmidt T, Yang C (2012) The hidden function of photosynthesis: a sensing system for environmental conditions that regulates plant acclimation responses. Protoplasma 249:125–135

Phillips R, McRoy C (eds) (1980) Handbook of seagrass biology: an ecosystem perspective. Garland STPM Press, New York

Pregnall AM, Smith RD, Kursar TA, Alberte RS (1984) Metabolic adaptation of *Zostera marina* (eelgrass) to diurnal periods of root anoxia. Mar Biol 83:141–147

Procaccini G, Beer S, Björk M, Olsen J, Mazzuca S, Santos R (2012) Seagrass ecophysiology meets ecological genomics: are we ready? Mar Ecol 33:522–527

Ralph P, Polk S, Moore K, Orth R, Smith W Jr (2002) Operation of the xanthophyll cycle in the seagrass *Zostera marina* in response to variable irradiance. J Exp Mar Biol Ecol 271:189–207

Raven J (2010) Inorganic carbon acquisition by eukaryotic algae: four current questions. Photosynth Res 106:123–134

Raven JA, Beardall J (2014) CO_2 concentrating mechanisms and environmental change. Aquat Bot 118:24–37

Retallack G (2001) A 300-million year record of atmospheric carbon dioxide from fossil plant cuticles. Nature 411:247–248

Sand-Jensen K (1977) Effect of epiphytes on eelgrass photosynthesis. Aquat Bot 3:55–63

Sand-Jensen K, Revsbech N, Jørgensen B (1985) Microprofiles of oxygen in epiphyte communities on submerged macrophytes. Mar Biol 89:55–62

Short F, Matso K, Hoven M, Whitten J, Burdick D, Short C (2001) Lobster use of eelgrass habitat in the Piscataqua River on the New Hampshire/Maine border, USA. Estuaries 24:277–284

Short F, Polidoro B, Livingstone S, Carpenter K, Bandiera S, Bujang J, Calumpong H, Carruthers T, Coles R, Dennison W, Erftmeijer P, Fortes M, Freeman A, Jagtap T, Kamal A, Kendrick G, Kenworthy J, La Nafie Y, Nasution I, Orth R, Prathep A, Sanciango J, van Tussebroek B, Vergara S, Waycott M, Zieman J (2011) Extinction risk assessment of the world's seagrass species. Biol Conserv 144(7):1961–1971

Short F, Wyllie-Echeverria S (1996) Natural and human-induced disturbance of seagrasses. Environ Conserv 23:17–27

Short FT, Neckles HA (1999) The effects of global climate change on seagrasses. Aquat Bot 63:169–196

Smith RD, Dennison WC, Alberte RS (1984) Role of seagrass photosynthesis in root aerobic processes. Plant Physiol 74:1055–1058

Smith RD, Pregnall AM, Alberte RS (1988) Effects of anaerobiosis on root metabolism of the seagrass *Zostera marina* L. (eelgrass). Mar Biol 98:131–141

Tomasko DA, Corbett CA, Greening HS, Raulerson GE (2005) Spatial and temporal variation in seagrass coverage in Southwest Florida: assessing the relative effects of anthropogenic nutrient load reductions and rainfall in four contiguous estuaries. Mar Pollut Bull 50:797–805

Trocine R, Rice J, Wells G (1981) Inhibition of seagrass photosynthesis by ultraviolet-B radiation. Plant Physiol 68:74–81

van Montrfrans J, Wetzel R, Orth R (1984) Epiphyte-grazer relationships in seagrass meadows: consequences for seagrass growth and production. Estuaries 7:289–309

van Tussenbroek BI, Villamil N, Márquez-Guzmán J, Wong R, Monroy-Velázquez LV, Solis-Weiss V (2016) Experimental evidence of pollination in marine flowers by invertebrate fauna. Nat Commun 7:12980

Voesenek LACJ, Bailey-Serres J (2015) Flood adaptive traits and processes: an overview. New Phytol 206:57–73

Woodward F (2002) Potential impacts of global elevated CO_2 concentrations on plants. Curr Opin Plant Biol 5:207–211

Xue X-G, Long S, Ort D (2010) Improving photosynthetic efficiency for greater yield. Annu Rev Mar Sci 61:235–261

Zimmerman R (2003) A biooptical model of irradiance distribution and photosynthesis in seagrass canopies. Limnol Oceanogr 48:568–585

Zimmerman R (2006) Light and photosynthesis in seagrass meadows. In: Larkum A, Orth R, Duarte C (eds) Seagrasses: biology, ecology and conservation. Springer, Dordrecht

Zimmerman R, Alberte R (1996) Effect of light/dark transition on carbon translocation in eelgrass *Zostera marina* seedlings. Mar Ecol Prog Ser 136:305–309

Zimmerman R, Cabello-Pasini A, Alberte R (1994) Modeling daily production of aquatic macrophytes from irradiance measurements: a comparative analysis. Mar Ecol Prog Ser 114:185–196

Zimmerman R, Caffrey J (2002) Chapter 8. Primary producers. In: Caffrey J, Brown M, Tyler W, Silberstein M (eds) Changes in a California estuary: a profile of elkhorn slough. Elkhorn Slough Foundation, Moss Landing, CA

Zimmerman R, Hill V, Celebi B, Jinuntuya M, Ruble D, Smith M, Cedeno T, Swingle W (2017) Experimental impacts of climate warming and ocean carbonation on eelgrass (*Zostera marina* L.) Mar Ecol Prog Ser 566:1–15

Zimmerman R, Hill V, Gallegos C (2015) Predicting effects of ocean warming, acidification and water quality on Chesapeake region eelgrass. Limnol Oceanogr 60:1781–1804

Zimmerman R, Kohrs D, Alberte R (1996) Top-down impact through a bottom-up mechanism: the effect of limpet grazing on growth, productivity and carbon allocation of *Zostera marina*. Oecologia 107:560–567

Zimmerman R, Kohrs D, Steller D, Alberte R (1997) Impacts of CO_2-enrichment on productivity and light requirements of eelgrass. Plant Physiol 115:599–607

Zimmerman R, Kohrs D, Steller DL, Alberte R (1995) Carbon partitioning in eelgrass. Regulation by photosynthesis and the response to daily light-dark cycles. Plant Physiol 108:1665–1671

Zimmerman R, Smith R, Alberte R (1987) Is growth of eelgrass nitrogen limited? A numerical simulation of the effects of light and nitrogen on the growth dynamics of *Zostera marina*. Mar Ecol Prog Ser 41:167–176

Zimmerman RC, Smith RD, Alberte RS (1989) Thermal acclimation and whole plant carbon balance in *Zostera marina* L. (eelgrass). J Exp Mar Biol Ecol 130:93–109

Zupo V, Buia M, Mazzella L (1997) A production model for *Posidonia oceanica* based on temperature. Estuar Coast Shelf Sci 44:483–492

Chapter 9
Gas and Liquid Chromatography-Mass Spectrometry-Based Metabolic Profiling of Marine Angiosperm *Zostera muelleri* (Alismatales, Zosteraceae)

Unnikrishnan Kuzhiumparambil, Manoj Kumar, and Peter Ralph

Abstract Seagrasses are monocotyledonous marine flowering plants that are considered lungs of the sea and are the most intense carbon sinks on the planet, delivering a range of ecologically and economically valuable biological services. In this study, we report the chemical fingerprint of *Zostera muelleri* using an untargeted metabolomic approach. High-performance liquid chromatography-mass spectrometry (HPLC-MS) and gas chromatography-mass spectrometry (GC-MS) were performed to study the metabolic profile of *Z. muelleri.* A total of 98 metabolites belonging to various chemical classes including flavonoids, phenolics, lipids, fatty acids, sugar alcohols and amino acids were identified, including two characteristic marker compounds of the genus, zosteric acid and rosmarinic acid. Chromatographic profiling yield a comprehensive map for the chemical constituents of *Z. muelleri*, and this method can be used as an effective and convenient approach to gain insights into the chemical composition of other seagrasses.

Keywords Seagrass • *Zostera muelleri* • Metabolomics • GC-MS • LC-MS • Primary and secondary metabolites

Contents

9.1 Introduction 190
9.2 Material and Methods 191
9.2.1 Chemicals 191
9.2.2 Extraction and Sample Preparation 191
9.2.3 Instrumentation 192
9.2.4 Chromatographic Peak Identification 192

Unnikrishnan Kuzhiumparambil and Manoj Kumar contributed equally to this work.

U. Kuzhiumparambil (✉) • M. Kumar • P. Ralph
Climate Change Cluster, Faculty of Science, University of Technology Sydney, Sydney, NSW, Australia
e-mail: Unnikrishnan.Kuzhiumparambil@uts.edu.au

© Springer International Publishing AG 2017

M. Kumar, P. Ralph (eds.), *Systems Biology of Marine Ecosystems*,
DOI 10.1007/978-3-319-62094-7_9

9.3 Results and Discussion 193
9.3.1 Identification of Rosmarinic Acid and Zosteric Acid 193
9.3.2 Profile of Z. muelleri 193
9.4 Conclusion 201
References 202

9.1 Introduction

Seagrasses are marine angiosperms that colonize coastal and estuarine sediments and play an important role in coastal food webs, carbon sequestration, sediment stabilization and nutrient cycling (Orth et al. 1976). Seagrasses are amongst the most productive aquatic autotrophic organisms on a global scale (Duarte and Chiscano 1999). *Zostera muelleri*, commonly known as eel grass, is a marine flowering plant that forms extensive meadows in sheltered soft-bottom habitats along North Atlantic and North Pacific temperate shores (Engle and Miller 2003).

Very few chemical and biological studies have been reported on *Z. muelleri* so far; however *Z. marina*, a related species, has been well studied on chemical and biological aspects. Pharmacological effects of *Z. marina* including antibiotic, antioxidant, nematicidal antimicrofoulant, antibacterial and anti-*Labyrinthula* properties have been reported (Kim et al. 2004; Custódio et al. 2016; Zidorn 2016). An in vitro cytotoxic activity of *Z. marina* to selectively reduce the viability of tumorous neuronal cells has also been recently demonstrated (Custódio et al. 2016). The most important constituent of *Zostera* spp. is zosteric acid, which is the most intensely studied natural product from seagrasses, to which the antimicrofoulant activity is attributed. Experimentally prepared silicone-based coatings containing zosteric acid have proven to delay microfouling. Nematicidal and antibacterial effects are attributed to rosmarinic acid (Achamlale et al. 2009). Other constituents reported in *Zostera* spp. include apigenin-7-O-β-D-glucoside, chrysoeriol, gallic acid, gentisic acid, p-hydroxybenzoic acid, vanillic acid, coumaric acid, caffeic acid and ferulic acid (Kim et al. 2004, Zidorn 2016, Grignon-Dubois and Rezzonico 2012, Quackenbush et al. 1986, Kawasaki et al. 1998, De Leeuw et al. 1995). Even though the presences of few other sulphated molecules have been reported, their molecular structure is still unknown (Zidorn 2016).

Big ambiguity remains on the chemical diversity amongst seagrass taxa. Due to the exposure of seagrasses to a number of unique environmental challenges, as well as submersed lifestyle, a high diversity of defence chemicals are expected. A comprehensive investigation on the composition of *Z. muelleri* with the aid of new analytical techniques is lacking. Lack of metabolite information makes it difficult to understand the pathways leading to the synthesis of secondary metabolites, their enzymes and regulation by external stimuli or their roles in plant defence mechanism (Farag et al. 2012).

Chromatography coupled with mass spectrometry has become an indispensable tool to elucidate the structure information without isolation and purification. Metabolic profiling using these techniques can resolve individual chemical component into separate peaks, enhancing the opportunity to uncover novel or minor metabolites and can provide advanced, simple and real-time means of metabolome analysis (Vaclavik et al. 2011). The objective of this study was to identify the metabolites in *Z. muelleri* using LC-Q-TOF (quadrupole time of flight)-MS and GC-MS.

9.2 Material and Methods

9.2.1 Chemicals

Methanol and acetonitrile (LC-MS grade) were obtained from Chem-Supply (Gilman, SA, Australia) and Merck (French Forest, NSW, Australia). Milli-Q grade water was used for all LC analysis. Formic acid, 2-amino anthracene and pentafluorobenzoic acid were obtained from Sigma Aldrich (Castle Hill, NSW, Australia).

9.2.2 Extraction and Sample Preparation

9.2.2.1 Gas Chromatography-Mass Spectrometry (GC-MS)

Metabolites were extracted using 500 μL extraction solution per 30 mg of lyophilized leaf powder (90% methanol with three internal standards, 100 μM labelled valine, 10 μM labelled sorbitol, 10 μM ribitol). Samples were incubated 20 min at 65 °C and 1250 rpm incubation in a thermomixer and then centrifuged for 3 min at 20,800xg at 65 °C. An aliquot of 60 μL of supernatant was transferred into a glass insert and evaporated to dryness in a vacuum evaporator for 3 h at ambient temperature. Glass inserts were transferred to a 2 mL GC vial and capped with magnetic crimp top.

Samples were derivatized using a CTC autosampler: 20 μL of 20 mg/mL methoxylamine in pyridine were added. Samples were incubated at 37 °C for 2 h while agitating at 750 rpm. Later, 20 μL N-methyl-N-(trimethylsilyl)trifluoroacetamide (MSTFA) was added and incubated at 37 °C for 30 min, while agitating at 750 rpm. Later, an aliquot of 1 μL was injected onto the GC-MS.

9.2.2.2 Liquid Chromatography-Mass Spectrometry (LC-MS)

Finely powdered freeze-dried material (20 mg) was weighed into a 1.5 mL Eppendorf tube. 400 μL of 100% MeOH solution containing internal standards was added and samples were vortexed for 3 min followed by incubation for 30 min at 25 °C. The samples were spun down in a microcentrifuge at 13,000 rpm for 10 min. Supernatant was collected and remaining pellet was re-extracted twice using 100% MeOH (400 μL) and 200 μL H_2O (Milli-Q), respectively, using the protocol as above. Combined extracts were stored at −80 °C until further analysis.

9.2.3 Instrumentation

9.2.3.1 GC-MS

The samples were run on a 7890A GC coupled to a 5975C MSD (Agilent); the column used was a VF 5 ms 30 m × 250 μm × 0.25 μm with a 10 m guard column (Agilent J&W). The carrier gas was helium and the column flow was 1 mLmin^{-1}. The temperature gradient of the oven was set initially at 70 °C for 1 min and then increased by 7 °C per minute to 325 °C. The inlet, thermal auxiliary, MS source and MS quadrupole temperatures were set to 250 °C, 280 °C, 230 °C and 150 °C, respectively. The scan range was m/z 50–600. Metabolites were identified in ADMIS version 2.72 (NIST) using the Metabolomics Australia in-house library and NIST 14 library version 2.2 for mass spectral comparison.

9.2.3.2 LC-MS

Instrument and LC-MS setup were as follows: LC-MS system (Agilent 6520 QTOF, Santa Clara, CA, USA) with a dual sprayer ESI source and attached to Agilent 1200 series HPLC system (Santa Clara, CA, USA) comprised of a vacuum degasser, binary pump, with a thermostated autosampler and column oven. The MS was operated in positive mode using the following conditions: nebulizer pressure 30 psi, gas flow rate 10 L/min, gas temperature 300 °C, capillary voltage 4000 V, fragmentor 150 and skimmer 65 V. The instrument was operated in the extended dynamic range mode with data collected in m/z range 70–1700. Chromatography was carried out using Zorbax Eclipse XDB-C18 column (2.1 × 100 mm, 1.8 μm column, Agilent)), maintained at 40 °C (± 1 °C) at a flow rate of 400 μL min^{-1} with a 20-min run time. A gradient LC method was used with mobile phases comprised of (A) 0.1% formic acid in deionized water and (B) 0.1% formic acid in acetonitrile. Gradient: A 5-min linear gradient from 5% solvent (B) to 30% solvent (B), followed by a 5-min linear gradient to 30% solvent (B) to 100% solvent (B), and then a 5-min hold at 100% solvent (B) and a 4-min re-equilibration at 5% solvent (B). Additionally, targeted MS/MS analysis was performed using Agilent 6510 Q TOF (Santa Clara, CA, USA) following the chromatographic conditions as detailed above. The MS parameters used were as follows: *m/z* 100–950 (MS/MS); capillary voltage, 3500 V; nebulizer pressure, 30 psig; gas flow, 5 L/min; gas temperature, 325 °C; fragmentor voltage, 160 V; collision energy, 15 and 25 eV; and skimmer voltage, 65 V.

9.2.4 Chromatographic Peak Identification

Molecular feature extraction was conducted using Agilent MassHunter Profinder (version B.06.00) and MassHunter Qualitative analysis (version B.07.00). Putative molecular feature identification was done by accurate mass (<10 ppm) and retention time (<0.3 min) matching using an in-house library maintained at Metabolomics

Australia. Additional peak identification was carried out by accurate mass (<10 ppm) matching using METLIN database, by manually interpreting the mass spectra, considering the data obtained from literature and fragmentation pattern of the compounds. Samples were analysed in triplicates and only those molecules identified in all the three samples are reported.

9.3 Results and Discussion

In this study, the metabolite profile of *Z. muelleri* was analysed. A total of 63 components were identified in the GC-MS analysis accounting for >90% of the total peak areas. This could be classified into four metabolite categories: organic acids (40%), sugar and sugar alcohols (25%), amino acids (23%), phytosterols and others (10%). A list of annotated molecules using GC-MS is listed in Table 9.1.

Forty-four metabolites were annotated using LC-MS analysis. Identified metabolites belonged to various classes including phenolic acids, amino acids, sugar alcohols, flavonoid glycosides/methyl ether derivatives and fatty acids. Based on metabolite identification, LC-MS chromatogram of *Z. muelleri* indicated three main regions, 0.6–1.5 min with eluted peaks linked to amino acids and sugar, 3.5–7.2 min for organic acid and flavonoids and 11–13 min for fatty acid and derivatives. The compounds annotated from *Z. muelleri* using LC-MS is summarized in Table 9.2.

9.3.1 Identification of Rosmarinic Acid and Zosteric Acid

Earlier studies have reported rosmarinic acid and zosteric acid as the marker compounds of *Zostera* species (Quackenbush et al. 1986). Rosmarinic acid, a dimer of caffeic acid, was characterized by the presence of molecular ion at m/z 361.0922 and fragments at m/z 343.0815 (M + H-18, loss of water molecule), 181.0494 (M + H-180, loss of caffeic acid and m/z 163.088 (M + H-198, loss of R(+)-β-(3,4-dihydroxyphenyl) lactic acid) (Barros et al. 2013; Taamalli et al. 2015). The molecular ion peak at m/z 245.0110 (Rt 3.48 min) was assigned to zosteric acid. Targeted QTOF analysis in negative ion mode showed product ions at m/z 198.991 and 163.038 corresponding to loss of carboxylic acid (–COOH) and sulphonic acid (–SO_3H), respectively, thus indicating the peak as zosteric acid.

9.3.2 Profile of Z. muelleri

In the last few decades, considerable progress has made in identifying secondary metabolites in seagrass using targeted and/or non-targeted metabolomic approaches. Several studies have demonstrated the high existence of natural secondary

Table 9.1 Metabolites annotated from LC-QTOF analysis. Molecular feature identification was done by accurate mass (<10 ppm) and retention time (<0.3 min) matching using (**a**) an in-house library maintained at Metabolomics Australia (MA) and (**b**) using METLIN database and mass spectral characteristics

Retention time	Putative metabolite name	Molecular mass	Formula	Mass Diff (ppm)	Metabolite class	Rt Match
9.08	Lauroyl carnitine	343.2717	$C_{19}H_{37}NO_4$	1.53	Acyl carnitine	y
0.86	Adenine	135.0549	$C_5H_5N_5$	−2.85	Amino acid	y
0.88	L-Proline	115.064	$C_5H_9NO_2$	−2.53	Amino acid	y
0.89	Urocanate	138.0434	$C_6H_6N_2O_2$	−3.31	Amino acid	y
0.89	3,4-Dihydroxy-L-phenylalanine	197.0684	$C_9H_{11}NO_4$	2.22	Amino acid	y
0.93	L-Tyrosine	181.0741	$C_9H_{11}NO_3$	−0.99	Amino acid	y
0.66	L-Carnitine	161.105	$C_7H_{15}NO_3$	1.29	Amino acid derivative	y
0.68	L-Glutamic acid	147.0535	$C_5H_9NO_4$	−2	Aminoacid	y
1.05	L-Isoleucine	131.0945	$C_6H_{13}NO_2$	0.92	Aminoacid	y
6.27	Benzaldehyde	106.0424	C_7H_6O	−4.76	Aromatic aldehyde	y
6.28	3-Hydroxycoumarin	162.0314	$C_9H_6O_3$	1.69	Benzopyrone	
0.93	Deoxycarnitine	145.1105	$C_7H_{15}NO_2$	−1.77	Deoxycarnitine	y
11.40	γ-Linolenic acid	278.2248	$C_{18}H_{30}O_2$	−0.73	Fatty acid	y
11.42	10,12,14-Octadecatrienoic acid	278.2255	$C_{18}H_{30}O_2$	−3.38	Fatty acid	
11.56	Myristic acid	228.2085	$C_{14}H_{28}O_2$	1.78	Fatty acid	y
11.59	Docosahexaenoic acid	328.2411	$C_{22}H_{32}O_2$	−2.73	Fatty acid	
11.71	Palmitoleic acid	254.227	$C_{16}H_{30}O_2$	−9.37	Fatty acid	
11.75	Palmitic amide	255.2558	$C_{16}H_{33}NO$	1.56	Fatty acid amide	
11.92	Oleamide	281.2717	$C_{18}H_{35}NO$	0.56	Fatty acid amide	
12.83	Stearamide	283.2874	$C_{18}H_{37}NO$	0.28	Fatty acid amide	
9.03	2E,4E,8E,10E-Dodecatetraenedioic acid	222.0896	$C_{12}H_{14}O_4$	−1.87	Fatty acyl	
11.29	9Z,12Z,15Z-Octadecatrienal	262.2298	$C_{18}H_{30}O$	−0.46	Fatty acyls	
7.16	Kaempferide	300.0638	$C_{16}H_{12}O_6$	−1.45	Flavanol	
6.29	Luteolin	286.048	$C_{15}H_{10}O_6$	−1.1	Flavone	
6.88	5,2′,5′-Trihydroxyflavone	270.0529	$C_{15}H_{10}O_5$	−0.36	Flavone	

6.34	Luteolin 4′-sulphate	366.0048	$C_{15}H_{10}O_9S$	−0.58	Flavone derivative	
6.88	Apigenin 7-sulphate	350.0099	$C_{15}H_{10}O_8S$	−0.65	Flavone derivative	
7.07	Luteolin 4′-methyl ether 3′-sulphate	380.0206	$C_{16}H_{12}O_9S$	−0.97	Flavone derivative	
3.58	Luteolin 4′-glucoside	448.1007	$C_{21}H_{20}O_{11}$	−0.34	Flavone glycoside	
4.31	Luteolin 5-(6″-malonylglucoside)	534.1014	$C_{24}H_{22}O_{14}$	−0.91	Flavone glycoside	
5.44	6-Hydroxyluteolin 5-rhamnoside	448.1006	$C_{21}H_{20}O_{11}$	−0.18	Flavone glycoside	
7.05	Diosmetin	300.0636	$C_{16}H_{12}O_6$	−0.56	Methoxy flavone	
0.89	Citric acid	192.0271	$C_6H_8O_7$	−0.49	Organic acid	y
1.00	L-Pipecolic acid	129.0794	$C_6H_{11}NO_2$	−2.87	Organic acid derivative	y
7.35	Methyl vanillate	182.0579	$C_8H_8O_4$	0.12	Organic acid derivative	y
3.67	3-(4-Hydroxyphenyl)lactate	182.0579	$C_9H_{10}O_4$	0.07	Organic ester	y
3.23	Ferulic acid	194.0578	$C_{10}H_{10}O_4$		Phenolic acid	
3.48	Zosteric acid	244.0042	$C_9H_8O_6S$	0.22	Phenolic acid	
4.43	Caffeic acid	180.0441	$C_9H_8O_4$	−10.41	Phenolic acid	y
4.83	Homovanillic acid	182.0577	$C_9H_{10}O_4$	0.99	Phenolic acid	y
6.27	Rosmarinic acid	360.0847	$C_{18}H_{16}O_8$	−0.6	Phenolic acid	
8.86	Biliverdin	582.2473	$C_{33}H_{34}N_4O_6$	0.92	Pigment	y
0.66	D-(+)-Raffinose	504.1667	$C_{18}H_{32}O_{16}$	4.53	Sugar	y
0.70	6-Phosphogluconic acid	276.0249	$C_6H_{13}O_{10}P$	−1.07	Sugar acid	y

Table 9.2 Metabolites annotated from GC-MS analysis. Metabolites were identified using the Metabolomics Australia in-house library and NIST 14 library version 2.2 for mass spectral comparison

Retention time	Putative metabolite name (Derivative)	MZ	Formula	Metabolite Class	Rt Match
7.23	Lactic acid	117.1	$C_3H_6O_3$	Organic acid	y
15.67	Aspartic acid	218.0	$C_4H_7NO_4$	Amino acid	y
14.07	β-alanine	248.1	$C_3H_7NO_2$	Amino acid	y
14.42	Homoserine	218.1	$C_4H_9NO_3$	Amino acid	y
8.00	L-Alanine	116.0	$C_3H_7NO_2$	Amino acid	y
18.16	L-Asparagine	231.0	$C_4H_8N_2O_3$	Amino acid	y
17.35	L-Glutamic acid	246.0	$C_5H_9NO_4$	Amino acid	y
19.82	L-Glutamine	156.0	$C_5H_{10}N_2O_3$	Amino acid	y
8.37	L-Glycine	102.1	$C_2H_5NO_2$	Amino acid	y
11.57	L-Isoleucine	158.1	$C_6H_{13}NO_2$	Amino acid	y
21.94	L-Lysine	317.2	$C_6H_{14}N_2O_2$	Amino acid	y
19.48	L-Ornithine	174.1	$C_5H_{12}N_2O_2$	Amino acid	y
17.52	L-Phenylalanine	192.0	$C_9H_{11}NO_2$	Amino acid	y
10.94	L-Serine	132.1	$C_3H_7NO_3$	Amino acid	y
16.29	L-Threonic acid	292.0	$C_4H_8O_5$	Amino acid	y
11.63	L-Threonine	219.0	$C_4H_9NO_3$	Amino acid	y
22.22	L-Tyrosine	218.1	$C_9H_{11}NO_3$	Amino acid	y
15.95	N-Acetylglutamic acid	276.1	$C_7H_{11}NO_5$	Amino acid	y
15.78	Pyroglutamic acid	258.0	$C_5H_7NO_3$	Amino acid	y
11.70	Proline	142.1	$C_5H_9NO_2$	Amino acid	y
10.12	L-Valine	144.0	$C_5H_{11}NO_2$	Amino acid	y
15.87	γ-aminobutyric acid	304.0	$C_4H_9NO_2$	Amino acid	y
22.33	4-Hydroxycinnamic acid	293.1	$C_9H_8O_3$	Aromatic acid	y
24.93	3,4-Dihydroxycinnamic acid	396.2	$C_9H_8O_4$	Aromatic acid	y
20.13	p-Coumaric acid	308.1	$C_9H_8O_3$	Aromatic acid	y

22.89	D-Gluconic acid	319.0	$C_6H_{12}O_7$	Carboxylic acid	y
19.38	Ribonic acid	292.0	$C_5H_9O_6$	Carboxylic acid	y
12.71	2-Pyrrolecarboxylic acid	240.1	$C_5H_5NO_2$	Carboxylic acid	y
20.29	Shikimic acid	204.0	$C_7H_{10}O_5$	Carboxylic acid	y
7.28	Oxalic acid	219.0	$C_2H_2O_4$	Dicarboxylic acid	y
12.00	Succinic acid	247.1	$C_4H_6O_4$	Dicarboxylic acid	y
11.81	Maleic acid	245.1	$C_4H_4O_4$	Dicarboxylic acid	y
9.92	Malonic acid	233.1	$C_3H_4O_4$	Dicarboxylic acid	y
12.68	Fumaric acid	245.0	$C_4H_4O_4$	Dicarboxylic acid	y
25.38	Phytol	143.1	$C_{20}H_{40}O$	Diterpene alcohol	y
13.15	3,4-Dihydroxydihydrofuran-2(3H)-one	247.1	$C_4H_6O_4$	Heterocyclic organic compound	y
11.14	Phosphoric acid	299.0	H_3PO_4	Inorganic acid	y
16.65	2-Ketoglutaric acid	198.0	$C_5H_6O_5$	Keto acid	y
19.57	L-Glycerol-3-phosphate	357.1	$C_3H_9O_6P$	Phosphoric ester of sugar alcohol	y
27.01	D-Glucose-6-phosphate	387.1	$C_6H_{13}O_9P$	Phosphorylated sugar	y
22.97	Sequoyitol	217.1	$C_7H_{14}O_6$	Poly alcohol	y
24.20	Myo-inositol	432.2	$C_6H_{12}O_6$	Poly alcohol	y
19.42	Cis-aconitic acid	280.1	$C_6H_6O_6$	Polycarboxylic acid	y
14.81	Citramalic acid	247.1	$C_5H_8O_5$	Polycarboxylic acid	y
20.42	Citric acid	347.0	$C_6H_8O_7$	Polycarboxylic acid	y
37.53	β-sitosterol	486.0	$C_{29}H_{50}O$	Sterol	y
36.80	Campesterol	472.0	$C_{28}H_{48}O$	Sterol	y
37.00	Stigmasterol	484.4	$C_{29}H_{48}O$	Sterol	y
32.84	Gentiobiose	361.1	$C_{12}H_{22}O_{11}$	Sugar	y
21.14	D-Fructose	307.0	$C_6H_{12}O_6$	Sugar	y

(continued)

Table 9.2 (continued)

Retention time	Putative metabolite name (Derivative)	MZ	Formula	Metabolite Class	Rt Match
18.71	β-D-1,6-anhydro-glucose	204.1	$C_6H_{10}O_5$	Sugar	y
21.51	D-Glucose	319.0	$C_6H_{12}O_6$	Sugar	y
27.51	D-Glycero-D-gulo-heptose	331.1	$C_7H_{14}O_7$	Sugar	y
30.53	Sucrose	451.0	$C_{12}H_{22}O_{11}$	Sugar	y
31.65	Trehalose	361.1	$C_{12}H_{22}O_{11}$	Sugar	y
17.49	Xylose	307.2	$C_5H_{10}O_5$	Sugar	y
20.20	Glucuronic acid	333.1	$C_6H_{10}O_7$	Sugar acid	y
12.23	Glyceric acid	292.0	$C_3H_6O_4$	Sugar acid	y
26.86	2-*O*-Glycerol-*α*-D-galactopyranoside	204.1	$C_9H_{18}O_8$	Sugar alcohol	y
22.14	Dulcitol	307.1	$C_6H_{14}O_6$	Sugar alcohol	y
21.96	Mannitol	319.1	$C_6H_{14}O_6$	Sugar alcohol	y
11.15	Glycerol	218.0	$C_3H_8O_3$	Sugar alcohol	y
15.39	Meso-erythritol	217.1	$C_4H_{10}O_4$	Sugar alcohol	y

metabolites wherein some have focused on their bioactive potential (Papenbrock 2012; Subhashini et al. 2013). However, considering almost 72 species of seagrasses distributed across the globe, only few have been examined for natural bioactives and their associated role. So far, seagrass *Posidonia oceanica* is the only most studied species with the identification of almost 51 natural compounds belonging to phenols, phenyl-methane/ethane/propane and their derivatives and flavonoids (Zidorn 2016; Subhashini et al. 2013). These secondary metabolites have been suggested to play crucial role not only for growth and development but also in defence from environmental stress, pathogen attack and herbivory (Kumar et al. 2016).

The importance of primary polar metabolites (such as sugars, TCA intermediates, amino acids, and/or polyalcohols) as taxonomical and physiological traits has also been well reported. In this study, 24 amino acids and 16 compounds related to "sugar and derivatives" were detected (Tables 9.1 and 9.2). Sugars, amino acids and organic acids are the major energy sources for photosynthesis and respiration in plants. Sugars, as a carbon source, provide the energy required for growth and development, and amino acids, as a nitrogen source, promote rapid growth.

In recent years, the role of primary metabolites in stress physiology of seagrasses has been reported. A metabolomic study on diurnal effects of anoxia in *Z. marina* identified varied expression levels of lactate, pyruvate, GABA (γ-aminobutyric acid), succinate, alanine, glutamate and glutamine. It was suggested that the function of the alanine-GABA shunt was to mitigate the toxic effects of lactate or ethanol produced during anoxic condition, by providing an alternative route for pyruvate metabolism. Also, alanine and GABA accumulation together with glutamate and glutamine allowed carbon and nitrogen storage during anoxia, and this would provide energy for metabolism upon re-establishment of normoxic conditions (Hasler-Sheetal et al. 2015). Another study on adaptive mechanism for chronic heat stress in *Z. marina* and *Z. noltii* have identified sucrose, fructose and *myo*-inositol as the most responsive metabolites (Gu et al. 2012). Given the osmoprotective function of sugars, the accumulation of *myo*-inositol in both species was suggested to act as a substrate to generate protein-stabilizing osmolytes such as di-*myo*-inositol phosphate. Further, differential expression of primary metabolites, particularly sugars such as glucose and fructose, has been reported in *Z. marina* under light stress conditions (Hasler-Sheetal et al. 2016). Therefore, identification of these metabolites and studying their regulation under stressful environment can provide insights on the acclimation and/or tolerance mechanism of Z. *muelleri* (an abundant species of Zosteraceae family in Australia) under the scenario of global climate change.

In both terrestrial and sea plants, shikimate pathway has been evident as the major pathway for the synthesis of phenylpropanoid and flavonoid derivatives forming the bulk of metabolites (Zidorn 2016; Papenbrock 2012). In the present study, phenolic acids such as caffeic, rosmarinic, zosteric and ferulic acid and 3-hydroxycoumarin were detected as major phenolic acids with homovanillic being identified for the first time in *Zostera* species. The identified phenolics were either hydroxybenzoic acid or hydroxycinnamic acid derivatives. An array of phenolic compounds such as coumaric acid, caffeic acid, ferulic acid, vanillic acids and rosmarinic acid have also been reported from leaf samples of *P. oceanica* and *Z. marina*

(Subhashini et al. 2013; Cuny et al. 1995; Haznedaroglu and Zeybek 2007). Zosteric acid is the marker bioactive isolated from various *Zostera* species and was first discovered in *Z. marina* (Todd et al. 1993). Zosteric acid is a phenylpropane derivative (*p*-(sulphoxy) cinnamic acid) with huge antifouling activities. Zosteric acid has been documented to inhibit colonization of bacteria and fungus thus hinders their biofilm formation and offers a zosteric acid as a promising broad spectrum antifoulant [hindering biofilm formation with zosteric acid]. Another major biomarker phenolic in *Zostera* species in rosmarinic acid identified in this study is caffeoyl ester, which has diverse applications from food preservatives to cosmetics due to its high antioxidant activity. Rosmarinic acid isolated from *Z. marina* and other species exhibited nematicidal, antibacterial and algicidal activities (Wang et al. 2012; Laabir et al. 2013). The identified phenolics in this study have been shown to exhibit anti-*Labyrinthula* activity isolated from *Z. marina* as natural defence against this pathogen (Vergeer and Develi 1997). In seagrasses, phenolics have been shown to accumulate considerably in most of the adverse conditions caused by environmental stress such as high light and salinity fluctuations. In contrast, recently a noticeable loss in phenolic substances (vanillin, coumaric and ferulic acid, proanthocyanidins) has been observed in seagrass *Cymodocea nodosa*, *Ruppia maritima and Potamogeton perfoliatus* with high grazing rate during ocean acidification conditions. These observations temper the recent predictions that seagrass would necessarily be "winners" in a high CO_2 world.

Flavonoids are another class of major metabolites identified in this study. In seagrasses, the polyphenolic flavonoids generally exist as flavones, flavonols, flavanons, flavanols and anthocyanidins and are synthesized from a common precursor – phenylalanine. In the present study, the identified flavonoids include flavonoid glycosides (luteolin 4'-glucoside and luteolin 5-(6"-malonylglucoside)), sulphated flavonoids (luteolin 4'-sulphate, luteolin 4'-methyl ether 3'-sulfate and apigenin 7-sulphate), flavone (luteolin and 5,2',5'-trihydroxyflavone) and flavanol (kaempferide). Sulphated flavonoids are of particular interest due to its role in plant physiology as a growth regulator and its ability to form stable complexes with other flavonoids. They are also well known for various biological activities such as anti-inflammatory, antiviral and antitumour activity (Subhashini et al. 2013). We reported here for the first time the existence of luteolin 3′-methyl ether 7-sulphate in *Z. muelleri*; however, this requires further confirmatory analysis. Seagrasses *Halophila*, *Enhalus*, *Thalassia* and *Zostera* species have been the most studied for the presence of sulphated flavonoids, with no record of such flavonoids in *Posidonia* sp. A considerable accumulation of flavonoids such as luteolin 7-sulphate and luteolin 7,3′-disulphate has been observed in *Z. marina* as a strategy of chemical defence to inhibit marine microorganism (Enerstvedt et al. 2016). In our study, luteolin 4′-glucoside and luteolin 5-(6″-malonylglucoside) are the only flavonoid glycosides that we observed. The existence of luteolin 7-O-glucoside, diosmetin-7-O-glucoside (*Z. marina* and *Z. nana*) and luteolin 7-O-glucoside (*Z. noltii*) suggests that *Zostera* spp. are strong enough to synthesize these flavonoid derivatives in stressful aquatic environment (Zidorn 2016; Subhashini et al. 2013). Similar to our findings, the presence of malonylated flavone glycoside derivatives has also been observed in *Halophila stipulacea* (Papenbrock 2012).

Recently, the presence of flavone glycoside thalassiolin B in significant concentrations from *T. testudinum* (partially characterized using 1H NMR and LC-MS) was reported which was capable of inhibiting the growth of *Labyrinthula* sp. (Vergeer and Develi 1997). Furthermore, various flavonoids such as luteolin, apigenin, luteolin-3'-glucuronide and luteolin-4-*O*-glucuronide have also been isolated and characterized by HPLC/MS^n and NMR from *T. testudinum* and *Enhalus acoroides* with anti-feedent and anti-larval activities against *Spodoptera litura* and *Bugula neritina* larvae (Vergeer and Develi 1997). Luteolin from leaf tissues of *T. testudinum* has also been shown to inhibit the settlement of motile zoospores of the protist *Schizochytrium aggregatum.*

The most abundant lipid classes defined in seagrass species are fatty acids, sterols and hydroxyl fatty acids. Free fatty acids and amides such as γ-linolenic acid, myristic acid, palmitoleic acid, docosahexaenoic acid, 10,12,14-octadecatrienoic acid, palmitic amide, oleamide and stearamide and steroidal compounds β-sitosterol, campesterol and stigmasterol were identified in this study. However, it should be noted that the extraction protocol in this study was not lipid specific and hence the fatty acid and derivatives identified in this study may not be conclusive. Many targeted and non-targeted GC-MS-based metabolomic approaches have revealed fluctuations in fatty acid composition in marine plants exposed to various biotic and abiotic stresses. An enhanced proportion of oleic acid and linoleic acid with a parallel decrease in palmitoleic acid was observed in marine macrophyte *Gracilaria corticata* and *Ectocarpus siliculosus* at hypersalinities in contrast to hyposalinity (Kumar et al. 2016). Higher PUFAs accumulation was suggested as an adaptive strategy to maintain greater membrane fluidity, to stabilize the protein complexes of PSII and to control the physicochemical properties of membranes, such as the increased activity of the Na^+/H^+ antiporter system of the plasma membrane in order to cope with hypersalinity stress (Kumar et al. 2016). Phytosterols play important role in plant adaptation to temperature and are also involved in the regulation of temperature involved membrane dynamics (Ribeiro et al. 2014). Sterols such as 24-ethylcholest-5-en-3*β*-ol, 24-ethylcholesta-5,22E-dien-3*β*-ol and 24-methylcholest-5-en-3*β*-ol have been previously reported from *Z. muelleri* (Gillan et al. 1984).

9.4 Conclusion

In this study, we demonstrate the potential of MS-based metabolite profiling for analysing a broad spectrum of metabolites in seagrasses. Information on metabolites is integral for linking genotype and phenotype and thus has a significant role in the development of system biology approaches in marine systems. The metabolite information of *Z. muelleri* attained in this study can provide a better insight into the biochemical composition of the species. However, it is highly likely that results of metabolomics when integrated with allied omic platform such as transcriptomics or proteomics will further deepen our understanding on system biology and to discover new biomolecules and metabolic pathways that are crucial for survival process of marine macrophytes in response to global climate change.

Acknowledgement LC-MS analysis and tentative identification of molecules using MA database was carried out by UK and Dr. Adrian Lutz, Metabolomics Australia, School of BioSciences, University of Melbourne. Tentative annotations using Metlin database and interpretation of mass spectra was carried out by UK. MK collected the seagrass samples and processed them for GC and LC-MS analysis. GC-MS profiling was carried out by Dorothee R Hahne, UWA Centre for Metabolomics, The University of Western Australia. MK is grateful to the Australian Research Council for awarding him Discovery Early Career Research Award (DECRA Fellowship, DE150100461-2015).

References

Achamlale S, Rezzonico B, Grignon-Dubois M (2009) Rosmarinic acid from beach waste: isolation and HPLC quantification in *Zostera detritus* from Arcachon lagoon. Food Chem 113(4):878–883

Barros L, Dueñas M, Dias MI, Sousa MJ, Santos-Buelga C, Ferreira ICFR (2013) Phenolic profiles of cultivated, in vitro cultured and commercial samples of *Melissa officinalis* L. infusions. Food Chem 136(1):1–8

Cuny P, Serve L, Jupin H, Boudouresque C-F (1995) Water soluble phenolic compounds of the marine phanerogam *Posidonia oceanica* in a Mediterranean area colonised by the introduced chlorophyte Caulerpa taxifolia. Aquat Bot 52(3):237–242

Custódio L, Laukaityte S, Engelen AH, Rodrigues MJ, Pereira H, Vizetto-Duarte C et al (2016) A comparative evaluation of biological activities and bioactive compounds of the seagrasses *Zostera marina* and *Zostera noltei* from southern Portugal. Nat Prod Res 30(6):724–728

De Leeuw JW, Irene W, Rijpstra C, Nienhuis PH (1995) Free and bound fatty acids and hydroxy fatty acids in the living and decomposing eelgrass *Zostera marina* L. Org Geochem 23(8):721–728

Duarte CM, Chiscano CL (1999) Seagrass biomass and production: a reassessment. Aquat Bot 65(1–4):159–174

Enerstvedt KH, Jordheim M, Andersen M (2016) Isolation and identification of flavonoids found in *Zostera marina* collected in Norwegian coastal waters. Am J Plant Sci 7(7):1163–1172

Engle JM, Miller KA (2003) Distribution and morphology of eelgrass (*Zoster marina* L.) at the California Channel Islands. In: Garcelon DK, Schwemm CA (eds) Sixth California Island symposium. Institute for Wildlife Studies, Arcata, pp 1584–1590

Farag MA, Porzel A, Wessjohann LA (2012) Comparative metabolite profiling and fingerprinting of medicinal licorice roots using a multiplex approach of GC–MS, LC–MS and 1D NMR techniques. Phytochemistry 76:60–72

Gillan FT, Hogg RW, Drew EA (1984) The sterol and fatty acid compositions of seven tropical seagrasses from North Queensland, Australia. Phytochemistry 23(12):2817–2821

Grignon-Dubois M, Rezzonico B (2012) First phytochemical evidence of chemotypes for the seagrass *Zostera noltii*. Plants 1(1):27–38

Gu J, Weber K, Klemp E, Winters G, Franssen SU, Wienpahl I et al (2012) Identifying core features of adaptive metabolic mechanisms for chronic heat stress attenuation contributing to systems robustness. Integr Biol 4(5):480–493

Hasler-Sheetal H, Castorani MCN, Glud RN, Canfield DE, Holmer M (2016) Metabolomics reveals cryptic interactive effects of species interactions and environmental stress on nitrogen and sulfur metabolism in seagrass. Environ Sci Technol 50(21):11602–11609

Hasler-Sheetal H, Fragner L, Holmer M, Weckwerth W (2015) Diurnal effects of anoxia on the metabolome of the seagrass *Zostera marina*. Metabolomics 11(5):1208–1218

Haznedaroglu MZ, Zeybek U (2007) HPLC determination of chicoric acid in leaves of *Posidonia oceanica*. Pharm Biol 45(10):745–748

Kawasaki W, Matsui K, Akakabe Y, Itai N, Kajiwara T (1998) Volatiles from *Zostera marina*. Phytochemistry 47(1):27–29

Kim JH, Cho YH, Park SM, Lee KE, Lee JJ, Lee BC et al (2004) Antioxidants and inhibitor of matrix metalloproteinase-1 expression from leaves of *Zostera marina* L. Arch Pharm Res 27(2):177–183

Kumar M, Kuzhiumparambil U, Pernice M, Jiang Z, Ralph PJ (2016) Metabolomics: an emerging frontier of systems biology in marine macrophytes. Algal Res 16:76–92

Laabir M, Grignon-Dubois M, Masseret E, Rezzonico B, Soteras G, Rouquette M et al (2013) Algicidal effects of *Zostera marina* L. and *Zostera noltii* Hornem. Extracts on the neuro-toxic bloom-forming dinoflagellate *Alexandrium catenella*. Aquat Bot 111:16–25

Orth RJ, Carruthers TJB, Dennison WC, Duarte CM, Fourqurean JW, Heck KL et al (1976) A global crisis for seagrass ecosystems. Bioscience 56(12):987–996

Papenbrock J (2012) Highlights in seagrasses phylogeny, physiology, and metabolism: what makes them special? ISRN Botany 2012:15.17

Quackenbush RC, Bunn D, Lingren W (1986) HPLC determination of phenolic acids in the water-soluble extract of *Zostera marina* L. (eelgrass). Aquat Bot 24(1):83–89

Ribeiro PR, Fernandez LG, de Castro RD, Ligterink W, Hilhorst HW (2014) Physiological and biochemical responses of *Ricinus communis* seedlings to different temperatures: a metabolomics approach. BMC Plant Biol 14(1):223

Subhashini P, Dilipan E, Thangaradjou T, Papenbrock J (2013) Bioactive natural products from marine angiosperms: abundance and functions. Nat Prod and Bioprospect 3(4):129–136

Taamalli A, Arráez-Román D, Abaza L, Iswaldi I, Fernández-Gutiérrez A, Zarrouk M et al (2015) LC-MS-based metabolite profiling of methanolic extracts from the medicinal and aromatic species *Mentha pulegium* and *Origanum majorana*. Phytochem Anal 26(5):320–330

Todd JS, Zimmerman RC, Crews P, Alberte RS (1993) The antifouling activity of natural and synthetic phenol acid sulphate esters. Phytochemistry 34(2):401–404

Vaclavik L, Lacina O, Hajslova J, Zweigenbaum J (2011) The use of high performance liquid chromatography–quadrupole time-of-flight mass spectrometry coupled to advanced data mining and chemometric tools for discrimination and classification of red wines according to their variety. Anal Chim Acta 685(1):45–51

Vergeer LHT, Develi A (1997) Phenolic acids in healthy and infected leaves of *Zostera marina* and their growth-limiting properties towards *Labyrinthula zosterae*. Aquat Bot 58(1):65–72

Wang J, Pan X, Han Y, Guo D, Guo Q, Li R (2012) Rosmarinic acid from eelgrass shows nematicidal and antibacterial activities against pine wood nematode and its carrying bacteria. Mar Drugs 10(12):2729–2740

Zidorn C (2016) Secondary metabolites of seagrasses (Alismatales and Potamogetonales; Alismatidae): chemical diversity, bioactivity, and ecological function. Phytochemistry 124:5–28

Part III

Chapter 10
Marine Microalgae: Systems Biology from 'Omics'

Justin Ashworth

Abstract Marine biological systems are vast, productive, long-lived, dynamic, and complex. The oceans are fed and balanced primarily through photosynthesis and nutrient cycling by ubiquitous marine microalgae, including cyanobacteria, found thriving even in vast nutrient-limited pelagic deserts, and larger eukaryotic phytoplankton, whose genetic and functional diversity appear substantially more complex and understudied than initially expected. The rapid complete sequencing of environmentally relevant genomes has provided the first precise molecular descriptions of complete biological systems. Metatranscriptomic data quickly and easily provide intercomparable system-wide and conditionally relevant functional information. Environmental proteomics are able to directly identify functional protein biomarkers of nutrient conditions. The integration of comprehensive molecular data (genomes, transcriptomes, proteomes, metabolomes) into models capable of rigorous hypothesis testing and prediction constitutes a new way to study the connections between genotype and phenotype, phenotype and environment, species and ecosystems, and interspecies evolution and adaptation. To date, marine microalgae are the first and most extensively studied marine organisms in terms of their functioning as coalescent molecular systems. The richness of data, systematic integration, and predictive models therein set a new example for the broad new study of marine life at unprecedented detail and comparability, promising answers to broad new scientific, statistical, and quantitative questions of critical concern for the present and future functioning and adaptability of the world's oceans.

Keywords Systems biology • Marine microeukaryotes • Transcriptomics • Diatoms • Microalgae • Data-driven science • Metagenomics • Metatranscriptomics • Gene expression • Environmental microbiology

---

J. Ashworth (✉)
Climate Change Cluster, University of Technology Sydney, Sydney, NSW, Australia
e-mail: Justin.Ashworth@uts.edu.au

© Springer International Publishing AG 2017

M. Kumar, P. Ralph (eds.), *Systems Biology of Marine Ecosystems*,
DOI 10.1007/978-3-319-62094-7_10

Contents

10.1 Introduction ... 208
10.2 Systems Biology ... 209
10.3 Marine Microalgal Genomics ... 210
10.4 Marine Microalgal Transcriptomics ... 212
10.5 Proteomics and Metabolomics ... 214
10.6 Integration and Meta-analysis ... 215
10.7 Prediction and Synthesis ... 216
References ... 217

10.1 Introduction

Marine systems are vast, productive, long-lived, dynamic, and complex. However, they are also finite and sensitive and can be overwhelmingly altered or impacted by strong short-term pressures of environmental or human origin. As these pressures and impacts proceed in the natural environment, it is crucial to consider questions about how these systems function and respond to change. What feeds and balances marine ecosystems? How do biooceanographic processes vary geographically, genetically, functionally, and in accordance with environmental change? What are the strengths and weaknesses of marine systems, and what are the ways in which they are likely to adapt to future ocean conditions? Can we transform detailed observations of past and present systems into valuable predictions and answers to present and future questions?

Marine ecosystems are fed and balanced primarily by the photosynthesis and nutrient cycling ubiquitous marine microalgae. The biology, photosynthesis, and life cycles of these organisms have adapted, evolved, diversified, and colonized all parts of the diverse and dynamic oceans, providing a food source and biogeochemical shuttle that sustains almost all other sea life. The ubiquity and critical function of robust primary productivity by marine microalgae are now wholly evident and fundamentally accepted (Falkowski 1997; Falkowski et al. 1998). However, the true complexity, variability, and adaptive paths that govern wild marine microbial systems now and in future scenarios are far from entirely measured or understood.

Broad and precise new capabilities now exist to measure and model the identities, distribution, dynamics, and functions of cellular processes operating in biological systems. More importantly, new questions can be addressed holistically based upon these increasing data (Karsenti et al. 2011). How do marine organisms balance their complex physiologies and cellular states in order to persist, thrive, and produce? What are the organized systems of genes and proteins that have specifically evolved to sense, respond, and catalyze these processes? How do these diverse and highly evolved unicellular machines manage to cope with dynamic and challenging conditions? What are the chemical species and processes that crucially mediate competitive and cooperative exchange? How do numerous evidently co-occurring species truly co-exist? How quickly can these functions adapt to new changes, and which features of biological systems are most subject to novel environmental and evolutionary selection pressures? What do these details mean for ecosystem dynamics and larger ocean processes now and in the future?

These and other important questions are now theoretically addressable through (**a**) well-reasoned scientific thinking and experimentation, (**b**) collection of relevant comprehensive molecular data (genomes, transcriptomes, proteomes, metabolomes), and (**c**) systematic, robust, and scientific methods of data interpretation, aggregation, analysis, modeling, prediction, and validation.

10.2 Systems Biology

The specific organization and intrinsic biological programs operating within living cells in their environments define their functional and ecological roles. Cast about randomly in a marine environment, the millions of highly specialized biomolecules inside a cell would accomplish very little, fail to self-replicate, and quickly cease to exist. Simply cataloging their presence individually in that case would mean equally little: it is the programmed and compartmentalised coordination of these biomolecules that govern their existence and functions. The systematic co-organization of nucleic acids, proteins, metabolites, and systems by millions of years of evolutionary selection results in irreducible and interconnected processes. This integrative nature is essential to their functions and relevance, and this thinking motivates and defines "systems biology" as a scientific discipline (Ideker et al. 2001). It is not simple however to accurately, rigorously, and scientifically measure, model, and study biological systems in an integrative or exhaustively detailed manner.

Biological systems can be defined at many levels: a single metabolic pathway, signaling cascade, or set of coevolving gene functions; each operates as a molecular system. The outward functions of these systems are defined (and selected upon) as much by their coalescent properties as a whole, as by the functions of their individual components. These small systems do not function or make sense except in the context of the additional information contained within their coevolving properties and interactions. No protein, gene, or environmental response mechanism operates in isolation; the operation of one biomolecule, phenotype, or gene depends on the functions and identities of others. Viruses, organelles, cells and tissues, organisms, populations, and ecosystems are all complex biological systems, whose relevance and operation similarly cannot be reduced to individual parts without missing critical information. Likewise, the relevance and specific effect or response of any single environmental variable can only be understood or predicted in the context of co-varying conditions. *The unique and irreducible information stored within the context and arrangement of a biological system in its dynamic environment is essential to understanding its nature and operation*—and the principles and theories therein are as important to study, measure, model, and predict as the properties of single genes, proteins, or metabolic components.

The importance of this "systems-level" thinking has long been recognized (von Bertalanffy 1968) but seldom easily or scientifically addressed. Billions of microorganisms teem in most marine waters, and the genomes of even the simplest among them encode for thousands of active biomolecules. The coalescent behavior of these numerous variables is often limited to outward observations, measurements, and qualitative classifications: cell size, growth rate, fluorometry, and elemental contents. However, the complex, dynamic, and interacting molecular and genetic parameters that give rise to emergent biological functions must be explicitly considered in order to reach sufficient or predictive answers to many fundamental questions. For example, certain species dominate others, depending on environmental factors. The cyanobacteria thrive in nutrient-limited pelagic waters (Zwirglmaier et al. 2008), while the large eukaryotic phytoplankton tend to dominate in areas abundant in macronutrients (de Vargas et al. 2015). But why is this? What combinations gene functions have critically coevolved into these divergently optimized systems-level properties? Which emergent properties of their molecular systems produce their respective advantages? What are the constraints and boundaries on their niches and adaptabilities? What are the minimum molecular and genetic differences (among many) required to explain and account for the outward differences between species? How have proliferative, competing, or cooperating species coevolved or co-opted advantages to become successful in each other's niches through evolution, gene transfer, and symbiosis? What are the possibilities and opportunities for this to continue and change in new and future environments?

The consideration of complex molecular, genetic, ecological, and environmental data and models becomes necessary in order to accurately and scientifically answer questions like these. The advent of efficient large-scale comprehensive molecular data collection (genomics, transcriptomics, proteomics, metabolomics, etc.) now offers broad and explicit information that can help to explicitly link genotype to phenotype, phenotype to environment, species to ecosystems, and intra- and inter-species evolution to adaptation. The amount of new data available is immense and is superseded in importance and opportunity only by a careful focus on biological questions, well-conceived hypotheses, and measurements and experiments designed to address them.

10.3 Marine Microalgal Genomics

The first comprehensive molecular data rapidly, efficiently, and completely collected for living systems were arguably the first microbial genomes (Fleischmann et al. 1995; Fraser et al. 1995, 1997, 1998; Bult et al. 1996). The rapid, systematic sequencing of genomes provided the first precise molecular descriptions of complete biological systems. Every functional biomolecule in the organism is encoded or imprinted in its genome, as well as the hardcoded portion of the biological program that controls its physiology, metabolism, life cycle, responses to change, potential for interactions with the extracellular environment, evolutionary history and relationships, and heritable capacity to genetically evolve new functions, adaptations, and emergent properties. The whole genome of an organism is necessary (but not sufficient) to fully understand and consider its biology and its current and future capabilities.

Prokaryotic Microalgae The most ubiquitous marine microalgae are the cyanobacteria, including the genera *Prochlorococcus* and *Synechococcus*. These are found thriving throughout marine systems, including throughout vast nutrient-limited pelagic deserts where organic nitrogen, phosphorus, iron, or silicate are insufficient for larger microalgae to thrive. The genomes of *Prochlorococcus* and *Synechococcus* are approximately 1.5–2.5 Mbp in length and contain between 1000 and 2500 genes (Dufresne et al. 2003; Palenik et al. 2003; Rocap et al. 2003), typically including all of the genes necessary for photosynthesis. These genomes appear remarkably streamlined in comparison to the 4.6Mbp genome of the *E. coli* bacterium, consisting of approximately 4300 genes (Blattner et al. 1997). It is hypothesized that the small genome sizes of cyanobacteria are optimized for nutrient-limited environments (Bentkowski et al. 2017).

Eukaryotic Microalgae The first fully sequenced marine microeukaryote genomes have revealed a surprising genetic and physiological complexity hidden within seemingly simple unicellular plankton. The genome of the siliceous "cosmopolitan" diatom, *Thalassiosira pseudonana*, the first diatom sequenced due to its ubiquity and compact genome size (32 Mbp), consists of 24 chromosomes and encodes an estimated 11,390 genes (Armbrust et al. 2004). Roughly half of these genes bear no confident similarity to any other genes of known function. The genome of *Phaeodactylum tricornutum*, the second diatom to be sequenced, similarly encodes over 10,000 genes, only about half of which bear detectable similarity to those found in *T. pseudonana*, despite only ~90 million years of divergent evolution between the two species. Both of these marine microeukaryote genomes imply a complexity greater than that of the first fully sequenced microeukaryote, the yeast *Saccharomyces cerevisiae*, the 12 Mbp genome of which consists of ~6000 genes (Goffeau et al. 1996). Similarly, the ~120 Mbp genome of the freshwater microalga *Chlamydomonas reinhardtii* harbors a surprisingly complex genome encoding ~16,700 genes (Merchant et al. 2007), less than 3000 of whose products bear confident similarity to those encoded in the genomes of *T. pseudonana* (Armbrust et al. 2004) or *P. tricornutum* (Bowler et al. 2008). The collection of microalgal genomes is rapidly expanding, including several additional diatom (Lommer et al. 2012; Traller et al. 2016; Mock et al. 2017; Basu et al. 2017) and dinoflagellate (Lin et al. 2015) genomes now available for research into these lesser understood phyla. In addition to this, at least hundreds of new genomes and millions of new putative protein coding genes have been cataloged by the latest exploratory and integrative oceanographic efforts (de Vargas et al. 2015).

Marine Microbial Metagenomes Within any water sample or marine ecosystem, there are multiple genomes present, numbering from the tens to trillions, depending on scale. The specific repertoires of variously encoded functions and alleles in any sample define the genetic potential of biological processes in that context and specify salient functions and features that are adapted to exist in the environment from which the sample was collected. The genomic information typically observed in marine environments is inexhaustibly complex and difficult to assemble or reduce (Gilbert and Dupont 2011; Iverson et al. 2012). Nevertheless, the sample-specific richness of taxonomy, diversity, allele distribution, and function gained through high-depth short-read sequencing of environmental samples quickly yields critical information to understand the aggregate nature and properties of marine microbial systems (Venter et al. 2004; Sunagawa et al. 2015). The genome, however, is largely passive and inactive and merely a source code in all living systems. In order to understand how the information stored and transmitted in genomes relates to biology itself, it is necessary to consider the messages and biomolecules transcribed from genomes to perform cellular functions.

10.4 Marine Microalgal Transcriptomics

The primary (and likely most ancient) dynamically encoded information in the cell is contained within RNAs conditionally and flexibly transcribed to and from the functionally inactive genome. Fortunately, this is also now relatively comprehensive and easy information to obtain through high-throughput sequencing. For microeukaryotes with large genome sizes and mixed cultures or in environmental samples, the size and complexity of the expressed transcriptome are significantly smaller in size than the corresponding whole genome space and by definition contain nearly all of genetically encoded functional information that is operating under a particular condition. These information, which are in many cases now readily comprehensive, provide a wealth of detailed information about crucially acting molecular processes and the emergent patterns of gene expression that give rise to various cellular states. While it is necessary but not sufficient to explain the operation of biological systems, transcriptomic data currently represent the broadest, most sensitive, and most easily obtainable and intercomparable system-wide functional information that can be collected for most organisms and biological systems.

The putative biological activities of the proteins produced by typically about half of all microalgal transcripts can be inferred bioinformatically, and while in some examples and species the expression levels of proteins can be uncorrelated to the expression levels of their transcripts, increases or decreases of mRNA transcript levels are often concordant with changes in the cellular levels of the proteins that they encode. mRNA sequencing of genetically variable species or environments also yields nucleotide polymorphisms in conserved and evolutionarily pressured protein-coding regions that may be linked to spatial, temporal, ecological, evolutionary, and functional divergence. For all of these reasons, a wealth of transcriptomic and metatranscriptomic data has been collected for marine microalgal species in a very short time. These data can be used to rapidly characterize informative gene regulatory profiles of biological systems in accordance with varying cellular states and environments.

Laboratory Studies The first comprehensive genome-wide expression studies in marine microalgae consisted of single-species experiments in which whole-transcriptome microarrays were used to measure changes in the expression of all genes under various conditions of relevance to the natural environment. These include tracking the cyanobacterium *Prochlorococcus* over its day/night cycle (Waldbauer et al. 2012), measuring the comprehensive responses of the diatom *T. pseudonana* to a panel of typical stresses and limitations (Mock et al. 2008), and interacting factors of the diel cycle, culture density, and nutrient exhaustion (Ashworth et al. 2013). For the diatoms in particular, abundant transcriptome-wide expression data have been collected using microarrays and mRNA sequencing under various laboratory conditions designed to simulate environmental variables. These include, among others: (1) *T. pseudonana*: silica, iron, and nitrogen limitation, low temperature and elevated pH (Mock et al. 2008), iron starvation (Thamatrakoln et al. 2012), silica starvation and re-supplementation (Shrestha et al. 2012; Smith et al. 2016b), diel growth from exponential to stationary phase (Ashworth et al. 2013), exposure to the pollutant benzo[*a*]pyrene (Carvalho et al. 2011), and growth at moderate and elevated CO_2 levels under moderate and elevated light and (2) *P. tricornutum*: silica limitation (Sapriel et al. 2009), acclimation to high light (Nymark et al. 2009), exposure to cadmium (Brembu et al. 2011), acclimation to light and dark cycles (Chauton et al. 2013), exposure to a panel of stresses and pollutants (Hook and Osborn 2012), darkness and re-illumination (Nymark et al. 2013), and growth in red, blue, and green light (Valle et al. 2014).

The comprehensive tracking of changes in gene expression over all of these various experimental conditions results in a rich and complex picture of transcriptome dynamics in these organisms (Ashworth et al. 2013, 2016; Levering et al. 2017), much of which is still yet to be sufficiently studied, understood, and fully applied to address fundamental questions and predictions with regard to marine systems and future change.

RNA Sequencing Whole transcriptome RNA sequencing at high depth has become affordable enough to simplify the process and information necessary to obtain transcriptomic profiles for any species or biological sample. The sequencing of transcribed RNA repertoires through reverse transcription and amplification is powerful and sensitive and can be informatively classified, quantified, and assembled even in the absence of corresponding genome sequences. For eukaryotes in particular, the functional and phylogenetic information density present in transcribed messenger RNA is high, and the amplification of cDNA improves signal detection—particularly in the case of poly-dT-primed strand synthesis and 3′-directed selective amplification (Xiong et al. 2017). Laboratory mRNA sequencing studies in marine microalgae include the profiling of diatoms to transcriptome dynamics at different levels of carbon dioxide (Hennon et al. 2015), during silica starvation (Smith et al. 2016b), diel cycling combined with iron limitation (Smith et al. 2016a), and an integrated range of several other environmental, nutrient, and chemical perturbations (Levering et al. 2017). Hundreds of additional new single-strain transcriptomes have been sequenced in order to cover dozens of completely new clades of marine microeukaryotes (Keeling et al. 2014), resulting in an atlas of transcribed and functional coding sequences of unprecedented comparative breadth and depth against which to integrate and contextualize individual new studies conducted in the laboratory and in the field.

Metatranscriptomic mRNA sequencing studies, or those conducted on samples collected in natural marine systems, are beginning to produce a deep and unprecedented richness of functional eukaryotic coding sequences operating in wild environments. The transcriptomic complexity and response of a wild phytoplankton community during iron supplementation experiments at sea revealed a rapid and iron-specific response encoded in native organisms that are evolutionarily and adaptively primed to cope with and take advantage of fluctuating iron levels in the North Pacific Ocean (Marchetti et al. 2012), confirming and expanding upon observations from related laboratory experiments. Environmentally responsive biological programs can be identified within these large data sets that help to explain acclimation and evolved survival mechanisms in unpredecented precision and detail (Marchetti et al. 2017). Meanwhile, in the Atlantic Ocean, transcriptomics are being used to more deeply observe and understand the intracellular behaviors and responses of phytoplankton with respect to nutrient conditions (Alexander et al. 2015). The ability to track detailed and specific intracellular programs in the natural environment provides an opportunity to deeply understand what native cellular communities are really doing as they occupy and experience different marine environments, and this may soon be broadly automated to study and predict the dynamics of communities in situ (Ottesen et al. 2013, 2014; Aylward et al. 2015).

10.5 Proteomics and Metabolomics

The majority of unique functional biomolecules in marine systems are proteins, and the majority of the remainder are metabolites and the products of enzymes. It would be deeply informative to know the identity and quantity of all proteins, metabolites, and biomaterials present in a cell or sample from the ecosystem—this could be used to model and predict metabolic flux—and would be more closely representative of true physiology and activity than transcript levels, from which the levels and functions of posttranscriptional, posttranslational, and metabolic products may importantly diverge.

Proteins are more complex polymers than nucleic acids, are complicated to uniquely separate and identify, and cannot be amplified; unfortunately, unlike for DNA and RNA, there is no simple way to simply sequence or quantify complex pools of proteins. Proteomic technologies based on peptide generation, automated multidimensional separation, and mass spectrometry are now able to finely sample and detect thousands of different peptides at high resolution, resulting in successful, albeit noncomprehensive proteomic analyses of microalgae in response to changing conditions (Nunn et al. 2009; Dyhrman et al. 2012; Nunn et al. 2013).

Quantification of complex protein pools is also a challenge and is best conducted using a pool of peptide standards matching the expected proteome. Exhaustive species-specific prototypic peptides have made this possible for the human proteome (Kusebauch et al. 2016), but this does not immediately translate to other species, and the theoretical pool of possible peptides in wild environments is prohibitively large for accurate and comprehensive de novo quantification. Nevertheless, environmental proteomics have been able to directly identify certain functional proteins that present and operate in accord with environmental conditions (Saito et al. 2014), and the proteomic detection and quantification of specifically validated and informative protein biomarkers may be an important tool for oceanography.

The broad measurement of metabolites and biomaterials in cells and natural systems similarly relies on adequate chemical separation, unique identification through mass spectrometry, and validated standards for quantification and thus are similarly challenging in terms of comprehensiveness and sensitivity compared to amplifiable nucleic acids. Nevertheless, deep and informative biochemical datasets are emerging that can be integrated with transcriptomic, proteomic, and environmental data to obtain cellular models that are predictive of basic emergent cellular properties.

10.6 Integration and Meta-analysis

The critical task and opportunity in systems biology and high-dimensional measurements (including omics) are to coherently integrate and apply these data into forms that are easily distilled and amenable to testable scientific questions. Some of these questions include

how do multiple experiments agree? Which aggregate patterns are robust and unique? Which features are condition specific? What are the informative trends and relationships between linked but orthogonal components: genome, transcriptome, proteome, metabolome, phenotype, community, and environment? What is the organizational and reactive information contained within the system, what are its constraints, and how is it most likely transmitted?

In the case of transcriptome-wide gene expression, coherent microarray or RNA sequencing data from many independent experiments can be readily integrated to discover patterns of co-expression and conditionality that are only evident in aggregation. In the case of microalgae, as in other organisms, this can more powerfully imply condition-specific units of implied co-regulation and function than individual experiments alone (Hennon et al. 2015), partition all genes into subgroups of statistical and conditional relatedness within and between species (Ashworth et al. 2016), and identify core features of apparent conditional metabolic control (Levering et al. 2017). The bioinformatics and integrative construction of metabolic models help to organize, explain, and predict the flow of metabolites in new microalgal species (Chang et al. 2011; Nogales et al. 2012; Levering et al. 2016; Kim et al. 2016). Layering additional data types together into multi-scale models is also in some sense simple, given adequately comprehensiveness and coherence (Karr et al. 2012).

This is now also imminently possible for microalgae. Increasingly, deep data collection and performance-optimised "data-driven" statistics must be intimately applied to justifiable biological hypotheses in order understand and compare profound and complex features evident in single cells or entire ecosystems. Modeling ocean processes has evidently little do to with the tidy normal distributions and convenient statistical models that easily obtained in the laboratory. In practice, assumptions about what to expect in environmental datasets must be very carefully considered, and designed uniquely to address new questions with statistically valid findings. The computational challenge alone rivals the prodigious stargazing efforts and brilliant scientific advances of the national space programs of the 20th century. The constellations of biological entities and processes occurring in the world's oceans still exceed our ability to fully observe them, or to accurately and predictively model all of the biological processes operating in balance with changing environments. This is a major new challenge for 21st century oceanography.

10.7 Prediction and Synthesis

The increasing depth and comprehensiveness available through high-throughput molecular data collection now parameterizes the operational details of living systems in typically overwhelming detail. This detail is necessary but not sufficient to constitute scientific knowledge and utility. The burden is on a [systems] biologist to demonstrate the soundness, practicality, and relevance of their data, models, and products to broader fields. How is high-data analysis and modeling *useful*, *informative*, and *predictive*? For what reasons were these complex data and models invested in and what is their readily transferable scientific or technological value?

Two applicable aims in this regard are (1) predictions of present and future ecosystem properties and dynamics and (2) predictive testing, optimization, and reengineering of cellular and system-wide properties. The use of complex "whole-cell" models with predictable aggregate properties could be used to deepen predictions of genetic and cellular functions that drive ecosystems (Bragg et al. 2010), and responses and adaptations of species and strains to ecological situations may soon be modelable as a function of complex, interacting, and adaptable molecular programs whose inputs, rules, and outputs are predictable. Complementary to pure new "data-driven" approaches are efficient and multiplexed laboratory experiments that are able to directly probe the functions and impacts of specific new gene candidates identified from conditional 'omics experiments. Two exciting examples of this are direct and thoroughly demonstrations of the evidently broad effects of transcription factors and regulatory systems on productive microalgal phenotypes. The direct experimental knockdowns of single transcription factors identified by targeted transcriptomics experiments in both *P. tricornutum* (Matthijs et al. 2017) and *N. gaditana* (Ajjawi et al. 2017) resulted in dramatic changes in metabolism and natural product profiles, demonstrating the genetic and regulatory flexibility of microalgal species with regard to microbial engineering.

The use of whole-cell modeling is also an attractive answer to the challenge microbial engineering that can probe the bounds and productive potential of microalgae. This may be crucial for pathways or cellular functions whose operation involves multigenic tuning, signaling and regulatory logic, subcellular organization, or large-scale bulk cell properties and phenotypes. Honest estimates of uncertainties, model assumptions, and validating tests are all critical to the scientific relevance of integrative analyses—as well as efficient and parsimonious algorithms that can be widely adopted and understood. True biotechnological gains from data-based modeling in microbial systems will require the contextualization and design of methods specifically to be predictive, parsimonious and practical engineering within larger real-world goals, constrains, and opportunities (Georgianna and Mayfield 2012).

Marine systems are vast, and they are ubiquitously populated, produced, and balanced by the contextual operation of diverse and complex microalgae. Understanding the molecular and genetic dynamics of present and future marine biological systems will be crucial to interpret large-scale observations and shifts in marine ecosystems. As cellular and biological systems often vary astonishingly in their distinctly varying modes of genome organization, regulation, metabolism, interactions, and environments, it will be crucial to begin modeling efforts by designing data collection, analyses, model, and algorithms to suit the essential biology at hand. Many systems may not simply conform to pre-existing assumptions, tools, or frameworks. As the depth and variety of available data and modeling approaches continue to increase, continued critical, honest, practical, efficient, and rigorous adaption of scientifically focused thinking, data collection, modeling, analysis, prediction, and validation methods will yield the most fit and fruitful and translatable products of systems-level scientific research.

References

Ajjawi I, Verruto J, Aqui M, Soriaga LB, Coppersmith J, Kwok K, Peach L, Orchard E, Kalb R, Xu W, Carlson TJ, Francis K, Konigsfeld K, Bartalis J, Schultz A, Lambert W, Schwartz AS, Brown R, Moellering ER (2017) Lipid production in Nannochloropsis gaditana is doubled by decreasing expression of a single transcriptional regulator. Nat Biotechnol 35(7):647–652

Alexander H, Jenkins BD, Rynearson TA, Dyhrman ST (2015) Metatranscriptome analyses indicate resource partitioning between diatoms in the field. Proc Natl Acad Sci 112:E2182–E2190. doi:10.1073/pnas.1421993112

Armbrust EV, Berges JA, Bowler C et al (2004) The genome of the diatom Thalassiosira pseudonana: ecology, evolution, and metabolism. Science 306:79–86. doi:10.1126/science.1101156

Ashworth J, Coesel S, Lee A et al (2013) Genome-wide diel growth state transitions in the diatom Thalassiosira pseudonana. Proc Natl Acad Sci U S A 110:7518–7523. doi:10.1073/pnas.1300962110

Ashworth J, Turkarslan S, Harris M et al (2016) Pan-transcriptomic analysis identifies coordinated and orthologous functional modules in the diatoms Thalassiosira pseudonana and Phaeodactylum tricornutum. Mar Genomics 26:21–28. doi:10.1016/j.margen.2015.10.011

Aylward FO, Eppley JM, Smith JM et al (2015) Microbial community transcriptional networks are conserved in three domains at ocean basin scales. Proc Natl Acad Sci U S A 112:5443–5448. doi:10.1073/pnas.1502883112

Basu S, Patil S, Mapleson D, Russo MT, Vitale L, Fevola C, Maumus F, Casotti R, Mock T, Caccamo M, Montresor M, Sanges R, Ferrante MI (2017) Finding a partner in the ocean: molecular and evolutionary bases of the response to sexual cues in a planktonic diatom. New Phytol 215(1):140–156

Bentkowski P, van Oosterhout C, Ashby B, Mock T (2017) The effect of extrinsic mortality on genome size evolution in prokaryotes. ISME J 11:1011–1018. doi:10.1038/ismej.2016.165

Blattner FR, Plunkett G, Bloch CA et al (1997) The complete genome sequence of Escherichia coli K-12. Science 277:1453–1462. doi:10.1126/science.277.5331.1453

Bowler C, Allen AE, Badger JH et al (2008) The Phaeodactylum genome reveals the evolutionary history of diatom genomes. Nature 456:239–244. doi:10.1038/nature07410

Bragg JG, Dutkiewicz S, Jahn O et al (2010) Modeling selective pressures on phytoplankton in the Global Ocean. PLoS One 5:e9569. doi:10.1371/journal.pone.0009569

Brembu T, Jørstad M, Winge P et al (2011) Genome-wide profiling of responses to cadmium in the diatom Phaeodactylum tricornutum. Environ Sci Technol 45:7640–7647. doi:10.1021/es2002259

Bult CJ, White O, Olsen GJ et al (1996) Complete genome sequence of the Methanogenic Archaeon, Methanococcus jannaschii. Science 273:1058–1073. doi:10.1126/science.273.5278.1058

Carvalho RN, Bopp SK, Lettieri T (2011) Transcriptomics responses in marine diatom Thalassiosira pseudonana exposed to the polycyclic aromatic hydrocarbon Benzo[a]pyrene. PLoS One 6:e26985. doi:10.1371/journal.pone.0026985

Chang RL, Ghamsari L, Manichaikul A et al (2011) Metabolic network reconstruction of Chlamydomonas offers insight into light-driven algal metabolism. Mol Syst Biol 7:518. doi:10.1038/msb.2011.52

Chauton MS, Winge P, Brembu T et al (2013) Gene regulation of carbon fixation, storage, and utilization in the diatom Phaeodactylum tricornutum acclimated to light/dark cycles. Plant Physiol 161:1034–1048. doi:10.1104/pp.112.206177

de Vargas C, Audic S, Henry N et al (2015) Eukaryotic plankton diversity in the sunlit ocean. Science 348:1261605. doi:10.1126/science.1261605

Dufresne A, Salanoubat M, Partensky F et al (2003) Genome sequence of the cyanobacterium Prochlorococcus marinus SS120, a nearly minimal oxyphototrophic genome. Proc Natl Acad Sci 100:10020–10025. doi:10.1073/pnas.1733211100

Dyhrman ST, Jenkins BD, Rynearson TA et al (2012) The Transcriptome and proteome of the diatom Thalassiosira pseudonana reveal a diverse phosphorus stress response. PLoS One 7:e33768. doi:10.1371/journal.pone.0033768

Falkowski PG (1997) Evolution of the nitrogen cycle and its influence on the biological sequestration of CO2 in the ocean. Nature 387:272–275. doi:10.1038/387272a0

Falkowski PG, Barber RT, Smetacek V (1998) Biogeochemical controls and feedbacks on ocean primary production. Science 281:200–206. doi:10.1126/science.281.5374.200

Fleischmann RD, Adams MD, White O et al (1995) Whole-genome random sequencing and assembly of Haemophilus influenzae Rd. Science 269:496–512. doi:10.1126/science.7542800

Fraser CM, Casjens S, Huang WM et al (1997) Genomic sequence of a Lyme disease spirochaete, Borrelia burgdorferi. Nature 390:580–586. doi:10.1038/37551

Fraser CM, Gocayne JD, White O et al (1995) The minimal Gene complement of mycoplasma genitalium. Science 270:397–404. doi:10.1126/science.270.5235.397

Fraser CM, Norris SJ, Weinstock GM et al (1998) Complete genome sequence of Treponema pallidum, the syphilis spirochete. Science 281:375–388. doi:10.1126/science.281.5375.375

Georgianna DR, Mayfield SP (2012) Exploiting diversity and synthetic biology for the production of algal biofuels. Nature 488:329–335. doi:10.1038/nature11479

Gilbert JA, Dupont CL (2011) Microbial Metagenomics: beyond the genome. Annu Rev Mar Sci 3:347–371. doi:10.1146/annurev-marine-120709-142811

Goffeau A, Barrell BG, Bussey H et al (1996) Life with 6000 genes. Science 274:546–567. doi:10.1126/science.274.5287.546

Hennon GMM, Ashworth J, Groussman RD et al (2015) Diatom acclimation to elevated CO2 via cAMP signalling and coordinated gene expression. Nat Clim Chang 5:761–765. doi:10.1038/nclimate2683

Hook SE, Osborn HL (2012) Comparison of toxicity and transcriptomic profiles in a diatom exposed to oil, dispersants, dispersed oil. Aquat Toxicol Amst Neth 124–125:139–151. doi:10.1016/j.aquatox.2012.08.005

Ideker T, Galitski T, Hood L (2001) A NEW APPROACH TO DECODING LIFE: systems biology. Annu Rev Genomics Hum Genet 2:343–372. doi:10.1146/annurev.genom.2.1.343

Iverson V, Morris RM, Frazar CD et al (2012) Untangling genomes from Metagenomes: revealing an uncultured class of marine Euryarchaeota. Science 335:587–590. doi:10.1126/science.1212665

Karr JR, Sanghvi JC, Macklin DN et al (2012) A whole-cell computational model predicts phenotype from genotype. Cell 150:389–401. doi:10.1016/j.cell.2012.05.044

Karsenti E, Acinas SG, Bork P et al (2011) A holistic Approach to marine eco-systems biology. PLoS Biol 9:e1001177. doi:10.1371/journal.pbio.1001177

Keeling PJ, Burki F, Wilcox HM et al (2014) The marine microbial eukaryote Transcriptome sequencing project (MMETSP): illuminating the functional diversity of eukaryotic life in the oceans through transcriptome sequencing. PLoS Biol 12:e1001889. doi:10.1371/journal.pbio.1001889

Kim J, Fabris M, Baart G et al (2016) Flux balance analysis of primary metabolism in the diatom Phaeodactylum tricornutum. Plant J 85:161–176. doi:10.1111/tpj.13081

Kusebauch U, Campbell DS, Deutsch EW et al (2016) Human SRMAtlas: a resource of targeted assays to quantify the complete human proteome. Cell 166:766–778. doi:10.1016/j.cell.2016.06.041

Levering J, Broddrick J, Dupont CL et al (2016) Genome-scale model reveals metabolic basis of biomass partitioning in a model diatom. PLoS One 11:e0155038. doi:10.1371/journal.pone.0155038

Levering J, Dupont CL, Allen AE, et al (2017) Integrated regulatory and metabolic networks of the marine diatom Phaeodactylum tricornutum predict the response to rising CO2 levels. mSystems 2:e00142-16. doi: 10.1128/mSystems.00142-16

Lin S, Cheng S, Song B, Zhong X, Lin X, Li W, Li L, Zhang Y, Zhang H, Ji Z, Cai M, Zhuang Y, Shi X, Lin L, Wang L, Wang Z, Liu X, Yu S, Zeng P, Hao H, Zou Q, Chen C, Li Y, Wang Y, Xu C, Meng S, Xu X, Wang J, Yang H, Campbell DA, Sturm NR, Dagenais-Bellefeuille S, Morse D (2015) The Symbiodinium kawagutii genome illuminates dinoflagellate gene expression and coral symbiosis. Science 350(6261):691–694

Lommer M, Specht M, Roy A-S, Kraemer L, Andreson R, Gutowska MA, Wolf J, Bergner SV, Schilhabel MB, Klostermeier UC, Beiko RG, Rosenstiel P, Hippler M, LaRoche J (2012) Genome and low-iron response of an oceanic diatom adapted to chronic iron limitation. Genome Biol 13(7):R66

Marchetti A, Schruth DM, Durkin CA et al (2012) Comparative metatranscriptomics identifies molecular bases for the physiological responses of phytoplankton to varying iron availability. PNAS 2012 109 (6) E317–E325. doi:10.1073/pnas.1118408109

Marchetti A, Moreno CM, Cohen NR, Oleinikov I, deLong K, Twining BS, Armbrust VE, Lampe RH, Müller K (2017) Development of a molecular-based index for assessing iron status in bloom-forming pennate diatoms. J Phycol 53(4):820–832

Merchant SS, Prochnik SE, Vallon O et al (2007) The Chlamydomonas genome reveals the evolution of key animal and Plant functions. Science 318:245–250. doi:10.1126/science.1143609

Mock T, Samanta MP, Iverson V et al (2008) Whole-genome expression profiling of the marine diatom Thalassiosira pseudonana identifies genes involved in silicon bioprocesses. Proc Natl Acad Sci U S A 105:1579–1584. doi:10.1073/pnas.0707946105

Mock T, Otillar RP, Strauss J, McMullan M, Paajanen P, Schmutz J, Salamov A, Sanges R, Toseland A, Ward BJ, Allen AE, Dupont CL, Frickenhaus S, Maumus F, Veluchamy A, Wu T, Barry KW, Falciatore A, Ferrante MI, Fortunato AE, Glöckner G, Gruber A, Hipkin R, Janech MG, Kroth PG, Leese F, Lindquist EA, Lyon BR, Martin J, Mayer C, Parker M, Quesneville H, Raymond JA, Uhlig C, Valas RE, Valentin KU, Worden AZ, Virginia Armbrust E, Clark MD,

Bowler C, Green BR, Moulton V, van Oosterhout C, Grigoriev IV (2017) Evolutionary genomics of the cold-adapted diatom Fragilariopsis cylindrus. Nature 541(7638):536–540

Matthijs M, Fabris M, Obata T, Foubert I, Franco-Zorrilla JM, Solano R, Fernie AR, Vyverman W, Goossens A (2017) The transcription factor bZIP14 regulates the TCA cycle in the diatom. EMBO J 36(11):1559–1576

Nogales J, Gudmundsson S, Knight EM et al (2012) Detailing the optimality of photosynthesis in cyanobacteria through systems biology analysis. Proc Natl Acad Sci 109:2678–2683. doi:10.1073/pnas.1117907109

Nunn BL, Aker JR, Shaffer SA et al (2009) Deciphering diatom biochemical pathways via whole-cell proteomics. Aquat Microb Ecol Int J 55:241–253. doi:10.3354/ame01284

Nunn BL, Faux JF, Hippmann AA et al (2013) Diatom proteomics reveals unique acclimation strategies to mitigate Fe limitation. PLoS One 8:e75653. doi:10.1371/journal.pone.0075653

Nymark M, Valle KC, Brembu T et al (2009) An integrated analysis of molecular acclimation to high light in the marine diatom Phaeodactylum tricornutum. PLoS One 4:e7743. doi:10.1371/journal.pone.0007743

Nymark M, Valle KC, Hancke K et al (2013) Molecular and photosynthetic responses to prolonged darkness and subsequent acclimation to re-illumination in the diatom Phaeodactylum tricornutum. PLoS One 8:e58722. doi:10.1371/journal.pone.0058722

Ottesen EA, Young CR, Eppley JM et al (2013) Pattern and synchrony of gene expression among sympatric marine microbial populations. Proc Natl Acad Sci U S A 110:E488–E497. doi:10.1073/pnas.1222099110

Ottesen EA, Young CR, Gifford SM et al (2014) Multispecies diel transcriptional oscillations in open ocean heterotrophic bacterial assemblages. Science 345:207–212. doi:10.1126/science.1252476

Palenik B, Brahamsha B, Larimer FW et al (2003) The genome of a motile marine Synechococcus. Nature 424:1037–1042. doi:10.1038/nature01943

Rocap G, Larimer FW, Lamerdin J et al (2003) Genome divergence in two Prochlorococcus ecotypes reflects oceanic niche differentiation. Nature 424:1042–1047. doi:10.1038/nature01947

Saito MA, McIlvin MR, Moran DM et al (2014) Multiple nutrient stresses at intersecting Pacific Ocean biomes detected by protein biomarkers. Science 345:1173–1177. doi:10.1126/science.1256450

Sapriel G, Quinet M, Heijde M et al (2009) Genome-wide Transcriptome analyses of silicon metabolism in Phaeodactylum tricornutum reveal the multilevel regulation of silicic acid transporters. PLoS One 4:e7458. doi:10.1371/journal.pone.0007458

Shrestha RP, Tesson B, Norden-Krichmar T et al (2012) Whole transcriptome analysis of the silicon response of the diatom Thalassiosira pseudonana. BMC Genomics 13:499. doi:10.1186/1471-2164-13-499

Smith SR, Gillard JTF, Kustka AB et al (2016a) Transcriptional orchestration of the global cellular response of a model Pennate diatom to Diel light cycling under iron limitation. PLoS Genet 12:e1006490. doi:10.1371/journal.pgen.1006490

Smith SR, Glé C, Abbriano RM et al (2016b) Transcript level coordination of carbon pathways during silicon starvation-induced lipid accumulation in the diatom Thalassiosira pseudonana. New Phytol 210:890–904. doi:10.1111/nph.13843

Sunagawa S, Coelho LP, Chaffron S et al (2015) Structure and function of the global ocean microbiome. Science 348:1261359. doi:10.1126/science.1261359

Traller JC, Cokus SJ, Lopez DA, Gaidarenko O, Smith SR, JP MC, Gallaher SD, Podell S, Thompson M, Cook O, Morselli M, Jaroszewicz A, Allen EE, Allen AE, Merchant SS, Pellegrini M, Hildebrand M (2016) Genome and methylome of the oleaginous diatom Cyclotella cryptica reveal genetic flexibility toward a high lipid phenotype. Biotechnol Biofuels 9(1):258

Thamatrakoln K, Korenovska O, Niheu AK, Bidle KD (2012) Whole-genome expression analysis reveals a role for death-related genes in stress acclimation of the diatom Thalassiosira pseudonana. Environ Microbiol 14:67–81. doi:10.1111/j.1462-2920.2011.02468.x

Valle KC, Nymark M, Aamot I et al (2014) System responses to equal doses of Photosynthetically usable radiation of blue, green, and red light in the marine diatom Phaeodactylum tricornutum. PLoS One 9:e114211. doi:10.1371/journal.pone.0114211

Venter JC, Remington K, Heidelberg JF et al (2004) Environmental genome shotgun sequencing of the Sargasso Sea. Science 304:66–74. doi:10.1126/science.1093857

von Bertalanffy L (1968) General system theory: foundations, development, applications. George Braziller, New York

Waldbauer JR, Rodrigue S, Coleman ML, Chisholm SW (2012) Transcriptome and proteome dynamics of a light-dark synchronized bacterial cell cycle. PLoS One 7:e43432. doi:10.1371/journal.pone.0043432

Xiong Y, Soumillon M, Wu J, et al (2017) A comparison of mRNA sequencing with random primed and 3′-directed libraries. bioRxiv 098905. doi: 10.1101/098905

Zwirglmaier K, Jardillier L, Ostrowski M et al (2008) Global phylogeography of marine Synechococcus and Prochlorococcus reveals a distinct partitioning of lineages among oceanic biomes. Environ Microbiol 10:147–161. doi:10.1111/j.1462-2920.2007.01440.x

Chapter 11
Application of 'Omics' Approaches to Microbial Oceanography

Deepa R. Varkey and Martina A. Doblin

Abstract Viruses, bacteria, archaea and single celled eukaryotes, collectively known as microbes, dominate the biomass and metabolism of ocean ecosystems. Marine microbes are highly abundant and critical to human survival, but the vast majority of taxa have not yet been cultured. The use of environmental nucleic acid sequencing as a cultivation-independent approach to microbial oceanography has therefore significantly expanded our understanding of the diversity, evolution, biogeography and important biogeochemical roles of marine microorganisms. Here we provide illustrative examples of how genomic, transcriptomic and proteomic approaches have been applied to marine microbes to advance our understanding of their ecology. A remaining challenge is the need to link phenotypes to their environment, requiring a better understanding of genomic features that influence transcription (e.g. promoters and methylation) as well as post-translational modifications, and how such regulatory processes are impacted by extracellular abiotic and biotic processes. In addition, the expansion of available protein and taxonomic databases will greatly increase our capacity to link microbial function to specific taxa.

Keywords Marine microbes • Microbial function • Genomics • Proteomics • Transcriptomics

Contents

11.1 Introduction 224
11.2 Influence of Oceanographic Processes on Microbial Distribution and Diversity 224
11.3 Influence of Oceanographic Processes on Microbial Gene Expression 227
11.4 Influence of Oceanographic Processes on Microbial Protein Expression 229
11.5 Conclusion 231
References 232

D.R. Varkey • M.A. Doblin (✉)
Climate Change Cluster, University of Technology Sydney,
PO Box 123 Broadway, Sydney, NSW 2007, Australia
e-mail: martina.doblin@uts.edu.au

© Springer International Publishing AG 2017

M. Kumar, P. Ralph (eds.), *Systems Biology of Marine Ecosystems*,
DOI 10.1007/978-3-319-62094-7_11

11.1 Introduction

Whilst oceanographers have been interested in microbially mediated ocean processes for more than a century, our understanding of the diversity and functional capacity of these organisms has really only been delivered since the development of molecular approaches in the 1980s. Prior to this, our knowledge about their roles in the ocean was primarily derived from culture-dependent approaches, whereby microbes were isolated and grown under different conditions and subsequently tested for their chemical and biological activity. However, as a consequence of the difficulty in cultivating many microbial species, these approaches only afforded insights into a very small proportion of the microbes in the ocean (Rappé and Giovannoni 2003).

In general, molecular methodologies rely on the fact that microbes share common genes that have changed relatively little throughout the evolutionary history of life on Earth. Analysing the differences in these conservative genes enables organisms to be classified into different species or operational taxonomic units (OTUs) in a way that reflects their evolutionary history (Giovannoni and Cary 1993). Moreover, DNA codes for a large variety of proteins which can then be used to infer microbial metabolism and function (DeLong 2009). The techniques that sequence DNA, examine its transcription, and the subsequent expression of proteins are known as genomics, transcriptomics and proteomics, respectively. Collectively, these methodologies are referred to as 'omic approaches.

Various 'omic techniques have been applied in broadscale ocean surveys such as the Global Ocean Sampling expedition (GOS), International Census of Marine Microbes (ICoMM) and Tara Oceans expedition, as well as long-term microbial observatories such as the Bermuda Atlantic Time-Series and Hawaii Ocean Time-series (BATS and HOT, respectively). Here we provide a focussed, but deliberately not exhaustive, set of laboratory and field-based examples which demonstrate the use of such 'omic approaches to characterise microbial processes in the ocean.

11.2 Influence of Oceanographic Processes on Microbial Distribution and Diversity

Early investigations of the diversity of marine microbial samples involved fingerprinting techniques (Avaniss-Aghajani et al. 1994) which separated ribosomal RNA (rRNA) fragments according to their length or nucleotide composition (e.g. terminal restriction fragment length polymorphism (T-RFLP), automated rRNA intergenic spacer analysis (ARISA)). Such methods provided no significant details on the identity of organisms, simply patterns of similarity between samples (Dorigo et al. 2005; Zinger et al. 2012). Alternatively, researchers identified microbes within natural samples by microscopy using fluorescence in situ hybridisation or created clone libraries using rRNA gene fragments, involving PCR amplification of a specific marker gene which was cloned into vectors, followed by sequencing (Dorigo et al. 2005). The construction of clone libraries is time-consuming, and identification was therefore restricted to 100 clones or less in most instances. However, with the advent of next-generation sequencing, it became possible to retrieve thousands of sequences from a single sample, significantly expanding our view of microbial diversity in the environment.

Over the last 30 years, multiple sequencing approaches have been used to undertake genomic studies of marine microbes, but the development of next-generation sequencing has led to increasing application of metabarcoding and metagenomics (Fig. 11.1) as rapid and effective methods for characterising ocean microbial communities. These are techniques that combine DNA-based identification with high-throughput DNA sequencing, enabling a high degree of parallel sequencing for the analyses of multiple samples or complex communities, or both. Amplicon sequencing (otherwise referred to as metabarcoding) is a technique that utilises the parallel sequencing power of next-generation sequencing to characterise microbial diversity based on a phylogenetic marker that is conserved across taxa (Fig. 11.1). The marker gene is PCR amplified and sequenced to determine the diversity and relative abundance of taxa present in the sample. Commonly used markers include the 16S rRNA

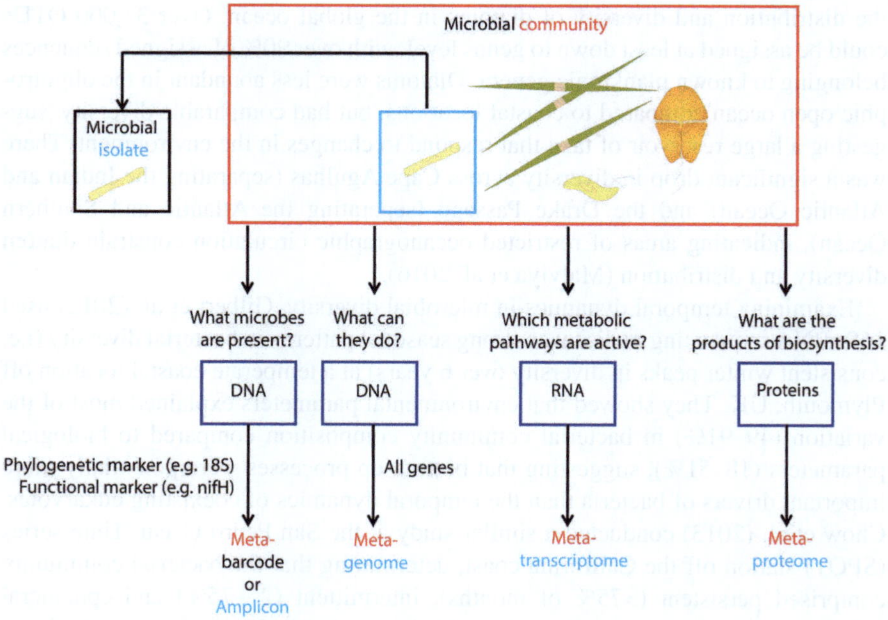

Fig. 11.1 Application of 'omics approaches to microbial communities (*red text*) and isolates (*blue text*). Genomic approaches using next-generation sequencing involve extraction of DNA and sequencing of fragments amplified from a single marker gene (amplicon sequencing or metabarcoding) or randomly generated fragments of DNA (shotgun metagenome sequencing), whilst transcriptomic and proteomic approaches examine the products of gene expression and their respective metabolic pathways

gene for prokaryotes (Pace et al. 1986), 18S and 28S rRNA genes for eukaryotes (Moon-van der Staay et al. 2000; Gong et al. 2013) and the internal transcribed spacer (ITS) region for both (Brown and Fuhrman 2005; Santoferrara et al. 2014). Such markers generally resolve taxa to family or genus level, but more specific genetic markers that provide higher taxonomic resolution have also been developed (e.g. *petB* that encodes the cytochrome b6 subunit of the cytochrome b6f complex (Mazard et al. 2012), *rbcL* that encodes the large subunit of the carbon fixation enzyme RuBisCO (Hamsher et al. 2011), *hsp90* that encodes a major heat shock protein (Hoppenrath and Leander 2010) and *nifH* that encodes the iron protein of nitrogenase enzyme (Zehr and Turner 2001). Amplicon sequencing has been applied to a range of marine environments across temporal and spatial scales and has been most insightful to microbial oceanographers when diversity of a specific microbial group has been assessed, coupled with investigations of their biogeochemical transformations (e.g. diversity of diazotrophs estimated using *nifH*, with measured rates of N_2 fixation; Messer et al. 2016).

To date, many studies have explored microbial community composition changes in response to oceanographic processes such as ocean currents, upwelling, estuarine outflows and the formation of mesoscale eddies (Treusch et al. 2009; Villar et al. 2015; Doblin et al. 2016; Zielinski et al. 2016). For example, Malviya et al. (2016) used the hypervariable V9 region of the 18S ribosomal marker gene to characterise the distribution and diversity of diatoms in the global ocean. Over 33,000 OTUs could be assigned at least down to genus level, with over 90% of assigned sequences belonging to known planktonic genera. Diatoms were less abundant in the oligotrophic open ocean compared to coastal locations, but had comparable diversity, suggesting a large reservoir of taxa that respond to changes in the environment. There was a significant drop in diversity across Cape Agulhas (separating the Indian and Atlantic Ocean) and the Drake Passage (separating the Atlantic and Southern Ocean), indicating areas of restricted oceanographic circulation constrain diatom diversity and distribution (Malviya et al. 2016).

Examining temporal dynamics in microbial diversity, Gilbert et al. (2012) used 16S rRNA sequencing to discover strong seasonal patterns in bacterial diversity (i.e. consistent winter peaks in diversity over 6 years) at a temperate coastal location off Plymouth, UK. They showed that environmental parameters explained most of the variation (49–91%) in bacterial community composition compared to biological parameters (18–51%), suggesting that bottom-up processes were potentially more important drivers of bacteria than the temporal dynamics of coexisting eukaryotes. Chow et al. (2013) conducted a similar study at the San Pedro Ocean Time-series (SPOT) station off the California coast, determining that the bacterial community comprised persistent (>75% of months), intermittent (25–75%) and ephemeral (<25%) taxa. There was a relatively stable core microbial community at both the surface and subsurface chlorophyll-a maximum (~30 m) such that OTUs were similar between samples taken days, weeks, months or years apart, with most pairs of samples having on average at least 36% similarity.

Shotgun metagenomics sequencing is another approach used to characterise microbial communities. No specific gene is targeted but rather the whole community DNA is fragmented and sequenced (Fig. 11.1), yielding thousands of sequences for multiple genes from multiple organisms present in the community. Metagenomics has been extensively applied to various environments across multiple spatial and temporal scales (see review Sharpton 2014; Thomas et al. 2012). This is a powerful technique that provides information about not just the microbial community composition and structure, but also their functional potential. This approach enables the identification of novel genes and biomarkers from uncultured organisms.

Allen et al. (2012) used a metagenomic approach to investigate microbial responses to upwelling in Californian coastal waters, demonstrating that oligotrophic waters were dominated by alpha-proteobacteria (primarily *Pelagibacter* sp.) and *Prochlorococcus* sp., with upwelled and aged upwelled water having distinctly different microbial communities. Sites from oligotrophic waters had a greater number of genomes and lower estimated genome size, whereas sites from upwelling regions had a smaller number of genomes and greater genome size (Allen et al. 2012). Upwelling sites were enriched in diatom sequences in the largest size class (3.0–200 μm), as well as the picoeukaryotic prasinophytes, *Micromonas* sp. and *Ostreococcus* sp. In contrast, offshore oligotrophic sites, particularly in the case of the largest size class, were enriched in sequences classified as dinoflagellates or other Alveolata. Within the picoeukaryotic size class (0.8–3.0 μm), the prasinophytes were more abundant at upwelling sites, whereas pelagophytes and ciliates were more abundant at oligotrophic sites.

Whilst such genomic datasets are more comprehensive, faster to produce and less dependent on taxonomic expertise, they rely on comparison with reference sequences from known taxa. This has necessitated sequencing of key microbial taxa to achieve representation of virus, bacteria and archaea genomes in reference databases, as well as a growing number of eukaryotes (Keeling et al. 2014).

Despite our increasing understanding of how microbes are distributed between and within ocean basins, across depth and over time, there is still no consensus on the role of environmental factors in regulating community assembly or whether biotic interactions are more important (Giovannoni and Vergin 2012). Furthermore, whilst genomic approaches provide a blueprint of the genetic diversity within communities, they lack information on gene expression dynamics and changes in microbial activity. To progress understanding of how microbes respond to different aspects of their growth environment, we therefore need to go beyond genomic approaches.

11.3 Influence of Oceanographic Processes on Microbial Gene Expression

Transcriptomics involves the exploration of genes that are expressed at a given time and place and how expression changes under different conditions. Whilst transcriptomics focuses on a single organism, metatranscriptomics is the sequencing of expressed genes within a whole community. Gene expression changes reveal active metabolic pathways and the molecular adjustment of cells in response to changing growth conditions. This approach is useful in determining the molecular mechanisms underpinning community dynamics and structure. Below we provide some illustrative examples of how these approaches have been used to characterise the responses of marine microbial communities to different oceanographic processes. Given that many oceanographic phenomena cause local and regional changes in resources for growth, many of these transcriptomic studies have explored responses to different nutrient conditions.

In a laboratory study, Mock et al. (2008) grew the widely distributed marine eukaryote *Thalassiosira pseudonana* axenically (clone CCMP 1335 for which the whole-genome sequence is available) to examine its transcriptome in response to different growth environments. Whole-genome expression profiling of this species revealed a set of 75 genes specifically upregulated during silicon limitation but not under low concentrations of nitrogen or iron, alkaline pH, or low temperatures (Mock et al. 2008). This study also found unexpectedly tight coupling of pathways initiated by iron (Fe) and silicon (Si) bioavailability (84 common genes upregulated by both Fe and Si limitation but no other treatments), suggesting that the in situ iron/silicon nutritional status of diatoms could be detected using molecular indicators, thereby enhancing our understanding of these important nutritional controls on diatom productivity.

Shi et al. (2011) compared the diversity and metabolic activity of a bacterioplankton community at discrete depths in a stratified water column using a metatranscriptomic approach. Comparison of the DNA and cDNA libraries suggested differential relative transcriptional activities per cell and revealed a transcriptionally active but less abundant *Prochlorococcus* population in the bottom of the photic zone. The metatranscriptomic dataset sheds light on the genes that were required to maintain function at different depths such as the high expression of oxidative stress-associated genes at the surface, carbon fixation and photosynthetic genes at DCM and ammonium assimilation genes at mesopelagic region (500 m depth). Functional information was used to associate individual taxa to biogeochemical processes including the role of *Roseobacter* in aerobic anoxygenic phototrophy (expression of proteorhodopsin) and *Crenarchaeota* in ammonia oxidation (expression of ammonium transporters and ammonium monooxygenase subunits). The study also highlighted the adjustment of gene expression based on nutrient availability and requirement, i.e. distinct profiles of transporters such as ABC-transporters, Na+/solute symporters and tripartite ATP-independent periplasmic transporters expressed by *Pelagibacter* populations at different depths (Shi et al. 2011).

To explore the role of iron in a natural phytoplankton community, Marchetti et al. (2012) undertook a microcosm experiment investigating the differential gene expression under conditions of iron limitation versus iron enrichment. Iron enrichment resulted in a community shift from small picophytoplankton to larger eukaryotic phytoplankton, mainly diatoms, haptophytes and chlorophytes, but there was also a distinct difference in expression within the three dominant phytoplankton groups. Under iron amendment, diatoms increased expression of genes involved in photosynthesis, N assimilation and carbohydrate storage, whilst haptophytes upregulated iron-containing proteins for light harvesting and photosynthesis with no change in nitrate assimilation genes. Notably, under iron-replete conditions, diatoms continued using iron-free versions of photosynthetic proteins (i.e. did not replace non-iron proteins with their more efficient iron-containing counterparts), selectively partitioning acquired iron towards nutrient assimilation rather than light harvesting. This could offer some explanation for why diatoms dominate low iron environments (Marchetti et al. 2012). Haptophytes, on the other hand, increased their expression of iron-containing proteins (Marchetti et al. 2012). Thus, application of a transcriptomic approach to examine nutrient amendment is not only useful to understand lifestyle strategies between taxonomic groups but also to explore nutrient partitioning within individual cells.

In a further example, Alexander et al. (2015) explored resource partitioning amongst coexisting diatoms using nutrient amendment and quantitative metatranscriptomics. This approach highlighted the specific pathways for nutrient acquisition and metabolism utilised by sympatric diatoms within a natural community. Samples for metatranscriptomes were collected from the field, and the Marine Microbial Eukaryotic Transcriptome Sequencing Project database was used as a reference to identify transcripts. Resource portioning was evident by the different transcriptional responses between *Skeletonema* spp. and *Thalassiosira rotula* in the same environment, enabling co-occurrence without competition, a first time observation. During the bloom, the dominant diatom was *Skeletonema* which had high expression of growth-related genes such as those involved in carbon, nitrogen, sulfur and lipid metabolism as well as nitrate and ammonia assimilation whilst the less abundant diatom, *Thalassiosira*, highly expressed genes associated with amino acid transporters as well as transport and repair (Alexander et al. 2015). This study highlighted that despite possessing similar nitrogen and phosphorous transport and metabolism genes, the two species differed temporally in their expression, suggesting resource partitioning.

11.4 Influence of Oceanographic Processes on Microbial Protein Expression

Whilst transcriptomics provides valuable insights into the metabolic potential of individual species and natural communities by pinpointing differentially expressed genes, the observed expression does not necessarily directly translate into synthesised proteins. Since there are multiple regulatory controls on genes and post-transcriptional modifications on expressed transcripts (Williams and Cavicchioli 2014), measures of expressed proteins (i.e. products of biosynthesis) provide verification of function over metabolic potential inferred from the genome or the transcriptome (Fig. 11.1). The high-throughput 'omic' approach, proteomics, determines the presence and abundance of expressed proteins, providing an additional level of information on microbial activity and function. Metaproteomics is the application of this technique to entire populations or communities without specific targets (Morris et al. 2010; Williams and Cavicchioli 2014). This requires reference genomes or metagenomic datasets to identify protein-coding genes. Complementing metagenomics with metaproteomics therefore has the most potential to decipher microbial community dynamics and function. The rapidly developing field of metaproteomics has had increasing application within marine systems to better understand microbial communities and their roles in biogeochemical functions (Wang et al. 2014; Williams and Cavicchioli 2014).

Two early studies by Sowell et al. (2009, 2011) elegantly demonstrated the utility of metaproteomics for microbial oceanography. Sowell et al. (2009) explored changes in protein expression under contrasting oceanographic conditions to examine how microbes inhabiting the upper ocean respond to their environmental conditions. Under stratified conditions in the oligotrophic Sargasso Sea, the heterotrophic marine bacterium SAR11 expressed high-affinity transporters for limiting nutrients (e.g. phosphate), providing a possible explanation for the dominance of this organism during nutrient-deplete summer months (Sowell et al. 2009). On the other hand, in the upwelling influenced and nutrient-rich coastal regions of the Oregon shelf, SAR11 alternatively displayed a higher expression of transporters associated with carbon and nitrogen uptake (Sowell et al. 2011). These comparative proteomic approaches therefore delivered a new understanding of how the ocean's most abundant bacterium adjusts its metabolic machinery according to resource availability, allowing it to succeed in both nutrient-limited and nutrient-replete conditions.

Morris et al. (2010) utilised metaproteomics to explore how nutrient utilisation strategies shape marine bacterial community structure along an oceanic nutrient gradient. They focussed on membrane proteins that were identified both functionally and taxonomically to provide insight into changes in both the composition of the community and the proteins they expressed. TonB-dependent transporters, which are membrane proteins that transport diverse nutrients across the outer membrane of Gram-negative bacteria using a proton motive force, were found to be more abundant in nutrient-rich coastal waters, whereas other transporters such as porins, permeases and major facilitator superfamily transporters, which transport a diverse range of substrates, were highly expressed in the nutrient-poor open ocean. Taxonomic assignment of these proteins highlighted that diverse taxa expressed TonB-dependent transporters whilst porins were mostly associated with the dominant open ocean microbes *Prochlorococcus* and SAR11 (Morris et al. 2010).

Proteomic approaches have also been used to examine microbial processes over time. Teeling et al. (2012) coupled metatranscriptomics and metagenomics to examine temporal dynamics of bacterioplankton substrate utilisation during the progression of a phytoplankton bloom. During the early stages of the bloom, there was a high expression of TonB-dependent transporters, which enable the utilisation of complex organic matter available at the start of the bloom and was linked to an initial dominance of *Bacteroidetes* (*Ulvibacter* and *Formosa*). Progression of the bloom and a change in algal exudate composition led to the dominance of copiotrophic opportunists that expressed transporters with broad substrate spectra, including members of the gammaproteobacteria, *Reinekea* spp. and alphaproteobacteria *Roseobacter* spp. This was followed by the rise of lineages of flavobacteria (*Polaribacter* and *Formosa*) and gammaproteobacteria (SAR92) which had expression profiles different to their predecessors. Thus in this study, metaproteomics allowed the investigation of the protein expression profiles to explain why different bacterioplankton taxa dominate across the progression of a phytoplankton bloom.

The study by Saito et al. (2014) demonstrated a different approach using proteomics, that is, targeted quantification of specific proteins of interest that were first identified using metaproteomics. This study explored how the cyanobacterium, *Prochlorococcus*, deals with multiple nutrient limitations to dominate nutrient-poor open ocean environments. Expression of targets identified from metaproteomes and culture-based studies, namely, flavodoxin (IdiA), global nitrogen response regulator (NtcA), nitrogen regulatory protein (P-II) and a urea transporter, was measured with high taxonomic resolution. The change in expression of these targets at stations in the Pacific Ocean was correlated with shifts in nutrient availability and involved increased expression of IdiA in regions of reduced iron availability and high expression of P-II in low-nitrogen regions. Thus using both targeted and non-targeted proteomic approaches, the influence of nutrient availability on specific taxa and the strategy adopted to tolerate nutrient limitations were explored.

11.5 Conclusion

Cultivation-independent 'omic' approaches have increased our understanding of the planet's most abundant organisms and their roles in the ocean. The application of genomic, transcriptomic and proteomic approaches to natural communities and microbial isolates have revealed previously undescribed microorganisms, new genes, and novel metabolic pathways. We now know that microscopic marine microbes are the most evolutionarily diverse organisms in the biosphere, are responsible for ~50% of global primary production (Field et al. 1998) and influence the Earth's climate by driving global biogeochemical cycles (Falkowski et al. 2008). These studies illustrate the immense potential that 'omic approaches have in understanding microbial communities beyond just the presence of species and their functional potential. By studying organisms at the gene expression level, we can begin to decipher how taxa adjust their molecular and physiological machinery to take advantage of changing environmental conditions and in turn shape microbial community structure, which ultimately determines ecosystem function. However, there is still much to learn about the responses of microbes to oceanographic processes, including the relative impact of physico-chemical changes in growth conditions versus biological interactions.

Acknowledgements We thank Justin Seymour, University of Technology Sydney and Martin Ostrowski, Macquarie University, for useful discussions and feedback, which greatly improved the content. This work was supported by the Climate Change Cluster, University of Technology Sydney and the Australian Research Council Discovery Scheme (DP14010134) to MD.

References

Alexander H, Jenkins BD, Rynearson TA et al (2015) Metatranscriptome analyses indicate resource partitioning between diatoms in the field. Proc Natl Acad Sci U S A 112:E2182–E2190

Allen LZ, Allen EE, Badger JH et al (2012) Influence of nutrients and currents on the genomic composition of microbes across an upwelling mosaic. ISME J 6:1403–1414

Avaniss-Aghajani E, Jones K, Chapman D et al (1994) A molecular technique for identification of bacteria using small subunit ribosomal RNA sequences. BioTechniques 17:144–146

Brown MV, Fuhrman JA (2005) Marine bacterial microdiversity as revealed by internal transcribed spacer analysis. Aquat Microb Ecol 41:15–23

Chow C-ET, Sachdeva R, Cram JA et al (2013) Temporal variability and coherence of euphotic zone bacterial communities over a decade in the Southern California Bight. ISME J 7:2259–2273

DeLong EF (2009) The microbial ocean from genomes to biomes. Nature 459:200–206

Doblin MA, Petrou K, Sinutok S et al (2016) Nutrient uplift in a cyclonic eddy increases diversity, primary productivity and iron demand of microbial communities relative to a western boundary current. PeerJ 4:e1973

Dorigo U, Volatier L, Humbert J-F (2005) Molecular approaches to the assessment of biodiversity in aquatic microbial communities. Water Res 39:2207–2218

Falkowski PG, Fenchel T, Delong EF (2008) The microbial engines that drive Earth's biogeochemical cycles. Science 320:1034–1039

Field CB, Behrenfeld MJ, Randerson JT et al (1998) Primary production of the biosphere: integrating terrestrial and oceanic components. Science 281:237–240

Gilbert JA, Steele JA, Caporaso JG et al (2012) Defining seasonal marine microbial community dynamics. ISME J 6:298–308

Giovannoni S, Cary SC (1993) Probing marine systems with ribosomal RNAs. Oceanography 6:95–104

Giovannoni SJ, Vergin KL (2012) Seasonality in ocean microbial communities. Science 335:671–676

Gong J, Dong J, Liu X et al (2013) Extremely high copy numbers and polymorphisms of the rDNA operon estimated from single cell analysis of oligotrich and peritrich ciliates. Protist 164:369–379

Hamsher SE, Evans KM, Mann DG et al (2011) Barcoding diatoms: exploring alternatives to COI-5P. Protist 162:405–422

Hoppenrath M, Leander BS (2010) Dinoflagellate phylogeny as inferred from heat shock protein 90 and ribosomal gene sequences. PLoS One 5:e13220

Keeling PJ, Burki F, Wilcox HM et al (2014) The Marine Microbial Eukaryote Transcriptome Sequencing Project (MMETSP): illuminating the functional diversity of eukaryotic life in the oceans through transcriptome sequencing. PLoS Biol 12:e1001889

Malviya S, Scalco E, Audic S et al (2016) Insights into global diatom distribution and diversity in the world's ocean. Proc Natl Acad Sci U S A 113(11):E1516–E1525

Marchetti A, Schruth DM, Durkin CA et al (2012) Comparative metatranscriptomics identifies molecular bases for the physiological responses of phytoplankton to varying iron availability. Proc Natl Acad Sci U S A 109:E317–E325

Mazard S, Ostrowski M, Partensky F et al (2012) Multi-locus sequence analysis, taxonomic resolution and biogeography of marine *Synechococcus*. Environ Microbiol 14:372–386

Messer LF, Mahaffey C, Robinson CM et al (2016) High levels of heterogeneity in diazotroph diversity and activity within a putative hotspot for marine nitrogen fixation. ISME J 10(6):1499–1513

Mock T, Samanta MP, Iverson V et al (2008) Whole-genome expression profiling of the marine diatom *Thalassiosira pseudonana* identifies genes involved in silicon bioprocesses. Proc Natl Acad Sci U S A 105:1579–1584

Moon-van der Staay SY, van der Staay GW, Guillou L et al (2000) Abundance and diversity of prymnesiophytes in the picoplankton community from the equatorial Pacific Ocean inferred from 18S rDNA sequences. Limnol Oceanogr 45:98–109
Morris RM, Nunn BL, Frazar C et al (2010) Comparative metaproteomics reveals ocean-scale shifts in microbial nutrient utilization and energy transduction. ISME J 4:673–685
Pace NR, Olsen GJ, Woese CR (1986) Ribosomal RNA phylogeny and the primary lines of evolutionary descent. Cell 45:325–326
Rappé MS, Giovannoni SJ (2003) The uncultured microbial majority. Annu Rev Microbiol 57:369–394
Saito MA, McIlvin MR, Moran DM et al (2014) Multiple nutrient stresses at intersecting Pacific Ocean biomes detected by protein biomarkers. Science 345:1173–1177
Santoferrara LF, Grattepanche JD, Katz LA et al (2014) Pyrosequencing for assessing diversity of eukaryotic microbes: analysis of data on marine planktonic ciliates and comparison with traditional methods. Environ Microbiol 16:2752–2763
Sharpton TJ (2014) An introduction to the analysis of shotgun metagenomic data. Front Plant Sci 5:209
Shi Y, Tyson GW, Eppley JM et al (2011) Integrated metatranscriptomic and metagenomic analyses of stratified microbial assemblages in the open ocean. ISME J 5:999–1013
Sowell SM, Abraham PE, Shah M et al (2011) Environmental proteomics of microbial plankton in a highly productive coastal upwelling system. ISME J 5:856–865
Sowell SM, Wilhelm LJ, Norbeck AD et al (2009) Transport functions dominate the SAR11 metaproteome at low-nutrient extremes in the Sargasso Sea. ISME J 3:93–105
Teeling H, Fuchs BM, Becher D et al (2012) Substrate-controlled succession of marine bacterioplankton populations induced by a phytoplankton bloom. Science 336:608–611
Thomas T, Gilbert J, Meyer F (2012) Metagenomics—a guide from sampling to data analysis. Microb Inform Exp 2:3
Treusch AH, Vergin KL, Finlay LA et al (2009) Seasonality and vertical structure of microbial communities in an ocean gyre. ISME J 3:1148–1163
Villar E, Farrant GK, Follows M et al (2015) Environmental characteristics of Agulhas rings affect interocean plankton transport. Science 348:1261447
Wang D-Z, Xie Z-X, Zhang S-F (2014) Marine metaproteomics: current status and future directions. J Proteome 97:27–35
Williams TJ, Cavicchioli R (2014) Marine metaproteomics: deciphering the microbial metabolic food web. Trends Microbiol 22:248–260
Zehr J, Turner P (2001) Nitrogen fixation: nitrogenase genes and gene expression. Method Microbiol 30:271–286
Zielinski BL, Allen AE, Carpenter EJ et al (2016) Patterns of transcript abundance of eukaryotic biogeochemically-relevant genes in the amazon river plume. PLoS One 11:e0160929
Zinger L, Gobet A, Pommier T (2012) Two decades of describing the unseen majority of aquatic microbial diversity. Mol Ecol 21:1878–1896

Chapter 12
Effects of Ocean Acidification and UV Radiation on Marine Photosynthetic Carbon Fixation

Kunshan Gao and Donat-P. Häder

Abstract The oceans absorb anthropogenically released CO_2 at a rate of more than one million tons per hour, which causes a pH decrease of seawater and results in ocean acidification (OA). The effect of OA and absorption of CO_2 via the biological carbon pump driven by marine photosynthesis has drawn increasing attentions. As a consequence, there are numerous studies on influences of OA on primary producers, and the effects on photosynthetic carbon fixation are still under debate. OA can promote the growth of diatoms at low PAR irradiances and inhibit it at high PAR. Besides, OA may influence metabolic pathways of phytoplankton, upregulating β-oxidation, and the tricarboxylic acid cycle, resulting in increased accumulation of toxic phenolic compounds. In parallel, phytoplankton cells in the upper mixed layer are affected by intense PAR and UV radiation (UVR). The calcareous layers of calcified algae, which have been shown to shield the organisms from UVR, are thinned due to OA, exposing the cells to increased UVR and further inhibiting the calcification. Therefore, effects of OA and UV on marine photosynthetic carbon fixation could be compounded. While the photosynthetic carbon fixation is controlled by other environmental stressors in addition to OA and UV, such as nutrients limitation and warming, combined effects of OA and UV have been less considered. In this review, we synthesize and analyze recent advances on effects of OA and UV and their combined effects, implying that future studies should pay special attentions to ecological and physiological effects of OA in the presence of solar UV irradiance to reflect more realistic implications. The ecophysiological effects of OA and/or UV and their mechanisms in complex environments should be further explored.

Keywords CO_2 • Ocean acidification • Photosynthesis • Primary producers • Solar UV radiation

K. Gao (✉)
State Key Laboratory of Marine Environmental Science, College of Ocean and Earth Sciences, Xiamen University, Xiang-An Campus, Xiamen 361102, China
e-mail: ksgao@xmu.edu.cn

D.-P. Häder
Department of Biology, Friedrich-Alexander University Erlangen-Nürnberg, Neue Str. 9, Erlangen 91096, Germany

© Springer International Publishing AG 2017

M. Kumar, P. Ralph (eds.), *Systems Biology of Marine Ecosystems*,
DOI 10.1007/978-3-319-62094-7_12

Contents

12.1 Introduction 236
12.2 Ocean Acidification and Its Effect on Photosynthetic Organisms 237
12.2.1 Ocean acidification 237
12.2.2 Responses of photosynthetic organisms to ocean acidification 239
12.3 UV and Its Effect on Marine Photosynthetic Carbon Fixation 242
12.3.1 UV radiation 242
12.3.2 The effect of UV on photosynthetic carbon fixation 244
12.4 The combined effects of OA and UV radiation 244
12.5 Conclusion 246
References 247

12.1 Introduction

Global environmental change has become a key factor influencing the sustainable development of humans, and increasing atmospheric CO_2 concentration is causing ocean acidification (OA), global warming, sea level rise, and altering water mixing dynamics. As a CO_2 sink, the oceans continuously absorb CO_2, playing an important role in mitigating global warming. Therefore, the oceanic absorption of CO_2 and its mechanism are key to study and predict the global environmental changes. Photosynthesis is the basic driver of the biological CO_2 pump. As a result, it is necessary to understand the relationship between photosynthesis and the change of marine chemical and physical environments to assess CO_2 uptake capacities in different regions.

Global warming caused by increasing atmospheric CO_2 concentrations results in rise of seawater temperature and shoaling of the upper mixed layer (UML) above the thermocline, which exposes the phytoplankton within this layer to higher integrated levels of UVR irradiances (Gao et al. 2012a). Meanwhile, the increased temperature on the ground and the decreased heat reflection from the Earth's surface through the atmosphere to the outer space lowers the temperature of the stratosphere and accelerates the depletion of stratospheric ozone. The release of CFCs (chlorofluorocarbons) has been reduced since the Montreal Protocol Agreement was signed; however, due to their long lifetimes (on the order of a century) in the stratosphere, these substances are still harmful for the stratospheric ozone (Bais et al. 2015). Besides, several new gases, harmful to the stratospheric ozone, are rapidly accumulating in the atmosphere (Laube et al. 2014). Thus, the ecophysiological effects of the enhanced UV-B radiation (280–315 nm) still gain much attention (Häder and Gao 2015).

In short, OA and increased UVR exposure negatively affect the efficiency of the marine biological CO_2 pump by influencing the physiological performance of marine primary producers in the UML. Therefore, their respective and combined effects on the photosynthetic carbon fixation should be examined in a context of marine environmental changing biology.

12.2 Ocean Acidification and Its Effect on Photosynthetic Organisms

12.2.1 Ocean Acidification

The exchange of CO_2 between the oceans and the atmosphere relies on seawater mixing intensity and surface ocean carbonate chemistry. The dissolved CO_2 in seawater remains in the upper layer for 6 years on average. The mixture between epipelagic and mesopelagic seawater (1000–4000 m) is relatively slow and needs hundreds of years (The Royal Society 2005). The oceans have absorbed more than 1/3 of the anthropogenically released CO_2 since the Industrial Revolution, which significantly mitigates global warming (Sabine et al. 2004). However, the atmospheric CO_2 concentration is still rising, with continuous oceanic absorption of CO_2, leading to a lowered alkalinity of the mesopelagic seawater. This process of increasing acidity of seawater caused by the rising atmospheric CO_2 concentration is termed "ocean acidification (OA)." More than half of the CO_2 absorbed from the atmosphere by the oceans remains at a depth of 0 to 400 m (Sabine et al. 2004); the longer CO_2 remains in the upper layer, the faster OA proceeds.

CO_2 is an acidic gas, and its dissolution into seawater leads to acidification:

$$CO_2 + H_2O = H_2CO_3 \tag{12.1}$$

$$H_2CO_3 = H^+ + HCO_3^- \tag{12.2}$$

$$HCO_3^- = H^+ + CO_3^{2-} \tag{12.3}$$

(The equilibrium constants of the above three reactions are dependent on temperature and salinity in surface oceans). As indicated by these reactions, when CO_2 dissolves into seawater, it forms carbonic acid. Then, carbonic acid dissociates and forms HCO_3^- and H^+. As the concentration of H^+ increases, reaction (12.3) is shifted to the right, causing the decrease of CO_3^{2-} concentration.

The seawater carbonate system of the upper oceans provides an inorganic carbon source for primary production. With the progressive pH decrease associated with OA, the concentration of inorganic carbon in the carbonate system and the concentration ratios of different inorganic carbons (CO_2, HCO_3^-, CO_3^{2-}) change, thus affecting the saturation Ω of $CaCO_3$ in seawater ($\Omega = Ca^{2+} \times CO_3^{2-} / K_c$, where K_c is the product of $Ca^{2+} \times CO_3^{2-}$ when the $CaCO_3$ solution is saturated, which is associated with the crystal type of $CaCO_3$ such as calcite and aragonite. Since the oceanic Ca^{2+} concentration is relatively constant (approximately 10 mM), the $CaCO_3$ saturation Ω mainly depends on the concentration of CO_3^{2-}. In general, HCO_3^- in the seawater accounts for more than 90% of the dissolved inorganic carbon (DIC), CO_3^{2-} for about 9%, and CO_2 for less than 1% (these percentages change at different latitudes or regions). Increasing atmospheric CO_2 results in increased concentrations of dissolved CO_2, HCO_3^-, and H^+, the decreased concentration of CO_3^{2-}, and consequently a decreased saturation of $CaCO_3$. Since the Industrial Revolution, the concentration of CO_3^{2-} in the epipelagic seawater has decreased approximately by 10% (Orr et al. 2005). If the atmospheric CO_2 concentration is doubled, the partial pressure of CO_2 ($_PCO_2$) of epipelagic seawater will also be doubled, the concentration of HCO_3^- increased by about 11%, DIC increased by 9%, and CO_3^{2-} decreased by 45% (Kleypas et al. 2006). It should be noted that the concentration of CO_3^{2-} varies in waters (dependent on temperature etc.). The concentration of CO_3^{2-} in polar waters is only 41% of that in tropical waters. By the end of this century, the concentration of CO_3^{2-} in tropical waters will decrease to 149 μmol/kg, while that of polar waters will decrease to 55 μmol/kg, which is 37% of that in tropical waters (Orr et al. 2005). In waters with relatively low temperature, OA will decrease the concentration of CO_3^{2-} to a larger extent. Therefore, OA has different effects on the chemical processes of different waters (Kleypas et al. 1999; Orr et al. 2005).

If utilization of fossil fuels by humans continues at the current rate, the atmospheric CO_2 concentration will rise up to 800–1000 ppmv before 2100, decreasing the pH of the upper layer by 0.3–0.4, which means the concentration of H^+ will increase by 100–150% (Zeebe and Wolf-Gladrow 2001; Caldeira and Wickett 2003; Gattuso et al. 2015). According to the predicted reserves of available fossil fuel, the anthropogenic CO_2 emissions will reach a maximum in 2150 and then decline. However, the high concentration of CO_2 will remain in the atmosphere for thousands of years. During this period, the oceans will continue to absorb CO_2, decreasing the pH of the upper ocean. Meanwhile, the CO_2 absorbed by the upper ocean gradually sinks to the deep ocean, decreasing the deep ocean pH as well. The absorbed CO_2 can sink even thousands of meters deep into the ocean (Caldeira and Wickett 2003). Thus, it can be seen that OA has influences on the chemical and biological processes not only in the euphotic zone but also in deeper layers. Even if humans stopped the CO_2 emission from now on, the tendency of OA would not be reversible in a short term (several hundreds of years) (The Royal Society 2005), since it takes thousands of years to complete the mixture of the upper and deep layers of the ocean. In the past, the concentration of atmospheric CO_2 changed slowly, which allowed mixing between upper and deeper layers to mitigate the pH drop of the upper ocean. But at present, OA proceeds much faster than ever in the past 300 million years (Hönisch et al. 2012) which is far beyond the mitigation capacity of geochemical processes.

OA changes the ocean carbonate system and other chemical processes (Millero 2007), which influence life processes relying on the chemical environment of seawater. Hence, potential effects of OA on organisms and ecosystems have drawn increasing attentions (Riebesell and Gattuso 2015). At the same time, in different waters and latitudes, progress of OA and its effects on ecosystems are controlled by other environmental factors (Riebesell and Gattuso 2015). In addition to visible light, temperature, and nutrients, UV is of special concern (Boyd 2011; Brennan and Collins 2015).

In summary, anthropogenically released CO_2 has already influenced fundamental marine chemical processes. How OA will affect biological processes, organisms, and biogeochemical process is of great significance for the scientific community and society.

12.2.2 Responses of Photosynthetic Organisms to Ocean Acidification

Chemical changes as a result of OA can influence physiological processes of organisms as well as their heredity and evolution. Marine phytoplankton (prokaryotic species and microalgae) as well as macroalgae can convert CO_2 into organic materials, which accounts for 50% of the global primary production (Field et al. 1998). Analysis of the ecological effects of OA on photosynthetic carbon fixation is key to understanding the relationship between OA and marine ecosystem and estimating the amount of CO_2 absorbed by the oceans.

A number of previous studies have shown that OA may either promote growth and photosynthesis of phytoplankton (Reibesell and Tortell 2011) or have no effect (Tortell et al. 2000; Kim et al. 2006; Gao and Campbell 2014) or have a negative effect on it, such as promoting mitochondrial respiration and photorespiration (Wu et al. 2010; Gao et al. 2012b). Shipboard studies in the South China Sea during several research cruises showed that the effect of OA on the photosynthetic carbon fixation by phytoplankton relies on the intensity of solar radiation and the depth distribution of phytoplankton. The increased concentration of CO_2 lowers the carbon fixation of phytoplankton communities and inhibits the growth of diatoms in shallow waters or under high irradiances while accelerating the growth of diatoms or augmenting carbon fixation in deeper layers or at low irradiances (Fig. 12.1). The biological mechanism behind this phenomenon is that the increased concentration of CO_2 downregulates the active uptake capacity of inorganic carbon (CO_2, HCO_3^-) of the cells, saving energy for CO_2-concentrating mechanisms, which promotes the growth of phytoplankton at low irradiances. Such energy saving interacts with increased acidity to stimulate light stress under intense levels of solar radiation, so that photorespiration can be enhanced (Gao et al. 2012a, b). On the other hand, in order to resist OA, phytoplankton cells increase synthesis of toxic phenolics and gain extra energy by degrading them, with mitochondrial respiration being enhanced. The underlying mechanism is that OA promotes several metabolic pathways such as β-oxidation, the tricarboxylic acid (TCA) cycle, glycolysis, etc. (Fig. 12.2), leading to a higher concentration of phenolics (toxic compounds) and a higher respiratory rate (Jin et al. 2015).

Nutrients also influence the physiological responses of phytoplankton to OA (Li and Gao 2012; Verspagen et al. 2014). The substrate of cyanobacterial and algal carboxylation is CO_2, while more than 90% of inorganic carbon in the surface ocean is HCO_3^- . Thus, if seawater carbonate chemistry changes (pCO_2 rises, pH decreases etc.), utilization of inorganic carbon and the energy to maintain an intracellular acid-base balance will be affected (Mackey et al. 2015). Furthermore, OA can influence some metabolic pathways of harmful red tide species with regard to toxin production. For example, under OA conditions, more toxic compounds were accumulated in the cells of *Pseudo-nitzschia* (Sun et al. 2011) and of toxic dinoflagellates (Hattenrath-Lehmann et al. 2015).

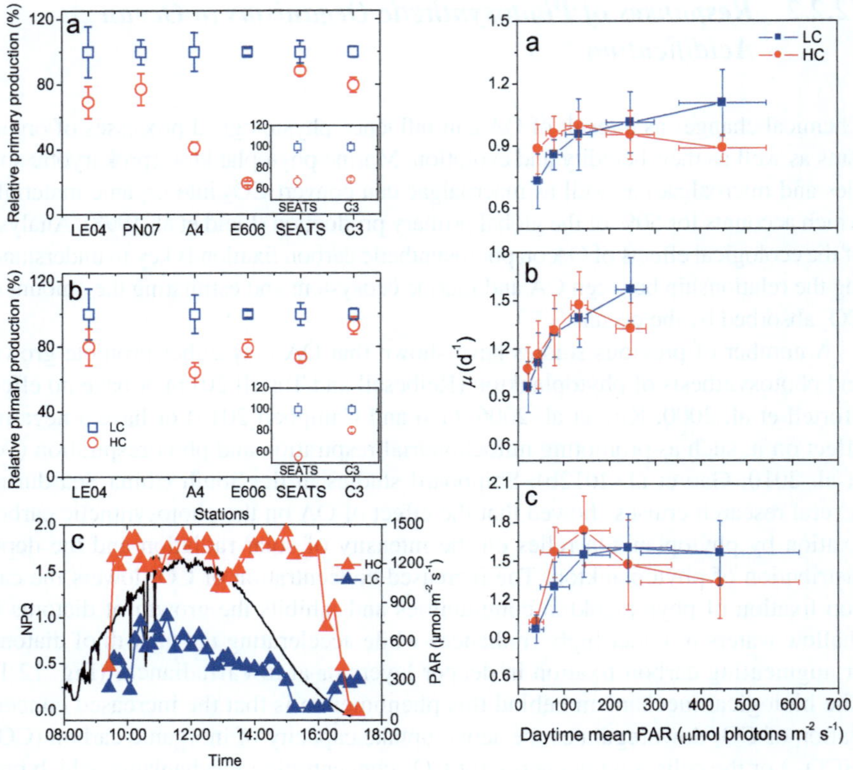

Fig. 12.1 The decrease of carbon fixation by phytoplankton in the upper layer of the South China Sea under acidification conditions (*red triangles*). *Left panels* **a** and **b** show the carbon fixation per volume of seawater and per chl *a*, respectively) and increased light stress (left panel c, non-photochemical quenching). *Right panels* **a**, **b** and **c**, respectively, represent the relationship between the irradiance and the growth rates of *Phaeodactylum tricornutum* (**a**), *Thalassiosira pseudonana* (**b**), and *Skeletonema costatum* (**c**) at low (390 μatm, *blue squares*) and high (1000 μatm, red circles) CO_2 concentrations showing growth rates as a function of irradiance. There is a coupling effect between OA and irradiance, as low radiation promotes the growth of diatoms while high radiation inhibits it (Gao et al. 2012a, b)

Oceanic photosynthetic carbon fixation is limited by the availability of nutrients and trace elements (such as Fe). In oligotrophic waters, biological nitrogen fixation by cyanobacteria can provide biologically usable nitrogen, which promotes growth of phytoplankton and increases photosynthetic carbon fixation. In this way, the marine diazotrophs mitigates the global warming indirectly (Michaels et al. 2001; Berthelot et al. 2015). OA can influence nitrogen fixation in cyanobacteria and also affects their photosynthetic carbon fixation directly and/or indirectly. For cyanobacteria with heterocysts, OA either promotes (Wannicke et al. 2012) or inhibits (Czerny et al. 2009) nitrogen fixation. In cyanobacteria without heterocysts, the extent of promotion varies among the species (Eichner et al. 2014). However, some research cruise studies show that OA caused by increased CO_2 concentrations

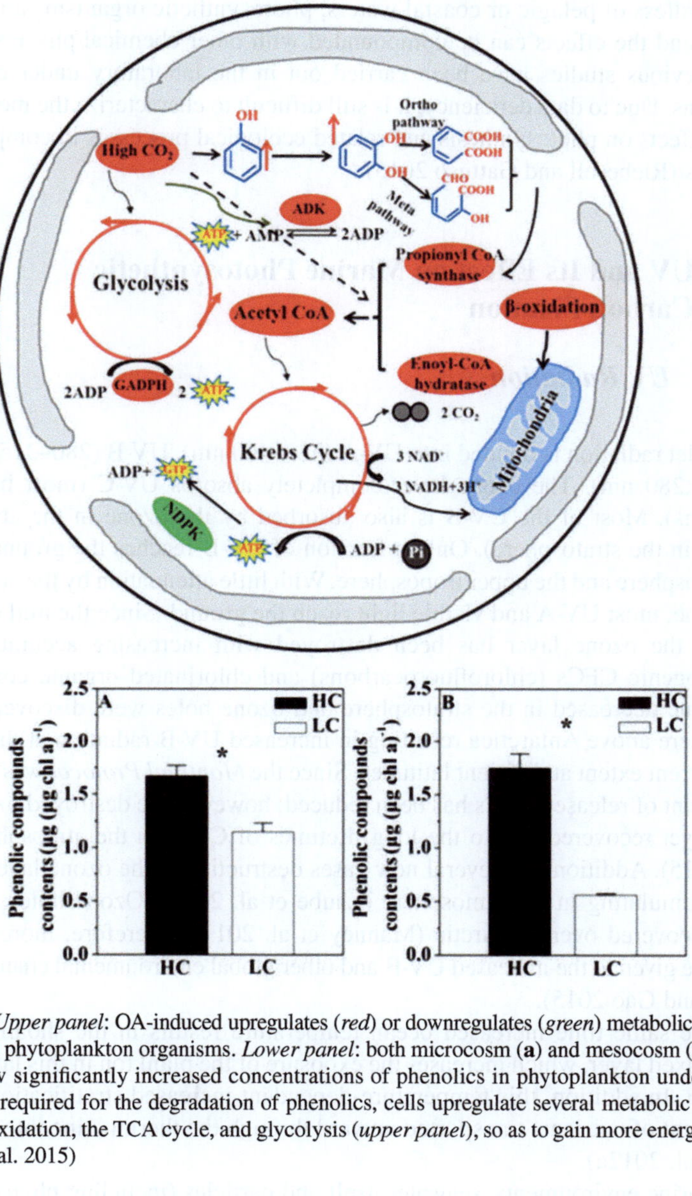

Fig. 12.2 *Upper panel*: OA-induced upregulates (*red*) or downregulates (*green*) metabolic pathways of calcified phytoplankton organisms. *Lower panel*: both microcosm (**a**) and mesocosm (**b**) experiments show significantly increased concentrations of phenolics in phytoplankton under acidification. As required for the degradation of phenolics, cells upregulate several metabolic pathways such as β-oxidation, the TCA cycle, and glycolysis (*upper panel*), so as to gain more energy to resist OA (Jin et al. 2015)

has no significant effect or inhibits nitrogen fixation in *Trichodesmium* (Böttjer et al. 2014; Gradoville et al. 2014). These different findings can be attributed to inherent physiological characteristics of species as well as different marine environments and experimental conditions. Due to different responses of different nitrogen-fixing organisms to OA, the structure of communities and biodiversity are bound to change with progressive OA (Hutchins et al. 2013; Gradoville et al. 2014).

Regardless of pelagic or coastal waters, photosynthetic organisms are affected by OA, and the effects can be compounded with other chemical/physical factors. Most previous studies have been carried out in the laboratory under controlled conditions. Due to data deficiency, it is still difficult to characterize the mechanisms of OA effects on photosynthesis and related ecological processes in complex environments (Riebesell and Gattuso 2015).

12.3 UV and Its Effect on Marine Photosynthetic Carbon Fixation

12.3.1 UV Radiation

Ultraviolet radiation is divided into UV-A (315–400 nm), UV-B (280–315 nm), and UV-C (<280 nm). The ozone layer completely absorbs UV-C (most harmful to organisms). Most of the UV-B is also absorbed by the ozone in the atmosphere (mainly in the stratosphere). Only a fraction of UV-B reaches the ground through the stratosphere and the upper troposphere. With little attenuation by the atmosphere and ozone, most UV-A and visible light reach the ground. Since the mid of the last century, the ozone layer has been destroyed with increasing accumulation of anthropogenic CFCs (chlorofluorocarbons) and chlorinated organic compounds. The ozone decreased in the stratosphere and ozone holes were discovered in the atmosphere above Antarctica resulting in increased UV-B radiation at the ground, to a different extent at different latitudes. Since the *Montreal Protocol* was enforced, the amount of released CFCs has been reduced; however, the destroyed ozone layer has not yet recovered due to the long lifetimes of CFCs in the atmosphere (Bais et al. 2015). Additionally, several new gases destructive to the ozone layer are rapidly accumulating in the atmosphere (Laube et al. 2014). Ozone holes have also been discovered over the Arctic (Manney et al. 2011). Therefore, more attention should be given to the increased UV-B and other global environmental change drivers (Häder and Gao 2015).

At the same time increased ocean temperature results in the shoaling of the upper mixed layer, which increases the exposure of the plankton in this layer to UV radiation. In addition, this temperature-dependent increased stratification reduces the amount of nutrients transported upward through the thermocline into the UML (Gao et al. 2012a).

In marine environments, seawater itself and particles (including phytoplankton) and dissolved matters absorb and scatter solar radiation. The degree of absorption and scattering depends on the wavelength. The solar radiation transmitted into the water undergoes exponential attenuation (Piazena and Häder 1997). The energy ratio of light with different wavelengths to the total light energy varies with depth (Hargreaves 2003). In estuary and coastal ecosystems, particulate organic matters

(POM) and chromophoric dissolved organic matter (CDOM) are the main factors resulting in the attenuation of solar irradiances. In offshore waters, due to high transparency, the attenuation of solar radiation mainly depends on the biomass density or degradation of the phytoplankton and the intake and excretion of inorganic/organic matter of zooplankton (Häder et al. 2011). The transmission depth varies greatly for light with different wavelengths in different waters. For example, in the Atlantic, UV-B (310 nm) can reach 30 m, 100 m for UV-A (380 nm), over 190 m for blue light (450–500 nm) and 20 m for other visible light wavelengths (600–700 nm) (Piazena et al. 2002). In the offshore waters of the South China Sea, the visible light can penetrate deeper than 80 m (the depth of the euphotic zone), 50 m for UV-A and 38 m for UV-B which accounts for 62% and 47% of the depth of the euphotic zone, respectively (Li et al. 2009). In the coastal waters of Shantou with large quantities of suspended particles, the visible light can only penetrate less than 10 m in the waters adjacent to the aquaculture area; UV-A and UV-B can only transmit less than 5 m and 3 m, respectively, accounting for 50% and 30% of the depth of the euphotic zone (Gao et al. 2007a; Li and Gao 2012). In the estuary of Jiulong River of Fujian Province, visible light penetrates even less (0.47–5 m), only 0.25–1.8 m for UV-A and 0.21–1.7 m for UV-B which accounts for 36–75% and 46–56% of the depth of the euphotic zone, respectively, of the transmission depth of visible light (Li et al. 2011b). Recently, it has been suggested that intensity and transmission depth of UV will also change with global ocean changes (Zepp et al. 2007; Häder et al. 2011). By affecting the phytoplankton biomass of the euphotic zone, global change will indirectly influence the attenuation of visible light and UV in the water column, which may increase the intensity and the transmission depth of UV (Zepp et al. 2007). On the other hand, dissolved organic carbon (DOC) and particulate organic carbon (POC) are the main factors attenuating UV-A and UV-B, respectively. Increased temperatures will accelerate bacterial decay and photodegradation of CDOM and increase the transmission depth of UV (Tzortziou et al. 2007; Zhang et al. 2009). Exposure to solar ultraviolet radiation is fluctuating due to vertical mixing within the epilimnion. Therefore, increased solar UVR (as a result of a shallower epilimnion and more frequent circulation near the surface), fluctuating irradiances (as a result of stronger vertical mixing due to increased wind stress), and attenuation of solar radiation in the water column will have a compound effect (Helbling et al. 2015). This new scenario for oceanic phytoplankton requires further studies.

UV-B irradiance at the earth surface also changes with latitude, ozone content in the stratosphere, and suspended particles in the atmosphere. Thus, the effect of UV on photosynthetic carbon fixation varies with different areas, waters, and depths. In the past, many marine in situ carbon sequestration studies and investigations have been carried out in ordinary transparent containers (glass or PC culture bottles) which block most of the UV radiation. As a result, the effects of UV radiation on carbon fixation and other biological processes have been ignored in many studies and investigations. A large number of reports have given insight into the effect of UV radiation on phytoplankton (see reviews Häder et al. 2014; Häder and Gao 2015).

12.3.2 The Effect of UV on Photosynthetic Carbon Fixation

UV can damage the DNA of phytoplankton (Häder and Gao 2015) and inhibit its repair (Gao et al. 2008; Rastogi et al. 2014) as well as negatively affect its physiological metabolism. Moreover, UV can also produce active free oxygen radicals indirectly which causes oxidative stress in the cells (Häder et al. 2014), which may lower the photosynthesis rate (Wu et al. 2005). Solar UV also changes the morphology of filamentous cyanobacteria and causes their death (Wu et al. 2005). Although in most areas UV-B accounts for less than 1% of the total solar energy, its harmful effect is usually stronger than that of UV-A, which accounts for about 6–8% of the total solar energy. However, when the water is mixed or solar radiation is fluctuating, the inhibition of photosynthetic carbon fixation by UV is significantly lower (Li et al. 2013). During a long evolution process, phytoplankton has acquired mechanisms to resist UV radiation, for example, the synthesis of the UV-shielding pigments such as mycosporine-like amino acids (MAAs), the removal mechanism of active oxygen free radicals, and the repair mechanisms of proteins and DNA (Häder et al. 2014; Rastogi et al. 2014). However, in the surface layer, phytoplankton cells are inevitably affected by UV, resulting in a decrease of carbon fixation (Helbling et al. 2003; Häder and Gao 2015). But in some cases, such as on cloudy days or in a relatively deeper layer, where the solar radiation is reduced to moderate or low levels, UV radiation may even stimulate carbon fixation (Fig. 12.3). It has been demonstrated that phytoplankton assemblages in coastal waters can utilize UV-A to drive their photosynthetic carbon fixation (Gao et al. 2007a, b). Further investigations showed that, after filtering out the visible light, a diatom-dominated phytoplankton community can still carry out photosynthetic carbon fixation, and the fixation rate steadily rises with increased UV-A intensity. However, the presence of UV-B decreases carbon fixation (Fig. 12.3a). The positive effect of UV-A is related to the intensity of solar radiation received by the phytoplankton and the size of the cells (Li et al. 2011a; Li and Gao 2013). Usually, phytoplankton species with larger cell sizes show higher capacity to utilize UV-A to fix carbon.

12.4 The Combined Effects of OA and UV Radiation

The oceanic upper mixed layer is facing the pressure of OA, rising temperatures and enhanced UV radiation, etc. Due to the lack of data related to effects of multiple global change drivers (Riebesell and Gattuso 2015), it is difficult to predict the trend of biological carbon fixation and the efficiency of the biological carbon pump. Although there is a large body of papers on the effects of OA or UV on the physiology of marine primary producers, little knowledge has been gained in understanding the compounding/coupling effects of OA and UV (Beardall et al. 2014).

The photosystems of plants show rapid responses to UV radiation. For example, the high radiation at noon lowers the PSII photochemical efficiency within a few minutes. Combined effects of OA and UVR (280–400 nm) significantly lower the PSII

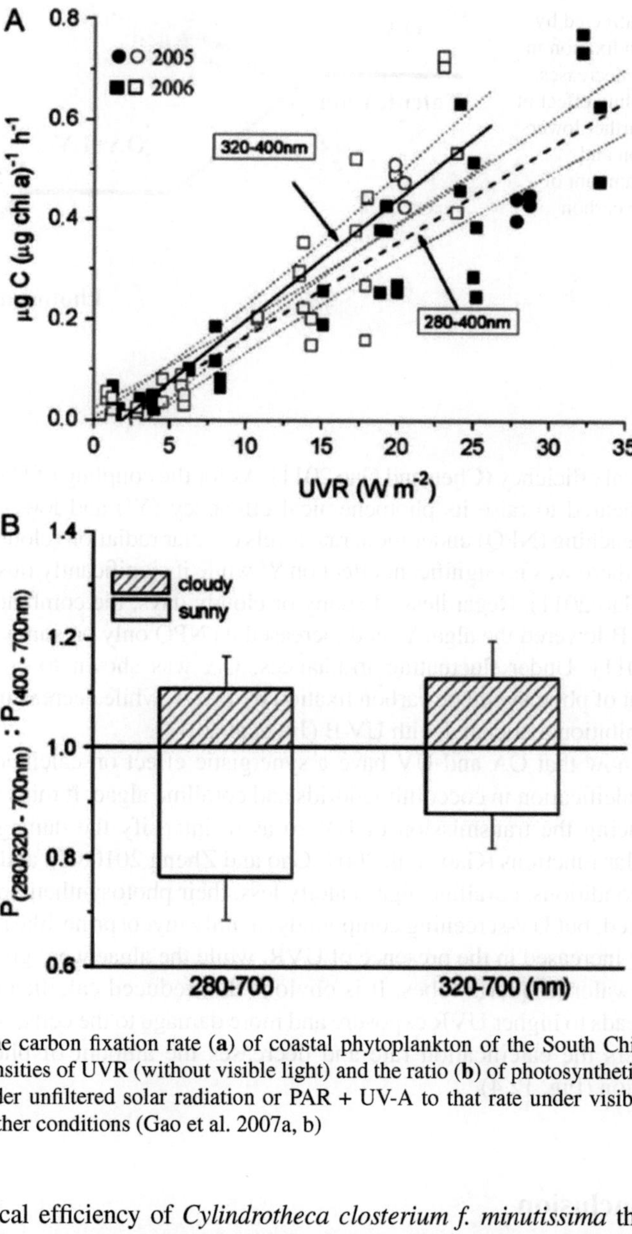

Fig. 12.3 The carbon fixation rate (**a**) of coastal phytoplankton of the South China Sea under different intensities of UVR (without visible light) and the ratio (**b**) of photosynthetic carbon fixation rates under unfiltered solar radiation or PAR + UV-A to that rate under visible light under different weather conditions (Gao et al. 2007a, b)

photochemical efficiency of *Cylindrotheca closterium f. minutissima* that is grown under at 1000 µatm CO_2 and solar radiation on the initial day of exposure; even when the organisms had been acclimated for 9 days, the coupling effect of OA and UV still significantly lowered its electron transport rate, though there was no significant difference in growth (Wu et al. 2012). For *Phaeocystis globosa* grown at 1000 µuatm CO_2 (with constant pH) and solar radiation, its photochemical efficiency varied with the intensity of solar radiation (a negative correlation with light intensity), indicating the lowest values at noon, and there was a coupling effect of UV and OA on its

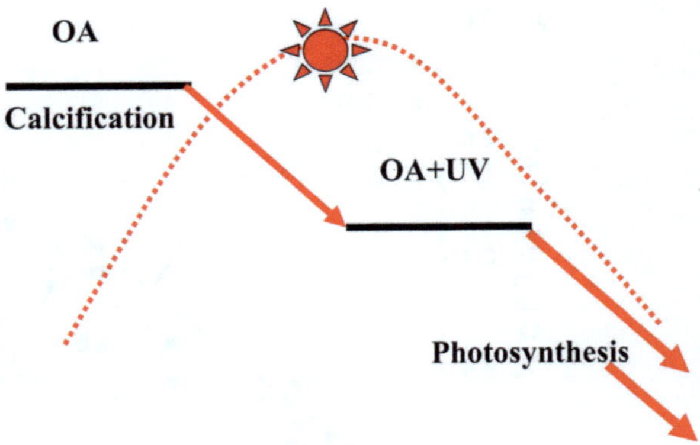

Fig. 12.4 As affected by OA, the carbon fixation in calcified algae decreases. The compounding effect of OA and UV further lowers the calcification and decreases the amount of photosynthetic carbon fixation

photochemical efficiency (Chen and Gao 2011). As for the coupling of UV-A and OA, this alga appeared to raise its photochemical efficiency (Y′) and lower non-photochemical quenching (NPQ) under moderate levels of solar radiation (cloudy days). On sunny days, there was no significant effect on Y′ while it significantly raised the NPQ (Chen and Gao 2011). Regardless of sunny or cloudy days, the combined effects of OA and UV-B lowered the algal Y′ and increased its NPQ only on sunny days (Chen and Gao 2011). Under fluctuating irradiances, OA was shown to counteract the enhancement of photosynthetic carbon fixation by UV-A, while decreasing the photochemical inhibition associated with UV-B (Jin et al. 2013).

Studies show that OA and UV have a synergistic effect on calcified algae. OA lowers the calcification in coccolithophorids and coralline algae. It thins the calcified layer, enhancing the transmission of UV so as to intensify the damage of UV to impair cellular functions (Gao et al. 2009; Gao and Zheng 2010; Xu and Gao 2014). Under OA conditions, coralline algae calcify less, their photosynthetic pigments are highly reduced, but UV-screening compounds (mainly mycosprine-like amino acids) significantly increased in the presence of UVR, while the algae were grown in flow-through seawater in quartz tubes. It is obvious that reduced calcification of these organisms leads to higher UVR exposure and more damage to the cells, which in turn further lowers the calcification rate and decreases the amount of photosynthetic carbon fixation (Fig. 12.4).

12.5 Conclusion

OA is a growing environmental issue in both pelagic and coastal waters, which appears to be more serious in Chinese waters, where eutrophication is intense with large-scaled sea farming. Since different regions would be affected by climate change to different extents, different organisms and marine ecosystems are exposed to multifactorial stresses associated with ocean climate changes. Therefore, it is necessary to study the effects of OA on marine photosynthetic carbon fixation under different scenarios. Though it would be ideal to investigate OA effects under natural conditions with fluctuating or changing abiotic factors, experimental logistics usually limit the combinations of several environmental drivers. Therefore, in investigations of photosynthetic responses to OA, phytoplankton, and/or macroalgae should be exposed to solar UV radiation. Mechanisms involved in the combined or compounded effects of OA and UV should be explored to establish a reliable theory toward understanding regional responses of marine primary producers to the changing ocean environment. Due to research activity logistics, under multiple environmental stressors, our knowledge is limited in our understanding of how the effect of OA or UV or UV plus OA changes. In some waters where the chemical environment is fluctuant, some regional problems such as the effect of OA on biological activities require further studies; the effect of OA on the process of UV photodegradation of organics will influence the amount of marine carbon fixation or carbon storage directly or indirectly.

References

Bais AF, McKenzie RL, Bernhard G, Aucamp PJ, Ilyas M, Madronich S, Tourpali K (2015) Ozone depletion and climate change: impacts on UV radiation. Photochem Photobiol Sci 14:19–52

Beardall J, Stojkovic S, Gao KS (2014) Interactive effects of nutrient supply and other environmental factors on the sensitivity of marine primary producers to ultraviolet radiation: implications for the impacts of global change. Aquat Biol 22:5–23

Berthelot H, Moutin T, L'Helguen S, Leblanc K, Hélias S, Grosso O, Leblond N, Charrière B, Bonnet S (2015) Dinitrogen fixation and dissolved organic nitrogen fueled primary production and particulate export during the VAHINE mesocosm experiment (New Caledonia lagoon). Biogeosciences 12:4099–4112

Böttjer D, Karl MD, Letelier MR et al (2014) Experimental assessment of diazotroph reponses to elevated seawater pCO_2 in the North Pacific Subtropical Gyre. Global Biogeochem Cycle 28(6):601–616

Boyd PW (2011) Beyond ocean acidification. Nat Geosci 4:273–274

Brennan G, Collins S (2015) Growth responses of a green alga to multiple environmental drivers. Nat Clim Chang 5:892–897

Caldeira K, Wickett ME (2003) Oceanography: anthropogenic carbon and ocean pH. Nature 425:365

Chen SW, Gao KS (2011) Solar ultraviolet radiation and CO_2-induced ocean acidification interacts to influence the photosynthetic performance of the red tide alga *Phaeocystis globosa* (Prymnesiophyceae). Hydrobiologia 675(1):105–117

Czerny J, Barcelos e Ramos J, Riebesell U (2009) Influence of elevated CO_2 concentrations on cell division and nitrogen fixation rates in the bloom-forming cyanobacterium *Nodularia spumigena*. Biogeosciences 6:1865–1875

Eichner M, Rost B, Kranz SA (2014) Diversity of ocean acidification effects on marine N_2 fixers. J Exp Mar Biol Ecol 457:199–207

Field CB, Behrenfeld MJ, Randerson JT et al (1998) Primary production of the biosphere: integrating terrestrial and oceanic components. Science 281:237–240

Gao KS, Campbell DA (2014) Photophysiological responses of marine diatoms to elevated CO_2 and decreased pH: a review. Funct Plant Biol 41:449–459

Gao KS, Zheng YQ (2010) Combined effects of ocean acidification and solar UV radiation on photosynthesis, growth, pigmentation and calcification of the coralline alga *Corallina sessilis* (Rhodophyta). Glob Change Biol 16:2388–2398

Gao KS, Wu Y, Li G, Wu H, Villafane VE, Helbling EW (2007a) Solar UV radiation drives CO_2 fixation in marine phytoplankton: a double-edged sword. Plant Physiol 144:54–59
Gao KS, Li G, Helbling EW, Villafane VE (2007b) Variability of UVR effects on photosynthesis of summer phytoplankton assemblages from a tropical coastal area of the South China Sea. Photochem Photobiol 83:802–809
Gao KS, Li P, Watanabe T, Helbling EW (2008) Combined effects of ultraviolet radiation and temperature on morphology, photosynthesis and DNA of *Arthrospira* (*Spirulina*) *platensis* (Cyanophyta). J Phycol 44:777–786
Gao KS, Ruan ZX, Villafane VE, Gattuso JP, Helbling EW (2009) Ocean acidification exacerbates the effect of UV radiation on the calcifying phytoplankter *Emiliania huxleyi*. Limnol Oceanogr 54:1855–1862
Gao KS, Xu JT, Gao G et al (2012a) Rising CO_2 and increased light exposure synergistically reduce marine primary productivity. Nat Clim Chang 2(7):519–523
Gao KS, Helbling EW, Häder DP, Hutchins DA (2012b) Responses of marine primary producers to interactions between ocean acidification, solar radiation, and warming. Mar Ecol Prog Ser 470:167–189
Gattuso JP, Magnan A, Bille R (2015) Contrasting futures for ocean and society from different anthropogenic CO_2 emissions scenarios. Science 349(6243):aac4722
Gradoville MR, White AE, Böttjer D, Church MJ, Letelier RM (2014) Diversity trumps acidification: lack of evidence for carbon dioxide enhancement of *Trichodesmium* community nitrogen or carbon fixation at Station ALOHA. Limnol Oceanogr 59:645–659
Häder DP, Gao KS (2015) Interactions of anthropogenic stress factors on marine phytoplankton. Front Environ Sci 3:1–14
Häder DP, Helbling EW, Smith RC et al (2011) Effects of UV radiation on aquatic ecosystems and interactions with climate change. Photochem Photobiol Sci 10:242–260
Häder DP, Williamson CE, Wängberg SÅ, Rautio M, Rose KC, Gao KS, Helbling EW, Sinha RP, Worrest R (2014) Effects of UV radiation on aquatic ecosystems and interactions with other environmental factors. Photochem Photobiol Sci 14:108–126
Hargreaves BR (2003) Water column optics and penetration of UVR. In: Helbling EW, Zagarese HE (eds) UV effects in aquatic organisms and ecosystems. The Royal Society of Chemistry, Cambridge, pp 59–105
Hattenrath-Lehmann TK, Smith JL, Wallace RB et al (2015) The effects of elevated CO_2 on the growth and toxicity of field populations and cultures of the saxitoxin-producing dinoflagellate, *Alexandrium fundyense*. Limnol Oceanogr 60(1):198–214
Helbling EW, Gao KS, Goncalves R et al (2003) Utilization of solar UV radiation by costal phytoplankton assemblages off SE China when exposed to fast mixing. Mar Ecol Prog Ser 259:59–66
Helbling EW, Banaszak AT, Villafañe VE (2015) Global change feed-back inhibits cyanobacterial photosynthesis. Sci Rep 5:14514
Hönisch B, Rdigwell A, Schmidt DN et al (2012) The geological record of ocean acidification. Science 335(6072):1058–1063
Hutchins DA, Fu FX, Webb EA, Walworth N, Tagliabue A (2013) Taxon-specific response of marine nitrogen fixers to elevated carbon dioxide concentrations. Nat Geosci 6:790–795
Jin P, Gao KS, Villafañe EV, Campbell AD, Helbling EW (2013) Ocean acidification alters the photosynthetic response of a coccolithophorid to fluctuating ultraviolet and visible radiation. Plant Physiol 162:2084–2094
Jin P, Wang TF, Liu NN, Dupont S, Beardall J, Boyd PW, Riebesell U, Gao KS (2015) Ocean acidification increases the accumulation of toxic phenolic compounds across trophic levels. Nat Commun 6:8714
Kim JM, Lee K, Shin K et al (2006) The effect of seawater CO_2 concentration on growth of a natural phytoplankton assemblage in a controled mesocosm experiment. Limnol Oceanogr 51:1629–1636
Kleypas JA, Buddemeier RW, Archer D et al (1999) Geochemical consequences of increased atmospheric carbon dioxide on coral reefs. Science 284(5411):118–120

Kleypas JA, Feely RA, Fabry VJ, et al (2006) Impacts of ocean acidification on coral reefs and other marine calcifiers: A guide for Future Research. Report of a workshop held 18–20 April 2005, St. Petersburg, FL, sponsored by NSF, NOAA, and the U.S. Geological Survey

Laube JC, Newland MJ, Hogan C et al (2014) Newly detected ozone-depleting substances in the atmosphere. Nat Geosci 7:266–269

Li G, Wu YP, Gao KS (2009) Effects of Typhoon Kaemi on coastal phytoplankton assemblages in the South China Sea, with special reference to the effects of solar UV radiation. J Geophys Res Biogeo 114:G04029

Li G, Gao KS, Gao G (2011a) Differential impacts of solar UV radiation on photosynthetic carbon fixation from the coastal to offshore surface waters in the South China Sea. Photochem Photobiol 87(2):329–334

Li G, Gao KS, Yuan DX et al (2011b) Relationship of photosynthetic carbon fixation with environmental changes in the Jiulong River estuary of the South China Sea, with special reference to the effects of solar UV radiation. Mar Pollut Bull 62(8):1852–1858

Li G, Che ZW, Gao KS (2013a) Photosynthetic carbon fixation by tropical coral reef phytoplankton assemblages: a UVR perspective. Algae 28:281–288

Li G, Gao KS (2013b) Cell size-dependent effects of solar UV radiation on primary production in coastal waters of the South China Sea. Estuar Coasts 36:728–736

Li W, Gao KS (2012) A marine secondary producer respires and feeds more in a high CO_2 ocean. Mar Pollut Bull 64:699–703

Mackey KRM, Morris JJ, Morel FMM et al (2015) Response of photosynthesis to ocean acidification. Oceanography 28(2):74–91

Manney GL, Santee ML, Rex M et al (2011) Unprecedented Arctic ozone loss in 2011. Nature 478:469–475

Michaels AF, Karl DM, Capone DG (2001) Element stoichiometry, new production and nitrogen fixation. Oceanography 14(4):68–77

Millero FJ (2007) The marine inorganic carbon cycle. Chem Rev 107:308–341

Orr JC, Fabry VJ, Aumont O et al (2005) Anthropogenic ocean acidification over the twenty-first century and its impact on calcifying organisms. Nature 437:681–686

Piazena H, Häder DP (1997) Penetration of solar UV and PAR into different waters of the Baltic Sea and remote sensing of phytoplankton. In: Häder DP (ed) The effects of ozone depletion on aquatic ecosystems. Academic Press, Austin, pp 45–96

Piazena H, Perez-Rodriguesb E, Häder DP et al (2002) Penetration of solar radiation into the water column of the central subtropical Atlantic Ocean-optical properties and possible biological consequences. Deep-Sea Res 49:3513–3528

Rastogi R, Singh S, Incharoensakdi A, Häder DP, Sinha R (2014) Ultraviolet radiation-induced generation of reactive oxygen species, DNA damage and induction of UV-absorbing compounds in the cyanobacterium *Rivularia* sp. HKAR-4. S Afr J Bot 90:163–169

Reibesell U, Tortell PD (2011) Effects of ocean acidification on pelagic organisms and ecosystems. In: Gattuso JP, Hansson L (eds) Ocean acidification. Oxford University Press, New York, pp 291–311

Riebesell U, Gattuso JP (2015) Lessons learned from ocean acidification research. Nat Clim Chang 5:12–14

Sabine CL, Feely RA, Gruber N et al (2004) The oceanic sink for anthropogenic CO_2. Science 305(5682):367–371

Sun J, Hutchins DA, Feng Y et al (2011) Effects of changing pCO2 and phosphate availability on domoic acid production and physiology of the marine harmful bloom diatom Pseudo-Nitzchia Multiseries. Limnol Oceanogr 56:829–840

The Royal Society (2005) Ocean acidification due to increasing atmospheric carbon dioxide. ISBN 0854036172

Tortell PD, Rau GH, Morel FMM (2000) Inorganic carbon acquisition in coastal Pacific phytoplankton communities. Limnol Oceanogr 45:1485–1500

Tzortziou M, Osburn CL, Neale PJ (2007) Photobleaching of dissolved organic material from a tidal marsh-estuarine system of the Chesapeake Bay. Photochem Photobiol 83:782–792

Verspagen JM, Van de Waal DB, Finke JF, Visser PM, Huisman J (2014) Contrasting effects of rising CO_2 on primary production and ecological stoichiometry at different nutrient levels. Ecol Lett 17:951–960

Wannicke N, Engel SEA, Grossart HP et al (2012) Response of Nodularia spumigena to PCO_2. Biogeosciences 9:2973–2988

Wu H, Gao KS, Ma ZL, Watanabe T (2005) Effects of solar ultraviolet radiation on biomass production and pigment contents of *Spirulina platensis* in commercial operations under sunny and cloudy weather conditions. Fish Sci 71:454–456

Wu XJ, Gao G, Giordano M et al (2012) Growth and photosynthesis of a diatom grown under elevated CO_2 in the presence of solar UV radiation. Fundam Appl Limnol 180(4):279–290

Wu Y, Gao KS, Riebesell U (2010) CO_2-induced seawater acidification affects physiological performance of the marine diatom *Phaeodactylum tricornutum*. Biogeosciences 7:2915–2923

Xu K, Gao KS (2014) Solar UV irradiances modulate effects of ocean acidification on the coccolithophorid *Emiliania huxleyi*. Photochem Photobiol 91:92–101

Zeebe RE, Wolf-Gladrow D (2001) CO_2 in seawater: equilibrium, kinetics, isotopes. Elsevier, Amsterdam, 346 pp

Zepp RG, Erickson DJ III, Paul ND et al (2007) Interactive effects of solar UV radiation and climate change on biogeochemical cycling. Photochem Photobiol Sci 6:286–300

Zhang Y, Liu M, Qin B et al (2009) Photochemical degradation of chromophoric dissolved organic matter exposed to simulated UV-B and natural solar radiation. Hydrobiologia 627:159–168

Chapter 13
Oxidative Stress-Induced Bioprospecting of Microalgae

Kaumeel Chokshi, Imran Pancha, Arup Ghosh, and Sandhya Mishra

Abstract Microalgae are sunlight-driven cell factories found in diverse marine and freshwater environments. With simple growth requirements (light, CO_2, N, P and K), microalgae produce various valuable products like carotenoids, antioxidants, fatty acids, enzymes, polymers, peptides, toxins and sterols. Their photosynthetic mechanism is similar to plants, but due to their simple cellular structure and submergence in an aqueous environment, in most cases, they have an efficient access to water, CO_2 and other nutrients. In addition, their growth is faster and photosynthetic efficiency is higher compared to terrestrial crop plants. Their shorter generation time allows production of lipids and carbohydrates in large amounts over short periods of time, which can be easily converted into biofuels. Due to these reasons, microalgae are considered as an alternative renewable feedstock for biofuel production. In any organism, fluctuating environmental conditions trigger a series of physiological processes and generation of reactive oxygen species (ROS) which are highly reactive and damage proteins, lipids, carbohydrates and DNA, ultimately resulting into cellular toxicity. Stress-induced ROS accumulation is counteracted by cellular enzymatic and non-enzymatic antioxidants. Excessive ROS damage the ability of the cells to readily detoxify the reactive intermediates or to repair the resulting damage, ultimately leading to oxidative stress conditions. Recent studies suggest that oxidative stress is a mediator for increased accumulation of lipid and various bioactive metabolites in microalgae. This chapter provides comprehensive

K. Chokshi • S. Mishra (✉)
Discipline of Salt and Marine Chemicals, CSIR—Central Salt and Marine Chemicals Research Institute, Bhavnagar 364002, India

Academy of Scientific and Innovative Research (AcSIR), CSIR—Central Salt and Marine Chemicals Research Institute, Bhavnagar 364002, India
e-mail: smishra@csmcri.res.in

I. Pancha
Discipline of Salt and Marine Chemicals, CSIR—Central Salt and Marine Chemicals Research Institute, Bhavnagar 364002, India

A. Ghosh
Academy of Scientific and Innovative Research (AcSIR), CSIR—Central Salt and Marine Chemicals Research Institute, Bhavnagar 364002, India

Discipline of Plants Omics, CSIR—Central Salt and Marine Chemicals Research Institute, Bhavnagar 364002, India

© Springer International Publishing AG 2017
M. Kumar, P. Ralph (eds.), *Systems Biology of Marine Ecosystems*,
DOI 10.1007/978-3-319-62094-7_13

information on bioprospecting of microalgae under oxidative stress conditions, mainly for their carotenoid accumulation and biofuel potential. An overview of omics platform including genomics, transcriptomics, proteomics and metabolomics is also provided in the context of better understanding the stress response of microalgae at cellular level and using these advanced approaches for the development of microalgal biofactory.

Keywords Microalgae • Bioprospecting • Oxidative stress • Reactive oxygen species • Biofuel • Carotenoids • Omics

Contents

13.1 Microalgae and an Overview of Microalgal Applications 252
13.2 Oxidative Stress and Reactive Oxygen Species 254
13.3 Role of Oxidative Stress in Carotenoid Accumulation in Microalgae 257
13.4 Link Between Oxidative Stress and Biofuel Potential of Microalgae 262
13.5 Other Recent Applications of Microalgae Linked with Oxidative Stress 266
13.6 Systems Biology and Omics Platform for Microalgal Biofactory 267
13.7 Conclusion and Future Perspectives 270
References 271

13.1 Microalgae and an Overview of Microalgal Applications

Microalgae are unicellular, microscopic (2–200 μm), polyphyletic, CO_2 evolving and sunlight-driven cell factories found in diverse environmental conditions and habitats such as lacustrine, brackish, freshwater, hyper-saline, wastewater ponds, dams, rivers, marine and coastal areas. In any habitat, microalgae have been shown to have successional tendencies due to variable nutrient availability, inclement weather and seasonal variations (Bernal et al. 2008). They are autotrophic organisms which grow by photosynthesis and are considered as eukaryotes, although the prokaryotic cyanobacteria are also included in this category (Greenwell et al. 2009). The enormous biodiversity of microalgae represent almost untapped resource. It has been estimated that about 20,000–800,000 microalgal species exist, of which only about 50,000 species are reported (Suganya et al. 2016).

The use of microalgae by humans started when the Chinese used *Nostoc* and few other cyanobacteria as a food source around 2000 years ago. After World War II, the USA, Germany and Japan started mass culture of microalgae to meet the increasing food demand (Ratha and Prasanna 2012). Since then microalgae have been explored for the control of water pollution, regeneration of atmosphere in biospheres, mitigation of greenhouse gases, etc. With simple growth requirements (light, CO_2, N, P and K), microalgae produce a wide range of potential metabolic products such as pigments, fatty acids, vitamins, carotenoids, antioxidants, enzymes, toxins, food supplements, pharmacological substances, polymers, etc. (Pancha et al. 2015a, b; Araujo et al. 2011). Microalgae are a great source of natural compounds which could be used as ingredients to enhance nutritional food content. Due to their high protein content, various microalgal species are considered as unconventional sources of protein. They also represent a valuable source of nearly all amino acids and essential vitamins. Microalgae are rich in pigments like chlorophyll (0.5–1% of dry weight), carotenoids (0.1–0.2% of dry weight) and phycobiliproteins (up to 8% of dry weight). Due to their high digestibility, microalgal carbohydrates (glucose, starch and polysaccharides) can be used in foods or feeds (Ratha and Prasanna 2012). Applications of various microalgal products are enlisted in Table 13.1.

Table 13.1 Applications of various products extracted from microalgae

Products	Microalgae	Applications
Pigments		
Phycocyanin	*Spirulina platensis*	Health food Cosmetics Pharmaceuticals
Phycoerythrin	*Porphyridium cruentum*	Fluorescent agent Biomedical research Diagnostic tool
Lutein Astaxanthin	*Haematococcus pluvialis* *Botryococcus braunii* *Chlorella zofingiensis* *Chlorococcum* sp.	Pigments Anti-inflammatory agents Cosmetics Provitamins
β-Carotene	*Dunaliella salina* *Vischeria stellata*	Pigments Food colourant Antioxidants Feed additives
Polyunsaturated fatty acids		
Eicosapentaenoic acid	*Chlorella minutissima* *Nannochloropsis* sp.	Nutraceuticals Food additives
Docosahexaenoic acid	*Crypthecodinium* sp. *Schizochytrium* sp.	
Linolenic acid	*Spirulina* sp.	
Vitamins		
Ascorbic acid (vitamin C) Biotin, α-tocopherol (vitamin E)	*Chlorella* sp. *Euglena gracilis*	Nutrition
Polysaccharides	*Porphyridium cruentum*	Antiviral agents Lubricants Flocculants
Lipids	*Botryococcus braunii* *Chlorella* sp. *Chlamydomonas reinhardtii* *Nannochloropsis* sp. *Scenedesmus* sp.	Biodiesel
Carbohydrates	*Arthrospira platensis* *Botryococcus braunii* *Chlorella* sp. *Dunaliella salina* *Haematococcus pluvialis*	Bioethanol Foods Feeds

Compiled from Chu (2012), Ratha and Prasanna (2012), Markou and Nerantzis (2013), Trivedi et al. (2015), Minhas et al. (2016)

More recently, microalgae have been considered as promising feedstocks for the generation of biofuels including biodiesel from oil, alcohols from carbohydrates and liquid or gaseous hydrocarbons from whole biomass (Bohutskyi et al. 2015).

The photosynthetic mechanism of microalgae is similar to plants, but due to their simple cellular structure and submergence in an aqueous environment, in most cases, they have an efficient access to water, CO_2 and other nutrients. They exhibit faster growth and higher photosynthetic efficiency compared to terrestrial crop plants (Chisti 2007). Their shorter generation time allows production of lipids and carbohydrates in large amounts over short periods of time, which can be easily converted into biodiesel and bioethanol, respectively. Moreover, microalgae can be cultivated in nonarable lands using seawater, brackish water or wastewater, due to which they do not compete with agricultural farmlands (Pancha et al. 2014). Under environmental stress conditions, they can accumulate substantial amounts of lipids and carbohydrates (50–70% of dry weight). Due to these reasons, microalgal biomass is now recognized as an alternative renewable feedstock for biofuel production. Further utilization of de-oiled (lipid extracted) biomass for various applications like bioethanol, biogas, animal feed, fertilizer, biosorption of dyes (Maurya et al. 2016), nanoparticle synthesis (Chokshi et al. 2016a, b), etc. reduces the overall cost of microalgal biofuel production, making it economically sustainable.

Bioprospecting comprises of searching and collecting of unique microbial strains for their potential applications. As microalgae are a potential source of bioactive compounds with pharmaceutical, biomedical and nutraceutical prospects, they could play an important role in producing biofuels and bio-based chemicals based on both, their natural components and refined products (Hu et al. 2013). Therefore, bioprospecting of microalgal strains is important to select the best strains that can produce higher amounts of desired metabolic products.

13.2 Oxidative Stress and Reactive Oxygen Species

In any organism, adverse environmental conditions trigger a series of physiological processes and generation of reactive oxygen species (ROS). These highly reactive ROS cause severe damage to proteins, lipids, carbohydrates and nucleic acids, often leading to alterations in cell structure, organelle dysfunction, mutagenesis (Halliwell and Gutteridge 1999) and oxidative cell injuries, ultimately resulting in cell death (Gill and Tuteja 2010). The production of ROS is an unavoidable consequence of aerobic life. It comprises both free radical (superoxide radicals, $O_2^{\bullet-}$; hydroxyl radicals, $OH^\bullet$; perhydroxy radicals, $HO_2^\bullet$; and alkoxy radicals, $RO^\bullet$) and non-radical (molecular) forms (hydrogen peroxide, H_2O_2 and singlet oxygen $^1O_2$). These are the partially reduced forms of atmospheric oxygen (O_2) resulted from the excitation of O_2 to form $^1O_2$. The single electron reduction of O_2 results in the generation of $O_2^{\bullet-}$, which inactivates several important enzymes having iron-sulphur clusters which are required for energy production and amino acid metabolism. This inactivation is caused by the oxidation of cluster leading to release of iron, which through Fenton chemistry produces highly reactive OH$^•$ radicals (Kumar et al. 2011a, b). Protonation of $O_2^{•-}$ produces $HO_2^•$. At low pH, dismutation of $O_2^{•-}$ is unavoidable, with one $O_2^{•-}$ giving up its added electron to another $O_2^{•-}$ forming O_2^{2-}. Protonation of O_2^{2-} results in the generation of H_2O_2. In the presence of transition metals like copper and iron, further reactions through Haber-Weiss cycle or Fenton reaction forms OH$^•$, which is the most reactive chemical species in the biological world (Mittler 2002; Gill and Tuteja 2010).

Stress-induced ROS accumulation is counteracted by an integral defence mechanism of the cell, which, under normal conditions, scavenges the excess oxidants and avoids the deleterious effects of ROS. This includes an array of enzymatic scavengers such as superoxide dismutase (SOD), catalase (CAT), ascorbate peroxidase (APX), glutathione reductase (GR), etc. and non-enzymatic antioxidant molecules such as pigments, proline, polysaccharides, polyphenols, carotenoids, flavonoids, etc. (Mittler et al. 2004; Cirulis et al. 2013). Generation and scavenging of ROS and their effects on cellular metabolites of microalgae are shown in Fig. 13.1. SOD is an important intracellular enzymatic antioxidant ubiquitous in all subcellular compartments of aerobic organisms prone to ROS-mediated oxidative stress (Gill and Tuteja 2010). It provides the first line of defence against toxic effects of ROS by catalysing dismutation of $O_2^{•-}$

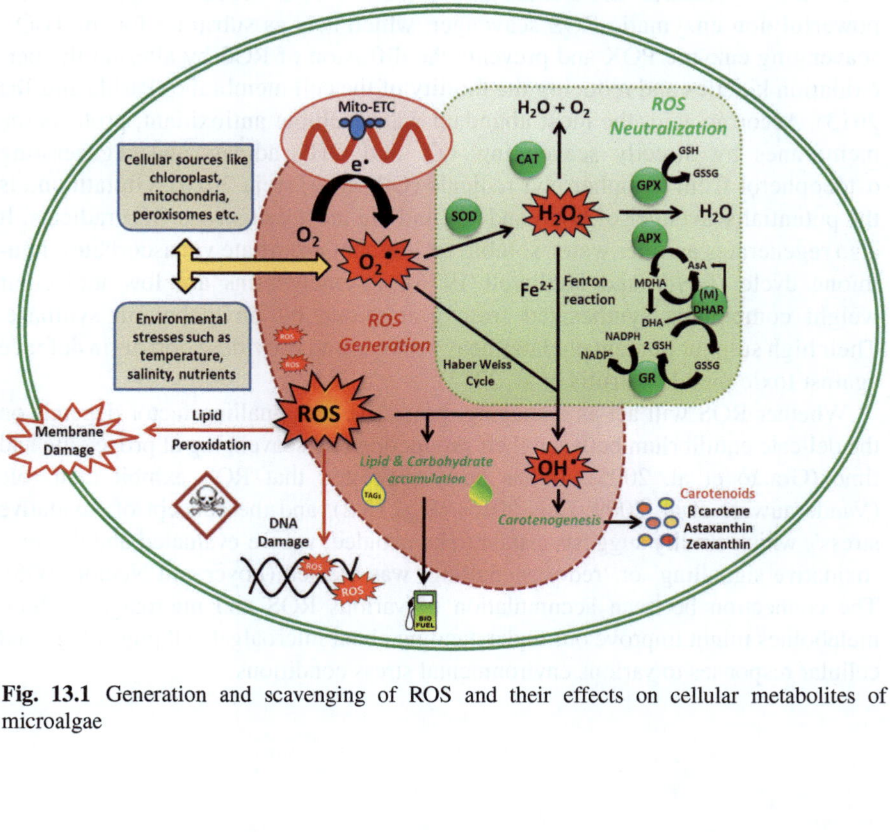

Fig. 13.1 Generation and scavenging of ROS and their effects on cellular metabolites of microalgae

(two $O_2^{\bullet-}$ are neutralized by the addition of two hydrogen ions) to H_2O_2 and O_2, thereby decreasing the risk of $OH^{\bullet}$ formation via Haber-Weiss-type reaction (Gill and Tuteja 2010). CAT is indispensable for ROS detoxification during stressed conditions (Garg and Manchanda 2009). It contains porphyrin haem active sites that directly dismutate H_2O_2 into H_2O and O_2. In 1 min, one molecule of CAT can convert approximately six million molecules of H_2O_2 into H_2O and O_2 (Gill and Tuteja 2010). APX is one of the peroxidases (POX) involved in the scavenging of H_2O_2 utilizing ascorbate as the electron donor. APX has higher affinity to H_2O_2 (μM range) than that of CAT (mM range) (Gill and Tuteja 2010). Besides these, several other enzymatic scavengers, viz. GR, glutathione S-transferases (GST), guaiacol peroxidase, glutathione peroxidase, dehydro- and monodehydroascorbate reductase, etc. also play an essential role in the cellular defence against ROS.

Various non-enzymatic antioxidants also play an important role in the mitigation of oxidative stress-induced cellular injuries. Carotenoids, the lipid soluble pigments, alleviate oxidative toxicity by direct quenching of $^1O_2$ or by dissipating the excess excitation energy from chlorophyll by direct transfer or via xanthophyll cycle and suppressing lipid peroxidation (Telfer 2005; Kumar et al. 2011a). Tocopherols, another lipid-soluble $^1O_2$ quencher, prevent the chain propagation step in lipid autoxidation (Ahmad et al. 2010). Proline, an osmoprotectant, is now considered an inhibitor of lipid peroxidation and the scavenger of $OH^{\bullet}$ and $^1O_2$ (Ashraf and Foolad 2007; Trovato et al. 2008). Polyphenol is a powerful non-enzymatic ROS scavenger, which acts as substrate for the H_2O_2-scavenging enzyme POX and prevents the diffusion of ROS by altering the peroxidation kinetics and reducing the fluidity of the cell membrane (Parida and Jha 2013). Ascorbic acid, the most abundant water-soluble antioxidant, protects the membranes by directly scavenging $O_2^{\bullet-}$ and $OH^{\bullet}$ radicals and regenerating α-tocopherol from tocopheroxyl radicals (Gill and Tuteja 2010). Glutathione is the potential scavenger of H_2O_2 and $^1O_2$ and the most dangerous $OH^{\bullet}$ radicals. It also regenerates another water-soluble antioxidant ascorbate via ascorbate-glutathione cycle (Foyer and Halliwell 1976). Phytochelatins are low molecular weight compounds synthesized from glutathione by phytochelatin synthase. Their high sulphur content chelates heavy metals and provides short-term defence against toxic metals (Cirulis et al. 2013).

Whether ROS will act as damaging, protective or signalling factor depends on the delicate equilibrium between their production and scavenging at proper site and time (Gratão et al. 2005). It has been suggested that ROS exhibit dual role (Vanderauwera et al. 2009; Luis 2015) (Fig. 13.2), and the concept of 'oxidative stress', which strictly suggests a state to be avoided, was re-evaluated and the term 'oxidative signalling' or 'redox signalling' was created (Foyer and Noctor 2005). The connection between accumulation of various ROS and microalgal cellular metabolites might improve our understanding about microalgal cell physiology and cellular responses to various environmental stress conditions.

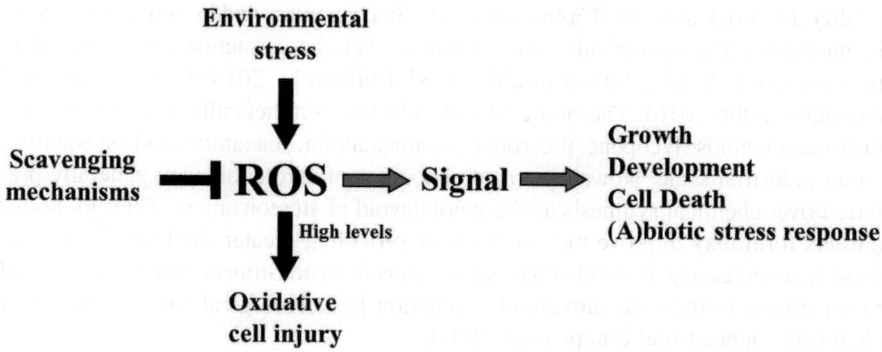

Fig. 13.2 Dual role of ROS [Reproduced from Vanderauwera et al. (2009)]

13.3 Role of Oxidative Stress in Carotenoid Accumulation in Microalgae

Carotenoids are among the most common, naturally occurring terpenoid pigments with a C40 methyl-branched hydrocarbon backbone which provides distinctive molecular structures and associated chemical properties including light-absorption features that are essential for photosynthesis (del Campo et al. 2007). There are over 600 carotenoids occurring in nature (Paliwal et al. 2016). Their colour varies from yellow to orange or red, depending on the number of conjugated double bonds of the polyene chain and corresponds to their ability to absorb photons in the blue and near-UV regions. Carotenoids are divided into carotenes (hydrocarbons containing no oxygen) such as α-carotene, β-carotene and lycopene and xanthophylls (hydrocarbons containing oxygen element) including lutein, zeaxanthin and violaxanthin (Zhang et al. 2014). Carotenoids can also be classified as primary carotenoids, which are functional and structural components of the photosynthetic apparatus and are essential for cell survival, and secondary carotenoids, which are accumulated as oil droplets in the plastids or cytoplasm, when exposed to specific environmental stimuli (via carotenogenesis) (Guedes et al. 2011) such as nutrient starvation, high salinity, high light, etc. Lutein is a major carotenoid as it acts as a primary carotenoid maintaining the integrity of cell membranes and protecting the cells from many forms of stress. Astaxanthin is considered a secondary carotenoid, present in lipid bodies outside the chloroplast, and has potential applications in human health (Minhas et al. 2016).

Carotenoids are synthesized de novo by all photosynthetic organisms, some bacteria and fungi. Vertebrates have to take up these essential molecules from their diet. Due to their physicochemical properties and high-added values, carotenoids are widely used by industries as natural food colourants to colour the flesh of fish or to enhance the colour of egg yolks in poultry industries; as feed additives in aquaculture, poultry, livestock and fish (Cazzonelli 2011); in cosmetics; and as active ingredients in medicinal pharmaceuticals. The global market for carotenoids is expected to increase from US $1.2 billion in 2010 to $1.4 billion by 2018 (D'Alessandro and Antoniosi Filho 2016). The nutraceutical industries synthetically manufacture five major carotenoids (lycopene, β-carotene, canthaxanthin, zeaxanthin and astaxanthin) on an industrial scale. Nowadays, natural sources of carotenoids are generally preferred over chemical synthesis as these are devoid of stereoisomers. Their mono- or diesters form may improve their stability by providing greater shelf life. Moreover, there is an increasing demand of natural products from consumers. This has promoted major efforts to improve carotenoid production from biological sources instead of chemical synthesis (del Campo et al. 2007).

Among various microalgae, *Haematococcus pluvialis*, *Dunaliella salina* and *Chlorella zofingiensis* are the extensively studied taxa as they are widely used in the commercial production of β-carotene and astaxanthin in medium- and large-scale cultures (Lemoine and Schoefs 2010). Carotenoid accumulation in *H. pluvialis* was first reported by Flotow (1844), which is well reviewed by Lemoine and Schoefs (2010). Under favourable environmental conditions, *H. pluvialis* grow as large flagellated green macrozoids. When conditions become unfavourable (e.g. light stress, deficiency of nitrogen, etc.), the cells become spherical, induce cyst formation and change colour from green to red. This resting stage corresponds to large red cells with a thick and heavy resistant cell wall. Tischer (1936) identified the pigment in resting cells and named 'haematochrom' as astaxanthin. Under oxidative stress, biosynthesis of astaxanthin enhances in these cyst cells (Kobayashi et al. 1993; 1997a, b) which is considered as a survival strategy developed by this organism (Boussiba 2000). Under stress conditions, *H. pluvialis* accumulates astaxanthin up to 2–3% (w/w) on a dry weight basis or 43 μg/g on a fresh weight basis (Boussiba et al. 1999; Lemoine and Schoefs 2010).

Tolerance to excessive ROS has been shown to be higher in astaxanthin-rich cysts than in vegetative cells (Kobayashi et al. 1997a, b; Li et al. 2010). In cysts, astaxanthin molecules are present as fatty acid esters and thus have both hydrophilic groups and hydrophobic esters at both β-ionone rings. Compared to nonesterified molecules, the esterified astaxanthin molecules have better capacity to detoxify $^1O_2$. This suggests their ability to function as stabilizers to maintain high antioxidant ability between hydrophilic and hydrophobic conditions (Hagen et al. 1993; Kobayashi et al. 1997a; Kobayashi and Sakamoto 1999). The proteomic analysis identified 70 proteins whose expression pattern changed in *H. pluvialis* following stress induction (Wang et al. 2004a, b). Proteins from the families SOD, CAT and POX were quickly upregulated within the first 12–48 h and then downregulated when astaxanthin molecules accumulated. This suggests primary enzymatic defence response of the cells that plays a critical role upon onset of stress and during the transition of green vegetative cells to cyst formation (Wang et al. 2004a, b). Kobayashi (2000) also observed decrease in the ROS formation in *H. pluvialis* with the increase in intracellular astaxanthin content. Compared to cyst cells, vegetative cells with no astaxanthin had higher ROS formation. Thus, vegetative cells might be more sensitive to ROS than cyst cells, suggesting the importance of morphological

Table 13.2 Summary of selected studies on oxidative stress-induced carotenoid production from microalgae

Products	Microalgae	Stress condition	Stress parameters	References
Astaxanthin, β-carotene, lutein, zeaxanthin, cryptoxanthin	*Spirulina platensis*	Addition of 2–8 mM H_2O_2	CAT, APX, SOD, ascorbic acid, glutathione, tocopherols	El-Baky et al. (2009)
Carotenoids	*Dunaliella salina*	Nitrogen depletion coupled with high light (240 μmol m^{-2} s^{-1})	Isoforms of SOD	Saha et al. (2013)
Astaxanthin, lutein, β-carotene	*Haematococcus* MT 2877	High light (250 μmol m^{-2} s^{-1}) induction	–	Hu et al. (2008)
Lutein	*Chlorella protothecoides*	Addition of 0.1 mM H_2O_2 and 0.01 mM NaClO plus 0.5 mM Fe^{2+}	–	Wei et al. (2008)
β-Carotene	*Dunaliella salina*	1–3 M NaCl coupled with addition of 1 mM n-propyl gallate	Ascorbate, H_2O_2, SOD, CAT, APX, MDA	Einali and Valizadeh (2015)
β-Carotene	*Dunaliella bardawil*	High light (500 W/m^2); addition of ROS-generating chemicals	SOD, CAT	Shaish et al. (1993)
Astaxanthin	*Chlorococcum* sp.	Addition of 0.1 mM H_2O_2	–	Ma and Chen (2001a, b)
Astaxanthin	*Chlorella zofingiensis*	Addition of 0.001–10 mM H_2O_2 and NaClO	–	Ip and Chen (2005)
Astaxanthin	*Haematococcus pluvialis*	Addition of 450 μM iron coupled with high temperature (30–36°C)	H_2O_2, $O_2^{\cdot-}$, $^1O_2$, $OH^{\cdot}$	Hong et al. (2015)

change from vegetative to cyst cells for their survival under adverse environmental conditions. When stress disappears, astaxanthin molecules and the precursors are used as carbon sources for restoring the cell growth (Qin et al. 2008).

Various stress conditions alter the accumulation of carotenoids in microalgae. Table 13.2 summarizes selected studies on oxidative stress-induced carotenoid production from microalgae. Hong et al. (2015) observed enhanced autotrophic astaxanthin production in *H. pluvialis* under high temperature via heat stress-driven Haber-Weiss reaction. High temperature of 30–36°C inhibited the accumulation of astaxanthin under photoautotrophic conditions, which was mainly due to depression of carotenogenesis attributed to excess levels of less ROS ($O_2^{\cdot-}$ and H_2O_2) generated under high temperature. Iron (450 μM) was added in the culture medium to convert these less ROS to more ROS ($^1O_2$ and OH•), thereby facilitating lipid peroxidation, which prompt the cells to synthesize fatty acids and astaxanthin to protect their lipid vesicles. After 18 days of photoautotrophic induction, the astaxanthin content of the cells dramatically increased by 75% at 30°C and 133% at 36°C, compared to that of cells exposed to heat stress alone. Saha et al. (2013) evaluated carotenoid production and differential response of various isoforms of SOD in *D. salina* under different stress conditions. The combination of nitrogen depletion coupled with high-light illumination (240 μmol m^{-2} s^{-1}) resulted into maximum carotenoid production during which the activity of Fe-SOD induced while that of Mn-SOD retained. The micronutrient Mn is one of the important minerals in algal photosynthesis which helps in enhancing inorganic mineral accumulation. This observation suggests essential requirement of Mn during growth of *D. salina* in stress conditions. Differential response of SODs is also reported as a short-term strategy of *H. pluvialis* to survive under oxidative stress conditions, and the organism adopt chronic molecular defence strategies by increasing carotenoid biosynthesis as long-term survival strategy (Wang et al. 2011). Einali and Valizadeh (2015) studied accumulation of β-carotene and response of cellular antioxidant scavengers in *D. salina* grown under salt stress. About 10% increase in β-carotene content was observed when cells were grown in 1 and 3 M NaCl. The H_2O_2 content decreased in cells grown at 1 M NaCl, whereas cells grown at 2 and 3 M NaCl had higher accumulation of H_2O_2. The activity of SOD increased with the increase in the salt concentration; APX activity was highest in cells grown in 2 M NaCl, while CAT activity decreased in all treatments. The total ascorbate and malondialdehyde (MDA, a natural biomarker of lipid peroxidation) contents also increased concomitantly with salt concentration.

Addition of various ROS agents to the culture media has been considered as a possible measure for improving carotenoid synthesis by microalgae. Ip and Chen (2005) used H_2O_2 and sodium hypochlorite (NaClO) to generate OH• and $^1O_2$ and studied carotenogenesis and astaxanthin production in heterotrophic culture of *C. zofingiensis* cultivated in the dark. Addition of 0.1 mM H_2O_2 enhanced secondary carotenoid biosynthesis, including astaxanthin, which was probably due to the formation of OH• radicals through the iron-catalysed Fenton reaction (Kobayashi et al. 1993). On the other hand, $^1O_2$ sharply reduced the secondary carotenoid content. Ma and Chen (2001a, b) observed increase in the astaxanthin formation from 5.8 to 6.5 mg/g in the mixotrophic culture of *Chlorococcum* sp. exposed to 0.1 mM H_2O_2 for 3 days. $^1O_2$ generated using methylene blue reduced the biomass and astaxanthin content. Even under heterotrophic conditions (Ma and Chen 2001a), addition of 0.1 mM H_2O_2 enhanced astaxanthin formation suggesting that light is not an obligatory enhancer for carotenogenesis. This suggests that suitable types of non-photochemical ROS generators could substitute light to stimulate the biosynthesis of secondary carotenoids in dark. It is known that ROS regulate astaxanthin accumulation by direct activation of biosynthetic latent enzymes GST (Aniya and Anders 1992) and GR (Miller and Claiborne 1991) or by activating the expression of genes coding for carotenogenesis enzymes (Bouvier et al. 1998).

El-Baky et al. (2009) studied the effect of H_2O_2 stress on carotenoid content and cellular antioxidants response of *Spirulina platensis*. The contents of hydrophilic antioxidants ascorbic acid and glutathione and lipophilic antioxidant tocopherol increased with the increase in the concentration of H_2O_2. The activities of enzymatic antioxidants CAT, POX and SOD also increased consistently. Carotenoid profiling showed increase in the accumulation of β-carotene, astaxanthin, lutein, zeaxanthin and cryptoxanthin with the increase in the concentration of H_2O_2. This suggests the role of different carotenoids and their derivatives as protectors in the dissipation of chloroplastic ROS (Bennoun 1998). Wei et al. (2008) studied the effects of oxidative stress on lutein production in heterotrophic *Chlorella protothecoides*. The addition of 0.1 mM H_2O_2 and 0.01 mM NaClO plus 0.5 mM Fe^{2+} into the culture led to the generation of $OH^{\bullet}$ radicals and enhanced the lutein content from 1.75 to 1.90 and 1.95 mg/g, respectively. The higher lutein content of 1.98 mg/g was observed when 0.01 mM H_2O_2 and 0.5 mM NaClO were used to generate $^1O_2$.

Oxidative stress caused by intense light illumination is also effective for inducing carotenoid accumulation in microalgae. The underlying mechanism is that excess photo-oxidation under high-light irradiance generates various ROS like $O_2^{\bullet-}$, $OH^{\bullet}$ and $^1O_2$. In the presence of ROS, the antioxidative carotenoids are produced, which could forestall damage of excessive irradiance by directly quenching triplet chlorophyll (^{3}Chl) or $^1O_2$ produced from photodynamic reactions (Krinsky 1979) and thus protect the cells against oxidative damage (Ip and Chen 2005). Hu et al. (2008) observed enhanced protection against photo-oxidative stress in an astaxanthin-overproduction mutant of *Haematococcus* MT 2877. After 3 days of high-light induction from 20 to 250 μmol m^{-2} s^{-1}, the astaxanthin content of mutant cells increased by twofolds compared to wild-type cells, whereas the lutein and β-carotene content remained almost constant. Regulation of astaxanthin biosynthesis might be most likely attributed to the 'secondary messengers'' role of ROS to transduce signalling pathways and altered expression of certain genes and antioxidant enzymes involved in carotenogenesis (Kullik and Storz 1994; Bouvier et al. 1998). Another mechanism of carotenoid photoprotection was described by Ben-Amotz et al. (1989) who concluded that large amount of β-carotene accumulate in the inter-thylakoid space of the green algae *Dunaliella bardawil*. This carotenoid could protect the cells from high irradiance injury by acting as a screen preventing excessive irradiance of blue light from reaching the antenna chlorophylls. This 'filter effect' of β-carotene is attributed to the large overlap between the absorption spectra of β-carotene and chlorophyll in the blue light regime (Boussiba 2000). Shaish et al. (1993) observed that addition of promoters of oxygen radicals greatly enhanced β-carotene synthesis, photodegradation of chlorophyll and inhibition of photosynthesis in *D. bardawil* under high irradiance and suggested that *D. bardawil* adapts to the oxidative stress caused by high irradiance by accumulation of β-carotene along with increasing the activities of CAT and SOD.

13.4 Link Between Oxidative Stress and Biofuel Potential of Microalgae

Biodiesel and bioethanol are the most important renewable fuels which can replace fossil-based fuels. Microalgal biomass has emerged as a potential alternative of plant-based biofuel feedstock due to its higher lipid and carbohydrate content (Chisti 2007; John et al. 2011). The productivities of microalgal biomass, lipid and carbohydrate are key parameters for the economic feasibility of microalgae-based biofuel. Microalgal lipid and carbohydrate can be enhanced by chemical stimuli like nutrient (nitrogen and phosphorous) starvation and salinity stress and physical stimuli like changes in culture pH, temperature, light intensity, photoperiods, etc. (Pancha et al. 2014; Chokshi et al. 2016a). Microalgal lipid is composed of neutral and glyco- and phospholipids. Neutral lipids are mainly composed of triacylglycerols (TAGs), which are important for the production of biodiesel; while glycolipids and phospholipids are the components of cell membranes (Olmstead et al. 2013). While studies on the oxidative stress-enhanced carotenoid accumulation in microalgae are widely reported, understanding of the connection between ROS and other cellular metabolites is still very poor. Therefore, increasing attempts are being made (Table 13.3) to understand the link between accumulation of ROS and synthesis of lipid and carbohydrate by microalgae.

Xin et al. (2011) firstly correlated cellular ROS level and lipid accumulation in *Scenedesmus* sp. grown at different temperature (10–30°C) and suggested that low cultivation temperature induces ROS accumulation in the cells and increases their lipid content. Chokshi et al. (2015) studied the effects of cultivation temperature on growth, biochemical composition and response of stress biomarkers in *Acutodesmus dimorphus*. Increase in the cultivation temperature from 35 to 38°C resulted in 28% higher neutral lipid accumulation after 15 days of growth. The H_2O_2 content of the cells increased by fourfolds, while the MDA level increased by twofolds. The activities of APX and CAT were also increased by 1.7- and 2.6-folds, respectively.

Ruiz-Domínguez et al. (2015) evaluated lipid accumulation and antioxidant activity in the acidophilic microalgae *Coccomyxa onubensis* under nutrient starvation. The total lipid content of the cultures lacking nitrogen, phosphorous or sulphur enhanced during the first 48 h of stress. Further increase in the stress duration increased the lipid accumulation only in the nitrogen-starved cultures. The activities of CAT, APX and GR increased with time suggesting induction of cellular antioxidant responses under nutrient deprivation. In another study, Yilancioglu et al. (2014) observed that low nitrogen condition increases lipid accumulation in the halophilic strain *D. salina*. There was higher level of lipid peroxidation (MDA level) and enhanced activation of SOD, CAT and APX enzymes. Further, the exogenous application of H_2O_2 induced oxidative stress and increased cellular lipid content up to 44% compared to control. This confirms that oxidative stress by itself can cause lipid accumulation and suggests that lipid accumulation under nitrogen depletion is mediated by oxidative stress.

Table 13.3 Summary of selected studies on biofuel potential of microalgae under oxidative stress conditions

Stress	Microalgae	Experimental condition	Stress parameters	References
Temperature	*Acutodesmus dimorphus*	25, 35 and 38°C	H_2O_2, MDA, APX, CAT	Chokshi et al. (2015)
	Scenedesmus sp.	10–30°C	ROS level	Xin et al. (2011)
Nutrients	*Chlamydomonas reinhardtii*	Nitrogen, sulphur, phosphorus and magnesium deprivation; nitrogen and zinc supplementation	Carotenoids, H_2O_2, SOD, CAT, APX, GR, Proline	Çakmak et al. (2015)
	Coccomyxa onubensis	Nitrogen, sulphur and phosphorus starvation	CAT, APX, GR	Ruiz-Domínguez et al. (2015)
	Dunaliella salina	0.05, 0.5 and 5 mM $NaNO_3$	MDA, SOD, CAT, APX	Yilancioglu et al. (2014)
	Chlorella sorokiniana	Nitrogen starvation	ROS, MDA, SOD, POD, CAT	Zhang et al. (2013)
	Chlorella vulgaris	Combination of nutrient excess and nitrogen starvation	$O_2^{\bullet-}$, $OH^{\bullet}$	Menon et al. (2013)
	Nitzschia closterium	932, 712, 491, 270 and 159 μM nitrogen	ROS level	Liu et al. (2012)
Salt	*Chlorella protothecoides*	10–50 g/L NaCl	ROS level, MDA, SOD, CAT, POX	Wang et al. (2016a, b)
	Scenedesmus sp.	0–400 mM NaCl	H_2O_2, APX, MDA, Proline	Pancha et al. (2015a)
Light	*Monoraphidium dybowskii Chlorella* sp.	40, 200 and 400 μmol m^{-2} s^{-1}	SOD, POD, CAT	He et al. (2015)
	Haematococcus pluvialis	20 and 400 μmol m^{-2} s^{-1}	GR, GPX, APX	Gwak et al. (2014)
Addition of chemicals	*Monoraphidium* sp.	5 mM glycine betaine	ROS level	Zhao et al. (2016)
	Dunaliella salina	0–150 mg/L phenol	MDA, SOD, CAT	Cho et al. (2016)
Wastewater	*Chlamydomonas debaryana* *Hindakia tetrachotoma* *Desmodesmus subspicatus* *Chlorella luteoviridis* *Parachlorella hussii*	Raw municipal wastewater secondary effluent; addition of 5–25 mol/m³ H_2O_2	ROS level, APX	Osundeko et al. (2013) Osundeko et al. (2014)

Nitrogen depletion has also been shown to correlate increased ROS accumulation with higher cellular lipid content in *Chlorella sorokiniana* C3 (Zhang et al. 2013). Higher oxidative stress indicated by increased level of membrane peroxidation was observed during oil droplet formation. The assays of SOD, POD and CAT suggested impaired ROS scavenging ability of the cells under nitrogen starvation. Menon et al. (2013) studied nutrient stress-induced lipid production and its relationship with specific ROS levels in *Chlorella vulgaris*. They observed 43-folds increase in the neutral lipid accumulation and 10–15-folds increase in the $O_2^{\bullet-}$ and $OH^{\bullet}$ radicals after shifting the cells from nutrient replete to starvation medium. From the experiments with different nutrient levels, a power-law correlation between specific intracellular levels of neutral lipid and $OH^{\bullet}$ radicals was observed suggesting a cascade effect of $OH^{\bullet}$ radicals on neutral lipids, possibly in the control of lipid synthesis. Changes in the intracellular level of ROS and neutral lipid accumulation in the marine diatom *Nitzschia closterium* cultivated under the range of nitrogen concentration (159–932 μM) was studied by Liu et al. (2012). The intracellular ROS content increased largely in all treatments during stationary phase (12 days) compared to the exponential phase (4 days). The neutral lipid accumulation increased with the decrease in nitrogen concentration in the medium after 8 and 12 days, which suggests that cells might accumulate TAG as a protective mechanism to survive under nitrogen-limited stress. It is known that ROS induced by nutrient starvation play a significant role in the regulation of starvation-induced autophagy (Goodson et al. 2011; Scherz-Shouval et al. 2007), which releases carbon moieties that might contribute to lipid synthesis (Goodson et al. 2011).

Çakmak et al. (2015) investigated the antioxidant response of *Chlamydomonas reinhardtii* under element deprivation (nitrogen, sulphur, phosphorus and magnesium) and supplementation (nitrogen and zinc). The total carotenoid content of the cells increased under all element regimes except magnesium deprivation. Element deprivation and zinc supplementation significantly increased H_2O_2 level and lipid peroxidation in *C. reinhardtii*. Confocal imaging showed induction in neutral lipid accumulation by cells under element stress. A dramatic decrease (70%) in the proline content was determined in nitrogen-deprived cells. A similar trend, but to a lesser extent, was also observed in sulphur-, phosphorus- and magnesium-deprived cells. Except for magnesium deprivation, oxygen radical absorbance capacity of the cells decreased under element deprivation, which confirms a state of oxidative damage, while a significant increase under nitrogen and zinc supplementation suggests sustained cellular homeostasis via increased production of antioxidants. Fluctuations in the activities of SOD, CAT, APX and GR were also observed under different element regimes, which refer to different metabolic sources of ROS production triggered by the absence or overabundance of the specific element.

Addition of salt in the growth medium is also reported to affect lipid accumulation and alter the cellular antioxidant system of microalgae. Wang et al. (2016a, b) studied salt stress (addition of 0–50 g/L NaCl)-induced lipid accumulation and oxidative response in heterotrophic culture of *C. protothecoides*. The highest lipid content of 41.5% was obtained after addition of 30 g/L NaCl in the culture medium. After 144 h of stress, compared to control, 4.4-folds higher ROS level was observed.

The MDA content also exhibited a positive correlation with the ROS production. The activities of SOD, CAT and POX increased by 1.66-, 3.6- and 2.98-folds, respectively than those of control. In another study (Pancha et al. 2015a), increase in the salinity of culture medium resulted into simultaneous increase in the lipid and carbohydrate accumulation by *Scenedesmus* sp. Addition of 400 mM NaCl increased the lipid and carbohydrate contents by 1.8- and 2-folds, respectively. There was over 9- and 1.5-folds increase in the H_2O_2 content and APX activity, respectively. The lipid peroxidation increased by 2-folds, while the proline content increased by 4.5-folds.

When the microalgal cells are subjected to high light, the excess electrons in photosynthetic electron transport chain generate numerous ROS. Lipids, especially neutral lipids, and carbohydrates are the preferred storage products in microalgae under various stress conditions as they have highly reduced states and could be used during adverse conditions for the survival of cells (Li et al. 2011). Along with various enzymatic and non-enzymatic antioxidants, synthesis of lipid could also serve as a receptor to dissipate excess electrons to overcome the oxidative damage. He et al. (2015) studied the effect of different light intensities (40, 200 and 400 μmol m^{-2} s^{-1}) on neutral lipid accumulation and changes in the activity of ROS scavenging enzymes in *Chlorella* sp. and *Monoraphidium dybowskii*. After 10 days of cultivation in high light of 400 μmol m^{-2} s^{-1}, there was about 3- and 5-folds increase in SOD activity, 4- and 4.6-folds increase in POD activity and 2.53- and 2.43-folds increase in CAT activity of *Chlorella* sp. and *M. dybowskii*, respectively. The protein and carbohydrate contents of the cells decreased and the chlorophyll degraded. The accumulation of total lipid and neutral lipid increased with the increase in light intensity. The transcriptomic differentiation of antioxidant defence of *H. pluvialis* revealed that an increase in ROS scavenging activity, a decrease in ROS production and the relaxation of over-reduction of photosynthetic electron transport chain work together to protect the cells against photo-oxidative stress (Gwak et al. 2014). Under high irradiance, accumulation of TAG was attributed to moderate upregulation of de novo fatty acid biosynthesis at the gene level as well as moderate elevation of the TAG assembly pathways. It is known that carbohydrate and lipid synthesis is a parallel phenomenon in microalgae, and reduced energy is first stored as carbohydrate and excess is converted into lipids (Fan et al. 2012). Under stress conditions, the ROS production might divert the provisions of carbon skeleton for amino acids and protein synthesis to TAG biosynthesis to resist the cell damage (Msanne et al. 2012).

Osundeko et al. (2013) studied oxidative stress tolerance of five different microalgal strains in raw municipal wastewater secondary effluent for their biofuel application. The ROS generation in all strains was about 130–170% higher relative to non-stressed conditions. There was significant reduction in cell density of non-adapted strains *Chlamydomonas debaryana*, *Hindakia tetrachotoma* and *Desmodesmus subspicatus* at 15 and 20 mol/m^3 H_2O_2, while *Chlorella luteoviridis* and *Parachlorella hussii* could completely survive and grow very well due to their substantial tolerance to oxidative stress generated by wastewater. These strains also exhibited high APX activity. In another study (Osundeko et al. 2014), 8 weeks of acclimation of various microalgae to wastewater resulted in significantly higher growth rate and biomass productivity. The acclimation to wastewater tolerance was correlated with higher accumulation of carotenoid and increased APX activity. These results suggest that acclimation of microalgae to wastewater environment involves increased oxidative stress tolerance activity, and oxidative stress-tolerant microalgae are highly efficient for biofuel feedstock production on wastewater. Higher oxidative stress tolerance is an indicator of many extremophiles adapted to toxic environments. *Chlamydomonas* sp. W80 and HS5 can grow in highly saline environment and tolerate oxidative damage induced by methyl viologen as shown by high expression of APX (Tanaka et al. 2011).

Addition of 5 mM glycine betaine (GB), a plant growth elicitor, to the culture medium increased the lipid content of *Monoraphidium* sp. QLY-1 by 29.06% (Zhao et al. 2016). Compared to control, neutral lipid content of the cells increased by 12.83%, while ROS accumulation increased by 29.74%, which suggests that GB may enhance the absorption of metal trace elements, including iron, to provoke oxidative stress in microalgae via iron-catalysed Haber-Weiss reaction (Hong et al. 2015). The high level of lipid accumulation may also correlate carbon fixation and lipid biosynthetic genes with GB induction (Wang et al. 2016a). Cho et al. (2016) studied the effects of phenol-induced oxidative stress and biodiesel production by marine microalgae *D. salina*. Increase in the phenol concentration (0–150 mg/L) in the culture medium significantly increased the MDA content and SOD activity in the cells after 10 days of exposure. The CAT activity also increased; however, it was not significantly correlated with phenol concentration. The total lipid accumulation in the cells was unaffected, but the overall lipid yield increased by approximately 26%, which might be due to increase in the biomass production by cells under different concentration of phenol.

13.5 Other Recent Applications of Microalgae Linked with Oxidative Stress

Apart from carotenoids and biofuel, microalgae also have the potential for various other applications. Biological hydrogen production is being evaluated as a promising substitute for carbonaceous fuels owing to its high conversion efficiency and high specific energy content (Saenz et al. 2015). Microalgae are among the groups of organisms able to produce clean, carbon-free energy as hydrogen through the expression of hydrogenase enzymes. Saenz et al. (2015) reported evidences of oxidative stress during hydrogen photoproduction in sulphur-deprived cultures of *C. reinhardtii*. At the beginning of the phase during which hydrogen production starts, i.e. at 48 h, the activity of CAT and APX was significantly lower than that at 0 h. However, during the peak of hydrogen production, their activity increased by 45% and 168%, respectively.

Heavy metal pollution is one of the serious environmental problems all around the world. Bioremediation is an effective and low-cost approach for the removal of heavy metals (Chehregani et al. 2009). Microalgae are capable of metal removal from the environment through bioadsorption, i.e. adhesion of metals to cellular surfaces, and bioaccumulation, i.e. intracellular accumulation. Recently, Belghith et al. (2016) reported metal removal capacity of *D. salina* under cadmium stress. As the metal concentration in the medium increased, the total removed cadmium by the cells also increased. Cadmium removal of 35.85% was observed at 75 mg/L concentration. Exposure of cadmium reduced the carbohydrate and increased the total metallothionein's protein content of the cells. The accumulation of carotenoid increased with the increase in the concentration of cadmium in the medium, while the amounts of polyphenols and flavonoids varied with cadmium concentration.

Rai et al. (2013) observed a concentration-dependent increase in the activity of SOD, CAT and APX and accumulations of carotenoid and MDA in *C. vulgaris* exposed to 0.01–100 μg/ml of chromium. Sabatini et al. (2009) studied oxidative stress and antioxidant response of *Scenedesmus vacuolatus* exposed to different concentrations of copper (6.2, 108, 210 and 414 μM) for 1 week. The increase in the concentration of copper in culture medium resulted in the significant increase of the total copper in *S. vacuolatus* along with their protein and MDA contents. Cells exhibited a significant increase in the CAT and SOD activity along with higher glutathione content in 414 μM copper.

13.6 Systems Biology and Omics Platform for Microalgal Biofactory

Systems biology is a comprehensive quantitative analysis of the manner in which all the components of a biological system interact functionally over time (Aderem 2005). The recent advances in systems biology utilizing the omics—genomics, transcriptomics, proteomics and metabolomics—approaches are now unlocking novel metabolic capabilities of microalgae and highlight them as a multiuse feedstock (Guarnieri and Pienkos 2015). A combinatorial approach using multiple omics platforms and the integration of their outcomes is now an effective strategy for clarifying the molecular systems that are integral for microalgal biofactory. Omics data sets can be used to construct in silico metabolic models of biochemical reaction networks for the analysis and prediction of cellular behaviour under genetic and environmental perturbations (Hong and Lee 2015), thus allowing the design of more efficient microalgae with the enhanced level of targeted products.

Genomics is considered as the generation and analyses of nucleotide sequences of the genome as well as cDNA collections. From sequences, individual genes and repeat elements are identified, the organization and arrangement of genes are analysed and comparisons are made among genomes (Jamers et al. 2009), which assists in phylogenetic analysis and can lead to the discovery of conserved genes (Hardison et al. 2003). Genomics plays a significant role in systems biology as genome annotation provides a framework for metabolic reconstruction. The best studied algal species is *C. reinhardtii* whose genome draft sequence was completed in 2003, and the whole genome sequence was published in 2007 (Merchant et al. 2007). The genomic information has been coupled with mRNA expression analysis of genes and is being used in the proteomic and metabolomic studies. The functional annotation of several genes in *C. reinhardtii* has led to elucidation of understanding of nitrate assimilation, photosynthesis, cell cycle regulation and various other metabolic pathways (Jamers et al. 2009). At present, whole genome sequences of various microalgae, viz. *C. reinhardtii*, *C. vulgaris*, *Coccomyxa subellipsoidea*, *Micromonas pusilla*, *Monoraphidium neglectum*, *Nannochloropsis gaditana*, *Ostreococcus tauri*, *Ostreococcus luminarius*, *Phaeodactylum tricornutum*, *Porphyridium purpureum*, etc., are available (Hong and Lee 2015), which might facilitate the scope of genetic improvement by transformation (Singh et al. 2016). Early genomic investigation of *C. reinhardtii* carotenoid biosynthesis identified the components of methylerythritol phosphate pathways and examined the light responsive nature of transcription of the carotenogenic machinery (Lohr et al. 2005).

Transcriptomics is the study of the functional importance of all genes that are expressed. It provides information on the presence and relative abundance of RNA transcripts and thus provides a better view of the active components in the cell than a genomic approach (Jamers et al. 2009). Transcriptomics can be used either as an actual identification tool or as a way to expand our fundamental knowledge and understanding of certain metabolic pathways. Among the first algalomic studies, to employ next-generation sequencing was a global transcriptomic examination of *C. reinhardtii* during nitrogen starvation (Miller et al. 2010), which altered carbon flux directly into fatty acid biosynthesis and downregulated genes involved in protein biosynthesis and photosynthesis. Similar transcriptional analysis by Msanne et al. (2012) reported immediate increase in the synthesis of starch, followed by TAG, in *C. reinhardtii* and *Coccomyxa* sp. C-169 under nutrient depletion. Transcriptomic analyses have identified 41 differentially regulated transcription factors in nitrogen-starved *C. reinhardtii* (Lv et al. 2013). With comparative transcriptome analysis in a *Neodesmus* sp., a recent study (Chang et al. 2016) suggested involvement of triose phosphate isomerase as a key enzyme regulating photosynthate partitioning between fatty acid and starch biosynthesis in green microalgae. Using transcriptomics data of *C. reinhardtii*, a microarray, i.e. a set of DNA sequences representing the entire set of genes of an organism, arranged in a grid pattern for use in genetic testing, has been developed, containing 10,000 oligonucleotide sequences, each presenting a unique gene and covering nearly the full genome of the algae (htttp://www.chlamy.org). Microarray analysis of *C. reinhardtii* exposed to copper revealed upregulation of glutathione peroxidase and a probable glutathione S-transferase, which are involved in oxidative stress defence mechanism (Jamers et al. 2006). Similarly, microarray analysis of *H. pluvialis* under astaxanthin-inducing culture conditions revealed decreased abundance of photosynthesis-related genes concurrent with higher abundance of stress-related and signal transduction genes (Eom et al. 2006).

Modern next-generation sequencing technology has allowed the use of transcriptomic data to construct metabolic networks of *Dunaliella tertiolecta* (Rismani-Yazdi et al. 2011) and *Botryococcus braunii* (Molnár et al. 2012). Transcriptomics have also been used to elucidate the transcriptional dynamics governing hydrogen production in microalgae. Transcriptomic regulation of hydrogen production by *C. reinhardtii* under sulphur deprivation revealed cellular reorganization of the photosynthetic apparatus and the differential abundance of transcripts coding for genes essential for energy-dependent quenching. These studies were complemented with functional gene knockouts, resulting in mutants with enhanced hydrogen production (Nguyen et al. 2008; Ghirardi et al. 2007; Mus et al. 2007). Sulphur-starved *C. reinhardtii* also exhibited increases in transcripts coding for genes involved in stress response and detoxification along with metabolic remodelling to generate increased reducing equivalents (Toepel et al. 2013).

Proteomic data provide information about peptide sequences, which can support incomplete genome annotation and network gap analysis. A recent study employed proteomic analyses to determine the effect of a dark stress on lipid biosynthesis in *P. tricornutum* (Bai et al. 2016) and reported induced expression of proteins in the biochemical pathways of glycolysis and the synthesis of fatty acids, potentially using excess carbon and nitrogen produced from protein breakdown, which resulted into 2.3-folds increase in total lipid content. Nguyen et al. (2011) reported proteomic profiling of oil bodies from *C. reinhardtii*, with focus on proteins involved in lipid metabolism, which can be used for further studies related to genetic engineering of the oil synthetic pathways of the strain. A sub-proteome analysis of *Chlamydomonas* (Terashima et al. 2010) has revealed the presence of proteins in the chloroplast exhibiting similarity to proteins found in the organisms that don't perform photosynthesis, thus suggesting the role of chloroplast enzymes in the capability of the cell to survive under heterotrophic conditions. Approximately 42 chloroplast proteins with unknown function were identified and showed that they do not have any similarity with other proteins from land plants and may be specific in microalgae species (Terashima et al. 2010). Proteomic approaches have also been used to evaluate the changes in protein abundance during photoautotrophic and mixotrophic growth of *Chlamydomonas*. The data supported the presence of carbonic anhydrases (Cah) in the microalgae being Cah1, Cah2, Cah3, Cah4 and Cah9 all detected during autotrophic growth, while only Cah1, Cah3 and Cah4 were detected during mixotrophic growth (Wienkoop et al. 2010). Proteomic approach was also used to study response of *H. pluvialis* to oxidative stress (Wang et al. 2004a) and to study the molecular basis of salinity tolerance in *D. salina* (Liska et al. 2004).

Metabolites are the end products of cellular regulatory processes, and their levels can be used as the ultimate response of cell to environmental changes (Jamers et al. 2009). Metabolomics is defined as the comprehensive and quantitative analysis of all (or a subset) metabolites in a biological system at a specific time point, reflecting a snapshot of all the regulatory events responding to the external environmental conditions (Kumar et al. 2016). Metabolomics produce a large compound set which can provide information about missing metabolic pathways in the organism. It shares distinct advantages with proteomics in terms of elucidating gene function, i.e. the total complement of proteins or metabolites changes according to the physiological, developmental or pathological state of a cell; unlike transcript analysis, proteins and metabolites are functional entities within the cell (Raamsdonk et al. 2001). Further, there are fewer metabolites than genes or gene products to be studied. Most of the metabolic analyses in microalgae have been focused on identification and quantification of economically valuable secondary metabolites like fatty acids, steroids, carotenoids, polysaccharides, lectins, toxins, etc. (Jamers et al. 2009). Integrated proteomic and metabolomic analysis in *C. reinhardtii* (Lee et al. 2012) reported changes in general stress responses and carbon assimilation process with increases in fatty acid and TAG biosynthetic machinery under nitrogen-limited conditions. Wase et al. (2014) examined and correlated the metabolic and proteomic response of *C. reinhardtii* under nitrogen stress. During nitrogen deprivation, there was a reduced abundance of proteins and metabolites involved in photosynthesis, carbon assimilation and chlorophyll biosynthesis and an increase in abundance of proteins of oxidative phosphorylation, nitrogen metabolism and biosynthesis of lipid and starch.

13.7 Conclusion and Future Perspectives

Microalgae have the potential to be used as a feedstock for biofuel production, a source for bioactive compounds, biomedical components and high value pigments. This has increased their demand in cosmetic and pharmaceutical industries. Microalgae constantly manage oxidative stress resulting from cellular metabolic reactions and environmental stress conditions. To manage oxidative stress, microalgae have mechanisms similar to plants that include cellular antioxidant scavengers. Generation of ROS and their subsequent signalling are key regulators in cell physiology and their responses to stress conditions. At high concentration, ROS impair cellular damage while at sublethal level they act as a 'signalling molecule' initiating defence genes and adaptive responses of the cell. However, there is much to learn about the initiation of ROS signalling, the sensing and response mechanisms and how the delicate balance between production and scavenging is controlled. It is also very important to know about the interactions between pathways mediated by ROS, cellular redox changes, hormonal changes, and other messenger molecules in microalgae. Although the role of oxidative stress in carotenogenesis is well studied, its link with the accumulation lipid and carbohydrate in microalgae is not well established yet. Further research is required to know how different ROS affect the synthesis of these metabolites under different stress conditions.

Algalomic analyses provide a wealth of information about the dynamics of genes, proteins and metabolites directly involved in various metabolic processes. Information gathered from omics studies can be used to direct genome editing of other microalgal species more amenable to genetic manipulation or possessing more desirable traits. As new algal species are identified and their genomes are sequenced, strategies must be developed to quickly annotate and characterize genes and proteins involved in lipid and starch metabolism. Principally omics projects should aim at different microalgae features such as evolution, adaptation and divergence compared with other species, gene, protein and metabolite information and their interaction. This would facilitate an understanding of the biology of microalgae in detail and the application of these concepts in the production of valuable products.

Acknowledgement CSIR-CSMCRI Registration Number: 115/2016. KC acknowledges AcSIR for his Ph.D. enrolment. KC and IP acknowledge CSIR-SRF and CSC-0203 for their funding support. Dr. Arvind Kumar, DC, SMC, is gratefully acknowledged for his motivation and constant support. Authors acknowledge the continuous support of all present and past laboratory members. Authors are also thankful to editors for providing an opportunity to write this chapter. The authors apologize to all researchers whose relevant work could not be cited due to space limitations.

References

Aderem A (2005) Systems biology: its practice and challenges. Cell 121:511–513

Ahmad P, Jaleel CA, Salem MA et al (2010) Roles of enzymatic and nonenzymatic antioxidants in plants during abiotic stress. Crit Rev Biotechnol 30:161–175

Aniya Y, Anders MW (1992) Activation of rat liver microsomal glutathione S-transferase by hydrogen peroxide: role for protein-dimer formation. Arch Biochem Biophys 296:611–616

Araujo GS, Matos LJBL, Gonçalves LRB et al (2011) Bioprospecting for oil producing microalgal strains: evaluation of oil and biomass production for ten microalgal strains. Bioresour Technol 102:5248–5250

Ashraf M, Foolad MR (2007) Roles of glycine betaine and proline in improving plant abiotic stress resistance. Environ Exp Bot 59:206–216

Bai X, Song H, Lavoie M et al (2016) Proteomic analyses bring new insights into the effect of a dark stress on lipid biosynthesis in *Phaeodactylum tricornutum*. Sci Rep 6:25494

Belghith T, Athmouni K, Bellassoued K et al (2016) Physiological and biochemical response of *Dunaliella salina* to cadmium pollution. J Appl Phycol 28:991–999

Ben-Amotz A, Shaish A, Avron M (1989) Mode of action of the massively accumulated β-carotene of *Dunaliella bardawil* in protecting the alga against damage by excess irradiation. Plant Physiol 91:1040–1043

Bennoun P (1998) Chlororespiration, sixteen years later. In: Rochais J, Goldschmidt M, Merchant S (eds) The molecular biology of chloroplasts and mitochondria in *Chlamydomonas*. Kluwer, Dordrecht, pp 675–683

Bernal CB, Vázquez G, Quintal IB, Bussy AL (2008) Microalgal dynamics in batch reactors for municipal wastewater treatment containing dairy sewage water. Water Air Soil Pollut 190:259–270

Bohutskyi P, Liu K, Nasr LK et al (2015) Bioprospecting of microalgae for integrated biomass production and phytoremediation of unsterilized wastewater and anaerobic digestion centrate. Appl Microbiol Biotechnol 99:6139–6154

Boussiba S (2000) Carotenogenesis in the green alga *Haematococcus pluvialis*: cellular physiology and stress response. Physiol Plant 108:111–117

Boussiba S, Bing W, Yuan JP et al (1999) Changes in pigments profile in the green alga *Haematococcus pluvialis* exposed to environmental stresses. Biotechnol Lett 21:601–604

Bouvier F, Backhaus RA, Camara B (1998) Induction and control of chromoplast-specific carotenoid genes by oxidative stress. J Biol Chem 273:30651–30659

Çakmak ZE, Ölmez TT, Çakmak T et al (2015) Antioxidant response of *Chlamydomonas reinhardtii* grown under different element regimes. Phycol Res 63:202–211

Cazzonelli CI (2011) Carotenoids in nature: insights from plants and beyond. Funct Plant Biol 38:833–847

Chang WC, Zheng HQ, Chen CNN (2016) Comparative transcriptome analysis reveals a potential photosynthate partitioning mechanism between lipid and starch biosynthetic pathways in green microalgae. Algal Res 16:54–62

Chehregani A, Noori M, Yazdi HL (2009) Phytoremediation of heavy-metal-polluted soils: screening for new accumulator plants in Angouran mine (Iran) and evaluation of removal ability. Ecotoxicol Environ Saf 72:1349–1353

Chisti Y (2007) Biodiesel from microalgae. Biotechnol Adv 25:294–306
Cho K, Lee CH, Ko K et al (2016) Use of phenol-induced oxidative stress acclimation to stimulate cell growth and biodiesel production by the oceanic microalga *Dunaliella salina*. Algal Res 17:61–66
Chokshi K, Pancha I, Ghosh T et al (2016a) Green synthesis, characterization and antioxidant potential of silver nanoparticles biosynthesized from de-oiled biomass of thermotolerant oleaginous microalgae *Acutodesmus dimorphus*. RSC Adv 6:72269–72274
Chokshi K, Pancha I, Maurya R et al (2016b) Growth medium standardization and thermotolerance study of the freshwater microalga *Acutodesmus dimorphus*-a potential strain for biofuel production. J Appl Phycol 28:2687–2696
Chokshi K, Pancha I, Trivedi K et al (2015) Biofuel potential of the newly isolated microalgae *Acutodesmus dimorphus* under temperature induced oxidative stress conditions. Bioresour Technol 180:162–171
Chu WL (2012) Biotechnological applications of microalgae. IeJSME 6:S24–S37
Cirulis JT, Scott JA, Ross GM (2013) Management of oxidative stress by microalgae. Can J Physiol Pharmacol 91:15–21
D'Alessandro EB, Antoniosi Filho NR (2016) Concepts and studies on lipid and pigments of microalgae: a review. Renew Sust Energ Rev 58:832–841
del Campo JA, García-González M, Guerrero MG (2007) Outdoor cultivation of microalgae for carotenoid production: current state and perspectives. Appl Microbiol Biotechnol 74:1163–1174
Einali A, Valizadeh J (2015) Propyl gallate promotes salt stress tolerance in green microalga *Dunaliella salina* by reducing free radical oxidants and enhancing β-carotene production. Acta Physiol Plant 37:1–11
El-Baky HHA, El Baz FK, El-Baroty GS (2009) Enhancement of antioxidant production in *Spirulina platensis* under oxidative stress. Acta Physiol Plant 31:623–631
Eom H, Lee C-G, Jin E (2006) Gene expression profile analysis in astaxanthin-induced *Haematococcus pluvialis* using a cDNA microarray. Planta 223:1231–1242
Fan J, Yan C, Andre C et al (2012) Oil accumulation is controlled by carbon precursor supply for fatty acid synthesis in *Chlamydomonas reinhardtii*. Plant Cell Physiol 53:1380–1390
Flotow JV (1844) Beobachtungen Über *Haematococcus pluvialis*. Novorum Actorum Acad Caesareae Leopoldinae-Carolinae Naturae Curiosorum 20:413–606
Foyer CH, Halliwell B (1976) The presence of glutathione and glutathione reductase in chloroplasts: a proposed role in ascorbic acid metabolism. Planta 133:21–25
Foyer CH, Noctor G (2005) Redox homeostasis and antioxidant signaling: a metabolic interface between stress perception and physiological responses. Plant Cell 17:1866–1875
Garg N, Manchanda G (2009) ROS generation in plants: boon or bane? Plant Biosyst 143:81–96
Ghirardi ML, Posewitz MC, Maness P-C et al (2007) Hydrogenases and hydrogen Photoproduction in oxygenic photosynthetic organisms. Annu Rev Plant Biol 58:71–91
Gill SS, Tuteja N (2010) Reactive oxygen species and antioxidant machinery in abiotic stress tolerance in crop plants. Plant Physiol Biochem 48:909–930
Goodson C, Roth R, Wang ZT, Goodenough U (2011) Structural correlates of cytoplasmic and chloroplast lipid body synthesis in *Chlamydomonas reinhardtii* and stimulation of lipid body production with acetate boost. Eukaryot Cell 10:1592–1606
Gratão PL, Polle A, Lea PJ, Azevedo RA (2005) Making the life of heavy metal-stressed plants a little easier. Funct Plant Biol 32:481–494
Greenwell HC, Laurens LML, Shields RJ et al (2009) Placing microalgae on the biofuels priority list: a review of the technological challenges. J R Soc Interface 7:703–726
Guarnieri MT, Pienkos PT (2015) Algal omics: unlocking bioproduct diversity in algae cell factories. Photosynth Res 123:255–263
Guedes AC, Amaro HM, Malcata FX (2011) Microalgae as sources of carotenoids. Mar Drugs 9:625–644
Gwak Y, Hwang YS, Wang B et al (2014) Comparative analyses of lipidomes and transcriptomes reveal a concerted action of multiple defensive systems against photooxidative stress in *Haematococcus pluvialis*. J Exp Bot 65:4317–4334

Hagen C, Braune W, Greulich F (1993) Functional aspects of secondary carotenoids in *Haematococcus lacustris* [Girod] Rostafinski (Volvocales) IV. Protection from photodynamic damage. J Photochem Photobiol B Biol 20:153–160

Halliwell B, Gutteridge J (1999) Free radicals in biology and medicine, 3rd edn. Oxford University Press, New York

Hardison RC, Alexandersson M, Cawley S et al (2003) Comparative genomics. PLoS Biol 1:e58

He Q, Yang H, Wu L, Hu C (2015) Effect of light intensity on physiological changes, carbon allocation and neutral lipid accumulation in oleaginous microalgae. Bioresour Technol 191:219–228

Hong ME, Hwang SK, Chang WS et al (2015) Enhanced autotrophic astaxanthin production from *Haematococcus pluvialis* under high temperature via heat stress-driven Haber–Weiss reaction. Appl Microbiol Biotechnol 99:5203–5215

Hong SJ, Lee CG (2015) Microalgal systems biology for biofuel production. In: Algal Biorefineries. Springer International Publishing, Cham, pp 3–21

Hu CW, Chuang LT, Yu PC, Chen CNN (2013) Pigment production by a new thermotolerant microalga *Coelastrella sp.* F50. Food Chem 138:2071–2078

Hu Z, Li Y, Sommerfeld M et al (2008) Enhanced protection against oxidative stress in an astaxanthin-overproduction *Haematococcus* mutant (Chlorophyceae). Eur J Phycol 43:365–376

Ip PF, Chen F (2005) Employment of reactive oxygen species to enhance astaxanthin formation in *Chlorella zofingiensis* in heterotrophic culture. Process Biochem 40:3491–3496

Jamers A, Blust R, De Coen W (2009) Omics in algae: paving the way for a systems biological understanding of algal stress phenomena? Aquat Toxicol 92:114–121

Jamers A, Van der Ven K, Moens L et al (2006) Effect of copper exposure on gene expression profiles in *Chlamydomonas reinhardtii* based on microarray analysis. Aquat Toxicol 80:249–260

John RP, Anisha GS, Nampoothiri KM (2011) Micro and macroalgal biomass: a renewable source for bioethanol. Bioresour Technol 102:186–193

Kobayashi M (2000) In vivo antioxidant role of astaxanthin under oxidative stress in the green alga *Haematococcus pluvialis*. Appl Microbiol Biotechnol 54:550–555

Kobayashi M, Kakizono T, Nagai S (1993) Enhanced carotenoid biosynthesis by oxidative stress in acetate-induced cyst cells of a green unicellular alga, *Haematococcus pluvialis*. Appl Environ Microbiol 59:867–873

Kobayashi M, Kakizono T, Nishio N et al (1997a) Antioxidant role of astaxanthin in the green alga *Haematococcus pluvialis*. Appl Microbiol Biotechnol 48:351–356

Kobayashi M, Kurimura Y, Kakizono T et al (1997b) Morphological changes in the life cycle of the green alga *Haematococcus pluvialis*. J Ferment Bioeng 84:94–97

Kobayashi M, Sakamoto Y (1999) Singlet oxygen quenching ability of astaxanthin esters from the green alga *Haematococcus pluvialis*. Biotechnol Lett 21:265–269

Krinsky NI (1979) Carotenoid protection against oxidation. Pure Appl Chem 51:649–660

Kullik I, Storz G (1994) Transcriptional regulators of the oxidative stress response in prokaryotes and eukaryotes. Redox Rep 1:23–29

Kumar M, Gupta V, Trivedi N et al (2011a) Desiccation induced oxidative stress and its biochemical responses in intertidal red alga *Gracilaria corticata* (Gracilariales, Rhodophyta). Environ Exp Bot 72:194–201

Kumar M, Gupta V, Kumari P et al (2011b) Assessment of nutrient composition and antioxidant potential of Caulerpaceae seaweeds. J Food Compos Anal 24:270–278

Kumar M, Kuzhiumparambil U, Pernice M et al (2016) Metabolomics: an emerging frontier of systems biology in marine macrophytes. Algal Res 16:76–92

Lee DY, Park J-J, Barupal DK, Fiehn O (2012) System response of metabolic networks in *Chlamydomonas reinhardtii* to total available ammonium. Mol Cell Proteomics 11:973–988

Lemoine Y, Schoefs B (2010) Secondary ketocarotenoid astaxanthin biosynthesis in algae: a multifunctional response to stress. Photosynth Res 106:155–177

Li Y, Han D, Sommerfeld M, Hu Q (2011) Photosynthetic carbon partitioning and lipid production in the oleaginous microalga *Pseudochlorococcum* sp. (Chlorophyceae) under nitrogen-limited conditions. Bioresour Technol 102:123–129

Li Y, Sommerfeld M, Chen F, Hu Q (2010) Effect of photon flux densities on regulation of carotenogenesis and cell viability of *Haematococcus pluvialis* (Chlorophyceae). J Appl Phycol 22:253–263
Liska AJ, Shevchenko A, Pick U, Katz A (2004) Enhanced photosynthesis and redox energy production contribute to salinity tolerance in *Dunaliella* as revealed by homology-based proteomics. Plant Physiol 136:2806–2817
Liu W, Huang Z, Li P et al (2012) Formation of triacylglycerol in *Nitzschia closterium* f. *minutissima* under nitrogen limitation and possible physiological and biochemical mechanisms. J Exp Mar Bio Ecol 418–419:24–29
Lohr M, Im C-S, Grossman AR (2005) Genome-based examination of chlorophyll and carotenoid biosynthesis in *Chlamydomonas reinhardtii*. Plant Physiol 138:490–515
Luis A (2015) ROS and RNS in plant physiology: an overview. J Exp Bot 66:2827–2837
Lv H, Qu G, Qi X et al (2013) Transcriptome analysis of *Chlamydomonas reinhardtii* during the process of lipid accumulation. Genomics 101:229–237
Ma RYN, Chen F (2001a) Induction of astaxanthin formation by reactive oxygen species in mixotrophic culture of *Chlorococcum* sp. Biotechnol Lett 23:519–523
Ma RYN, Chen F (2001b) Enhanced production of free trans-astaxanthin by oxidative stress in the cultures of the green microalga *Chlorococcum* sp. Process Biochem 36:1175–1179
Markou G, Nerantzis E (2013) Microalgae for high-value compounds and biofuels production: a review with focus on cultivation under stress conditions. Biotechnol Adv 31:1532–1542
Maurya R, Paliwal C, Ghosh T et al (2016) Applications of de-oiled microalgal biomass towards development of sustainable biorefinery. Bioresour Technol 214:787–796
Menon KR, Balan R, Suraishkumar GK (2013) Stress induced lipid production in *Chlorella vulgaris*: relationship with specific intracellular reactive species levels. Biotechnol Bioeng 110:1627–1636
Merchant SS, Prochnik SE, Vallon O et al (2007) The Chlamydomonas genome reveals the evolution of key animal and plant functions. Science 318:245–250
Miller H, Claiborne A (1991) Peroxide modification of monoalkylated glutathione reductase. Stabilization of an active-site cysteine-sulfenic acid. J Biol Chem 266:19342–19350
Miller R, Wu G, Deshpande RR et al (2010) Changes in transcript abundance in *Chlamydomonas reinhardtii* following nitrogen deprivation predict diversion of metabolism. Plant Physiol 154:1737–1752
Minhas AK, Hodgson P, Barrow CJ, Adholeya A (2016) A review on the assessment of stress conditions for simultaneous production of Microalgal lipids and carotenoids. Front Microbiol 7:546
Mittler R (2002) Oxidative stress, antioxidants and stress tolerance. Trends Plant Sci 7:405–410
Mittler R, Vanderauwera S, Gollery M, Van Breusegem F (2004) Reactive oxygen gene network of plants. Trends Plant Sci 9:490–498
Molnár I, Lopez D, Wisecaver JH et al (2012) Bio-crude transcriptomics: Gene discovery and metabolic network reconstruction for the biosynthesis of the terpenome of the hydrocarbon oil-producing green alga, *Botryococcus braunii* race B (Showa)*. BMC Genomics 13:576
Msanne J, Xu D, Konda AR et al (2012) Metabolic and gene expression changes triggered by nitrogen deprivation in the photoautotrophically grown microalgae *Chlamydomonas reinhardtii* and *Coccomyxa* sp. C-169. Phytochemistry 75:50–59
Mus F, Dubini A, Seibert M et al (2007) Anaerobic acclimation in *Chlamydomonas reinhardtii*: anoxic gene expression, hydrogenase induction, and metabolic pathways. J Biol Chem 282:25475–25486
Nguyen HM, Baudet M, Cuiné S et al (2011) Proteomic profiling of oil bodies isolated from the unicellular green microalga *Chlamydomonas reinhardtii*: with focus on proteins involved in lipid metabolism. Proteomics 11:4266–4273

Nguyen AV, Thomas-Hall SR, Malnoë A et al (2008) Transcriptome for photobiological hydrogen production induced by sulfur deprivation in the green alga *Chlamydomonas reinhardtii*. Eukaryot Cell 7:1965–1979

Olmstead ILD, Hill DRA, Dias DA et al (2013) A quantitative analysis of microalgal lipids for optimization of biodiesel and omega-3 production. Biotechnol Bioeng 110:2096–2104

Osundeko O, Davies H, Pittman JK (2013) Oxidative stress-tolerant microalgae strains are highly efficient for biofuel feedstock production on wastewater. Biomass Bioenergy 56:284–294

Osundeko O, Dean AP, Davies H, Pittman JK (2014) Acclimation of microalgae to wastewater environments involves increased oxidative stress tolerance activity. Plant Cell Physiol 55:1848–1857

Paliwal C, Ghosh T, George B et al (2016) Microalgal carotenoids: potential nutraceutical compounds with chemotaxonomic importance. Algal Res 15:24–31

Pancha I, Chokshi K, George B et al (2014) Nitrogen stress triggered biochemical and morphological changes in the microalgae *Scenedesmus* sp. CCNM 1077. Bioresour Technol 156:146–154

Pancha I, Chokshi K, Ghosh T et al (2015a) Bicarbonate supplementation enhanced biofuel production potential as well as nutritional stress mitigation in the microalgae *Scenedesmus* sp. CCNM 1077. Bioresour Technol 193:315–323

Pancha I, Chokshi K, Maurya R et al (2015b) Salinity induced oxidative stress enhanced biofuel production potential of microalgae *Scenedesmus* sp. CCNM 1077. Bioresour Technol 189:341–348

Parida AK, Jha B (2013) Inductive responses of some organic metabolites for osmotic homeostasis in peanut (*Arachis hypogaea* L.) seedlings during salt stress. Acta Physiol Plant 35:2821–2832

Qin S, Liu G-X, Hu Z-Y (2008) The accumulation and metabolism of astaxanthin in *Scenedesmus obliquus* (Chlorophyceae). Process Biochem 43:795–802

Raamsdonk LM, Teusink B, Broadhurst D et al (2001) A functional genomics strategy that uses metabolome data to reveal the phenotype of silent mutations. Nat Biotechnol 19:45–50

Rai UN, Singh NK, Upadhyay AK, Verma S (2013) Chromate tolerance and accumulation in *Chlorella vulgaris* L.: role of antioxidant enzymes and biochemical changes in detoxification of metals. Bioresour Technol 136:604–609

Ratha SK, Prasanna R (2012) Bioprospecting microalgae as potential sources of "green energy"-challenges and perspectives (review). Appl Biochem Microbiol 48:109–125

Rismani-Yazdi H, Haznedaroglu BZ, Bibby K, Peccia J (2011) Transcriptome sequencing and annotation of the microalgae *Dunaliella tertiolecta*: pathway description and gene discovery for production of next-generation biofuels. BMC Genomics 12:148

Ruiz-Domínguez MC, Vaquero I, Obregón V et al (2015) Lipid accumulation and antioxidant activity in the eukaryotic acidophilic microalga *Coccomyxa* sp. (strain *onubensis*) under nutrient starvation. J Appl Phycol 27:1099–1108

Sabatini SE, Juárez AB, Eppis MR et al (2009) Oxidative stress and antioxidant defenses in two green microalgae exposed to copper. Ecotoxicol Environ Saf 72:1200–1206

Saenz ME, Bisova K, Touloupakis E et al (2015) Evidences of oxidative stress during hydrogen photoproduction in sulfur-deprived cultures of *Chlamydomonas reinhardtii*. Int J Hydrog Energy 40:10410–10417

Saha SK, Moane S, Murray P (2013) Effect of macro- and micro-nutrient limitation on superoxide dismutase activities and carotenoid levels in microalga *Dunaliella salina* CCAP 19/18. Bioresour Technol 147:23–28

Scherz-Shouval R, Shvets E, Fass E et al (2007) Reactive oxygen species are essential for autophagy and specifically regulate the activity of Atg4. EMBO J 26:1749–1760

Shaish A, Avron M, Pick U, Ben-Amotz A (1993) Are active oxygen species involved in induction of β-carotene in *Dunaliella bardawil*? Planta 190:363–368

Singh P, Kumari S, Guldhe A et al (2016) Trends and novel strategies for enhancing lipid accumulation and quality in microalgae. Renew Sust Energ Rev 55:1–16

Suganya T, Varman M, Masjuki HH, Renganathan S (2016) Macroalgae and microalgae as a potential source for commercial applications along with biofuels production: a biorefinery approach. Renew Sust Energ Rev 55:909–941

Tanaka S, Ikeda K, Miyasaka H et al (2011) Comparison of three *Chlamydomonas* strains which show distinctive oxidative stress tolerance. J Biosci Bioeng 112:462–468

Telfer A (2005) Too much light? How β-carotene protects the photosystem II reaction centre. Photochem Photobiol Sci 4:950–956

Terashima M, Specht M, Naumann B, Hippler M (2010) Characterizing the anaerobic response of *Chlamydomonas reinhardtii* by quantitative proteomics. Mol Cell Proteomics 9:1514–1532

Tischer J (1936) Über das Euglenarhodon und andere Carotinoide einer roten Euglene. (Carotinoide der Süßwasseralgen, I. Teil.). Hoppe-Seyler's Zeitschrift für Physiol. Chemie 239:257–269

Toepel J, Illmer-Kephalides M, Jaenicke S et al (2013) New insights into *Chlamydomonas reinhardtii* hydrogen production processes by combined microarray/RNA-seq transcriptomics. Plant Biotechnol J 11:717–733

Trivedi J, Aila M, Bangwal DP et al (2015) Algae based biorefinery-how to make sense? Renew Sust Energ Rev 47:295–307

Trovato M, Mattioli R, Costantino P (2008) Multiple roles of proline in plant stress tolerance and development. Rend Lincei 19:325–346

Vanderauwera S, Hoeberichts FA, Van Breusegem F (2009) Hydrogen peroxide-responsive genes in stress acclimation and cell death. In: Reactive oxygen species in plant signaling. Springer, Berlin Heidelberg, pp 149–164

Wang J, Sommerfeld M, Hu Q (2011) Cloning and expression of isoenzymes of superoxide dismutase in *Haematococcus pluvialis* (Chlorophyceae) under oxidative stress. J Appl Phycol 23:995–1003

Wang S-B, Chen F, Sommerfeld M, Hu Q (2004a) Proteomic analysis of molecular response to oxidative stress by the green alga *Haematococcus pluvialis* (Chlorophyceae). Planta 220:17–29

Wang S-B, Hu Q, Sommerfeld M, Chen F (2004b) Cell wall proteomics of the green alga *Haematococcus pluvialis* (Chlorophyceae). Proteomics 4:692–708

Wang T, Ge H, Liu T et al (2016a) Salt stress induced lipid accumulation in heterotrophic culture cells of *Chlorella protothecoides*: mechanisms based on the multi-level analysis of oxidative response, key enzyme activity and biochemical alteration. J Biotechnol 228:18–27

Wang Y, He B, Sun Z, Chen Y-F (2016b) Chemically enhanced lipid production from microalgae under low sub-optimal temperature. Algal Res 16:20–27

Wase N, Black PN, Stanley BA, DiRusso CC (2014) Integrated quantitative analysis of nitrogen stress response in *Chlamydomonas reinhardtii* using metabolite and protein profiling. J Proteome Res 13:1373–1396

Wei D, Chen F, Chen G et al (2008) Enhanced production of lutein in heterotrophic *Chlorella protothecoides* by oxidative stress. Sci China Ser C Life Sci 51:1088–1093

Wienkoop S, Weiß J, May P et al (2010) Targeted proteomics for *Chlamydomonas reinhardtii* combined with rapid subcellular protein fractionation, metabolomics and metabolic flux analyses. Mol BioSyst 6:1018–1031

Xin L, Hong-ying H, Yu-ping Z (2011) Growth and lipid accumulation properties of a freshwater microalga *Scenedesmus* sp. under different cultivation temperature. Bioresour Technol 102:3098–3102

Yilancioglu K, Cokol M, Pastirmaci I et al (2014) Oxidative stress is a mediator for increased lipid accumulation in a newly isolated *Dunaliella salina* strain. PLoS One 9:e91957

Zhao Y, Li D, Ding K et al (2016) Production of biomass and lipids by the oleaginous microalgae *Monoraphidium* sp. QLY-1 through heterotrophic cultivation and photo-chemical modulator induction. Bioresour Technol 211:669–676

Zhang YM, Chen H, He CL, Wang Q (2013) Nitrogen starvation induced oxidative stress in an oil-producing green alga *Chlorella sorokiniana* C3. PLoS One 8:e69225

Zhang J, Sun Z, Sun P et al (2014) Microalgal carotenoids: beneficial effects and potential in human health. Food Funct 5:413–425

Part IV

Chapter 14
Bioactive Small Molecules Mediate Microalgal-Bacterial Interactions

Leen Labeeuw, Anna R. Bramucci, and Rebecca J. Case

Abstract Microalgae are a diverse group of photosynthetic microorganisms found throughout the eukaryote tree. Although unicellular, they have complex relationships with the bacteria that surround them. These interactions can range from obligate symbiosis, where the bacterium is required for host survival, to pathogenic, where the bacterial pathogen can kill the host alga. The nature of these algal-bacterial interactions appear to be tightly regulated by both algal and bacterial bioactive molecules, creating a complex system of chemical interactions through which these different species can chemically communicate with each other and directly alter the other physiology. In this way the bacterium is able to exploit (and manipulate) its host to become a more conducive habitat (e.g. algal phycosphere, aquatic biofilms, etc.) for bacterial survival. However, the identity of many of these small molecules and the mechanisms by which they control these exchanges are often overlooked or misunderstood. The ability to eavesdrop on the chemical cross talk occurring between algae and bacteria may open up a vast potential for new knowledge, relating to understanding bacterial-algal relationships, evolution and possibly hijacking this communication to better control microbes in commercial systems. This chapter outlines some of the known bioactive chemicals that mediate these microalgal-bacterial interactions, highlighting what is currently known about these systems and areas that need further investigation.

Keywords Microalgae • Bacteria • Microbial behaviour/signalling • Microbial ecology • Bioactive small molecules

L. Labeeuw
Department of Biological Sciences, University of Alberta, Edmonton, AB T6G 2E9, Canada

Plant Functional Biology and Climate Change Cluster (C3), University of Technology, Sydney, Broadway, NSW 2007, Australia

A.R. Bramucci • R.J. Case (✉)
Department of Biological Sciences, University of Alberta, Edmonton, AB T6G 2E9, Canada
e-mail: rcase@ualberta.ca

© Springer International Publishing AG 2017

M. Kumar, P. Ralph (eds.), *Systems Biology of Marine Ecosystems*,
DOI 10.1007/978-3-319-62094-7_14

Contents

14.1 Introduction 280
14.2 Beneficial Interactions 282
14.2.1 Mutualism 282
14.2.2 Commensalism 287
14.3 Negative Interactions 287
14.3.1 Parasitism 287
14.3.2 Pathogenicity 288
14.4 Possible Applications 291
14.5 Conclusion 292
References 292

14.1 Introduction

Phytoplankton are a diverse group of photosynthetic microorganisms, or primary producers, including both eukaryotic algae and cyanobacteria, which live planktonically in the water column (Litchman et al. 2015). Phytoplankton are important players in global biogeochemical cycles, accounting for approximately 40–50% of the world's carbon fixation and driving the marine carbon pump (Falkowski 1994). In other words, phytoplankton contribute almost half of the global net primary productivity, despite accounting for less than 1% of the global photosynthetic biomass (Field et al. 1998; Falkowski 2012). In addition, phytoplankton make up the base of the marine food web, are major players in driving ecosystem functioning and diversification and act as a habitat-forming species in the open ocean, where they form massive blooms (Falkowski 2012; Stevenson 2014).

Phytoplankton also represent most of the diversity in algae, a term applied to a large and diverse group of photosynthetic eukaryotic organisms only distantly related across the phylogenetic tree of life (Fig. 14.1) (Archibald 2009; Keeling 2010). A billion years ago, the establishment of a cyanobacterium as an endosymbiont in a non-photosynthetic eukaryote created the chloroplast (Gray 1999; Yoon et al. 2004), leading to the evolution of the ancestor of all extant archaeplastids (primary endosymbiosis). However, the subsequent understanding of plastid evolution becomes murky due to successive transfers to other lineages through secondary and tertiary endosymbiotic events, as well as multiple loss events (Keeling 2010). This has led to algae being present in nearly every one of the major eukaryotic supergroups, namely, Archaeplastida, Excavata, the SAR (Stramenopiles-Alveolates-Rhizaria) group and the currently unaffiliated haptophytes and cryptophytes (Fig. 14.1). Although these groups are clearly distinct phylogenetically, there is a continuing debate on their relationship to each other (Keeling 2013; Burki 2014; Derelle et al. 2015).

In their natural environment, microalgae are surrounded by bacteria. Their interactions can be shaped by the complex exchange of small bioactive molecules. These bacterial and algal bioactive molecules may be involved in communication, behavioural modification, or as weaponry between bacteria and eukaryotes, with possible biotechnology applications. For example, algae are a rich source of natural products (such as antibiotics, lipids, fatty acids, toxins, nutraceuticals and pharmaceuticals) (Bhatnagar and Kim 2010), with phytoplankton accounting for approximately 15%

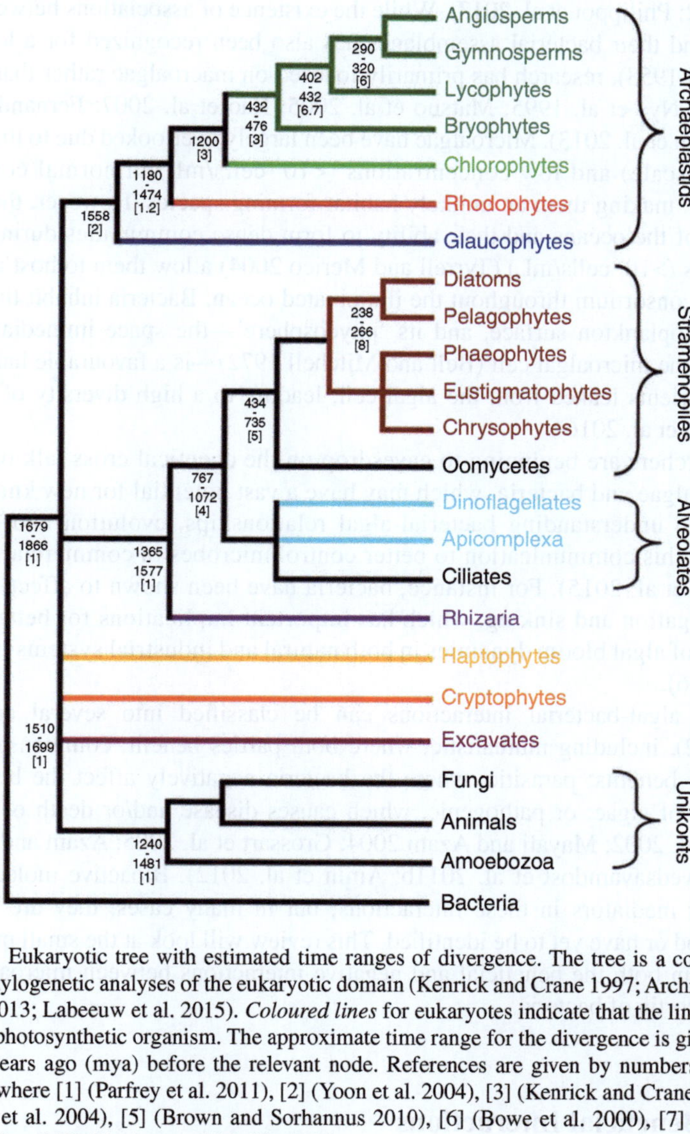

Fig. 14.1 Eukaryotic tree with estimated time ranges of divergence. The tree is a consensus of current phylogenetic analyses of the eukaryotic domain (Kenrick and Crane 1997; Archibald 2009; Keeling 2013; Labeeuw et al. 2015). *Coloured lines* for eukaryotes indicate that the lineage has at least one photosynthetic organism. The approximate time range for the divergence is given in millions of years ago (mya) before the relevant node. References are given by numbers in square brackets, where [1] (Parfrey et al. 2011), [2] (Yoon et al. 2004), [3] (Kenrick and Crane 1997), [4] (Douzery et al. 2004), [5] (Brown and Sorhannus 2010), [6] (Bowe et al. 2000), [7] (Schneider et al. 2004) and [8] (Kooistra and Medlin 1996)

of the discovered marine natural products (Hu et al. 2011). All too often however, novel natural products are sought for their therapeutic and antibiotic uses, while their role in mediating chemical communication between microbes is considered secondary (Yim et al. 2007; Bhatnagar and Kim 2010).

In terrestrial environments, plant-microbe interactions occurring within the plant rhizosphere—the area immediately surrounding plant roots and surrounding bacteria—have been extensively explored (Jones 1998; van Loon et al. 1998; Doornbos et al. 2012; Philippot et al. 2013). While the existence of associations between macroalgae and their bacterial assemblages has also been recognized for a long time (Provasoli 1958), research has primarily focused on macroalgae rather than microalgae (De Nys et al. 1995; Matsuo et al. 2005; Rao et al. 2007; Fernandes et al. 2011; Egan et al. 2013). Microalgae have been largely overlooked due to their small size (μm scale) and low concentrations ($<10^3$ cells/mL) in normal conditions (Li 2002), making them an unlikely habitat-forming species. However, the expansiveness of the oceans and their ability to form dense communities during bloom conditions ($>10^6$ cells/mL) (Tyrrell and Merico 2004) allow them to host a diverse bacterial consortium throughout the illuminated ocean. Bacteria inhabit the immediate phytoplankton surface, and its 'phycosphere'—the space immediately surrounding the microalgal cell (Bell and Mitchell 1972)—is a favourable habitat due to the nutrients leaked from the algal cell, leading to a high diversity of bacteria (Ramanan et al. 2016).

Researchers are beginning to eavesdrop on the chemical cross talk occurring between algae and bacteria, which may have a vast potential for new knowledge, relating to understanding bacterial-algal relationships, evolution and possibly hijacking this communication to better control microbes in commercial systems (Demuez et al. 2015). For instance, bacteria have been shown to affect microalgal aggregation and sinking, which has important implications for better understanding of algal bloom dynamics in both natural and industrial systems (Grossart et al. 2006).

These algal-bacterial interactions can be classified into several categories (Fig. 14.2), including mutualistic, where both parties benefit; commensal, where one party benefits; parasitic, where the bacteria negatively affect the health and fecundity of algae; or pathogenic, which causes disease and/or death of the host (Joint et al. 2002; Mayali and Azam 2004; Grossart et al. 2005; Azam and Malfatti 2007; Seyedsayamdost et al. 2011b; Amin et al. 2012). Bioactive molecules are important mediators in these interactions; but in many cases, they are not well understood or have yet to be identified. This review will look at the small molecules involved in both the beneficial and negative interactions between microalgae and their consortia of bacteria.

14.2 Beneficial Interactions

14.2.1 Mutualism

There are many potential advantages for bacteria in living near or attaching to an algal host. Many bacteria can rapidly sense and respond to chemical gradients using chemotaxis, allowing them to navigate towards the algal host (Miller et al. 2004; Stocker and Seymour 2012). Attachment to the host gives the bacteria a ready supply of organic compounds and an environment to interact closely with other bacteria (Geng and Belas 2010). Bacteria have been shown to chemotax to small diffusible

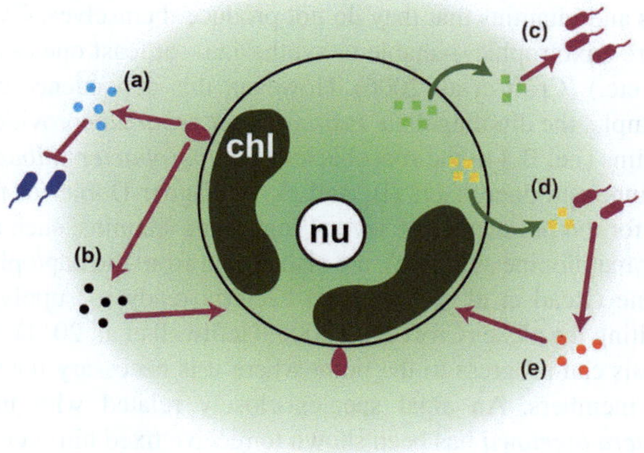

Fig. 14.2 Model of interactions between bacteria and microalgae mediated by bioactive molecules. Host algal cell with cell wall (*black outline*), chloroplasts (*chl*) and nucleus (*nu*). The phycosphere is depicted in *green* around the algal cell, representing the algal exudates. These exudates include bioactive molecules capable of diffusing, while hydrophobic molecules will concentrate at the cell surface, where microalgal-bacterial interactions occur. Bacteria (*pink*) are depicted around the algal cell, both attached and free-living. *Lines* indicate the source and direct of the molecule towards its target with positive (*pointed arrow*) and negative (*flat-ended arrow*) interactions occurring. Bioactives produced by bacteria (*circles*) and algae (*squares*). Bacterial symbionts release antibiotics, such as tropodithietic acid (TDA), which inhibit other bacteria (**a**). Bacterial symbionts can also produce bioactives that increase algal health or fecundity, such as growth-promoting hormones or essential vitamins (**b**). Algal hosts release metabolites (dissolved organic matter (DOM) and trace nutrients), which can act as nutrients or sensory cues (**c**), such as the chemotactic cue and metabolite dimethylsulfoniopropionate (DSMP). Some algal metabolites released during senescence, such as *p*-coumaric acid, act as signals to bacteria (**d**). Such signals can cue the production of antibiotics (**e**) that target the aging host, such as roseobacticides

molecules released by phytoplankton such as sugars, amino acids and algal metabolites which are beneficial to their growth (Miller et al. 2004; Seymour et al. 2010; Mandal et al. 2011). There are many cases whereby both the algae and the bacteria benefit from an exchange of nutrients to which they might otherwise have limited access (Goecke et al. 2010). For example, some bacteria forming close associations with algae can modify the type and abundance of siderophores—organic molecules that bind to iron and increase its solubility—they produce, such that their algal partner can easily use them to scavenge iron (Amin et al. 2009a). One specific form of siderophore, vibrioferrin, was isolated from several species of *Marinobacter*, a common symbiont of dinoflagellates such as *Gymnodinium catenatum.* While vibrioferrin is a relatively weak scavenger of iron, it is particularly sensitive to light and undergoes an irreversible photolytic reaction producing iron (Fe(III)) that is more readily taken up by both the bacteria and the surrounding dinoflagellates. The bacteria therefore produce a weaker siderophore to promote sharing scavenged iron with their algal symbionts (Amin et al. 2009b).

Bacteria and algae can also mutually benefit from a close interaction by gaining trace nutrients and vitamins that they do not produce themselves. Over half of surveyed algae are auxotrophic—unable to synthesize—at least one essential vitamin (e.g. B_1, B_{12}, etc.) (Croft et al. 2006). However, this dependence can work both ways; for example, the dinoflagellate *Prorocentrum minimum* provides organic carbon and vitamins (i.e. B_3) to the roseobacter *Dinoroseobacter shibae*, in return for vitamins it cannot produce itself (B_1 and B_{12}) (Wagner-Döbler et al. 2010). The genes coding for the production of several important vitamins, such as vitamin B_{12} (necessary for methionine synthase), are also absent from the haptophyte *Emiliania huxleyi* genome (Read et al. 2013), with bacteria ready to supply this vitamin, thereby benefiting the algae (Croft et al. 2005; Helliwell et al. 2011). Such nutrient-based symbiosis can progress to the point where it is necessary for the survival of one or both members. An algal species closely related with the haptophyte *Braarudosphaera bigelowii* has been shown to receive fixed nitrogen from its symbiont, the cyanobacterium UCYN-A, in exchange for organic carbon. What is remarkable about this system is that the cyanobacterium has lost photosystem II and the tricarboxylic acid (TCA) cycle in its genome, while the alga was shown to virtually always carry the bacterium, which is suggestive of this being an obligate symbiosis (Thompson et al. 2012; Cabello et al. 2015).

The exchange of nutrients is not the only benefit of symbiosis. Symbiotic interactions can also be critical in the development and behaviour of organisms. Such behavioural and developmental interactions have yet to be identified in phytoplankton. However, certain types of macro green and red algae depend on bacteria to control (or at least help determine) their growth and morphology throughout their life cycle (Goecke et al. 2010), and they can be reduced to a unicellular form in the absence of symbionts (Matsuo et al. 2005). In *Monostroma oxyspermum*, a bacterially produced secondary metabolite, thallusin, alters the multicellularity and differentiation of this macro green alga from loose aggregates of single cells to the final differentiated, leafy morphology (Matsuo et al. 2005). The unicellular motile zoospore stage of the macro green alga *Ulva* is attracted to acylated homoserine lactones (AHLs) produced by the bacterium *Vibrio anguillarum*, impacting the selection of surface sites for permanent attachment by the alga (Joint et al. 2002; Tait et al. 2005). *Ulva* development was also found to be tightly regulated to induce the characteristic morphology by a dual interaction of two bacterial species: *Cytophaga* MS6 and one of the three identified *Proteobacteria* (either *Roseobacter* MS2, *Sulfitobacter* MS3 and *Halomonas* MS1) (Spoerner et al. 2012). Together, these bacteria release molecules that resemble (but cannot be functionally replaced by) plant hormones, cytokinins and auxins that allow for normal growth of the macro green alga *Ulva mutabilis* compared to undifferentiated growth in axenic cultures (Spoerner et al. 2012; Wichard 2015). Some of the bacteria were found to have chemotaxis towards algal cell wall components, as well as a high affinity for the algal metabolites (Spoerner et al. 2012). However, it remains to be seen if bacteria can play a role in cell differentiation within the life cycle of microalgae.

Small molecules produced in the context of bacterial-algal interactions can also affect marine biogeochemical cycles. The unicellular haptophyte *E. huxleyi* is a key driver of its local ecosystem and has been found to have numerous chemically mediated antagonistic interactions with microbes, such as grazers, viruses and bacteria (Wolfe et al. 1997; Seyedsayamdost et al. 2011b; Vardi et al. 2012; Bidle 2015). However, it produces one key metabolite that promotes beneficial relationships: dimethylsulphoniopropionate (DMSP). This sulphurous compound is used as an antioxidant, osmoregulator and cryoprotector (Stefels 2000; Burkill et al. 2002; Sunda et al. 2002). Roseobacters, which are numerically dominant in *E. huxleyi* blooms (Green et al. 2015), play a critical role in converting DMSP into dimethyl sulphide (DMS), an important source of carbon and sulphur for the bacteria (González et al. 2000; Kiene et al. 2000; Moran et al. 2003; Malmstrom et al. 2004; Miller and Belas 2004; Geng and Belas 2010; Seymour et al. 2010). DMS forms the basis of cloud condensation nuclei, thereby affecting weather and having global relevance in affecting Earth's climate (Howard et al. 2006; Dickschat et al. 2010).

DMSP may be supplemented with polyhydroxyalkanoate (PHA) as a carbon source, as shown for the bacterium *Dinoroseobacter shibae* when in coculture with the dinoflagellate *Prorocentrum minimum*, with differential usage during light and dark periods observed (Wang et al. 2014b). One roseobacter, *Ruegeria pomeroyi* DSS-3, was shown to have upregulated catabolism of another sulphur compound produced by its algal host *Thalassiosira pseudonana*, 2,3-dihydroxypropane-1-sulphonate (DHPS). The algal host did not upregulate the production of DHPS in the presence of the bacterium, so the molecule is not released in response to the coculture, but rather the bacteria seems to take advantage of the presence of this molecule (Durham et al. 2015). In both examples, the bacteria provided the vitamin B_{12} to the auxotrophic algae (Wang et al. 2014b; Durham et al. 2015).

Some bacteria can also control the community composition of other bacteria in the phycosphere through their production of antibiotics. In the most extreme case, this can give the bacteria exclusive access to the alga's secreted nutrients while providing the alga with a defence mechanism against fouling agents (other bacteria) that might otherwise be harmful (Rao et al. 2007). Bacteria in the roseobacter clade are thought to produce several antibiotics of interest (Cude et al. 2012; Bentzon-Tilia and Gram 2017). One such antibiotic that is controlled by quorum sensing (QS) is tropodithietic acid (TDA), a potent antibiotic produced by several roseobacters such as *Phaeobacter* and *Ruegeria* (Brinkhoff et al. 2004; Bruhn et al. 2005). In *Phaeobacter gallaeciensis*, TDA production is controlled by AHLs, and the TDA can itself act as an autoinducer for TDA production (Berger et al. 2011). However, QS is only one of the regulatory systems involved in TDA synthesis, as *P. gallaeciensis* QS mutants show delayed TDA production, not complete lack of TDA synthesis (Prol García et al. 2013).

Some algae have evolved mechanisms to turn this bacterial antimicrobial biosynthesis to their advantage by manipulating their bacterial symbionts. For example, some algae secrete compounds that mimic QS signalling molecules naturally produced by bacteria. QS molecules are autoinducers released by bacteria as a function of their population, and when they reach the minimum concentration (quorum), expression of specific genes is induced (Waters and Bassler 2005). Algae are known to manipulate this process by producing analogues to block QS-regulated virulence in algal pathogens (Rao et al. 2007; Case et al. 2011; Harder et al. 2012), and they can also produce mimics. Riboflavin (vitamin B_2) and its derived version lumichrome are secreted by a chlorophyte, *Chlamydomonas*. These compounds can mimic the QS N-acyl-homoserine lactone (AHL) molecules, thereby prematurely inducing the bacteria (e.g. *Pseudomonas* or *Vibrio* spp.) to initiate production of antibacterials to protect the algal host, *Chlamydomonas*, against possible pathogens, even when the bacteria are at low concentrations (Teplitski et al. 2004; Rajamani et al. 2008).

Plant hormones (phytohormones) are well characterized and are known to play a role in plant-microbe interactions. Algae have been suggested to produce a range of plant hormones (Tarakhovskaya et al. 2007; Lu and Xu 2015), although their presence in algae is debated (Lau et al. 2009; Ross and Reid 2010). Plant hormones may play a role in an alga's stress response, as suggested for the chlorophyte *Klebsormidium crenluatum* (Holzinger and Becker 2015). Auxins are an important class of plant hormones, and the most abundant form, indole-3-acetic acid (IAA), was shown to be produced by an axenic brown macroalgae, *Ectocarpus siliculosus*, with an effect on the growth of the alga (Le Bail et al. 2010). However, subsequent research indicated that an un-culturable microbe associated with the alga might be responsible for the IAA production (Dittami et al. 2014). As genes for the initial steps in IAA biosynthesis have been identified in the alga's genome and genes involved in the final steps found in its bacterial partner's genome, it has been hypothesized that they cooperate to produce this compound (Dittami et al. 2014). Some bacteria, including roseobacters (Ashen et al. 1999; Fernandes et al. 2011; Amin et al. 2015; Labeeuw et al. 2016) and other marine groups (Maruyama et al. 1989; Gutierrez et al. 2009), are also capable of producing IAA on their own (Kudoyarova et al. 2015). This suggests that IAA is an important bioactive in the cross talk between algae and marine bacteria, similar to their function between plants and their bacteria (Patten and Glick 2002; Spaepen et al. 2007). Bacterially produced auxin has been implicated in gall formation on the (macro) rhodophyte *Prionitis lanceolata* (Ashen et al. 1999) and bud induction in another (macro) rhodophyte, *Gracilaria dura* (Singh et al. 2011). The question of whether IAA, which impacts cell differentiation and growth, would affect microalgae has been investigated. While it is unlikely to play a role in cellular differentiation, as it does in multicellular phototrophs, it has been shown to play a role in regulating growth and cell division (Amin et al. 2015; Borowitzka 2016). Exogenous application of IAA has been shown to impact microalgal chlorophyte growth (Jin et al. 2008; Stirk et al. 2013; Salama et al. 2014). It has also been shown to stimulate growth in the chlorophyte *Chlorella vulgaris* when cocultured with an IAA-producing bacterium (De-Bashan et al. 2008). This was further demonstrated in the diatom *Pseudo-nitzschia multiseries*, which provided its associated roseobacter *Sulfitobacter* sp. SA11 with tryptophan, which the bacteria converted into IAA, which in turn impacted the growth of the diatom (Amin et al. 2015). A similar model was proposed between the haptophyte *E. huxleyi* and the roseobacter *Phaeobacter inhibens* (Segev et al. 2016). Recently, production of IAA in response to stimulation with tryptophan was demonstrated in the axenic coccolith-bearing strain of *E. huxleyi*, while it was not produced in the bald diploid strain. Although the addition of IAA did not cause a morphological change in the coccolith-bearing strain, it did cause increased size and membrane permeability in the bald strain, indicating a signalling role between cell types of the alga (Labeeuw et al. 2016).

14.2.2 Commensalism

The ability of bacteria to switch from a mutualistic to pathogenic interaction is attributed to environmental factors (Valiente-Banuet and Verdú 2008; Ramanan et al. 2016). One theory is that there is a continuum of interactions and bacteria pass through a stage of commensalism on the way to pathogenicity (Zapalski 2011). Another theory put forth is that the end stage of many mutualists and parasites is naturally as commensals (Sachs and Simms 2006). The study of commensals is complicated, as they are difficult to distinguish from mutualists. There are often 'satellite' or epiphytic bacterial communities described on algal hosts, but these studies have not investigated how the algae might be benefiting from the relationship (Schäfer and Abbas 2002; Tujula et al. 2010). Experiments have shown that adding in natural bacterial communities to axenic algal strains can have a distinct effect on the host growth and survival (Grossart et al. 2005; Bolch et al. 2011). There are many examples of stable communities of bacteria living with algae, with the bacteria benefiting from the leaked nutrients from the algae; however, it is difficult to prove the absence of some algal benefit, thereby proving them to be true commensals rather than mutualists (Cole 1982; Bratbak and Thingstad 1985).

14.3 Negative Interactions

14.3.1 Parasitism

Parasitism, whereby one species benefits at the expense of the overall health and fecundity of its host, is found between algae and bacteria, or between algal species, as algae can often act parasitically to other algae. For example, approximately 10% of red algae are parasites of other, free-living, red algae (Hancock et al. 2010). Fungi, amoebae, alveolates as well as other protists and other algae species (often of the same phylum) are known to be parasites of microalgae as well (Park et al. 2004; Carney and Lane 2014). Parasitism can lead to the premature demise of the algal bloom population (Grami et al. 2011). Parasites are important for the resilience of a system against perturbation and ecosystem functioning (Gachon et al. 2010). However, the effects of bacterial parasitism and the chemical signals and metabolites underpinning these interactions are largely unresolved and an area for future studies.

14.3.2 Pathogenicity

Bacteria have a suite of chemical defences and mechanisms for defence and antagonism against other bacteria. However, the chemical signals between algal and bacterial pathogenic interactions are less well known (Wietz et al. 2013). One advantage to bacteria in killing their algal host is that it leads to access of metabolites and nutrients within their host. As such, proximity to the dying host should allow for more direct benefit. This is the case for some bacteria, in which close association is a necessity for pathogenicity, with some bacterial pathogens even residing directly in the algal cell wall (Wang et al. 2010). However, it has been suggested that only a minority of bacterial pathogens require direct cell-to-cell contact to negatively affect their host (Mayali and Azam 2004; Demuez et al. 2015). Chemotaxis has been shown to be an important feature in bacteria obtaining the dissolved organic matter (DOM) released from lytic algal cells (Stocker et al. 2008; Smriga et al. 2016). Some bacteria may capitalize on this and induce lysis of their algal hosts (Wang et al. 2010).

A few bacteria have been proposed to have the ability to switch from a mutualistic to pathogenic phase, induced by either molecule made due to aging of the algal host, the presence of a high nutrient media, accumulation of bacterial QS molecules or other unknown signals (Seyedsayamdost et al. 2011b; Wang et al. 2014b). Chemotaxis towards algal products such as DMSP is important for mutualists, as it can lead to the area around the algae containing higher concentrations. However, chemotaxis towards DMSP is also important in achieving the quorum of bacteria needed to achieve the QS signal concentration required to switch to a pathogenic lifestyle (Lovejoy et al. 1998; Nakashima et al. 2006; Wang and Yuan 2014). Environmental conditions and nutrient levels are an important factor in generating an algicidal response in bacteria, as a threshold concentration of bacteria and nutrients is required to stimulate lytic responses (Mayali and Doucette 2002; Amaro et al. 2005). One example is the bacteria *Kordia algicida* which has targeted algicidal effects against selected diatom species, as it only releases the algicidal proteases once a quorum has been reached (Paul and Pohnert 2011).

Effects of algicidal bacteria can include decreased chlorophyll and photosynthesis, induction of caspase-like activity and loss of cell wall integrity (Fu et al. 2012; Mayers et al. 2016). Pathogenicity seems to be a targeted affair, as algicidal bacteria are often found to kill one strain or species but not another (Mayali and Azam 2004; Demuez et al. 2015). However, unlike viral killing of algae (Bidle and Vardi 2011), the main mechanisms and compounds used by the bacteria remain mostly uncharacterized (Mayali and Azam 2004; Seyedsayamdost et al. 2011b). The algicidal bacteria *Alteromonas* sp. and *Thalassobius aestuarii* sp. release enzymes that specifically target the cell wall of the *Alexandrium tamarense*, including chitinase or β-glucosidase (Wang et al. 2010). Proteases induced in the stationary phase of *Pseudoalteromonas* sp. were implicated in the death of *Skeletonema costatum* (Lee et al. 2000; Mitsutani et al. 2001).

Specific algicidal molecules have been harder to identify, although recently the QS messenger 2-heptyl-4-quinolone (HHQ) (Diggle et al. 2007) produced by

Pseudoalteromonas piscicida was found to be a potent algicidal molecule causing the death of *E. huxleyi* (Harvey et al. 2016). The bacterium *Alteromonas* sp. was isolated from a harmful algal bloom, and four compounds (2-undecen-1′-yl-4-quinolone, 2-undecyl-4-quinolone, 3-hexyl-6-pentyl-4-hydroxyl-2H-pyran-2-one and 6-heptyl-3-hexyl-4-hydroxyl-2H-pyran-2-one) were isolated which had specific activity against harmful dinoflagellates but had a lower activity against other microalgae (Cho 2012). Two algicidal compounds, prolyl-methionine and hypoxanthine, were identified from a bacterium *Bacillus* sp. which was isolated from a bloom of the haptophyte *Phaeocystis globosa*; they were found to target antioxidant systems within the alga (Yang et al. 2015). Lactones produced by *Ruegeria pomeroyi* were found to target and harm only the algae (*Chlorella fusca*) when tested against other bacteria, fungi and the alga (Riclea et al. 2012). In addition to algicidal molecules, bacteria can also produce molecules that negatively affect the behaviour of algae. Anatoxin-a and microcystin-LR are two toxins produced by cyanobacteria that can inhibit motility in the chlorophyte *Chlamydomonas reinhardtii* and cause increased settling and sinking of the alga (Kearns and Hunter 2001).

Algae have developed a suite of small molecules that can act as chemical defences, which they can deploy to inhibit colonization, organisms competing for available resources or predators (Hay 1996; Wolfe et al. 1997; Potin et al. 2002; Steinberg and de Nys 2002). For example, (macro) brown algae produce phlorotannins which are potent antimicrobial metabolites against a range of bacteria, fungi and other algae (Eom et al. 2012). Likewise, microalgae have been shown to produce antibiotic compounds. Various diatoms produce polyunsaturated aldehydes (PUA) which can inhibit various bacteria (conversely, it was also found to stimulate a few specific species of bacteria as well) (Ribalet et al. 2008).

One of the most studied chemical defence systems has been in the (macro) red algae *Delisea pulchra*, which is subject to bleaching due to various bacterial pathogens (Case et al. 2011; Kumar et al. 2016). The macroalga can release furanones, similar in structure to the QS AHL molecules, thereby inhibiting the bacterial QS response and preventing them from expressing phenotypes controlled by this system such as biofilm formation or virulence (Case et al. 2011; Harder et al. 2012; Gardiner et al. 2015). Signals for activation of these defences can be components of the alga themselves. When the degradation products of one of the main components of the algal cell wall, agar, is detected, the macro red algae *Gracilaria conferta* responds with oxidative bursts and halogenating activity against any potentially pathogenic bacteria (Weinberger et al. 1999). Interestingly, one of the causative agents of bleaching disease in *Delisea*, *Ruegeria* sp. R11, requires glutathione peroxidase to be virulent: presumably this enzyme negates the algal defence mechanism of oxidative bursts (Gardiner et al. 2015). This pathogen was isolated from *D. pulchra*; however, it has a wider host range as it is able to kill two of the three unicellular cell types of *E. huxleyi* through the same mechanism of bleaching at elevated temperatures (Mayers et al. 2016). Microalgae such as the chlorophyte *Chlorella saccharophila* have also been found to disrupt QS gene expression in bacteria as well (Natrah et al. 2011), indicating that a similar role may be present in microalgae.

Previous research has proposed that the roseobacter *Phaeobacter gallaeciensis* displays a fascinating switch from mutualism to pathogenicity in response to algal senescence molecules. The bacterium is thought to detect aging in its algal host through production of cell wall breakdown products, which trigger production of compounds called roseobacticides that can act as potent algicides (Seyedsayamdost et al. 2011b). Another roseobacter, *Silicibacter* sp. strain TM1040, is also thought to produce an algicidal molecule known as roseobacter motility inducer (RMI), which causes death in the chlorophyte *Tetraselmis* (Sule and Belas 2013). The production of these compounds is stimulated by the proposed senescence molecule *p*-coumaric acid (*p*-CA) (Seyedsayamdost et al. 2011a, b, 2014; Sule and Belas 2013), an intermediary of lignin biosynthesis (Schaefer et al. 2008; Weng and Chapple 2010). Roseobacticides are assembled from the bacterially produced phenylacetic acid (PAA), algal *p*CA and cysteine derived from algal DMSP (Sule and Belas 2013; Seyedsayamdost et al. 2014), linked to TDA biosynthesis and regulated by QS AHL signals (Wang et al. 2016). These roseobacticides are of interest as they demonstrate a highly specific activity. When tested against selected haptophytes, a green alga, a diatom and a cryptomonad, only the *E. huxleyi* culture was completely killed by roseobacticides, and a third of the cryptomonad *Rhodomonas salina* cells was killed. The diatom *Chaetoceros muelleri* demonstrated morphological changes in response to the roseobacticide (Seyedsayamdost et al. 2011b). This targeted activity, such as cell lysis resulting in the release of internal compounds, can be useful for commercial processing.

The fact that *P. gallaeciensis* responds to algal *p*-CA, a lignin intermediary, is an evolutionary puzzle; lignin is a complex and highly recalcitrant form of carbon often thought to be one of the key evolutionary advancements allowing the movement of plants from marine habitats to terrestrial ecosystems, essential for structural support and water retention (Boerjan et al. 2003; Weng and Chapple 2010). The discovery of lignin and its intermediates in algae presents the intriguing possibility that these intermediates may have an alternate role as signalling molecules (Schaefer et al. 2008; Martone et al. 2009; Seyedsayamdost et al. 2011b; Goiris et al. 2014; Labeeuw et al. 2015). This is consistent with the theory that lignin and its intermediates may form an ancient microbial defence system of plants and algae against bacteria (Boudet 2000; Tronchet et al. 2010; Labeeuw et al. 2015).

Segev et al. (2016) propose that IAA is another metabolite involved in both the symbiotic and pathogenic phase of the *P. inhibens* DSM 17395 and *E. huxleyi* interaction. However, it should be noted that the concentration at which IAA becomes algicidal (1000 μM) is much higher than any marine bacteria which is currently known to be produced (e.g. 10 nM) (Maruyama et al. 1989; Xie et al. 1996; Fernandes et al. 2011). Concentrations of 100 μM have been shown to be detrimental to the health of algae (Bajguz and Piotrowska-Niczyporuk 2013; Labeeuw et al. 2016), while even in plants, concentrations of IAA above 200 μM are lethal (Baker and Ray 1965). It remains to be demonstrated that IAA is algicidal at biologically relevant concentrations. The IAA precursor tryptophan was also shown to be algicidal at that high concentration (1000 μM) (Labeeuw et al. 2016). Interestingly, IAA production was lower in the coculture of the pathogenic *Ruegeria* sp. R11 on the coccolith-bearing strain of *E. huxleyi* (which was shown to produce IAA in axenic cultures) in the presence of tryptophan, which may indicate a faster uptake of IAA in the coculture, hence its lower detection rate in the supernatant, or that the tryptophan is being diverted into other pathways in the coculture (Labeeuw et al. 2016). However, tryptophan did increase the virulence of the bacteria on the algae, indicating a role for tryptophan in the virulence of the bacteria (Labeeuw et al. 2016; Segev et al. 2016).

14.4 Possible Applications

Algae are grown commercially for a number of reasons, such as food supplements, for animal feedstocks, or biofuels (Spolaore et al. 2006; Chisti 2007; Borowitzka 2013). Processing of the algae is an important step when the aim is to collect valuable algal metabolites that can be used and/or sold for additional profit, with harvesting and extraction being identified as one of the key bottlenecks currently (Greenwell et al. 2010). The current harvesting methods can be problematic to scale up. One area that can be further investigated for improvement is flocculation, which often precedes other harvesting steps. The use of bioflocculants—flocculent-producing bacteria—to coagulate the algae is an attractive alternative to possibly toxic chemical flocculants (Oh et al. 2001; Gutzeit et al. 2005; Wang et al. 2012; González-Fernández and Ballesteros 2013). Removing bacteria from algal species has been shown to drastically reduce their flocculating ability, and further research in bacterial species involved in algal flocculation would likely help improve harvesting (Lee et al. 2013; Ramanan et al. 2016).

This also demonstrates the importance of understanding the natural communities present in algal systems and the possibilities in artificially creating a community ('synthetic ecology') to increase the robustness and productivity of a commercial algal system (Kazamia et al. 2012; Cho et al. 2015). Using more complex communities to produce biofuels could prevent crashes in the algal population from occurring, which causes delays in production and increases in cost. Understanding the underlying interactions occurring in this community, as well as the bioactive molecules involved, will allow for control over the systems, including increased lipid production (Keshtacher-Liebso et al. 1995; Lenneman et al. 2014; Cho et al. 2015). Alternatively, addition of bacteria, or bioactive molecules that cause a change in the symbiosis of the bacteria towards the algae, could allow for timed death within the system. The impact and possibilities of bacteria in commercial systems have been greatly underestimated, and only recently is the potential of bacteria starting to be recognized (Wang et al. 2014a), although the role of the metabolites involved is still greatly overlooked (Franz et al. 2013; Natrah et al. 2014; Demuez et al. 2015). The addition of a single chemical compound instead of bacteria to induce a specific desired change raises unique challenges and possibilities in biotechnological advancement of algae.

14.5 Conclusion

Bacteria and algae have a diverse range of possible interactions, some of which have been well studied and reviewed for macroalgae (Egan et al. 2013, 2014; Singh and Reddy 2014) and microalgae (Mayali and Azam 2004; Azam and Malfatti 2007; Geng and Belas 2010; Natrah et al. 2014; Ramanan et al. 2016). However, the identity of the causative bioactive compound specific to each interaction has often not been elucidated. Some research has been done on understanding the chemical signalling between algal species and within populations (Legrand et al. 2003; Borowitzka 2016), interactions with algae and their predators (Tillmann 2004; Pohnert et al. 2007), but less is known about the chemical underpinnings of the algal-bacterial interactions (Demuez et al. 2015; Hom et al. 2015). Understanding the cross talk between the algal host and their bacterial symbionts allows us to better understand and manipulate the environment.

More research is needed on how the various organisms communicate with each other in a complex community rather than the regulated (and often one-on-one) cocultures regularly studied in the lab. More complex communities will have beneficial and harmful interactions occurring between various organisms within the community, which will not be encompassed by a single model system. Further complicating the system, some bacteria may break down products produced by other members of the community, which will have novel impacts on the system being studied. As a final level of complexity, the environmental conditions may affect the interactions (Grossart 1999). Small-scale systems can be used to successfully monitor long-term interactions between microalgae and bacteria (Bramucci et al. 2015). However, translating the micro-scale experiments into larger scales is also an important field that needs further work, as the larger scales can affect the algal characteristics, as the turbulence, light penetration, nutrient availability and the bacterial assemblage present become less homogenous, and this heterogeneity can make extrapolations difficult from smaller scale (Grossart 1999; Sing et al. 2013; Lohrer et al. 2015). While recent advances in omics reveal a new level of detail in our understanding in microbial ecology (Jansson et al. 2012; Cooper and Smith 2015), they can be used in conjunction with lab systems or to better inform new hypothesizes for new experiments (Amin et al. 2015; Hom et al. 2015). As researchers continue to probe into the underlying novel functions of bioactive molecules, it is likely that even more new insights about their complex roles in shaping and controlling microbial communities will arise.

Acknowledgement This work was supported by Natural Sciences and Engineering Research Council of Canada (grant 402105) to RJC.

References

Amaro AM, Fuentes MS, Ogalde SR et al (2005) Identification and characterization of potentially algal-lytic marine bacteria strongly associated with the toxic dinoflagellate Alexandrium catenella. J Eukaryot Microbiol 52:191–200

Amin SA, Green DH, Hart MC, et al (2009a) Photolysis of iron-siderophore chelates promotes bacterial, algal mutualism. Proc Natl Acad Sci U S A 106:17071–17076
Amin SA, Green DH, Küpper FC, et al (2009b) Vibrioferrin, an unusual marine siderophore: iron binding, photochemistry, and biological implications. Inorg Chem 48:11451–11458
Amin SA, Hmelo LR, van Tol HM, et al (2015) Interaction and signalling between a cosmopolitan phytoplankton and associated bacteria. Nature 522:98–101
Amin SA, Parker MS, Armbrust EV (2012) Interactions between diatoms and bacteria. Microbiol Mol Biol Rev 76:667–684
Archibald JM (2009) The puzzle of plastid evolution. Curr Biol 19:R81–R88
Ashen JB, Cohen JD, Goff LJ (1999) GC-SIM-MS detection and quantification of free indole-3-acetic acid in bacterial galls on the marine alga *Prionitis lanceolata* (Rhodophyta). J Phycol 35:493–500
Azam F, Malfatti F (2007) Microbial structuring of marine ecosystems. Nat Rev Microbiol 5:782–791
Bajguz A, Piotrowska-Niczyporuk A (2013) Synergistic effect of auxins and brassinosteroids on the growth and regulation of metabolite content in the green alga *Chlorella vulgaris* (Trebouxiophyceae). Plant Physiol Biochem 71:290–297
Baker DB, Ray PM (1965) Relation between effects of auxin on cell wall synthesis and cell elongation. Plant Physiol 40:360–368
Bell W, Mitchell R (1972) Chemotactic and growth responses of marine bacteria to algal extracellular products. Mar Biol Lab 143:265–277
Bentzon-Tilia M, Gram L (2017) Biotechnological applications of the Roseobacter clade. In: Paterson R, Lima N (eds) Bioprospecting: success. Potential and Constraints. Springer International Publishing, Cham, pp 137–166
Berger M, Neumann A, Schulz S et al (2011) Tropodithietic acid production in *Phaeobacter gallaeciensis* is regulated by N-acyl homoserine lactone-mediated quorum sensing. J Bacteriol 193:6576–6585
Bhatnagar I, Kim SK (2010) Immense essence of excellence: marine microbial bioactive compounds. Mar Drugs 8:2673–2701
Bidle KD (2015) The molecular ecophysiology of programmed cell death in marine phytoplankton. Annu Rev Mar Sci 7:341–375
Bidle KD, Vardi A (2011) A chemical arms race at sea mediates algal host-virus interactions. Curr Opin Microbiol 14:449–457
Boerjan W, Ralph J, Baucher M (2003) Lignin biosynthesis. Annu Rev Plant Biol 54:519–546
Bolch CJS, Subramanian TA, Green DH (2011) The toxic dinoflagellate Gymnodinium Catenatum (Dinophyceae) requires marine bacteria for growth. J Phycol 47:1009–1022
Borowitzka MA (2016) Chemically-mediated interactions in microalgae. In: Borowitzka MA (ed) The physiology of microalgae. Springer International Publishing, Switzerland, pp 321–357
Borowitzka MA (2013) High-value products from microalgae-their development and commercialisation. J Appl Phycol 25:743–756
Boudet A-M (2000) Lignins and lignification: selected issues. Plant Physiol Biochem 38:81–96
Bowe LM, Coat G, DePamphilis CW (2000) Phylogeny of seed plants based on all three genomic compartments: extant gymnosperms are monophyletic and *Gnetales*' closest relatives are conifers. Proc Natl Acad Sci U S A 97:4092–4097
Bramucci AR, Labeeuw L, Mayers TJ et al (2015) A small volume bioassay to assess bacterial/phytoplankton co-culture using WATER-pulse-amplitude-modulated (WATER-PAM) fluorometry. J Vis Exp 97:e52455
Bratbak G, Thingstad TF (1985) Phytoplankton-bacteria interactions: an apparent paradox? Analysis of a model system with both competition and commensalism. Mar Ecol Prog Ser 25:23–30
Brinkhoff T, Bach G, Heidorn T et al (2004) Antibiotic production by a Roseobacter clade-affiliated species from the German Wadden Sea and its antagonistic effects on indigenous isolates. Appl Environ Microbiol 70:2560–2565

Brown JW, Sorhannus U (2010) A molecular genetic timescale for the diversification of autotrophic stramenopiles (Ochrophyta): substantive underestimation of putative fossil ages. PLoS One 5:e12759. doi:10.1371/journal.pone.0012759
Bruhn JB, Nielsen KF, Hjelm M et al (2005) Ecology, inhibitory activity, and morphogenesis of a marine antagonistic bacterium belonging to the Roseobacter clade. Appl Environ Microbiol 71:7263–7270
Burki F (2014) The eukaryotic tree of life from a global phylogenomic perspective. Cold Spring Harb Perspect Biol 6:a016147
Burkill PH, Archer SD, Robinson C et al (2002) Dimethyl sulphide biogeochemistry within a coccolithophore bloom (DISCO): an overview. Deep Res Part II Top Stud Oceanogr 49:2863–2885
Cabello AM, Cornejo-Castillo FM, Raho N et al (2015) Global distribution and vertical patterns of a prymnesiophyte–cyanobacteria obligate symbiosis. ISME J 10:693–706
Carney LT, Lane TW (2014) Parasites in algae mass culture. Front Microbiol 5:1–8
Case RJ, Longford SR, Campbell AH et al (2011) Temperature induced bacterial virulence and bleaching disease in a chemically defended marine macroalga. Environ Microbiol 13:529–537
Chisti Y (2007) Biodiesel from microalgae. Biotechnol Adv 25:294–306
Cho DH, Ramanan R, Heo J et al (2015) Enhancing microalgal biomass productivity by engineering a microalgal-bacterial community. Bioresour Technol 175:578–585
Cho JY (2012) Algicidal activity of marine *Alteromonas* sp. KNS-16 and isolation of active compounds. Biosci Biotechnol Biochem 76:1452–1458
Cole J (1982) Interactions between bacteria and algae in aquatic ecosystems. Annu Rev Ecol Syst 13:291–314
Cooper MB, Smith AG (2015) Exploring mutualistic interactions between microalgae and bacteria in the omics age. Curr Opin Plant Biol 26:147–153
Croft MT, Lawrence AD, Raux-Deery E et al (2005) Algae acquire vitamin B12 through a symbiotic relationship with bacteria. Nature 438:90–93
Croft MT, Warren MJ, Smith AG (2006) Algae need their vitamins. Eukaryot Cell 5:1175–1183
Cude WN, Mooney J, Tavanaei AA et al (2012) Production of the antimicrobial secondary metabolite indigoidine contributes to competitive surface colonization by the marine roseobacter *Phaeobacter* sp. strain Y4I. Appl Environ Microbiol 78:4771–4780
De-Bashan LE, Antoun H, Bashan Y (2008) Involvement of indole-3-acetic acid produced by the growth-promoting bacterium *Azospirillum Spp.* in promoting growth of *Chlorella Vulgaris*. J Phycol 44:938–947
De Nys R, Steinberg PD, Willemsen P et al (1995) Broad spectrum effects of secondary metabolites from the red alga *Delisea pulchra* in antifouling assays. Biofouling 8:259–271
Demuez M, González-Fernández C, Ballesteros M (2015) Algicidal microorganisms and secreted algicides: new tools to induce microalgal cell disruption. Biotechnol Adv 33:1615–1625
Derelle R, Torruella G, Klimeš V et al (2015) Bacterial proteins pinpoint a single eukaryotic root. Proc Natl Acad Sci U S A 112:E693–E699
Dickschat JS, Zell C, Brock NL (2010) Pathways and substrate specificity of DMSP catabolism in marine bacteria of the Roseobacter clade. Chembiochem 11:417–425
Diggle SP, Matthijs S, Wright VJ et al (2007) The *Pseudomonas aeruginosa* 4-quinolone signal molecules HHQ and PQS play multifunctional roles in quorum sensing and iron entrapment. Chem Biol 14:87–96
Dittami SM, Barbeyron T, Boyen C et al (2014) Genome and metabolic network of "*Candidatus* Phaeomarinobacter ectocarpi" Ec32, a new candidate genus of Alphaproteobacteria frequently associated with brown algae. Front Genet 5:1–13
Doornbos RF, van Loon LC, Bakker PAHM (2012) Impact of root exudates and plant defense signaling on bacterial communities in the rhizosphere. A review. Agron Sustain Dev 32:227–243
Douzery EJP, Snell EA, Bapteste E et al (2004) The timing of eukaryotic evolution: does a relaxed molecular clock reconcile proteins and fossils? Proc Natl Acad Sci U S A 101:15386–15391
Durham BP, Sharma S, Luo H et al (2015) Cryptic carbon and sulfur cycling between surface ocean plankton. Proc Natl Acad Sci 112:453–457

Egan S, Fernandes ND, Kumar V et al (2014) Bacterial pathogens, virulence mechanism and host defence in marine macroalgae. Environ Microbiol 16:925–938
Egan S, Harder T, Burke C et al (2013) The seaweed holobiont: understanding seaweed-bacteria interactions. FEMS Microbiol Rev 37:462–476
Eom SH, Kim YM, Kim SK (2012) Antimicrobial effect of phlorotannins from marine brown algae. Food Chem Toxicol 50:3251–3255
Falkowski P (2012) The power of plankton. Science 483:7–10
Falkowski PG (1994) The role of phytoplankton photosynthesis in global biogeochemical cycles. Photosynth Res 39:235–258
Fernandes N, Case RJ, Longford SR et al (2011) Genomes and virulence factors of novel bacterial pathogens causing bleaching disease in the marine red alga *Delisea pulchra*. PLoS One 6:e27387
Field CB, Behrenfeld MJ, Randerson JT et al (1998) Primary production of the biosphere: integrating terrestrial and oceanic components. Science 281:237–240
Franz AK, Danielewicz MA, Wong DM et al (2013) Phenotypic screening with oleaginous microalgae reveals modulators of lipid productivity. ACS Chem Biol 8:1053–1062
Fu L, Li D, Wu C et al (2012) Effects of algicidal bacterium BS03 (*Microbulbifer sp.*) on the growth and antioxidant systems of *Alexandrium tamarense*. Acta Microbiol Sin 52:784–790
Gachon CMM, Sime-Ngando T, Strittmatter M et al (2010) Algal diseases: spotlight on a black box. Trends Plant Sci 15:633–640
Gardiner M, Thomas T, Egan S (2015) A glutathione peroxidase (GpoA) plays a role in the pathogenicity of *Nautella italica* strain R11 towards the red alga *Delisea pulchra*. FEMS Microbiol Ecol 91:1–5
Geng H, Belas R (2010) Molecular mechanisms underlying roseobacter-phytoplankton symbioses. Curr Opin Biotechnol 21:332–338
Goecke F, Labes A, Wiese J et al (2010) Chemical interactions between marine macroalgae and bacteria. Mar Ecol Prog Ser 409:267–299
Goiris K, Muylaert K, Voorspoels S et al (2014) Detection of flavonoids in microalgae from different evolutionary lineages. J Phycol 50:483–492
González-Fernández C, Ballesteros M (2013) Microalgae autoflocculation: an alternative to high-energy consuming harvesting methods. J Appl Phycol 25:991–999
González JM, Simó R, Massana R et al (2000) Bacterial community structure associated with a dimethylsulfoniopropionate-producing North Atlantic algal bloom. Appl Environ Microbiol 66:4237–4246
Grami B, Rasconi S, Niquil N et al (2011) Functional effects of parasites on food web properties during the spring diatom bloom in lake pavin: a linear inverse modeling analysis. PLoS One. doi:10.1371/journal.pone.0023273
Gray MW (1999) Evolution of organellar genomes. Curr Opin Genet Dev 9:678–687
Green DH, Echavarri-Bravo V, Brennan D et al (2015) Bacterial diversity associated with the coccolithophorid algae *Emiliania huxleyi* and *Coccolithus pelagicus* f. Braarudii. Biomed Res Int 2015:194540
Greenwell HC, Laurens LML, Shields RJ et al (2010) Placing microalgae on the biofuels priority list: a review of the technological challenges. J R Soc Interface 7:703–726
Grossart H-P (1999) Interactions between marine bacteria and axenic various conditions in the lab. Aquat Microb Ecol 19:1–11
Grossart H-P, Levold F, Allgaier M et al (2005) Marine diatom species harbour distinct bacterial communities. Environ Microbiol 7:860–873
Grossart HP, Czub G, Simon M (2006) Algae-bacteria interactions and their effects on aggregation and organic matter flux in the sea. Environ Microbiol 8:1074–1084
Gutierrez CK, Matsui GY, Lincoln DE et al (2009) Production of the phytohormone indole-3-acetic acid by estuarine species of the genus vibrio. Appl Environ Microbiol 75:2253–2258
Gutzeit G, Lorch D, Weber A et al (2005) Bioflocculent algal-bacterial biomass improves low-cost wastewater treatment. Water Sci Technol 52:9–18

Hancock L, Goff L, Lane C (2010) Red algae lose key mitochondrial genes in response to becoming parasitic. Genome Biol Evol 2:897–910

Harder T, Campbell AH, Egan S et al (2012) Chemical mediation of ternary interactions between marine holobionts and their environment as exemplified by the red alga *Delisea pulchra*. J Chem Ecol 38:442–450

Harvey EL, Deering RW, Rowley DC et al (2016) A bacterial quorum-sensing precursor induces mortality in the marine coccolithophore, *Emiliania huxleyi*. Front Microbiol 7:1–12

Hay ME (1996) Marine chemical ecology: what's known and what's next? J Exp Mar Bio Ecol 200:103–134

Helliwell KE, Wheeler GL, Leptos KC et al (2011) Insights into the evolution of vitamin B12 auxotrophy from sequenced algal genomes. Mol Biol Evol 28:2921–2933

Holzinger A, Becker B (2015) Desiccation tolerance in the streptophyte green alga *Klebsormidium*: the role of phytohormones. Commun Integr Biol 8:e1059978

Hom EFY, Aiyar P, Schaeme D et al (2015) A chemical perspective on microalgal-microbial interactions. Trends Plant Sci 20:689–693

Howard EC, Henriksen JR, Buchan A et al (2006) Bacterial taxa that limit sulfur flux from the ocean. Science 314:649–652

Hu G-P, Yuan J, Sun L et al (2011) Statistical research on marine natural products based on data obtained between 1985 and 2008. Mar Drugs 9:514–525

Jansson JK, Neufeld JD, Moran MA et al (2012) Omics for understanding microbial functional dynamics. Environ Microbiol 14:1–3

Jin Q, Scherp P, Heimann K et al (2008) Auxin and cytoskeletal organization in algae. Cell Biol Int 32:542–545

Joint I, Tait K, Callow ME et al (2002) Cell-to-cell communication across the prokaryote-eukaryote boundary. Science 298:1207

Jones DL (1998) Organic acids in the rhizosphere—a critical review. Plant Soil 205:25–44

Kazamia E, Aldridge DC, Smith AG (2012) Synthetic ecology—a way forward for sustainable algal biofuel production? J Biotechnol 162:163–169

Kearns KD, Hunter MD (2001) Toxin-producing *Anabaena flos-aquae* induces settling of *Chlamydomonas reinhardtii*, a competing motile alga. Microb Ecol 42:80–86

Keeling PJ (2010) The endosymbiotic origin, diversification and fate of plastids. Philos Trans R Soc Lond Ser B Biol Sci 365:729–748

Keeling PJ (2013) The number, speed, and impact of plastid endosymbioses in eukaryotic evolution. Annu Rev Plant Biol 64:583–607

Kenrick P, Crane P (1997) The origin and early evolution of plants on land. Nature 389:33–39

Keshtacher-Liebso E, Hadar Y, Chen Y (1995) Oligotrophic bacteria enhance algal growth under iron-deficient conditions. Appl Environ Microbiol 61:2439–2441

Kiene RP, Linn LJ, Bruton JA (2000) New and important roles for DMSP in marine microbial communities. J Sea Res 43:209–224

Kooistra WHCF, Medlin LK (1996) Evolution of the diatoms (Bacillariophyta). Mol Phylogenet Evol 6:391–407

Kudoyarova GR, Arkhipova TN, Melentev AI (2015) Role of bacterial phytohormones in plant growth regulation and their development. In: Maheshwari DK (ed) Bacterial metabolites in sustainable Agroecosystem. Springer, Switzerland

Kumar V, Zozaya-Valdes E, Kjelleberg S et al (2016) Multiple opportunistic pathogens can cause a bleaching disease in the red seaweed Delisea Pulchra. Environ Microbiol 18:3962–3975

Labeeuw L, Khey J, Bramucci AR et al (2016) Indole-3-acetic acid is produced by *Emiliania huxleyi* coccolith-bearing cells and triggers a physiological response in bald cells. Front Microbiol 7:1–16

Labeeuw L, Martone PT, Boucher Y et al (2015) Ancient origin of the biosynthesis of lignin precursors. Biol Direct 10:1–21

Lau S, Shao N, Bock R et al (2009) Auxin signaling in algal lineages: fact or myth? Trends Plant Sci 14:182–188

Le Bail A, Billoud B, Kowalczyk N et al (2010) Auxin metabolism and function in the multicellular brown alga *Ectocarpus siliculosus*. Plant Physiol 153:128–144

Lee J, Cho DH, Ramanan R et al (2013) Microalgae-associated bacteria play a key role in the flocculation of *Chlorella vulgaris*. Bioresour Technol 131:195–201

Lee S, Kato J, Takiguchi N et al (2000) Involvement of an extracellular protease in algicidal activity of the marine bacterium *Pseudoalteromonas* sp. strain A28. Appl Environ Microbiol 66:4334–4339

Legrand C, Rengefors K, Fistarol GO et al (2003) Allelopathy in phytoplankton—biochemical, ecological and evolutionary aspects. Phycologia 42:406–419

Lenneman EM, Wang P, Barney BM (2014) Potential application of algicidal bacteria for improved lipid recovery with specific algae. FEMS Microbiol Lett 354:102–110

Li WKW (2002) Macroecological patterns of phytoplankton in the northwestern North Atlantic Ocean. Nature 419:154–157

Litchman E, de Tezanos PP, Edwards KF et al (2015) Global biogeochemical impacts of phytoplankton: a trait-based perspective. J Ecol 103:1384–1396

Lohrer AM, Thrush SF, Hewitt JE et al (2015) The up-scaling of ecosystem functions in a heterogeneous world. Sci Rep 5:10349

Lovejoy C, Bowman JP, Hallegraeff GM (1998) Algicidal effects of a novel marine *Pseudoalteromonas* isolate (class *Proteobacteria*, gamma subdivision) on harmful algal bloom species of the genera *Chattonella, Gymnodinium*, and *Heterosigma*. Appl Environ Microbiol 64:2806–2813

Lu Y, Xu J (2015) Phytohormones in microalgae: a new opportunity for microalgal biotechnology? Trends Plant Sci 20:273–282

Malmstrom RR, Kiene RP, Kirchman DL (2004) Identification and enumeration of bacteria assimilating dimethylsulfoniopropionate (DMSP) in the North Atlantic and Gulf of Mexico. Limnol Oceanogr 49:597–606

Mandal SK, Singh RP, Patel V (2011) Isolation and characterization of exopolysaccharide secreted by a toxic dinoflagellate, *Amphidinium carterae* Hulburt 1957 and its probable role in harmful algal blooms (HABs). Microb Ecol 62:518–527

Martone PT, Estevez JM, Lu F et al (2009) Discovery of lignin in seaweed reveals convergent evolution of cell-wall architecture. Curr Biol 19:169–175

Maruyama A, Maeda M, Simidu U (1989) Distribution and classification of marine bacteria with the ability of Cytokinin and Auxin production. Bull Japanese Soc Microb Ecol 5:1–8

Matsuo Y, Imagawa H, Nishizawa M et al (2005) Isolation of an algal morphogenesis inducer from a marine bacterium. Science 307:1598

Mayali X, Azam F (2004) Algicidal bacteria in the sea and their impact on algal blooms. J Eukaryot Microbiol 51:139–144

Mayali X, Doucette GJ (2002) Microbial community interactions and population dynamics of an algicidal bacterium active against *Karenia brevis* (Dinophyceae). Harmful Algae 1:277–293

Mayers TJ, Bramucci AR, Yakimovich K et al (2016) A bacterial pathogen displaying temperature-enhanced virulence of the microalga *Emiliania huxleyi*. Front Microbiol. doi:10.3389/fmicb.2016.00892

Miller TR, Belas R (2004) Dimethylsulfoniopropionate metabolism by *Pfiesteria*-associated Roseobacter spp. Appl Environ Microbiol 70:3383–3391

Miller TR, Hnilicka K, Dziedzic A et al (2004) Chemotaxis of *Silicibacter* sp. strain TM1040 toward dinoflagellate products. Appl Environ Microbiol 70:4692–4701

Mitsutani A, Yamasaki I, Kitaguchi H et al (2001) Analysis of algicidal proteins of a diatom-lytic marine bacterium *Pseudoalteromonas sp.* strain A25 by two-dimensional electrophoresis. Phycologia 40:286–291

Moran MA, González JM, Kiene RP (2003) Linking a bacterial taxon to sulfur cycling in the sea: studies of the marine Roseobacter group. Geomicrobiol J 20:375–388

Nakashima T, Miyazaki Y, Matsuyama Y et al (2006) Producing mechanism of an algicidal compound against red tide phytoplankton in a marine bacterium *γ-proteobacterium*. Appl Microbiol Biotechnol 73:684–690

Natrah FMI, Bossier P, Sorgeloos P et al (2014) Significance of microalgal-bacterial interactions for aquaculture. Rev Aquac 6:48–61
Natrah FMI, Kenmegne MM, Wiyoto W et al (2011) Effects of micro-algae commonly used in aquaculture on acyl-homoserine lactone quorum sensing. Aquaculture 317:53–57
Oh H-M, Lee SJ, Park M-H et al (2001) Harvesting of *Chlorella vulgaris* using a bioflocculant from *Paenibacillus sp.* AM49. Biotechnol Lett 23:1229–1234
Parfrey LW, Lahr DJG, Knoll AH et al (2011) Estimating the timing of early eukaryotic diversification with multigene molecular clocks. Proc Natl Acad Sci U S A 108:13624–13629
Park MG, Yih W, Coats DW (2004) Parasites and phytoplankton, with special emphasis on dinoflagellate infections. J Eukaryot Microbiol 51:145–155
Patten CL, Glick BR (2002) Role of *Pseudomonas putida* indoleacetic acid in development of the host plant root system. Appl Environ Microbiol 68:3795–3801
Paul C, Pohnert G (2011) Interactions of the algicidal bacterium Kordia algicida with diatoms: regulated protease excretion for specific algal lysis. PLoS One 6:e21032
Philippot L, Raaijmakers JM, Lemanceau P et al (2013) Going back to the roots: the microbial ecology of the rhizosphere. Nat Rev Microbiol 11:789–799
Pohnert G, Steinke M, Tollrian R (2007) Chemical cues, defence metabolites and the shaping of pelagic interspecific interactions. Trends Ecol Evol 22:198–204
Potin P, Bouarab K, Salaün J-PP et al (2002) Biotic interactions of marine algae. Curr Opin Plant Biol 5:308–317
Prol García MJ, D'Alvise PW, Gram L (2013) Disruption of cell-to-cell signaling does not abolish the antagonism of *Phaeobacter gallaeciensis* toward the fish pathogen *Vibrio anguillarum* in algal systems. Appl Environ Microbiol 79:5414–5417
Provasoli L (1958) Nutrition and ecology of protozoa and algae. Annu Rev Microbiol 12:279–308
Rajamani S, Bauer WD, Robinson JB et al (2008) The vitamin riboflavin and its derivative lumichrome activate the LasR bacterial quorum-sensing receptor. Mol Plant-Microbe Interact 21:1184–1192
Ramanan R, Kim BH, Cho DH et al (2016) Algae-bacteria interactions: evolution, ecology and emerging applications. Biotechnol Adv 34:14–29
Rao D, Webb JS, Holmström C et al (2007) Low densities of epiphytic bacteria from the marine alga Ulva Australis inhibit settlement of fouling organisms. Appl Environ Microbiol 73:7844–7852
Read BA, Kegel J, Klute MJ et al (2013) Pan genome of the phytoplankton *Emiliania* underpins its global distribution. Nature 499:209–213
Ribalet F, Intertaglia L, Lebaron P et al (2008) Differential effect of three polyunsaturated aldehydes on marine bacterial isolates. Aquat Toxicol 86:249–255
Riclea R, Gleitzmann J, Bruns H et al (2012) Algicidal lactones from the marine Roseobacter clade bacterium *Ruegeria pomeroyi*. Beilstein J Org Chem 8:941–950
Ross JJ, Reid JB (2010) Evolution of growth-promoting plant hormones. Funct Plant Biol 37:795–805
Sachs JL, Simms EL (2006) Pathways to mutualism breakdown. Trends Ecol Evol 21:585–592
Salama E-S, Kabra AN, Ji M-K et al (2014) Enhancement of microalgae growth and fatty acid content under the influence of phytohormones. Bioresour Technol 172C:97–103
Schaefer AL, Greenberg EP, Oliver CM et al (2008) A new class of homoserine lactone quorum-sensing signals. Nature 454:595–599
Schäfer H, Abbas B (2002) Genetic diversity of "satellite" bacteria present in cultures of marine diatoms. FEMS Microbiol Ecol 42:25–35
Schneider H, Schuettpelz E, Pryer KM et al (2004) Ferns diversified in the shadow of angiosperms. Nature 428:553–557
Segev E, Wyche TP, Kim KH et al (2016) Dynamic metabolic exchange governs a marine algal-bacterial interaction. elife 5:e17473
Seyedsayamdost MR, Carr G, Kolter R et al (2011a) Roseobacticides: small molecule modulators of an algal-bacterial symbiosis. J Am Chem Soc 133:18343–18349
Seyedsayamdost MR, Case RJ, Kolter R et al (2011b) The Jekyll-and-Hyde chemistry of *Phaeobacter gallaeciensis*—supplementary information. Nat Chem 3:331–335

Seyedsayamdost MR, Wang R, Kolter R et al (2014) Hybrid biosynthesis of roseobacticides from algal and bacterial precursor molecules. J Am Chem Soc 136:15150–15153

Seymour JR, Simo R, Ahmed T et al (2010) Chemoattraction to dimethylsulfoniopropionate throughout the marine microbial food web. Science 329:342–345

Sing SF, Isdepsky A, Borowitzka MA et al (2013) Production of biofuels from microalgae. Mitig Adapt Strateg Glob Chang 18:47–72

Singh RP, Bijo AJ, Baghel RS et al (2011) Role of bacterial isolates in enhancing the bud induction in the industrially important red alga *Gracilaria dura*. FEMS Microbiol Ecol 76:381–392

Singh RP, Reddy CRK (2014) Seaweed-microbial interactions: key functions of seaweed-associated bacteria. FEMS Microbiol Ecol 88:213–230

Smriga S, Fernandez VI, Mitchell JG et al (2016) Chemotaxis toward phytoplankton drives organic matter partitioning among marine bacteria. Proc Natl Acad Sci U S A 113:1576–1581

Spaepen S, Vanderleyden J, Remans R (2007) Indole-3-acetic acid in microbial and microorganism-plant signaling. FEMS Microbiol Rev 31:425–448

Spoerner M, Wichard T, Bachhuber T et al (2012) Growth and thallus morphogenesis of *Ulva mutabilis* (Chlorophyta) depends on a combination of two bacterial species excreting regulatory factors. J Phycol 48:1433–1447

Spolaore P, Joannis-Cassan C, Duran E et al (2006) Commercial applications of microalgae. J Biosci Bioeng 101:87–96

Stefels J (2000) Physiological aspects of the production and conversion of DMSP in marine algae and higher plants. J Sea Res 43:183–197

Steinberg PD, de Nys R (2002) Chemical mediation of colonization of seaweed surfaces. J Phycol 38:621–629

Stevenson J (2014) Ecological assessments with algae: a review and synthesis. J Phycol 50:437–461

Stirk WA, Ördög V, Novák O et al (2013) Auxin and cytokinin relationships in 24 microalgal strains. J Phycol 49:459–467

Stocker R, Seymour JR (2012) Ecology and physics of bacterial chemotaxis in the ocean. Microbiol Mol Biol Rev 76:792–812

Stocker R, Seymour JR, Samadani A et al (2008) Rapid chemotactic response enables marine bacteria to exploit ephemeral microscale nutrient patches. Proc Natl Acad Sci U S A 105:4209–4214

Sule P, Belas R (2013) A novel inducer of Roseobacter motility is also a disruptor of algal symbiosis. J Bacteriol 195:637–646

Sunda W, Kieber DJ, Kiene RP et al (2002) An antioxidant function for DMSP and DMS in marine algae. Nature 418:317–320

Tait K, Joint I, Daykin M et al (2005) Disruption of quorum sensing in seawater abolishes attraction of zoospores of the green alga *Ulva* to bacterial biofilms. Environ Microbiol 7:229–240

Tarakhovskaya ER, Maslov YI, Shishova MF (2007) Phytohormones in algae. Russ J Plant Physiol 54:163–170

Teplitski M, Chen H, Rajamani S et al (2004) *Chlamydomonas reinhardtii* Secretes compounds that mimic bacterial signals and interfere with quorum sensing regulation in bacteria. Plant Physiol 134:137–146

Thompson AW, Foster RA, Krupke A et al (2012) Unicellular cyanobacterium symbiotic with a single-celled eukaryotic alga. Science 337:1546–1550

Tillmann U (2004) Interactions between planktonic microalgae and protozoan grazers. J Eukaryot Microbiol 51:156–168

Tronchet M, Balagué C, Kroj T et al (2010) Cinnamyl alcohol dehydrogenases-C and D key enzymes in lignin biosynthesis, play an essential role in disease resistance in *Arabidopsis*. Mol Plant Pathol 11:83–92

Tujula NA, Crocetti GR, Burke C et al (2010) Variability and abundance of the epiphytic bacterial community associated with a green marine *Ulvacean* alga. ISME J 4:301–311

Tyrrell T, Merico A (2004) *Emiliania huxleyi*: bloom observation and the conditions that induce them. In: Thierstein HR, Young JR (eds) Coccolithophores: from molecular processes to global impact. Coccolithophores: from molecular processes to global impact. Springer-Verlag, Berlin, pp 75–98

Valiente-Banuet A, Verdú M (2008) Temporal shifts from facilitation to competition occur between closely related taxa. J Ecol 96:489–494

van Loon LC, Bakker PAHM, Pieterse CMJ (1998) Systemic resistance induced by rhizosphere bacteria. Annu Rev Phytopathol 36:453–483

Vardi A, Haramaty L, Van Mooy BAS et al (2012) Host-virus dynamics and subcellular controls of cell fate in a natural coccolithophore population. Proc Natl Acad Sci U S A 109:19327–19332

Wagner-Döbler I, Ballhausen B, Berger M et al (2010) The complete genome sequence of the algal symbiont *Dinoroseobacter shibae*: a hitchhiker's guide to life in the sea. ISME J 4:61–77

Wang H, Hill RT, Zheng T et al (2014a) Effects of bacterial communities on biofuel-producing microalgae: stimulation, inhibition and harvesting. Crit Rev Biotechnol 36:1–12

Wang H, Laughinghouse HD, Anderson MA et al (2012) Novel bacterial isolate from Permian groundwater, capable of aggregating potential biofuel-producing microalga *Nannochloropsis oceanica* IMET1. Appl Environ Microbiol 78:1445–1453

Wang H, Tomasch JJ, Jarek M et al (2014b) A dual-species co-cultivation system to study the interactions between Roseobacters and dinoflagellates. Front Microbiol 5:1–11

Wang M, Yuan W (2014) Bacterial lysis of microalgal cells. J Sustain Bioenergy Syst 4:243–248

Wang R, Gallant É, Seyedsayamdost R (2016) Investigation of the genetics and biochemistry of roseobacticide production in the Roseobacter clade bacterium *Phaeobacter inhibens*. MBio 7:1–10

Wang X, Li Z, Su J et al (2010) Lysis of a red-tide causing alga, Alexandrium tamarense, caused by bacteria from its phycosphere. Biol Control 52:123–130

Waters CM, Bassler BL (2005) Quorum sensing: communication in bacteria. Annu Rev Cell Dev Biol 21:319–346

Weinberger F, Friedlander M, Hoppe H (1999) Oligoagars elicit a physiological response in *Gracilaria Conferta* (Rhodophyta). J Phycol 35:747–755

Weng J-K, Chapple C (2010) The origin and evolution of lignin biosynthesis. New Phytol 187:273–285

Wichard T (2015) Exploring bacteria-induced growth and morphogenesis in the green macroalga order Ulvales (Chlorophyta). Front Plant Sci 6:86

Wietz M, Duncan K, Patin NV et al (2013) Antagonistic interactions mediated by marine bacteria: the role of small molecules. J Chem Ecol 39:879–891

Wolfe GV, Steinke M, Kirst GO (1997) Grazing-activated chemical defense in a unicellular marine alga. Nature 387:894–897

Xie H, Pasternak JJ, Glick BR (1996) Isolation and characterization of mutants of the plant growth promoting rhizobacterium pseudomonas putida GR12-2 that overproduce indoleacetic-acid. Curr Microbiol 32:67–71

Yang Q, Chen L, Hu X et al (2015) Toxic effect of a marine bacterium on aquatic organisms and its algicidal substances against *Phaeocystis globosa*. PLoS One 10:1–16

Yim G, Wang HH, Davies J (2007) Antibiotics as signalling molecules. Philos Trans R Soc Lond Ser B Biol Sci 362:1195–1200

Yoon HS, Hackett JD, Ciniglia C et al (2004) A molecular timeline for the origin of photosynthetic eukaryotes. Mol Biol Evol 21:809–818

Zapalski MK (2011) Is absence of proof a proof of absence? Comments on commensalism. Palaeogeogr Palaeoclimatol Palaeoecol 302:484–488

Chapter 15
Exploring the Complexity of Macroalgal-Bacterial Interactions Through Interkingdom Signalling System

Ravindra Pal Singh, Ramesh Kothari, and Suhelen Egan

Abstract Macroalgae belong to a diverse group of photosynthetic organisms that play an essential role in marine ecosystems. These ecosystem engineers contribute significantly to global primary production and are the major habitat formers on rocky shores in temperate waters, providing food and shelter for marine life. Macroalgae harbour a rich diversity of associated microorganisms with varied functions related to host performance and defence. In particular, epiphytic bacterial communities have been reported as essential for normal morphological development of the algal host. Moreover, bacteria with antifouling properties are thought to protect chemically undefended macroalgae from detrimental, secondary colonization by other microscopic and macroscopic epibiota. This tight relationship suggests that macroalgae and epiphytic bacteria interact as an integrated functional entity or "holobiont". Many of these interactions are controlled via chemical signalling systems in a type of interkingdom communication. Indeed recent studies have demonstrated that chemical signalling molecules from bacteria regulate important functions in green algae such as reproduction and host defence.

Keywords Macroalgae • Quorum sensing • Morphology • Macroalgal-microbial interaction • Next-generation sequencing

Contents

15.1 Introduction 302
15.2 Quorum Sensing and Cross Kingdom Communication with Macroalgae 304
15.3 Host-Associated Bacterial Diversity and Next-Generation Sequencing 306
15.4 Conclusion 309
References 309

R.P. Singh (✉) • R. Kothari
Department of Biosciences, Saurashtra University, Rajkot 360005, Gujarat, India
e-mail: ravindrapal.1441@gmail.com

S. Egan
School of Biological, Earth and Environmental Science, Centre for Marine Bio-Innovation, The University of New South Wales, Sydney, NSW, Australia

© Springer International Publishing AG 2017
M. Kumar, P. Ralph (eds.), *Systems Biology of Marine Ecosystems*,
DOI 10.1007/978-3-319-62094-7_15

15.1 Introduction

Macroalgae secrete a variety of organic nutrients in surrounding environment that are utilized as a nutrient source and surface settlement cues by diverse microorganisms (Bengtsson and Ovreas 2010; Goecke et al. 2010; Bengtsson et al. 2012; Godinho et al. 2013). Microbial communities living on the macroalgal surface are highly dynamic and include bacteria, fungi, diatoms, protozoa, spores and larvae of marine invertebrates (Holmstrom et al. 2002; Tujula et al. 2010; Burke et al. 2011a). Macroalgae encompass intra- or extracellular bacteria and eukaryotes that constitute the endo- and ectophycosphere, respectively (Hollants et al. 2011, 2013; Singh et al. 2015). These associated bacteria can exhibit a beneficial interaction with their host by assisting with morphogenesis, growth and reproduction (Matsuo et al. 2005; Marshall et al. 2006; Singh et al. 2011a, b; Spoerner et al. 2012). The first evidence for a role of bacteria in algal development was reported in the middle of the twentieth century when researchers observed that axenic cultures of the green alga, *Ulva*, species displayed an abnormal morphology (Provasoli and Pintner 1953, 1980; Provasoli 1958). These observations have since been repeated in other algal species. For example, in the case of the green macroalga *Monostroma oxyspermum*, a specific marine bacterial strain, YM2-23, has been shown to be essential for the normal morphological development of the host (Matsuo et al. 2005). Likewise, the morphogenesis and growth of *Gracilaria dura* are impaired unless Gram-positive bacteria (e.g. *Bacillus* sp.) are present (Singh et al. 2011a).

Associated microbial communities can also provide vitamins and essential nutrients to the algal host, thereby benefiting their growth. For example, nitrogen fixation activity of some associated bacterial strains significantly influences growth of the green and red seaweeds (Chisholm et al. 1996; Singh et al. 2011a). Bacterial biofilm and their extracellular polymeric substance and chemical compounds are found to be important for settling zoospores in Ulvaceae (Tait et al. 2005; Singh et al. 2013) and releasing spores in Gracilariaceae (Weinberger et al. 2007; Singh et al. 2015). Moreover, bacteria are thought to protect macroalgal surfaces from biofouling pressure via the production of both general and specific biological active chemical metabolites (i.e. bioactives) (Wahl et al. 2012). Among other chemical compounds, bacterial signalling or quorum sensing (QS) systems have gained attention for their involvement in the bidirectional communication between bacteria and hosts (Venturi and Fuqua 2013).

QS systems of bacteria modulate the coordinated expression of genes involved in a diverse set of bacterial phenotypes including biofilm formation, motility, antibiotic production, virulence gene expression and exchange of genetic material (Williams 2007). Gram-negative bacteria use a variety of different QS signalling molecules; among them, *N*-acyl homoserine lactone (AHL) is arguably the most widely studied molecule that controls several physiological responses of bacteria (Fuqua et al. 2001; Venturi and Fuqua 2013). Investigation of AHL-mediated QS began with the study of bioluminescence in the Hawaiian bobtail squid *Euprymna scolopes*. One bacterium, *Aliivibrio fischeri* (formally *Vibrio fischeri*), is present in the light organ of *E. scolopes* where, at high concentrations, it expresses the luciferase enzyme resulting in a visible bioluminescent phenotype (Visick et al. 2000). Luciferase expression in the bacterium is controlled by the paradigm of QS, LuxI/LuxR or the autoinducer 1 (AI1) system, in which the LuxI is responsible for the production of the AHL signalling molecule (in this case a 3-oxo-C6-HSL) and LuxR acts as the response regulator protein (Engebrecht and Silverman 1984). The LuxI-LuxR type QS system is not restricted to *A. fischeri*, and this now classical QS system has been described to control a variety of phenotypes in a range of Gram-negative bacteria (Whitehead et al. 2001; Galloway et al. 2011; Rutherford and Bassler 2012).

In addition to controlling a range of bacterial phenotypes via cell-to-cell communication, QS signalling has been proposed to facilitate communication between microorganisms and their hosts such as human, plant and algae (Hughes and Sperandio 2008). Over the last decade, some unique types of LuxR receptor proteins have been reported that respond to bacterial AHLs produced by the same or other bacterial cells as well as low molecular weight compounds produced by hosts, as shown in Fig. 15.1 (Fuqua 2006; Subramoni and Venturi 2009a; Patel et al. 2013).

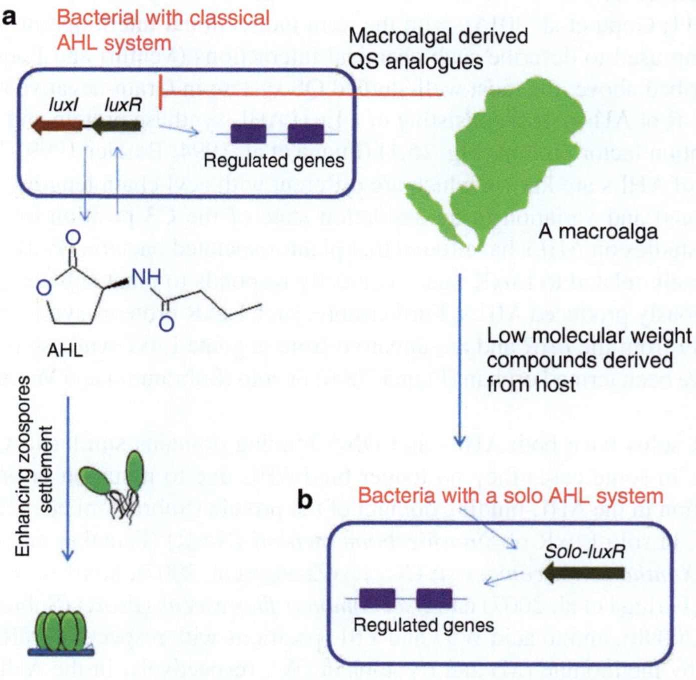

Fig. 15.1 Complex interkingdom quorum sensing (QS) signalling system between bacteria and macroalgae. (**a**) Classical AHL system of bacteria. (**b**) Solo-LuxR (lacking adjacent *luxI*) responding to low molecular weight compounds derived from host macroalga. AHLs are known to modulate zoospore settlement in *Ulva* species. Macroalgae also produce AHL mimic compounds (QS analogues) that attenuate QS system of bacteria

Subramoni et al. (2015) performed a bioinformatic survey of LuxR solo regulators and identified such receptors in diverse host-associated bacteria and found that some played an important function during host-bacterial interaction and are causing diseases. The aim of the chapter is to unravel the complexity of macroalgal microbiome with a particular focus on the role of chemical interactions via QS systems. We will further discuss the opportunities afforded by implementing next-generation sequencing (NGS) technologies in future studies aimed at understanding these complex microbial-host symbioses.

15.2 Quorum Sensing and Cross Kingdom Communication with Macroalgae

Many plant-associated bacteria undergo chemical signalling with the plant host via yet unidentified low molecular weight compounds (Subramoni and Venturi 2009a). There is now growing evidence that plant- and algal-derived compounds interact with bacterial regulatory proteins that either inhibit or activate bacterial QS (Manefield et al. 1999; Bodini et al. 2009; Patankar and Gonzalez 2009a, b; Lee et al. 2011; Gopu et al. 2015), with the term bidirectional interkingdom signalling now being used to describe such chemical interactions (Venturi and Fuqua 2013). As described above, the most well-studied QS system in Gram-negative bacteria is the LuxI-R or AI1 system consisting of a LuxI AHL synthase protein and the LuxR transcription factor protein (Fig. 15.1) (Fuqua et al. 1994; Bassler 1999). To date, a number of AHLs are known which are different with acyl chain lengths (from 4 to 18 carbons) and variation in the oxidation state of the C3 position of the chain. Further studies on AHLs have found that plant-associated bacteria (PAB) have proteins closely related to LuxR that specifically responds to plant signals rather than endogenously produced AHLs. Furthermore, such LuxR proteins synthesized independently from the *luxR* and are unpaired from cognate LuxI synthase (Fig. 15.1), thus, have been termed orphan (Fuqua 2006) or solo (Subramoni and Venturi 2009a) receptor.

LuxR solos have both AHL- and DNA-binding domains similar to QS LuxR; however, in some cases they no longer bind AHL due to mutation or amino acid substitution in the AHL-binding domain of the protein (Subramoni et al. 2015). For instance, in solo LuxR of *Sinorhizobium meliloti* (NesR) (Patankar and Gonzalez 2009a), *Xanthomonas campestris* (XccR) (Zhang et al. 2007), *Xanthomonas oryzae* (OryR) (Ferluga et al. 2007) and *Pseudomonas fluorescens* (PsoR) (Subramoni and Venturi 2009b), amino acid W57 and Y61 (positions with respect to TraR) are substituted by methionine (M) and tryptophan (W), respectively, in the AHL-binding domain. Interestingly, these specific amino acid substitutions indicated that the solo LuxR proteins were binding to low molecular weight compounds produced by plant rather than AHLs (Patankar and Gonzalez 2009a). Further study on the function of solo LuxR in *Xanthomonas oryzae* pv. *oryzae* found that these LuxR regulate virulence of the plant pathogen (Ferluga and Venturi 2009). Interestingly VarR, a type of solo LuxR, is also identified from macroalgae-associated bacterial pathogen *Nautella italica* R11 (Fernandes et al. 2011). The strain has been identified to induce the bleaching disease in vivo and in vitro in *Delisea pulchra* (Campbell et al. 2011; Fernandes et al. 2011). Via a combination of allelic exchange mutagenesis, physiological characterization and high-throughput proteomics, Gardiner et al. (2015) found evidence for *varR* to act as a regulator of colonization and virulence in this organism. Particularly, the study observed that 3.4% of the predicted proteome of *N. italica* R11 was differentially expressed between planktonic and biofilm conditions of *varR* mutant as compared to wild type. Thus, a subset of biofilm-associated protein of *N. italica* R11 is controlled by solo *varR* indicating their importance in attachment, biofilm maturation and infection in *D. pulchra*.

Cross kingdom communication through QS exists between bacteria and macroalgae. AHL molecules increase zoospore settlement in green macroalgae (Joint et al. 2002) and promote carpospore liberation in some red macroalgae (Weinberger et al. 2007; Singh et al. 2015). Joint et al. (2002) demonstrated that zoospores of *Enteromorpha* species respond to biofilm of the *Vibrio anguillarum*. *V. anguillarum* is a Gram-negative bacterium and can be able to produce C6-HSL, 3-hydroxy-C6-HSL and 3-oxo-C10-HSL (Milton et al. 1997). Tait et al. (2005) determined the rational effect of these AHLs on zoospore settlement and observed that longer *N*-acyl side-chains tended to result in increasing zoospore settlement. Furthermore, availability of 3-hydroxy and 3-oxo at C3 of acyl chain makes them more active AHLs than C6-HSL and C10-HSL towards zoospore settlement. When a lactonase-coding gene (that degrades AHLs) *aiiA* was expressed into *V. anguillarum*, the recombinant strain lost the activity of enhancing zoospore settlement, providing strong support for the involvement of AHLs (Joint et al. 2002). Wheeler et al. (2006) used wild and *vanM* mutant types of *V. anguillarum* and defined that the orientation of zoospores does not change during settlement and their swimming speed decreases more rapidly on the wild-type *V. anguillarum* biofilms as compared to *vanM* mutant. Thus, it is presumed that chemokinesis mechanism operates instead of chemotactic in which zoospore swimming speed rapidly decreases in the presence of AHLs. Further, experiments on this system suggested that AHLs influence Ca^{2+} influx in the zoospores which preferentially induced the settlement of them on bacterial biofilms producing AHLs (Wheeler et al. 2006; Joint et al. 2007).

Intriguingly, whilst AHLs promote zoospore settlement in the *Ulva*, they have a negative impact on germination and early growth stages of settled zoospores (Twigg et al. 2014). Using both native AHL producers (i.e. *Sulfitobacter* spp. 376 and *Shewanella* spp. 79) and synthetic AHLs, Twigg et al. (2014) could show that germling length was significantly reduced in the presence of AHLs or AHL-producing biofilms compared to the controls. The authors hypothesised that this apparent paradox (i.e. AHL induction of settlement yet retardation of germination and early growth) could be explained by the fact that slower growing algae can outcompete fast growers in nutrient-limited environments (Twigg et al. 2014). Thus, these results provide further evidence of the complexity of bacterial-macroalgal interactions and the ability of AHLs to modulate multiple aspects of green algal reproduction and early growth.

The effect of AHL is not only restricted to green algae but has also been observed in red algal species such as *Gracilaria* and *Acrochaetium* species. In these macroalgae, C_4- and C_6-HSLs have been shown to control carpospore liberation (Weinberger et al. 2007; Singh et al. 2015). Furthermore, sodium dodecyl sulphate-polyacrylamide gel electrophoresis analysis of total protein of C_4- and C_6-HSL-treated cystocarps of *G. dura* revealed the induction of specific polypeptide bands of approximate molecular weights 50 and 60 kDa (Singh et al. 2015). The finding suggested that these unidentified proteins may have a role in carpospores releasing from cystocarp. Nevertheless, identifying interconnection of these proteins and their expression under AHL treatment would enhance knowledge about carpospore liberation from *G. dura*.

15.3 Host-Associated Bacterial Diversity and Next-Generation Sequencing

There are a growing number of studies that have assessed the diversity of the epiphytic bacterial communities associated with macroalgal hosts (Staufenberger et al. 2008; Goecke et al. 2010; Burke et al. 2011a; Hollants et al. 2013). These studies highlight that bacterial communities associated with macroalgal whilst distinct from the bacteria found in the surrounding seawater vary across different seasons (Tujula et al. 2010; Burke et al. 2011b), geographical locations (Tait et al. 2009; Lachnit et al. 2011; Bondoso et al. 2014), different parts of the thallus (Staufenberger et al. 2008) and different pools based on host trait (Campbell et al. 2015).

Several studies have suggested that bacteria within the phyla *Proteobacteria* and *Firmicutes* are among the most abundant taxa associated with the surface of macroalgal hosts as summarized previously (Goecke et al. 2010; Singh and Reddy 2014). Therefore one might predict that these bacterial groups have particular characteristics that have enabled their adaptation to marine hosts, including but not limited to (1) the ability to cope with multiple stress parameters such as ionic, osmotic, chaotropic, hydrophobic and other activities of solutes, (2) providing direct benefit to the host (see below) and (3) having high-efficiency energy-generation systems (Burke et al. 2011b; Wahl et al. 2012; Cray et al. 2013). With respect to host benefits, many of these bacterial communities are reported to possess various biological activities such as antibacterial, antisettlement, antilarval and antifungal activities (Egan et al. 2008; Penesyan et al. 2009) that also secure their abundance over the host surface.

Early studies of bacterial communities associated with macroalgae employed culture-dependent or microscopy methods with fewer studies employing culture-independent molecular methods such as restriction fragment length polymorphism and terminal restriction fragment length polymorphism of DNA for identifying bacterial communities of the macroalgal hosts (Burr and West 1971; Meusnier et al. 2001, 2002; Longford et al. 2007; Lachnit et al. 2011; Bengtsson et al. 2012). Later molecular approaches removed some of the bias associated with culturing techniques, although they had limitations with respect to their ability to explore functional relationship of associated bacterial communities and their hosts.

Broad access to -omics technology (i.e. genomics, transcriptomics and proteomics) in sequencing is fostering a great deal of interest across many areas of biology (Turnbaugh et al. 2007; Yang and Li 2012; Kostic et al. 2013; Ursell and Knight 2013). In particular, these technologies are now making it possible to study microbial communities in unprecedented detail in order to understand how the microbiome impacts physiology and propensity to disease in diverse hosts (Adesemoye et al. 2009; Bakker et al. 2012; Knief et al. 2012). These advancements are beginning to unravel the complex interactions between the environment, host genetics and microbiome in diverse systems of the ocean, soil, invertebrate animals (corals, sponges, insects, etc.), algae, plants and humans (Turnbaugh et al. 2007; Berg and Smalla 2009; Turner et al. 2013; Krishnan et al. 2014; Yang and Jobin 2014; Ainsworth et al. 2015; Ding et al. 2015). These studies have highlighted the close association of microorganisms with their host and environment where they play essential role in host life cycle. In the case of human, a large focus has been the use of -omics technologies to study the gut microbiome (Cenit et al. 2014; Sun and Chang 2014; Nakayama et al. 2015; Wang et al. 2015). Several studies have been performed on the healthy and diseased intestines revealing the variation in community composition and their functional capacity correlating with host state (Turnbaugh et al. 2006; Atarashi et al. 2011; Ridlon et al. 2014). Some of the gut microbial communities degrade dietary fibres and convert them into small chain fatty acid which are warrant to the well-being of human (Byrne et al. 2015). Those beneficial microbial communities have used to modulate diseased gut and subsequently promoted health of the gut (Neyrinck et al. 2012). Furthermore, modulation of gut microbiomes may be carried out by using probiotic bacteria, diet supplement, antimicrobial compound and faecal microbiota transplant (Walsh et al. 2014). Such studies are providing mechanistic insights into the host-microbiome interaction and are leading to the development not only of new diagnostic methods but also treatment of a variety of human diseases based on the detection of gut microbiomes (Li et al. 2008; Rajpal and Brown 2013). These understanding could be useful for promoting growth of the macroalgal host through modulating some of the beneficial microbial communities as below.

Highly studied -omic approaches have also been used to assess microbial diversity and in a number of marine hosts. For example, there are now several research articles that describe omic analysis of microbial communities associated with marine sponges (Gurgui and Piel 2010; Hentschel et al. 2012; Trindade-Silva et al. 2012; O'Connor-Sanchez et al. 2014). These data not only demonstrate that the microbiome of different sponge species varies in their degree of host specificity at the phylogenetic level but also these studies have begun to identify some of the functional genes that are characteristic of sponge communities. For example, metatranscriptomics study of the marine sponge *Geodia barretti* revealed that bacterial communities belonging to *Chloroflexi*, *Poribacteria* and *Acidobacteria* were most abundant and some of the functional characteristics of genes were identified that were related to membrane transport, nitrification and related biological process (Radax et al. 2012). Another study of Yang and Li (2012) demonstrated that nitrogen cycling genes (ammonia oxidizing) only present in the endosome tissue rather than cortex region of the sponge (*Astrosclera willeyana*).

Microbiomes of macroalgae have not been as extensively investigated as compared to other hosts (sponges and human gut). However, few studies have used metagenomic approach to identify composition and function of associated microbial communities (Burke et al. 2011a; de Oliveira et al. 2012; Dittami et al. 2014; Martin et al. 2014; Campbell et al. 2015; Marzinelli et al. 2015). Martin et al. (2014) studied functional metagenomics of bacterial communities associated with the brown alga *Ascophyllum nodosum* and identified 13 novel putative esterase loci and 2 glycoside hydrolase loci. Marzinelli et al. (2015) studied bacterial and archaeal communities associated with 260 samples of the kelp *Ecklonia radiata* from different biogeographical regions across the Australian continent and found extremely stable microbial communities that were influenced by host state more than location. Metagenomic studies of microbial communities of the *Ulva australis* indicate that microbial communities on seaweeds are established based on a conservation of functional traits rather than microbial species composition (Burke et al. 2011a). Core groups of functional gene predicted to be important for algal association have been repeatedly identified in microbial communities of *U. australis* including those related to nitrate reduction, motility, QS systems, osmoregulation, cell differentiation, virulence and defence. Of particular relevance to microbial communities associated with macroalgae included stress responses relevant to protection against oxidative stress, desiccation as well as degradation of host-secreted metabolites. de Oliveira et al. (2012) have used transcriptomics analysis on the red macroalga (*Laurencia dendroidea*) and their associated microbiome. Dominant bacterial groups belonging to nitrogen-fixing *Cyanobacteria* and aerobic heterotrophic *Proteobacteria* were identified. Comparative analysis of transcripts revealed an abundance of transcripts related to glycolysis, polysaccharide and lipid breakdown. Other features of the metatranscriptomics included the expression of genes related to recognition of macroalgal surface, biofilm formation and mechanism for host defence resistance and stress (biosynthesis of terpenoid backbone) (de Oliveira et al. 2012). Transcripts related to bacterial QS signalling were also detected in the meta-transcriptome analysis of *L. dendroidea* microbiome, including a high abundance for genes coding for the enzyme *S*-adenosylmethionine synthetases, an important precursor for the synthesis of bacterial AHLs (Hanzelka and Greenberg 1996) among other biological reactions (Takusagawa et al. 1996). Whilst many functional activities of the microbiome of *L. dendroidea* are congruent with those associated with that of the *U. australis* (Burke et al. 2011a), the studies to date represent a snapshot of these complex communities. There is a need not only to further validate these studies but to assess the stability of the microbial community and the microbiome response to different environmental conditions commonly exposed to the host. With the increased accessibility to next-generation sequencing (NGS) technologies including methods for community profiling via the sequencing of phylogenetic marker genes (e.g. 16S rRNA gene amplicon sequencing); whole community genome (metagenomics) or meta-mRNA sequencing (metatranscriptomics) for analysis of gene expression via Roche 454, Illumina or more recently PacBio (Caporaso et al. 2012; Merriman and Rothberg 2012; Knief 2014); and metaproteomics, it is now possible to address these and more questions related to macroalgal-microbial interactions. The results obtained from larger-scale NGS projects will allow for understating complexity of biological function of macroalgal microbiomes. Understanding complexity of this interaction will help to increase the gross productivity of economical important macroalgae and mitigating several diseases via microbiome modulation as similarly to what approaches have been applied on the higher plant.

15.4 Conclusion

Bacterial communities associated with the diverse macroalgae are essential for normal life cycle of the host, in which they determine the morphogenesis, growth and reproduction in different ways. There are growing evidences for interkingdom chemical communication between macroalgae and their associated bacteria where they modulate several phenomena of each other. Advances in NGS technologies will improve our understanding of their composition and functions. So far, metagenomic and metatranscriptomics techniques have successfully employed to analyse the microbiomes of human, plants, insects, animals and marine life (sponges and macroalgae) revealing details of their biology and evolution of not known prior to these technological developments. The application of these sensitive NGS techniques to the macroalgal holobiont will invariably provide much needed information regarding the role of microbial associations for macroalgal health and function. Once identified, specific core members or functional microbial groups that benefit host health may then be used, for example, as probiotics in order to enhance macroalgal production in aquaculture or for developing an early warning system for macroalgal diseases.

References

Adesemoye AO, Torbert HA, Kloepper JW (2009) Plant growth promoting rhizobacteria allow reduced application rates of chemical fertilizers. Microb Ecol 58(4):921–929. doi:10.1007/s00248-009-9531-y

Ainsworth DT, Krause L, Bridge T, Torda G, Raina JB, Zakrzewski M et al (2015) The coral core microbiome identifies rare bacterial taxa as ubiquitous endosymbionts. ISME J 9(10):2261–2274. doi:10.1038/ismej.2015.39

Atarashi K, Tanoue T, Shima T, Imaoka A, Kuwahara T, Momose Y et al (2011) Induction of colonic regulatory T cells by indigenous *Clostridium* species. Science 331(6015):337–341. doi:10.1126/science.1198469

Bakker MG, Manter DK, Sheflin AM, Weir TL, Vivanco JM (2012) Harnessing the rhizosphere microbiome through plant breeding and agricultural management. Plant Soil 360(1-2):1–13. doi:10.1007/s11104-012-1361-x

Bassler BL (1999) How bacteria talk to each other: regulation of gene expression by quorum sensing. Curr Opin Microbiol 2(6):582–587

Bengtsson MM, Ovreas L (2010) Planctomycetes dominate biofilms on surfaces of the kelp *Laminaria hyperborea*. BMC Microbiol 10:261. doi:10.1186/1471-2180-10-261

Bengtsson MM, Sjotun K, Lanzen A, Ovreas L (2012) Bacterial diversity in relation to secondary production and succession on surfaces of the kelp *Laminaria hyperborea*. ISME J 6(12):2188–2198. doi:10.1038/ismej.2012.67

Berg G, Smalla K (2009) Plant species and soil type cooperatively shape the structure and function of microbial communities in the rhizosphere. FEMS Microbiol Ecol 68(1):1–13. doi:10.1111/j.1574-6941. 2009.00654.x

Bodini SF, Manfredini S, Epp M, Valentini S, Santori F (2009) Quorum sensing inhibition activity of garlic extract and p-coumaric acid. Lett Appl Microbiol 49(5):551–555. doi:10.1111/j.1472-765X.2009.02704.x

Bondoso J, Balague V, Gasol JM, Lage OM (2014) Community composition of the *Planctomycetes* associated with different macroalgae. FEMS Microbiol Ecol 88(3):445–456. doi:10.1111/1574-6941.12258

Burke C, Steinberg P, Rusch D, Kjelleberg S, Thomas T (2011a) Bacterial community assembly based on functional genes rather than species. Proc Natl Acad Sci U S A 108(34):14288–14293. doi:10.1073/pnas.1101591108

Burke C, Thomas T, Lewis M, Steinberg P, Kjelleberg S (2011b) Composition, uniqueness and variability of the epiphytic bacterial community of the green alga *Ulva australis*. ISME J 5(4):590–600. doi:10.1038/ismej.2010.164

Burr FA, West JA (1971) Protein bodies in *Bryopsis hypnoides*: their relationship to wound-healing and branch septum development. J Ultrastruct Res 35(5):476–498

Byrne CS, Chambers ES, Morrison DJ, Frost G (2015) The role of short chain fatty acids in appetite regulation and energy homeostasis. Int J Obes 39(9):1331–1338. doi:10.1038/ijo. 2015.84

Campbell AH, Harder T, Nielsen S, Kjelleberg S, Steinberg PD (2011) Climate change and disease: bleaching of a chemically defended seaweed. Glob Chang Biol 17(9):2958–2970. doi:10.1111/j.1365-2486.2011.02456.x

Campbell AH, Marzinelli EM, Gelber J, Steinberg PD (2015) Spatial variability of microbial assemblages associated with a dominant habitat-forming seaweed. Front Microbiol 6:230. doi:10.3389/fmicb.2015.00230

Caporaso JG, Lauber CL, Walters WA, Berg-Lyons D, Huntley J, Fierer N et al (2012) Ultra-high-throughput microbial community analysis on the Illumina HiSeq and MiSeq platforms. ISME J 6(8):1621–1624. doi:10.1038/ismej.2012.8

Cenit MC, Matzaraki V, Tigchelaar EF, Zhernakova A (2014) Rapidly expanding knowledge on the role of the gut microbiome in health and disease. Biochim Biophys Acta 1842(10):1981–1992. doi:10.1016/j.bbadis.2014.05.023

Chisholm JRM, Dauga C, Ageron E, Grimont PAD, Jaubert JM (1996) Roots in mixotrophic algae. Nature 381(6583):565–565. doi:10.1038/381565b0

Cray JA, Bell AN, Bhaganna P, Mswaka AY, Timson DJ, Hallsworth JE (2013) The biology of habitat dominance; can microbes behave as weeds? Microb Biotechnol 6(5):453–492. doi:10.1111/1751-7915.12027

de Oliveira LS, Gregoracci GB, Silva GG, Salgado LT, Filho GA, Alves-Ferreira M et al (2012) Transcriptomic analysis of the red seaweed *Laurencia dendroidea* (Florideophyceae, Rhodophyta) and its microbiome. BMC Genomics 13:487. doi:10.1186/1471-2164-13-487

Ding X, Peng XJ, Jin BS, Xiao M, Chen JK, Li B et al (2015) Spatial distribution of bacterial communities driven by multiple environmental factors in a beach wetland of the largest freshwater lake in China. Front Microbiol 6:129. doi:10.3389/fmicb.2015.00129

Dittami SM, Eveillard D, Tonon T (2014) A metabolic approach to study algal-bacterial interactions in changing environments. Mol Ecol 23(7):1656–1660. doi:10.1111/mec.12670

Egan S, Thomas T, Kjelleberg S (2008) Unlocking the diversity and biotechnological potential of marine surface associated microbial communities. Curr Opin Microbiol 11(3):219–225. doi:10.1016/j.mib.2008.04.001

Engebrecht J, Silverman M (1984) Identification of genes and gene products necessary for bacterial bioluminescence. Proc Natl Acad Sci U S A 81(13):4154–4158

Ferluga S, Venturi V (2009) OryR is a LuxR-family protein involved in interkingdom signaling between pathogenic *Xanthomonas oryzae* pv. *oryzae* and rice. J Bacteriol 191(3):890–897. doi:10.1128/JB.01507-08

Ferluga S, Bigirimana J, Hofte M, Venturi V (2007) A LuxR homologue of *Xanthomonas oryzae* pv. *oryzae* is required for optimal rice virulence. Mol Plant Pathol 8(4):529–538. doi:10.1111/j.1364-3703.2007.00415.x

Fernandes N, Case RJ, Longford SR, Seyedsayamdost MR, Steinberg PD, Kjelleberg S et al (2011) Genomes and virulence factors of novel bacterial pathogens causing bleaching disease in the marine red alga *Delisea pulchra*. PLoS One 6(12):e27387. doi:10.1371/journal.pone.0027387

Fuqua C (2006) The QscR quorum-sensing regulon of *Pseudomonas aeruginosa*: an orphan claims its identity. J Bacteriol 188(9):3169–3171. doi:10.1128/JB.188.9.3169-3171.2006

Fuqua C, Parsek MR, Greenberg EP (2001) Regulation of gene expression by cell-to-cell communication: acyl-homoserine lactone quorum sensing. Annu Rev Genet 35:439–468. doi:10.1146/annurev.genet.35.102401.090913

Fuqua WC, Winans SC, Greenberg EP (1994) Quorum sensing in bacteria - the Luxr-Luxi family of cell density-responsive transcriptional regulators. J Bacteriol 176(2):269–275

Galloway WR, Hodgkinson JT, Bowden SD, Welch M, Spring DR (2011) Quorum sensing in Gram-negative bacteria: small-molecule modulation of AHL and AI-2 quorum sensing pathways. Chem Rev 111(1):28–67. doi:10.1021/cr100109t

Gardiner M, Fernandes ND, Nowakowski D, Raftery M, Kjelleberg S, Zhong L et al (2015) VarR controls colonization and virulence in the marine macroalgal pathogen *Nautella italica* R11. Front Microbiol 6:1130. doi:10.3389/fmicb.2015.01130

Godinho VM, Furbino LE, Santiago IF, Pellizzari FM, Yokoya NS, Pupo D et al (2013) Diversity and bioprospecting of fungal communities associated with endemic and cold-adapted macroalgae in Antarctica. ISME J 7(7):1434–1451. doi:10.1038/ismej.2013.77

Goecke F, Labes A, Wiese J, Imhoff JF (2010) Chemical interactions between marine macroalgae and bacteria. Mar Ecol Prog Ser 409:267–299. doi:10.3354/Meps08607

Gopu V, Kothandapani S, Shetty PH (2015) Quorum quenching activity of *Syzygium cumini* (L.) Skeels and its anthocyanin malvidin against *Klebsiella pneumoniae*. Microb Pathog 79:61–69. doi:10.1016/j.micpath.2015.01.010

Gurgui C, Piel J (2010) Metagenomic approaches to identify and isolate bioactive natural products from microbiota of marine sponges. Methods Mol Biol 668:247–264. doi:10.1007/978-1-60761-823-2_17

Hanzelka BL, Greenberg EP (1996) Quorum sensing in *Vibrio fischeri*: evidence that S-adenosylmethionine is the amino acid substrate for autoinducer synthesis. J Bacteriol 178(17):5291–5294

Hentschel U, Piel J, Degnan SM, Taylor MW (2012) Genomic insights into the marine sponge microbiome. Nat Rev Microbiol 10(9):641–U675. doi:10.1038/nrmicro2839

Hollants J, Decleyre H, Leliaert F, De Clerck O, Willems A (2011) Life without a cell membrane: challenging the specificity of bacterial endophytes within Bryopsis (Bryopsidales, Chlorophyta). BMC Microbiol 11:255. doi:10.1186/1471-2180-11-255

Hollants J, Leliaert F, Verbruggen H, Willems A, De Clerck O (2013) Permanent residents or temporary lodgers: characterizing intracellular bacterial communities in the siphonous green alga *Bryopsis*. Proc Biol Sci 280(1754):20122659. doi:10.1098/rspb.2012.2659

Holmstrom C, Egan S, Franks A, McCloy S, Kjelleberg S (2002) Antifouling activities expressed by marine surface associated *Pseudoalteromonas* species. FEMS Microbiol Ecol 41(1):47–58. doi:10.1111/j.1574-6941.2002.tb00965.x

Hughes DT, Sperandio V (2008) Inter-kingdom signalling: communication between bacteria and their hosts. Nat Rev Microbiol 6(2):111–120. doi:10.1038/nrmicro.1836

Joint I, Tait K, Callow ME, Callow JA, Milton D, Williams P et al (2002) Cell-to-cell communication across the prokaryote-eukaryote boundary. Science 298(5596):1207–1207. doi:10.1126/science.1077075

Joint I, Tait K, Wheeler G (2007) Cross-kingdom signalling: exploitation of bacterial quorum sensing molecules by the green seaweed *Ulva*. Philos Trans R Soc Lond Ser B Biol Sci 362(1483):1223–1233. doi:10.1098/rstb.2007.2047

Knief C (2014) Analysis of plant microbe interactions in the era of next generation sequencing technologies. Front Plant Sci 5:216. doi:10.3389/fpls.2014.00216

Knief C, Delmotte N, Chaffron S, Stark M, Innerebner G, Wassmann R et al (2012) Metaproteogenomic analysis of microbial communities in the phyllosphere and rhizosphere of rice. ISME J 6(7):1378–1390. doi:10.1038/ismej.2011.192

Kostic AD, Howitt MR, Garrett WS (2013) Exploring host-microbiota interactions in animal models and humans. Genes Dev 27(7):701–718. doi:10.1101/gad.212522.112

Krishnan M, Bharathiraja C, Pandiarajan J, Prasanna VA, Rajendhran J, Gunasekaran P (2014) Insect gut microbiome—an unexploited reserve for biotechnological application. Asian Pac J Trop Biomed 4(Suppl 1):S16–S21. doi:10.12980/APJTB.4.2014C95

Lachnit T, Meske D, Wahl M, Harder T, Schmitz R (2011) Epibacterial community patterns on marine macroalgae are host-specific but temporally variable. Environ Microbiol 13(3):655–665. doi:10.1111/j.1462-2920.2010.02371.x

Lee KM, Lim J, Nam S, Yoon MY, Kwon YK, Jung BY et al (2011) Inhibitory effects of broccoli extract on *Escherichia coli* O157:H7 quorum sensing and in vivo virulence. FEMS Microbiol Lett 321(1):67–74. doi:10.1111/j.1574-6968.2011.02311.x

Li M, Wang B, Zhang M, Rantalainen M, Wang S, Zhou H et al (2008) Symbiotic gut microbes modulate human metabolic phenotypes. Proc Natl Acad Sci U S A 105(6):2117–2122. doi:10.1073/pnas.0712038105

Longford SR, Tujula NA, Crocetti GR, Holmes AJ, Holmstrom C, Kjelleberg S et al (2007) Comparisons of diversity of bacterial communities associated with three sessile marine eukaryotes. Aquat Microb Ecol 48(3):217–229. doi:10.3354/Ame048217

Manefield M, de Nys R, Kumar N, Read R, Givskov M, Steinberg P et al (1999) Evidence that halogenated furanones from *Delisea pulchra* inhibit acylated homoserine lactone (AHL)-mediated gene expression by displacing the AHL signal from its receptor protein. Microbiology 145:283–291

Marshall K, Joint I, Callow ME, Callow JA (2006) Effect of marine bacterial isolates on the growth and morphology of axenic plantlets of the green alga *Ulva linza*. Microb Ecol 52(2):302–310. doi:10.1007/s00248-006-9060-x

Martin M, Biver S, Steels S, Barbeyron T, Jam M, Portetelle D et al (2014) Identification and characterization of a halotolerant, cold-active marine endo-beta-1,4-glucanase by using functional metagenomics of seaweed-associated microbiota. Appl Environ Microbiol 80(16):4958–4967. doi:10.1128/AEM.01194-14

Marzinelli EM, Campbell AH, Zozaya Valdes E, Verges A, Nielsen S, Wernberg T et al (2015) Continental-scale variation in seaweed host-associated bacterial communities is a function of host condition, not geography. Environ Microbiol 17(10):4078–4088. doi:10.1111/1462-2920.12972

Matsuo Y, Imagawa H, Nishizawa M, Shizuri Y (2005) Isolation of an algal morphogenesis inducer from a marine bacterium. Science 307(5715):1598. doi:10.1126/science.1105486

Merriman B, Rothberg JM (2012) Progress in ion torrent semiconductor chip based sequencing. Electrophoresis 33(23):3397–3417. doi:10.1002/elps.201200424

Meusnier I, Olsen JL, Stam WT, Destombe C, Valero M (2001) Phylogenetic analyses of *Caulerpa taxifolia* (Chlorophyta) and of its associated bacterial microflora provide clues to the origin of the Mediterranean introduction. Mol Ecol 10(4):931–946

Meusnier I, Valero M, Destombe C, Gode C, Desmarais E, Bonhomme F et al (2002) Polymerase chain reaction-single strand conformation polymorphism analyses of nuclear and chloroplast

DNA provide evidence for recombination, multiple introductions and nascent speciation in the *Caulerpa taxifolia* complex. Mol Ecol 11(11):2317–2325

Milton DL, Hardman A, Camara M, Chhabra SR, Bycroft BW, Stewart GS et al (1997) Quorum sensing in *Vibrio anguillarum*: characterization of the vanI/vanR locus and identification of the autoinducer N-(3-oxodecanoyl)-L-homoserine lactone. J Bacteriol 179(9):3004–3012

Nakayama J, Watanabe K, Jiang J, Matsuda K, Chao SH, Haryono P et al (2015) Diversity in gut bacterial community of school-age children in Asia. Sci Rep 5:8397. doi:10.1038/srep08397

Neyrinck AM, Possemiers S, Verstraete W, De Backer F, Cani PD, Delzenne NM (2012) Dietary modulation of clostridial cluster XIVa gut bacteria (*Roseburia* spp.) by chitin-glucan fiber improves host metabolic alterations induced by high-fat diet in mice. J Nutr Biochem 23(1):51–59. doi:10.1016/j.jnutbio. 2010.10.008

O'Connor-Sanchez A, Rivera-Dominguez AJ, De los Santos-Briones C, Lopez-Aguiar LK, Pena-Ramirez YJ, Prieto-Davo A (2014) Acidobacteria appear to dominate the microbiome of two sympatric Caribbean Sponges and one Zoanthid. Biol Res 47:67. doi:10.1186/0717-6287-47-67

Patankar AV, Gonzalez JE (2009a) An orphan LuxR homolog of *Sinorhizobium meliloti* affects stress adaptation and competition for nodulation. Appl Environ Microbiol 75(4):946–955. doi:10.1128/AEM.01692-08

Patankar AV, Gonzalez JE (2009b) Orphan LuxR regulators of quorum sensing. FEMS Microbiol Rev 33(4):739–756. doi:10.1111/j.1574-6976.2009.00163.x

Patel HK, Suarez-Moreno ZR, Degrassi G, Subramoni S, Gonzalez JF, Venturi V (2013) Bacterial LuxR solos have evolved to respond to different molecules including signals from plants. Front Plant Sci 4:447. doi:10.3389/fpls.2013.00447

Penesyan A, Marshall-Jones Z, Holmstrom C, Kjelleberg S, Egan S (2009) Antimicrobial activity observed among cultured marine epiphytic bacteria reflects their potential as a source of new drugs. FEMS Microbiol Ecol 69(1):113–124. doi:10.1111/j.1574-6941.2009.00688.x

Provasoli L (1958) Effect of plant hormones on *Ulva*. Biol Bull 114(3):375–384. doi:10.2307/1538992

Provasoli L, Pintner IJ (1953) Ecological implications of in vitro nutritional requirements of algal flagellates. Ann N Y Acad Sci 56(5):839–851

Provasoli L, Pintner IJ (1980) Bacteria induced polymorphism in an axenic laboratory strain of *Ulva lactuca* (Chlorophyceae). J Phycol 32:479–482

Radax R, Rattei T, Lanzen A, Bayer C, Rapp HT, Urich T et al (2012) Metatranscriptomics of the marine sponge *Geodia barretti*: tackling phylogeny and function of its microbial community. Environ Microbiol 14(5):1308–1324. doi:10.1111/j.1462-2920.2012.02714.x

Rajpal DK, Brown JR (2013) Modulating the human gut microbiome as an emerging therapeutic paradigm. Sci Prog 96(Pt 3):224–236

Ridlon JM, Kang DJ, Hylemon PB, Bajaj JS (2014) Bile acids and the gut microbiome. Curr Opin Gastroenterol 30(3):332–338. doi:10.1097/MOG.0000000000000057

Rutherford ST, Bassler BL (2012) Bacterial quorum sensing: its role in virulence and possibilities for its control. Cold Spring Harb Perspect Med 2(11). doi:10.1101/cshperspect.a012427

Singh RP, Reddy CRK (2014) Seaweed-microbial interactions: key functions of seaweed-associated bacteria. FEMS Microbiol Ecol 88(2):213–230. doi:10.1111/1574-6941. 12297

Singh RP, Bijo AJ, Baghel RS, Reddy CRK, Jha B (2011a) Role of bacterial isolates in enhancing the bud induction in the industrially important red alga *Gracilaria dura*. FEMS Microbiol Ecol 76(2):381–392. doi:10.1111/j.1574-6941.2011.01057.x

Singh RP, Mantri VA, Reddy CRK, Jha B (2011b) Isolation of seaweed-associated bacteria and their morphogenesis-inducing capability in axenic cultures of the green alga *Ulva fasciata*. Aquat Biol 12(1):13–21. doi:10.3354/Ab00312

Singh RP, Shukla MK, Mishra A, Reddy CRK, Jha B (2013) Bacterial extracellular polymeric substances and their effect on settlement of zoospore of *Ulva fasciata*. Colloids Surf B Biointerfaces 103:223–230. doi:10.1016/j.colsurfb.2012.10.037

Singh RP, Baghel RS, Reddy CRK, Jha B (2015) Effect of quorum sensing signals produced by seaweed-associated bacteria on carpospore liberation from *Gracilaria dura*. Front Plant Sci:6. doi:10.3389/Fpls.2015.00117

Spoerner M, Wichard T, Bachhuber T, Stratmann J, Oertel W (2012) Growth and thallus morphogenesis of *Ulva mutabilis* (Chlorophyta) depends on a combination of two Bacterial species excreting regulatory factors. J Phycol 48(6):1433–1447. doi:10.1111/j.1529-8817.2012.01231.x

Staufenberger T, Thiel V, Wiese J, Imhoff JF (2008) Phylogenetic analysis of bacteria associated with *Laminaria saccharina*. FEMS Microbiol Ecol 64(1):65–77. doi:10.1111/j.1574-6941.2008.00445.x

Subramoni S, Venturi V (2009a) LuxR-family 'solos': bachelor sensors/regulators of signalling molecules. Microbiology 155(Pt 5):1377–1385. doi:10.1099/mic.0.026849-0

Subramoni S, Venturi V (2009b) PpoR is a conserved unpaired LuxR solo of *Pseudomonas putida* which binds N-acyl homoserine lactones. BMC Microbiol 9:125. doi:10.1186/1471-2180-9-125

Subramoni S, Florez Salcedo DV, Suarez-Moreno ZR (2015) A bioinformatic survey of distribution, conservation, and probable functions of LuxR solo regulators in bacteria. Front Cell Infect Microbiol 5:16. doi:10.3389/fcimb.2015.00016

Sun J, Chang EB (2014) Exploring gut microbes in human health and disease: pushing the envelope. Genes Dis 1(2):132–139. doi:10.1016/j.gendis.2014.08.001

Tait K, Joint I, Daykin M, Milton DL, Williams P, Camara M (2005) Disruption of quorum sensing in seawater abolishes attraction of zoospores of the green alga *Ulva* to bacterial biofilms. Environ Microbiol 7(2):229–240. doi:10.1111/j.1462-2920.2004.00706.x

Tait K, Williamson H, Atkinson S, Williams P, Camara M, Joint I (2009) Turnover of quorum sensing signal molecules modulates cross-kingdom signalling. Environ Microbiol 11(7):1792–1802. doi:10.1111/j.1462-2920.2009.01904.x

Takusagawa F, Kamitori S, Markham GD (1996) Structure and function of S-adenosylmethionine synthetase: crystal structures of S-adenosylmethionine synthetase with ADP, BrADP, and PPi at 28 angstroms resolution. Biochemistry 35(8):2586–2596. doi:10.1021/bi952604z

Trindade-Silva AE, Rua C, Silva GG, Dutilh BE, Moreira AP, Edwards RA et al (2012) Taxonomic and functional microbial signatures of the endemic marine sponge *Arenosclera brasiliensis*. PLoS One 7(7):e39905. doi:10.1371/journal.pone.0039905

Tujula NA, Crocetti GR, Burke C, Thomas T, Holmstrom C, Kjelleberg S (2010) Variability and abundance of the epiphytic bacterial community associated with a green marine *Ulvacean alga*. ISME J 4(2):301–311. doi:10.1038/ismej.2009.107

Turnbaugh PJ, Ley RE, Mahowald MA, Magrini V, Mardis ER, Gordon JI (2006) An obesity-associated gut microbiome with increased capacity for energy harvest. Nature 444(7122):1027–1031. doi:10.1038/Nature05414

Turnbaugh PJ, Ley RE, Hamady M, Fraser-Liggett CM, Knight R, Gordon JI (2007) The human microbiome project. Nature 449(7164):804–810. doi:10.1038/nature06244

Turner TR, James EK, Poole PS (2013) The plant microbiome. Genome Biol 14(6):209. doi:10.1186/gb-2013-14-6-209

Twigg MS, Tait K, Williams P, Atkinson S, Camara M (2014) Interference with the germination and growth of *Ulva zoospores* by quorum-sensing molecules from *Ulva*-associated epiphytic bacteria. Environ Microbiol 16(2):445–453. doi:10.1111/1462-2920.12203

Ursell LK, Knight R (2013) Xenobiotics and the human gut microbiome: metatranscriptomics reveal the active players. Cell Metab 17(3):317–318. doi:10.1016/j.cmet.2013.02.013

Venturi V, Fuqua C (2013) Chemical signaling between plants and plant-pathogenic bacteria. Annu Rev Phytopathol 51:17–37. doi:10.1146/annurev-phyto-082712-102239

Visick KL, Foster J, Doino J, McFall-Ngai M, Ruby EG (2000) *Vibrio fischeri* lux genes play an important role in colonization and development of the host light organ. J Bacteriol 182(16):4578–4586

Wahl M, Goecke F, Labes A, Dobretsov S, Weinberger F (2012) The second skin: ecological role of epibiotic biofilms on marine organisms. Front Microbiol 3. doi:10.3389/Fmicb. 2012.00292

Walsh CJ, Guinane CM, O'Toole PW, Cotter PD (2014) Beneficial modulation of the gut microbiota. FEBS Lett 588(22):4120–4130. doi:10.1016/j.febslet. 2014.03.035

Wang WL, Xu SY, Ren ZG, Tao L, Jiang JW, Zheng SS (2015) Application of metagenomics in the human gut microbiome. World J Gastroenterol 21(3):803–814. doi:10.3748/wjg.v21.i3.803

Weinberger F, Beltran J, Correa JA, Lion U, Pohnert G, Kumar N et al (2007) Spore release in *Acrochaetium* sp (Rhodophyta) is bacterially controlled. J Phycol 43(2):235–241. doi:10.1111/j.1529-8817.2007.00329.x

Wheeler GL, Tait K, Taylor A, Brownlee C, Joint I (2006) Acyl-homoserine lactones modulate the settlement rate of zoospores of the marine alga *Ulva intestinalis* via a novel chemokinetic mechanism. Plant Cell Environ 29(4):608–618

Whitehead NA, Barnard AM, Slater H, Simpson NJ, Salmond GP (2001) Quorum-sensing in Gram-negative bacteria. FEMS Microbiol Rev 25(4):365–404

Williams P (2007) Quorum sensing, communication and cross-kingdom signalling in the bacterial world. Microbiology 153(Pt 12):3923–3938. doi:10.1099/mic.0.2007/012856-0

Yang Y, Jobin C (2014) Microbial imbalance and intestinal pathologies: connections and contributions. Dis Model Mech 7(10):1131–1142. doi:10.1242/dmm.016428

Yang Z, Li Z (2012) Spatial distribution of prokaryotic symbionts and ammoxidation, denitrifier bacteria in marine sponge *Astrosclera willeyana*. Sci Rep 2:528. doi:10.1038/srep00528

Zhang L, Jia Y, Wang L, Fang R (2007) A proline iminopeptidase gene upregulated in planta by a LuxR homologue is essential for pathogenicity of *Xanthomonas campestris* pv. *campestris*. Mol Microbiol 65(1):121–136. doi:10.1111/j.1365-2958.2007.05775.x

Chapter 16
Role of Bacteria in Coral Ecosystem

Neha P. Patel, Sweta B. Kumar, and S. Haldar

Abstract Coral reefs are the most diverse and valuable of all marine ecosystems on earth. This ecosystem is often called the rainforests of the sea due to its diversity. It provides enormous benefits to the human and its surrounding community. But the increase in industrialization, urbanization, and mechanization has led to the decline in coral health which ultimately affects the productivity and sustainability of the reef ecosystem. Thus, there is an increasing interest in the field of coral ecology in the last few decades driven by the desire to understand the biology of coral animal for its protection. Remarkable progress has been made in understanding the coral holobiont of the reef ecosystem. The role of dinoflagellate in coral holobiont has been studied very well, but the role of coral-associated bacteria in holobiont is not clearly understood in spite many scientific studies. Bacterial diversity associated with coral holobionts is known to act both as drivers and indicators of disturbances in the coral reef ecosystem due to their short generation time. Conversely, to the health benefits the bacteria provide to the coral host, certain bacteria in conditions of environmental stress cause coral bleaching and other diseases. Thus, understanding the bacteriology of coral holobiont may help in preventing the worldwide destruction of coral reefs. Moreover, these studies may lead to the discovery of novel product with commercial importance. For this reason, the chapter aims to review the abundance and diversity of bacteria associated with healthy and diseased corals and their role in coral health. We are also discussing the recent technologies used in the study of coral-associated bacterial diversity and their role in structuring coral ecosystem.

Keywords Coral bacteriology • Coral bleaching • Dinoflagellates • Holobiont • DGGE • qPCR

N.P. Patel • S.B. Kumar • S. Haldar (✉)
Marine Biotechnology and Ecology Division, CSIR- Central Salt and Marine, Chemicals Research Institute (CSIR-CSMCRI), Gijubhai Badheka Marg, Bhavnagar 364002, Gujarat, India

Academy of Scientific and Innovative Research (AcSIR), CSIR-Central Salt and Marine Chemicals Research Institute (CSIR-CSMCRI), Council of Scientific and Industrial Research (CSIR), Gijubhai Badheka Marg, Bhavnagar 364002, Gujarat, India
e-mail: shaldar@csmcri.res.in

© Springer International Publishing AG 2017
M. Kumar, P. Ralph (eds.), *Systems Biology of Marine Ecosystems*,
DOI 10.1007/978-3-319-62094-7_16

Contents

16.1 Introduction 318
16.2 Coral and Its Structure 320
16.3 Bacterial Diversity Associated with Coral Reef Ecosystem 321
16.4 Role of Coral-Associated Bacteria 323
16.4.1 Coral-Associated Bacteria as a Companion of Coral Holobiont 323
16.4.1.1 Bacteria as a Key Nutrient Supplier 324
16.4.1.2 Bacteria Required for Metamorphosis and Settlement of Coral Larva 325
16.4.1.3 Coral-Associated Bacteria Contribute to Host Defense Against Unwanted Colonization and Biofouling 326
16.4.2 Coral-Associated Bacteria as a Pathogen 327
16.5 Recent Technologies to Study the Role of Bacteria in Coral Ecosystem 327
16.5.1 PCR-Based Methods 328
16.5.1.1 Denaturing Gradient Gel Electrophoresis (DGGE) 330
16.5.1.2 Quantitative Real-Time PCR (qRT-PCR) 330
16.5.2 Nucleic Acid Hybridization-Based Techniques 331
16.5.2.1 Fluorescent In Situ Hybridization (FISH) 331
16.5.2.2 Microarray 333
16.5.2.3 Membrane-Based Methods 333
16.5.3 Western Blot: A Protein-Based Molecular Technique 335
16.5.4 Next-Generation Tools 336
16.5.5 Omics Tools 336
16.5.5.1 Metaproteomics 336
16.5.5.2 Metabolomics 337
References 337

16.1 Introduction

Coral reefs are the most diverse and valuable of all marine ecosystems on earth. This ecosystem is often called the rainforests of the sea due to the diversity it harbors (Connell 1978). It supports more species per unit area than any other marine ecosystem. According to a recent survey, it harbors approximately 4000 species of fish, 800 species of hard corals, and hundreds of other species. Scientists have estimated that there may be another one to eight million undiscovered species of living organisms in the reef ecosystem (Reaka-Kudla 1997). The estimated total net benefit per year from the reef ecosystem of the world is $29.8 billion mainly derived from the ecosystem services such as tourism and recreation, coastal protection, fisheries, and biodiversity (Brooke and Ross 2014) (Fig. 16.1). In recent years, it has also become a valuable resource for the modern drug discovery (Rosenberg and Falkovitz 2004). Therefore, there is an increasing interest in the field of coral ecology in the last few decades, mainly driven by the desire to harvest the benefits that it provide to the human and its neighboring community. However, for the productivity and sustainability of reef ecosystem, healthy corals are crucial, but in this day and age, they are on the verge of degradation due to various anthropogenic activities such as blast fishing and coastal water pollution (Faden 2008). These activities alter

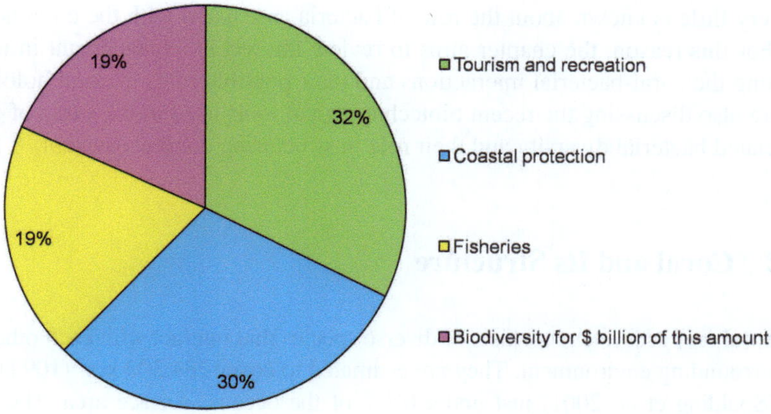

Fig. 16.1 Economic value of coral reef ecosystem

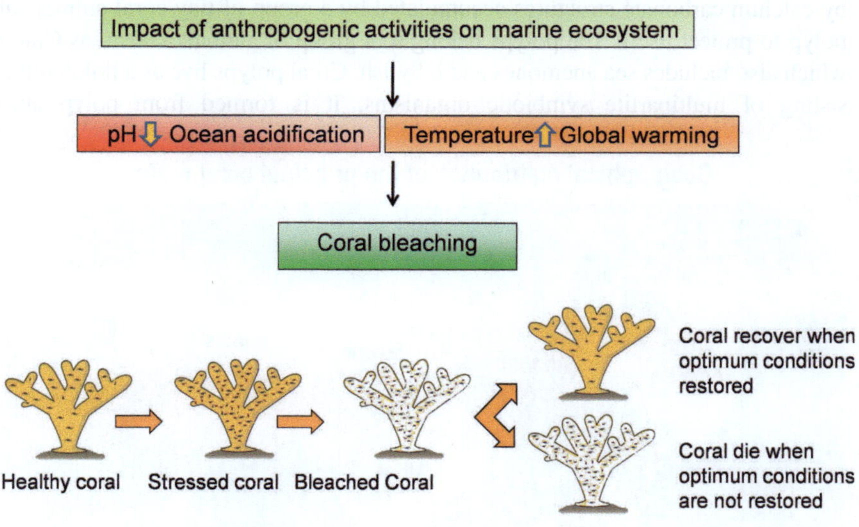

Fig. 16.2 Human impacts on coral ecosystem

the abiotic factors such as ocean temperature and pH to facilitate biotic factors like pathogens such as bacteria, viruses, and algae to cause coral disease (Ritchie 2006; Teplitski and Ritchie 2009; Mao-Jones et al. 2010) (Fig. 16.2). Therefore, there is an urgent need of the hour to protect this reef ecosystem from destruction. But before designing any strategy for its protection, it is important to understand the biology of the corals and the cause of its degradation. Impressive progress has been made in understanding the coral holobiont of the reef ecosystem. The symbiotic relationship between the coral animal and the dinoflagellates is studied very well, but very little is known about the role of bacteria associated with the coral holobiont. For this reason, the chapter aims to review the recent advancement in understanding the coral-bacterial interactions and their possible roles in coral holobiont. We are also discussing the recent biotechnological tools used in the study of coral-associated bacterial diversity and their role in structuring coral ecosystem.

16.2 Coral and Its Structure

The coral reef ecosystem consists of diverse species that interact with each other and the surrounding environment. They are estimated to cover 284,300 km^2 (109,800 sq. mi) (Spalding et al. 2001) just under 0.1% of the oceans' surface area. The Indo-Pacific region that include the Red Sea, Indian Ocean, Southeast Asia, and the Pacific) accounts for 91.9%, Southeast Asia accounts for 32.3%, and the Pacific including Australia accounts for 40.8%, while Atlantic and Caribbean coral reefs account for 7.6% of the total (Spalding et al. 2001) (Fig. 16.3). The coral reef ecosystem is held by calcium carbonate structures accumulated by a group of tiny coral animal called polyp to protect itself. The polyps belong to a group of animals known as Cnidaria, which also includes sea anemones and jellyfish. Coral polyps live as a holobiont consisting of multipartite symbiotic organisms. It is formed from polyp animal,

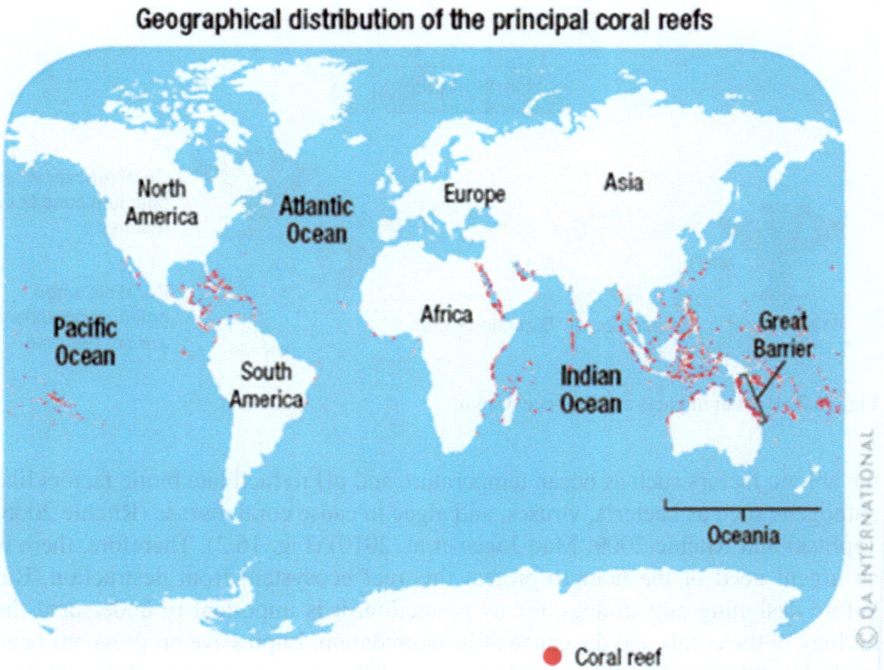

Fig. 16.3 Geographical distributions of the principle coral reefs

endosymbiotic dinoflagellates, bacterial and viral associates of polyp, and dinoflagellates (Rohwer et al. 2002). Coral polyps are composed of two layers of cells, the epidermis and gastrodermis, covered by a surface mucus layer and connected to a large, porous calcium carbonate skeleton (Rosenberg et al. 2007). The symbiotic dinoflagellate from the genus *Symbiodinium* is a photosynthetic associate that resides inside the membrane-bound vacuoles within specialized cells of the polyp. They are the main source of carbon nutrition as they translocate approximately 60–80% of their photosynthate to the coral host, allowing the holobiont to thrive in nutrient-poor waters (Tremblay et al. 2012). These dinoflagellate endosymbionts contribute significantly to the physiological attributes of the coral holobiont as these associations are dynamic and flexible. Corals can expel their dinoflagellate symbionts and acquire new strains (or even clades) of *Symbiodinium* that may be more effective in aiding holobiont to respond under environmental stress conditions (Little et al. 2004). The flexible associations of the dinoflagellates with coral encouraged researchers to investigate flexibility of the interactions between corals and the bacteria. With the incorporation of new high-throughput techniques like metagenomics, DGGE in coral microbiology, the knowledge about the nature of coral-microbe interactions is growing by leaps and bounds to address questions about the makeup and dynamics of coral microbial assemblages.

16.3 Bacterial Diversity Associated with Coral Reef Ecosystem

Bacteria are omnipresent and corals are no exception. They have coexisted ever since the early stages of evolution. They represent all conceivable modes of interactions—mutualism, parasitism, and commensalism. These interactions together influence the ecosystem directly or indirectly. It is well known that corals and bacteria synergistically affect each other's physiology and metabolism. Their interactions are important for the primary productivity of the most ecosystems.

In general, corals provide mainly three niches to bacteria each of which has a distinct bacterial population. These include the surface mucus layer, the coral tissue (including the gastrodermis cavity), and the calcium carbonate skeleton (Bourne and Munn 2005; Koren and Rosenberg 2006; Rosenberg et al. 2007) (Fig. 16.4). The diversity of coral-associated bacteria has been reasonably well documented in these niches using both culture-dependent and culture-independent techniques (Rohwer et al. 2002; Bourne and Munn 2005; Wegley et al. 2007; Dinsdale et al. 2008; Thurber et al. 2009; Cook et al. 2013). These studies indicate that it is common for a single coral to house many of the known divisions of bacteria such as *Gammaproteobacteria*, *Alphaproteobacteria*, *Cyanobacteria*, *Firmicutes*, and *Bacteroidetes* (Littman et al. 2009; Bayer et al. 2013). Moreover, the various coral niches are reported to be colonized by distinctive species (Sweet et al. 2011; Agostini et al. 2012), and environmental disturbances can alter coral microbial communities

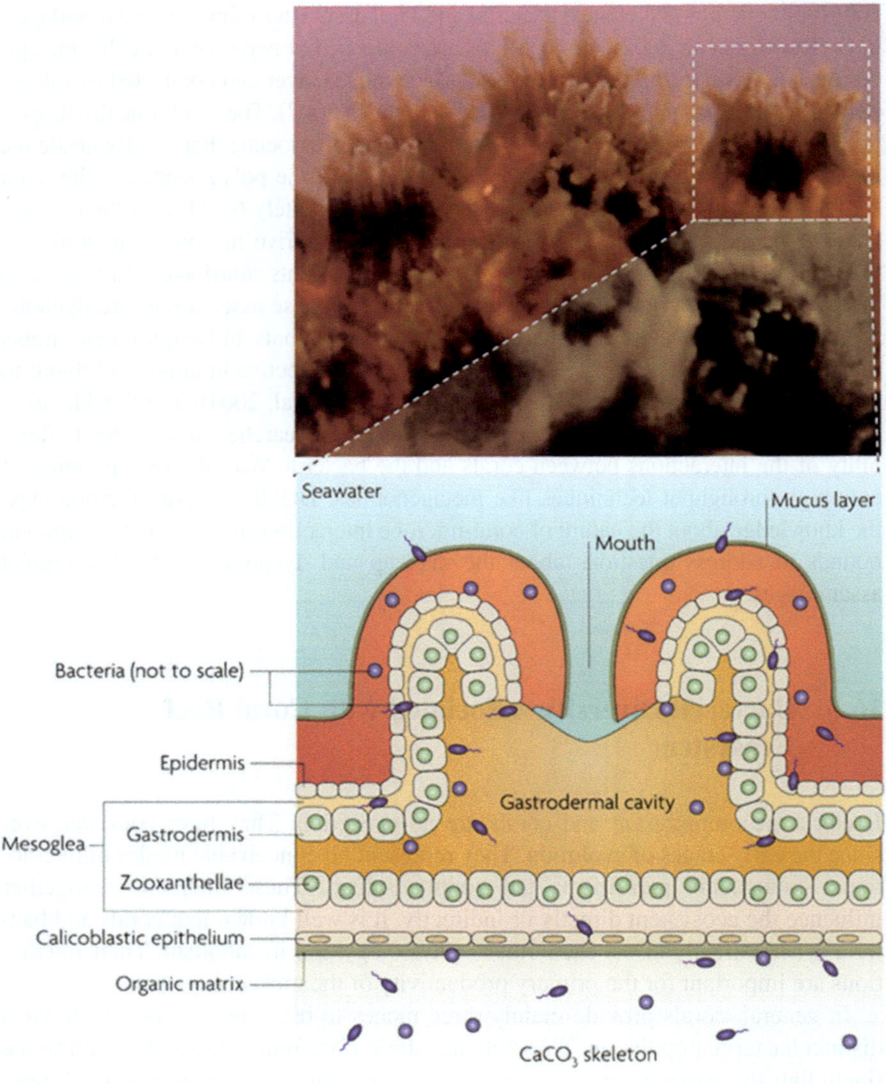

Fig. 16.4 Structure of coral holobiont

(Bourne et al. 2008) and functions (Thurber et al. 2009). However, studies are going on to acquire knowledge about how coral-bacterial relationships are affected by variations in spatial, temporal, and ecological conditions of coral species (Kimes et al. 2013).

Whether the association of bacteria with corals is specific or nonspecific is the topic of debate among scientific communities. Many studies report that bacteria associated with corals are host specific across geographically distant sites (Rohwer et al. 2002; Chen et al. 2011), suggesting that the coral host plays some role in recruiting the bacteria within the holobiont. Whereas some other studies have revealed that coral-associated microbial species display site specificity with community composition that differs in location rather than coral species (Littman et al. 2009; Barott et al. 2011) which suggests that ecological factors have some role in influencing the coral-associated-bacterial communities. It is not clear how bacteria are associated with corals—whether corals recruit their own microbiota or microbes choose to colonize the host. But in either case, the hypothetical mechanism behind coral-bacterial association involves motility/chemotaxis (Tout et al. 2014) and quorum sensing (Ransome et al. 2014a).

Now, scientists are also focusing on microbial functioning rather than phylogeny to define the diversity of microbial communities on the different coral host. Functional studies based on metagenomics, transcriptomics, proteomics, and metabolomics will provide us the knowledge related to the diversity and functionality of symbiotically associated bacteria rather than opportunistic bacteria on the coral surface (Burke et al. 2011). For example, Tout et al. (2014) studied microbial community composition and function within a coral reef ecosystem. They observed substantially different microbial compositions and metabolic functions between the four niches across Heron Island Reef, within the Great Barrier Reef. The microbial composition of seawater and sandy substrate is dominated by the genes associated with core housekeeping processes such as lipid, carbohydrate, protein, and nucleic acid metabolism, whereas the metagenome from the coral surface had an enhanced occurrence of genes associated with dynamic processes such as motility and chemotaxis. These studies suggest that metabolic pathways and functional capabilities define the "core" microbiota more accurately than the phylogenic diversity on the coral host (Krediet et al. 2013). Thus, a scientist should combine both functional-based and phylogenetic-based studies to define the diversity of coral-associated bacteria.

16.4 Role of Coral-Associated Bacteria

Bacteria have a beneficial as well as a detrimental role in coral holobiont. Understanding this role and mechanism behind them will probably help us in protecting the reef ecosystem, for instance, by the application of probiotic bacteria on diseased corals. Moreover, we could find promising bacteria having commercial application in the medical fields such as high antibiotic resistance and ROS scavenging properties. Maybe in the future, these bacteria can be taken as probiotic by health-compromised patients.

16.4.1 Coral-Associated Bacteria as a Companion of Coral Holobiont

The beneficial role of coral-associated bacteria involves nutrient supplying and cycling such as carbon, nitrogen, and sulfur, larval metamorphosis and settlement, host resilience, and disease resistance, by providing antibiotic defense against

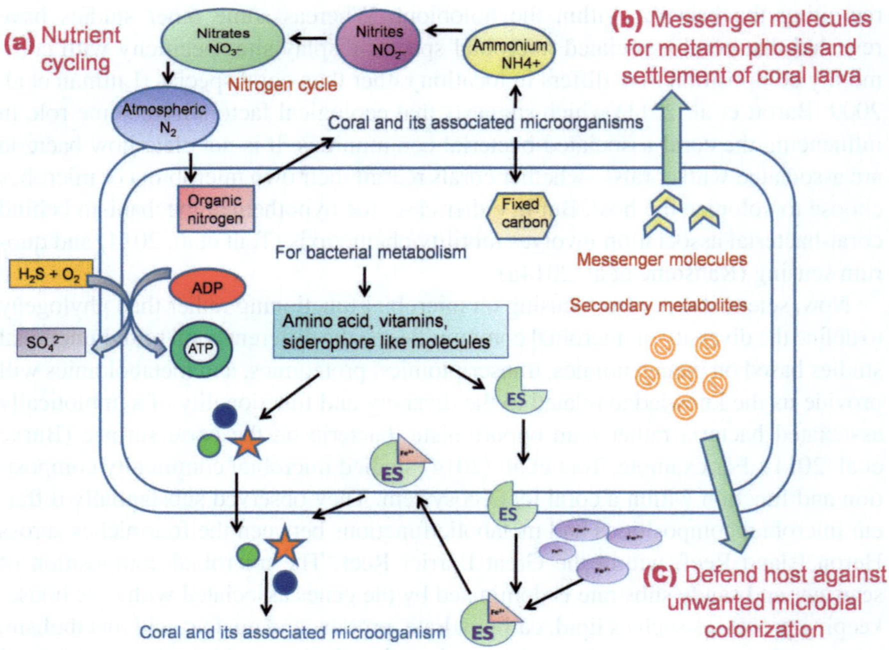

Fig. 16.5 Role of bacteria in coral holobiont as a companion

unwanted microbes and competing with pathogens for nutrients such as by producing siderophore (Fig. 16.5).

16.4.1.1 Bacteria as a Key Nutrient Supplier

The productivity of coral reef ecosystem is largely dependent on the role of coral-associated bacteria. They play a fundamental role in biogeochemical cycling of essential nutrients such as nitrogen and sulfur and deliver these recycled nutrients and trace elements to the coral holobiont (Azam and Malfatti 2007; Falkowski et al. 2008). They also metabolize organic matter for the coral holobiont. Many scientific studies have supported the association of nitrogen-fixing bacteria with corals. For example, a study by Lesser et al. 2007 showed the presence of a nitrogen-fixing *Cyanobacterium* within the cells of the Caribbean coral *Montastraea cavernosa* that fulfill the nitrogen demand of the host. Moreover, the abundance of *Symbiodinium* has been associated with the abundance of nitrogen-fixing bacteria such as those belonging to rhizobia taxa (Olson and Lesser 2013) which may provide a substantial nutritional benefit to all members of the coral holobiont (Lema et al. 2012). *Vibrio harveyi* and *V. alginolyticus* have shown the capability of nitrogen fixation in coral mucus and dominate the culturable nitrogen-fixing bacteria of the Brazilian coral *Mussismilia hispida* (Chimetto et al. 2008).

There are several pieces of evidence that support the role of bacteria in sulfur cycle. For instance, in a study by Raina et al. (2009), they suggested the role of coral-associated bacteria in providing sulfur to the host by degrading sulfur compounds such as dimethylsulfoniopropionate (DMSP), dimethyl sulfide (DMS), and acrylic acid. Moreover, these sulfur compounds may have a role in structuring bacterial communities beneficial for the health of corals.

Coral-associated bacteria also participate in scavenging limiting nutrients such as iron by siderophore production, amino acids, and vitamins for the host. The microorganisms in the coral gastrovascular cavity may have roles in food digestion and nutrient absorption like the gut microorganism of other organisms (Thompson et al. 2014). A study by Agostini et al. (2012) measured the physical properties of a gastric cavity of *Galaxea fascicularis* at different depths by using fiber-optic microsensors and observed elevated levels of vitamin B12, phosphate, and nitrogen species. They hypothesized that vitamin B12 may be produced by bacteria as a nutrient for coral or *Symbiodinium*. Some coral-associated bacteria are photosynthetic in nature that may provide the photosynthetic products in the absence of their photosynthetic partner such as during coral bleaching to help the coral animal to rejuvenate. For example, a study has shown the presence of *Cyanobacteria* in bleached *Oculina patagonica* that supply photosynthesized products to the coral (Fine and Loya 2002, 2004). Moreover, these bacteria are also the direct source of nutrition to corals through bacterivory (Kushmaro and Kramarsky-Winter 2004).

16.4.1.2 Bacteria Required for Metamorphosis and Settlement of Coral Larva

Metamorphosis and settlement are tightly coupled processes in corals wherein a mobile planula larva transforms into a sedentary polyp animal. Emerging evidence shows that bacteria play an important role in this transformation process of the coral animal. These studies have reported that the coral-associated bacteria provide protection to the coral spawn or newly hatched larvae through their antimicrobial activity (Marquis et al. 2005). A study by Negri et al. (2001) showed an increase in metamorphosis and larval settlement on crustose coralline algae (CCA) surfaces that harbors a consortium of a complex community of bacteria. Moreover, antibiotic treatment of larval cultures and rock surface showed sufficient inhibition for a larval settlement that signifies the importance of bacterial activity (Huggett et al. 2006; Vieira et al. 2016). Various researches are enduring to understand the mechanism behind the induction of metamorphosis and settlement of coral larva. Scientist suggests that this event may be induced by diffusible or potentially contact-mediated signals (Hadfield 2011; Dobretsov et al. 2013; Shikuma et al. 2014). For example, Tebben et al. (2011) isolated tetrabromopyrrole (TBP) from various *Pseudoalteromonas* strains isolated from CCA surfaces, known to stimulate metamorphosis in *Acropora millepora*. In contrast, a recent study reported that both natural and synthetic TBP induce metamorphosis as well as a coral larval settlement in Caribbean corals (Sneed et al. 2014).

Thus, it suggests that TBP may have divergent impacts on net larval recruitment that depends on the timing of larval settlement.

Bacteria are also known to guide the newly coral polyp to settle near the proximity of adult coral colonies or other coral competitors such as macro- and turf algae (Smith et al. 2006; Marhaver et al. 2013).

Despite several studies investigating the potential role of bacteria in the induction of larval settlement and metamorphosis, a thorough understanding of all the players involved during spawning and their mechanism is not clearly understood, and thus it remains an important avenue for ongoing studies.

16.4.1.3 Coral-Associated Bacteria Contribute to Host Defense Against Unwanted Colonization and Biofouling

Bacteria associated with corals guard the host from various pathogenic microorganisms. There are various mechanisms by which these bacteria provide protection to the coral host. For instance, secondary metabolite production such as antibiotics against opportunistic harmful bacteria on the coral surface has been reported by various scientific studies (Mansson et al. 2011; Raina et al. 2016). A study by Ritchie (2006) demonstrated that coral mucus has a role in structuring beneficial coral-associated microbial communities through antibacterial activity. They may also interrupt cell-to-cell communication among pathogens and compete with pathogens for limiting nutrients and space on host surfaces (Shnit-Orland and Kushmaro 2009). In addition to functioning as antimicrobials, these bacteria have the ability to inhibit pathogen's catabolic enzymes required by the pathogens for establishing themselves on the host. For example, microorganisms associated with *A. palmate* are found to interfere with the ability of white pox pathogen *Serratia marcescens* and to use coral mucus by blocking the induction of the glycosidase (Krediet et al. 2013).

Besides these, coral-associated bacteria have high ROS scavenging activities such as production of antioxidant compound sand enzymes which enable them to survive on oxidative-stressed coral surface and also protect the host and its associated microorganisms by the detrimental consequences of these radicals such as photo-inhibition which ultimately results in coral bleaching and death (Reshef et al. 2006). Some of these antioxidants are reported to associate with virulence such as superoxide dismutase enzyme (Munn et al. 2008; Banin et al. 2003). So, the researchers have hypothesized that this ability of coral bacteria combined with host property of ROS production prevents unwanted colonization of opportunistic and pathogenic bacteria that are responsible for coral diseases.

Moreover, siderophore-producing bacteria are found on the coral surface which suggest that these bacteria have strong biocontrol abilities against pathogenic bacteria (Solanki et al. 2014). Thus, this area required further scientific studies to establish the beneficial role of bacteria in coral health. Moreover, these inhibitory interactions among coral-associated bacteria on the coral surface could be of immense importance for searching secondary metabolite-producing bacteria having pharmaceutical application.

16.4.2 Coral-Associated Bacteria as a Pathogen

Over the past several decades, the decline in coral reef has been observed. Various bacterial pathogens have been found responsible for this decline. These pathogens cause diseases such as coral bleaching and black band disease, to name a few (Rosenberg et al. 2007; Pratte 2013). Coral diseases are linked to various environmental stresses such as temperature increase, water pollution, and many other anthropogenic activities which altogether make coral animal susceptible to infections by pathogens (Roder et al. 2014). For instance, the decline in the Caribbean populations of the elkhorn coral *A. palmata* is due to environmental stress resulting in bleaching, and the disease has been reported. They showed a decrease in antibiotic activity of bacterial isolates from the mucus and tissue of *A. palmata* during a summer bleaching event and were dominated by members of the genus *Vibrio* which are the common pathogens of coral diseases (Ritchie 2006). Another study by Rypien et al. (2010) observed that members of the *Gammaproteobacteria* particularly *Vibrionales* and *Alteromonadales* had high antagonistic activity against other coral bacteria. Other than antimicrobial activity against coral native bacteria, there are several ways by which these bacteria cause coral diseases such as by producing high radical oxygen species (ROS) scavenging compounds and enzymes which enable them to thrive on the oxidative-stressed coral surface resulting from the metabolic activities of zooxanthellae and associated microbes. Moreover, according to several studies, these antioxidant enzymes are related to virulence such as superoxide dismutase in *V. shiloi*, which aid in its pathogenicity (Munn et al. 2008). Siderophore-producing bacteria are also known to cause coral disease by competing with the host and its associated microorganisms for iron which is a limiting nutrient present in very low concentration in the marine environment. Siderophore is known to play a dual role in the infectious process. These are by enhancing the growth of the invading pathogen and inhibiting host defense system (Autenrieth et al. 1991). However, in the middle of the negative role of coral pathogens, a study by Marhaver et al. (2013) suggested the importance of pathogenic and opportunistic bacteria in structuring the reef community by mortality of nearby conspecific coral recruits (Fig. 16.6).

16.5 Recent Technologies to Study the Role of Bacteria in Coral Ecosystem

Various biotechnological techniques are required to study the role of bacteria in the coral ecosystem. These techniques include biochemical methods, microbiological methods, and molecular techniques. With recent advancement and easy accessibility, molecular methods based on targeting DNA, RNA, and proteins have gained much popularity among researchers to explore diverse bacteria and their functional role in coral holobiont. Below, we reviewed some of the advance molecular techniques used in coral-bacterial research.

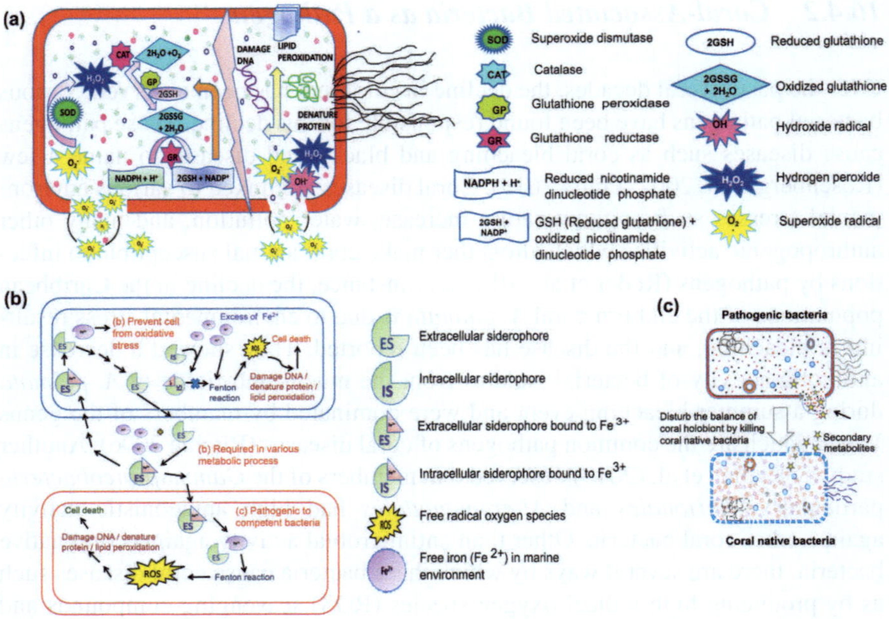

Fig. 16.6 Mechanism behind pathogenecity. (**a**) ROS production and ROS scavenging system for survival on the coral host. (**b**) Siderophore production for iron scavenging and pathogenecity to host and its associated microorganisms. (**c**) Secondary metabolite against coral and its associated microorganisms

16.5.1 PCR-Based Methods

Polymerase chain reaction (PCR) has myriad of applications in coral ecology in combination with other molecular techniques such as denaturing gradient gel electrophoresis (DGGE) and qRT-PCR. This technique is generally used for phylogeny or functional analysis of genes present in microbial communities in a varied ecosystem. It is developed by Kary Mullis in the 1980s (Hills et al. 1987). PCR is a DNA-based technique that generates thousands to multiple copies of a single particular segment of DNA template. It involves thermal cycling of repetitive heating and cooling of the reaction mixture containing buffer with mg^{2+}, dNTPs, primers (both forward and reverse), Taq polymerase, and dsDNA template. This process results in denaturation of dsDNA into ssDNA and its replication by Taq polymerase enzyme.

In coral bacteriology, it is used in combination with other molecular techniques to study diversity and abundance of coral-associated bacteria and their identification and phylogenetic relationship. This knowledge will provide us the key to exploring the role of bacteria in coral holobiont. Below, we are describing the most admired PCR-based techniques used in coral bacteriology (Fig. 16.7).

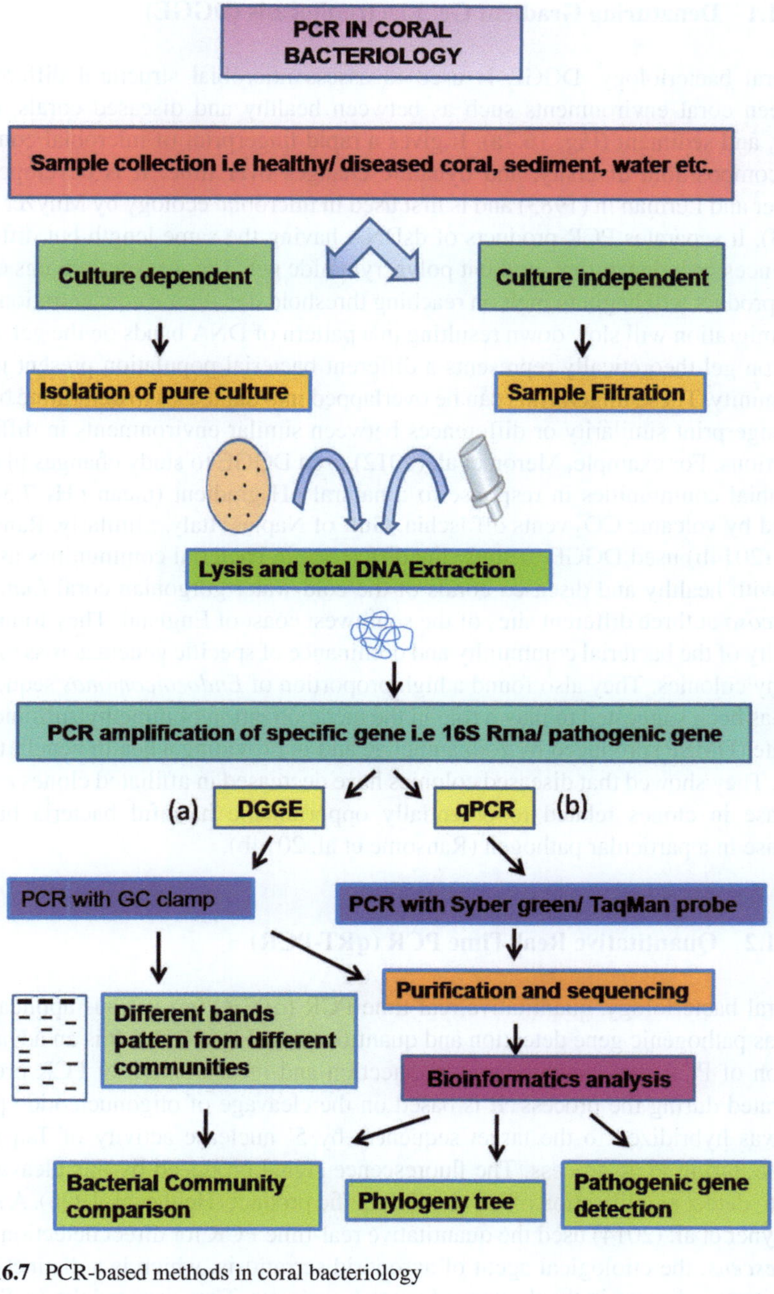

Fig. 16.7 PCR-based methods in coral bacteriology

16.5.1.1 Denaturing Gradient Gel Electrophoresis (DGGE)

In coral bacteriology, DGGE is used to assess microbial structural differences between coral environments such as between healthy and diseased corals, coral water, and sediment (Fig. 16.7a). It gives a rapid fingerprint of microbial community composition, diversity, and dynamic changes over time. It is developed by Fischer and Lerman in (1983) and is first used in microbial ecology by Muyzer et al. (1993). It separates PCR products of dsDNA having the same length but different sequences on a denaturant gradient polyacrylamide gel. The weaker domains of the PCR product will begin to melt on reaching threshold denaturant concentration, and their migration will slow down resulting in a pattern of DNA bands on the gel. Each band on gel theoretically represents a different bacterial population present in the community. These fingerprints can be overlapped into databases to determine bacterial fingerprint similarity or differences between similar environments in different conditions. For example, Meron et al. (2012) used DGGE to study changes in coral microbial communities in response to a natural pH gradient (mean pH_T 7.3–8.1) caused by volcanic CO_2 vents off Ischia, Gulf of Naples, Italy. Similarly, Ransome et al. (2014b) used DGGE to study the difference in bacterial communities associated with healthy and diseased corals of the cold-water gorgonian coral *Eunicella verrucosa* at three different sites of the southwest coast of England. They found the stability of the bacterial community and dominance of specific genera across visibly healthy colonies. They also found a high proportion of *Endozoicomonas* sequences that has been suggested to play a role in the metabolization of dimethylsulfoniopropionate (DMSP) produced by zooxanthellae and in providing a health benefit to the coral. They showed that diseased colonies have decreased in affiliated clones and an increase in clones related to potentially opportunistic harmful bacteria but no increase in a particular pathogen (Ransome et al. 2014b).

16.5.1.2 Quantitative Real-Time PCR (qRT-PCR)

In coral bacteriology, quantitative real-time PCR (qPCR) has various applications such as pathogenic gene detection and quantification (Fig. 16.7b). It is an advanced version of PCR that enables reliable detection and measurement of PCR product generated during the process. It is based on the cleavage of oligonucleotide probe that was hybridized to the target sequence, by 5′ nuclease activity of Taq polymerase during PCR process. The fluorescence signal produced by this cleavage is used to detect amplification of the target-specific product (Heid et al. 1996). A study by Joyner et al. (2014) used the quantitative real-time PCR for direct detection of *S. marcescens*, the etiological agent of acroporid serratiosis, which is a distinct form of white pox disease in the threatened coral *A. palmate*. They targeted the *luxS* gene to distinguish *S. marcescens* from other *Serratia* species with a reliable quantitative limit for the detection of ten cell equivalents (CE) per reaction.

16.5.2 Nucleic Acid Hybridization-Based Techniques

Nucleic acid hybridization-based techniques are popularly used in coral-bacterial studies such as FISH, microarrays, ribotyping, and colony hybridization (Fig. 16.8). Nucleic acid hybridization simply relies on Watson-Crick base pairing of a single-stranded nucleic acid to the radiolabeled nucleic acid probe. This technique has led to the detection, quantification, and purification of a specific segment of DNA/RNA, their cytogenetic localization in cells, and comparative gene expression analysis that may help in understanding coral bacteriology.

16.5.2.1 Fluorescent In Situ Hybridization (FISH)

Fluorescent in situ hybridization (FISH) is a molecular technique based on the use of fluorescently labeled, rRNA-targeted oligonucleotide probes which penetrate inside the cell and specifically hybridize to the target rRNA sequences in the ribosomes and are visualized under epifluorescence microscopy (Fig. 16.8a). If there are no target sequences in the ribosomes, the probes will not hybridize and will be eliminated by a subsequent washing step. During FISH hybridization, each single probe will theoretically hybridize with a single molecule of rRNA. Therefore, it is expected that the larger the number of ribosomes in the cell, the stronger the fluorescence will be. As a result, the intensity of the FISH signal can be taken as an indicator of cellular metabolic activity. Moreover, each cell will show a nearly

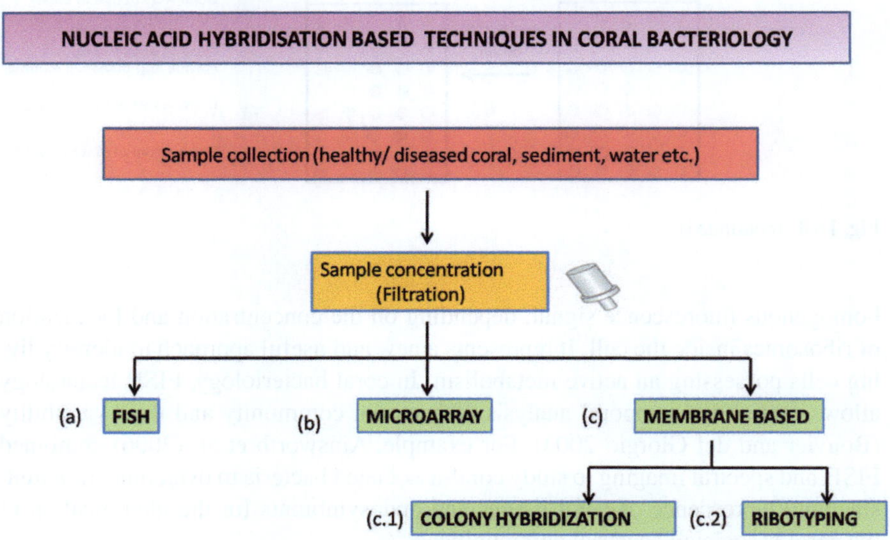

Fig. 16.8 Nucleic acid hybridization-based techniques in coral bacteriology. **(a)** Schematic diagram showing the principle of FISH. **(b)** Schematic diagram showing the principle of microarray

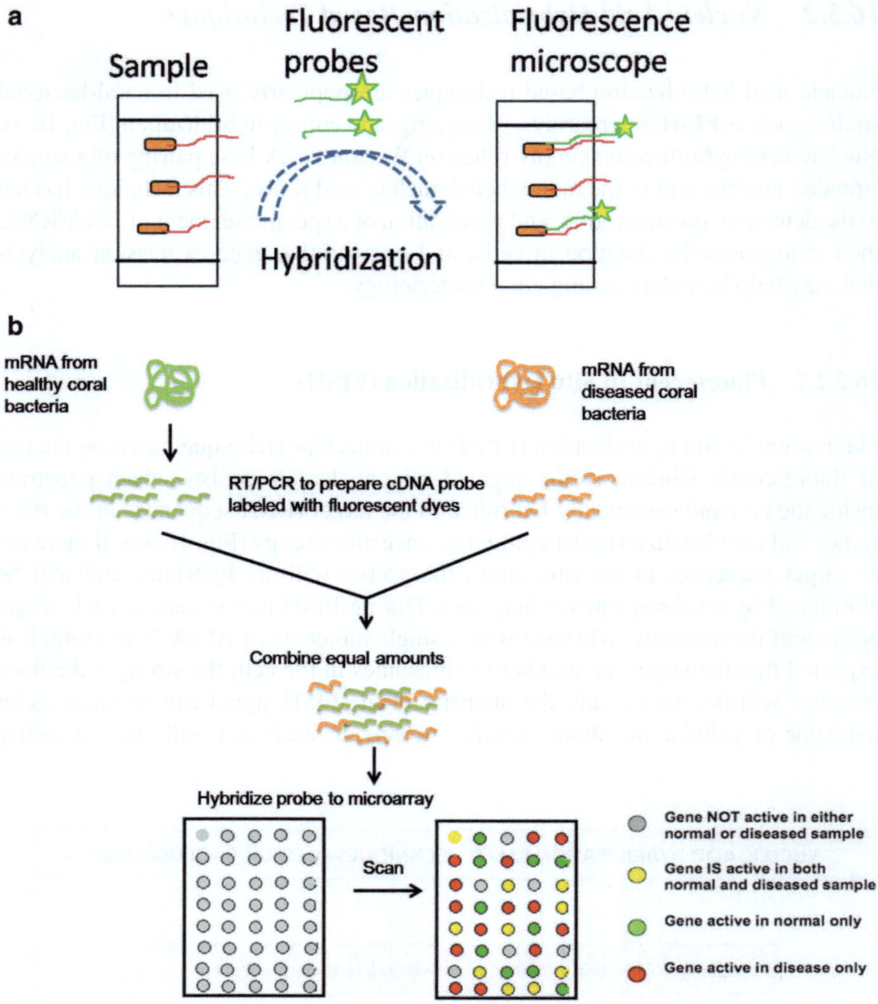

Fig. 16.8 (continued)

homogenous fluorescence signal, depending on the concentration and localization of ribosomes inside the cell. It represents a new and useful approach to identify living cells possessing an active metabolism. In coral bacteriology, FISH technology allows spatial and temporal analysis of bacterial community and their variability (Bouvier and del Giorgio 2003). For example, Ainsworth et al. (2006) combined FISH and spectral imaging to study coral-associated bacteria to overcome the extensive autofluorescence of coral tissues and endosymbionts for the identification of the coral-associated bacterial communities.

16.5.2.2 Microarray

Microarray has gained much popularity in coral-bacterial studies. It is a nucleic acid hybridization-based technique in which nucleic acid is hybridized to a very large set of oligonucleotide probes immobilized on a solid support generally called gene chip (Fig. 16.8b). It is mainly used to study gene expression and genetic variation in different samples. A typical microarray experiment involves extraction of total RNA separately from the different samples. Then, these RNA samples are reverse transcribed into cDNAs and labeled with different fluorescent dyes. Samples are mixed in equal proportion and hybridized with the cloned sequences on the gene chip. Hybridization of cDNA with the DNA on gene chip results into production of a fluorescent signal that indicates the expression level of the various genes. In coral bacteriology, 16s rRNA gene microarray is used to assay differences in bacterial assemblages of healthy and diseased colonies. Recent studies found increased bacterial richness in diseased samples highlighting the role of opportunistic conditions in structuring microbial community patterns during disease. They observed that host transcriptome under yellow-band disease (YBD) showed a reduced cellular expression of defense- and metabolism-related processes (Closek et al. 2014; Roder et al. 2014).

16.5.2.3 Membrane-Based Methods

Membrane-based techniques such as colony hybridization and ribotyping are also gaining momentum in coral-bacterial studies (Fig. 16.8c):

a. *Colony hybridization* is a molecular approach for microbial community analysis. The technique utilizes a membrane usually nylon or nitrocellulose on which blotted colonies are lysed to release nucleic acids which are denatured and hybridized with a labeled probe (Fig. 16.8 (c1)). The membrane is then visualized by UV or autoradiography. It can be used for screening clones or bacterial isolates based on probes utilized in the experiment. The probes can be a fragment of DNA or RNA of variable length constructed on the basis of specific genes. It's a versatile technique and can be used for a large number of samples. Bourne et al. (2008) have used probes Vib-sp1 and Vib-GV to screen for *Vibrionaceae*-positive clones. As it identifies a target gene, this technique can be very useful for the identification of potential pathogens from the corals. The limitation of this method is that there is a probability of cross-hybridization and misidentification of closely related species (Cerdà-Cuéllar and Blanch 2002).
b. *Ribotyping* is another membrane-based technique which entails digestion of bacterial genomic DNA with specific restriction enzymes followed by gel electrophoresis and then the subsequent transfer of the DNA fragments onto nylon or nitrocellulose membranes (Fig. 16.8 (c2)). The DNA on the membrane is then hybridized with a labeled 16S or 23S rRNA probe which can be visualized and compared with reference organisms from the available database. Basically, this method has been used for the identification of pathogenic species (Prevost et al. 1992; Carson et al. 2001; Gerner-Smidt 1992). Among pathogens, also there are many clones and it is difficult

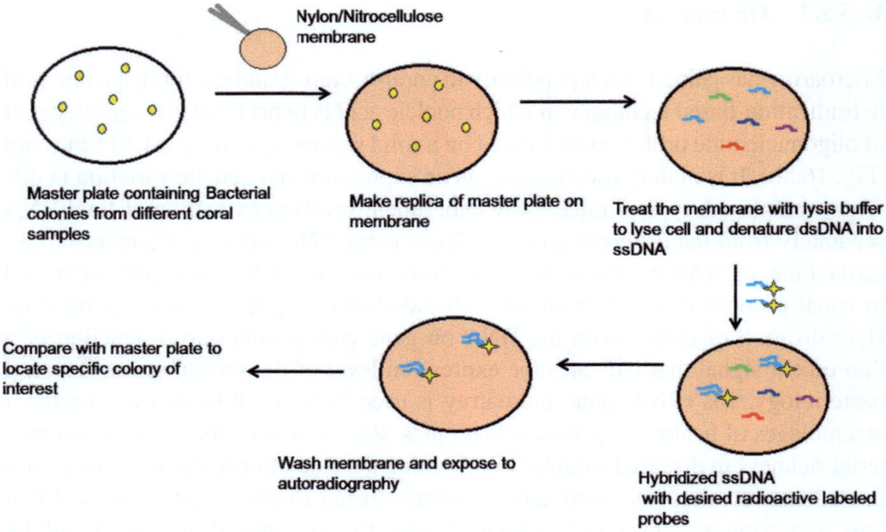

Fig. 16.8 (c1) Schematic diagram showing the principle of colony hybridization

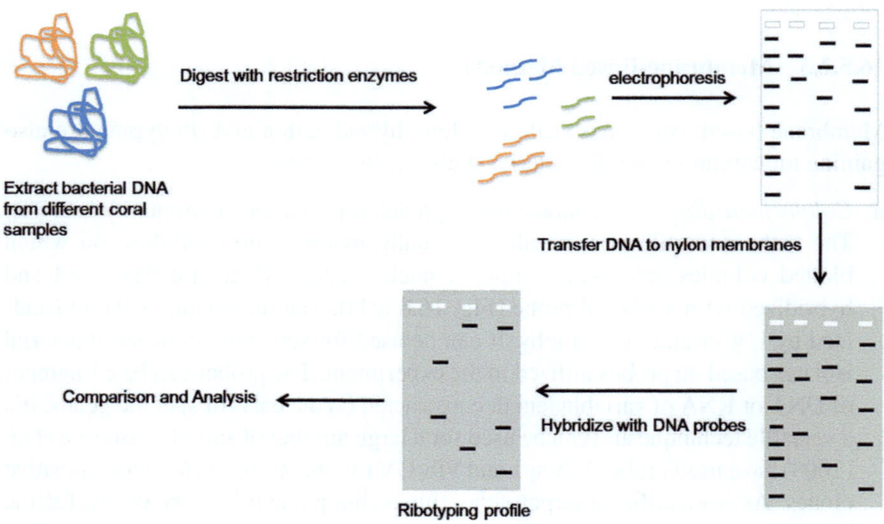

Fig. 16.8 (c2) Schematic diagram showing the principle of ribotyping

to identify the clones which are virulent. Therefore, this targeted gene-based method can identify the virulent clone among the pathogens which can harm the corals. This fingerprinting technique was used in the taxonomy of *vibrios* (Ilboudo et al. 2016). In addition to that, closely related *Vibrio* species can be effectively differentiated with the help of ribotyping (Austin et al. 1995). Genomic diversity of environmental *Vibrio* strains associated with fish and oysters has also been studied by Austin et al. (1997)

with the aid of ribotyping. Though it is a robust method for the determination of molecular epidemiology of bacterial pathogens, the time requirement for blotting, use of radioactivity in detection and its lengthiness, makes this technique less attractive and tedious as well (Kostman et al. 1992). But this technique is very less explored for the study of coral-associated bacteria. Hence, ribotyping can be efficiently adapted for exploring the diversity of coral-associated bacteria and detection of potentially pathogenic microbes.

16.5.3 Western Blot: A Protein-Based Molecular Technique

Proteins profiling in coral-bacterial ecology may provide us the information related to the function of bacteria and their mechanism in coral holobiont. Western blot is the technique for developing protein profiles of different bacterial samples from coral. It involves blotting of protein samples on the membrane and their hybridization with specific radiolabeled protein probe (Fig. 16.9). Thus, sometimes it is also called protein immunoblot. It is a very sensitive method widely applied in coral-bacterial studies. For example, Lesser et al. (2004) have reported that immunoblots of coral homogenates challenged with a polyclonal antibody against phycoerythrin revealed a positive cross-reaction with the β-polypeptide of phycoerythrin produced by marine *Cyanobacteria* which evidently shows that intracellular *Cyanobacteria* are associated with the coral. In another study, sponge-bacteria associations are being explored through Western blotting technique (Böhm et al. 2001). Additionally, it can be used very effectively to detect toxin released by bacteria which is detrimental to coral health.

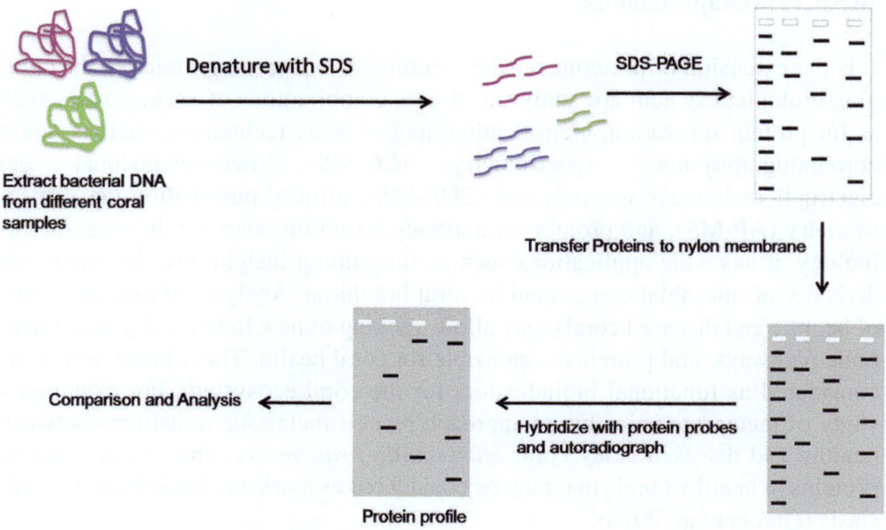

Fig. 16.9 Schematic diagram showing the principle of Western blotting

16.5.4 Next-Generation Tools

With the advancement of sequencing technology that offers larger data at much lower cost, the focus has shifted on the next-generation tools. These include Illumina (Solexa), 454 pyrosequencing, SOLiD sequencing, and Ion Torrent: Proton/PGM sequencing. These tools have given access to fully sequenced and annotated genomes that enable identification of genes, genomic regions, or alleles corelated to factors such as disease susceptibility, stress tolerance, growth rates, and metabolism. Further, access to complete genome sequences will facilitate the development of molecular markers that differentiate selective and random population divergence. NGS has already been applied in analyzing microbial community and their function in coral holobiont. For instance, a study found significant different microbial community among coral compartments using a combination of DGGE with next-generation sequencing and electron microscopy. This study also revealed the role of bacteria in nitrogen fixing. These bacteria were belonging to the *Rhodobacteraceae* and *Vibrionaceae* families that form part of *O. patagonica* tissues of core microbiome. Furthermore, they found sequences of coral pathogens, *V. mediterranei* and *V. coralliilyticus*, in both bleached and healthy corals (Yang et al. 2016) which suggest that these bacteria may turn pathogenic in adverse conditions. Thus, this advance sequencing technology would also be helpful in studying pathogenic bacteria in coral holobiont.

16.5.5 Omics Tools

16.5.5.1 Metaproteomics

It is the extension of proteomics where complete expressed proteins in the complex biological system are analyzed. It uses combinations of various tools such as for protein separation, identification, and/or assay techniques, such as liquid chromatography-mass spectrometry (LC-MS), two-dimensional gel electrophoresis-mass spectrometry (2DE-MS), affinity purification-mass spectrometry (AP-MS), and protein- or antibody-based microarrays. In coral microbiology, it has wide applications such as for gaining insight into the functional diversity of microbial component in coral holobiont. Analysis of metaproteome of healthy and diseased corals may allow tracking of new functional genes, metabolic pathways, and proteins responsible for coral health. These proteins may be considered as functional bioindicators for the coral ecosystem. For example, a study of metaproteomics-based approach reveals metabolic transitions between healthy and diseased stony coral *Mussismilia braziliensis.* They found a set of proteins in healthy corals that may be considered as markers of holobiont homeostasis (Garcia et al. 2016).

16.5.5.2 Metabolomics

It is the study of metabolites and their interactions within a biological system. Among all other "omics" measures, metabolomics is a powerful approach because metabolites and their concentrations directly reflect the underlying biochemical activity. Thus, metabolomics best represents the molecular phenotype. There are two main approaches in metabolomics studies. Untargeted approach measures as many metabolites as possible from a range of biological samples without any (intended) bias, whereas targeted approach measures a set of metabolites. For coral microbiology studies, mainly untargeted metabolomics approach is used. For example, a study on metabolomics of reef benthic interactions reveals a bioactive lipid involved in coral defense. These were platelet-activating factor (PAF) and lyso-PAF, similar to central inflammatory modulators present in mammals. This shows that self and non-self-recognition among some of the oldest extant holobionts involve bioactive lipids identical to those in highly derived taxa-like humans (Quinn et al. 2016).

Acknowledgment NPP and SBK sincerely thank CSIR and DBT for providing financial assistance by awarding JRF and SRF. This study is a part of the ongoing CSIR Network project VACCIN. This book chapter has been assigned to CSIR-CSMCRI—121/2016 registration number.

References

Agostini S, Suzuki Y, Higuchi T et al (2012) Biological and chemical characteristics of the coral gastric cavity. Coral Reefs 31:147–156

Ainsworth TD, Fine M, Blackall LL et al (2006) Fluorescence in situ hybridization and spectral imaging of coral-associated bacterial communities. Appl Environ Microbiol 72:3016–3020

Austin B, Alsina M, Austin D et al (1995) Identification and typing of *Vibrio anguillarum*: a comparison of different methods. Syst Appl Microbiol 18:285–302

Austin B, Austin D, Blanch A et al (1997) A comparison of methods for the typing of fish-pathogenic *Vibrio* spp. Syst Appl Microbiol 20:89–101

Autenrieth I, Hantke K, Heesemann J (1991) Immunosuppression of the host and delivery of iron to the pathogen: a possible dual role of siderophores in the pathogenesis of microbial infections? Med Microbiol Immunol 180:135–141

Azam F, Malfatti F (2007) Microbial structuring of marine ecosystems. Nat Rev Microbiol 5(10):782–791

Banin E, Vassilakos D, Orr E et al (2003) Superoxide dismutase is a virulence factor produced by the coral bleaching pathogen *Vibrio shiloi*. Curr Microbiol 46:418–422

Barott KL, Rodriguez-Brito B, Janouškovec J et al (2011) Microbial diversity associated with four functional groups of benthic reef algae and the reef-building coral *Montastraea annularis*. Environ Microbiol 13:1192–1204

Bayer T, Neave MJ, Alsheikh-Hussain A et al (2013) The microbiome of the red sea coral *Stylophora pistillata* is dominated by tissue-associated endozoicomonas bacteria. Appl Environ Microbiol 79:4759–4762

Böhm M, Hentschel U, Friedrich A et al (2001) Molecular response of the sponge *Suberites domuncula* to bacterial infection. Mar Biol 139:1037–1045

Bourne DG, Munn CB (2005) Diversity of bacteria associated with the coral *Pocillopora damicornis* from the Great Barrier Reef. Environ Microbiol 7:1162–1174

Bourne D, Iida Y, Uthicke S, Smith-Keune C (2008) Changes in coral-associated microbial communities during a bleaching event. ISME J 2112:350–363

Bouvier T, del Giorgio PA (2003) Factors influencing the detection of bacterial cells using fluorescence in situ hybridization (FISH): a quantitative review of published reports. FEMS Microbiol Ecol 44:3–15

Brooke S, Ross SW (2014) First observations of the cold-water coral *Lophelia pertusa* in mid-Atlantic canyons of the USA. Deep Sea Res Part 2 Top Stud Oceanogr 104:245–251

Burke C, Steinberg P, Rusch D et al (2011) Bacterial community assembly based on functional genes rather than species. Proc Natl Acad Sci 108:14288–14293

Carson CA, Shear BL, Ellersieck MR (2001) Identification of fecal *Escherichia coli* from humans and animals by ribotyping identification of fecal *Escherichia coli* from humans and animals by ribotyping. Appl Environ Microbiol 67:1503–1507

Cerdà-Cuéllar M, Blanch AR (2002) Detection and identification of *Vibrio scophthalmi* in the intestinal microbiota of fish and evaluation of host specificity. J Appl Microbiol 93:261–268

Chen CP, Tseng CH, Chen CA et al (2011) The dynamics of microbial partnerships in the coral *Isopora palifera*. ISME J 5:728–740

Chimetto LA, Brocchi M, Thompson CC et al (2008) Vibrios dominate as culturable nitrogen-fixing bacteria of the Brazilian coral *Mussismilia hispida*. Syst Appl Microbiol 31:312–319

Closek CJ, Sunagawa S, DeSalvo MK et al (2014) Coral transcriptome and bacterial community profiles reveal distinct Yellow Band Disease states in *Orbicella faveolata*. ISME J 8:2411–2422

Connell JH (1978) Diversity in tropical rain forests and coral reefs. Science 199:1302–1310

Cook GM, Rothenberger JP, Sikaroodi M et al (2013) A comparison of culture-dependent and culture-independent techniques used to characterize bacterial communities on healthy and white plague-diseased corals of the *Montastraea annularis* species complex. Coral Reefs 32:375–388

Dinsdale EA, Edwards RA, Hall D et al (2008) Functional metagenomic profiling of nine biomes. Nature 452:629–632

Dobretsov S, Abed RMM, Teplitski M (2013) Mini-review: inhibition of biofouling by marine microorganisms. Biofouling 29:423–441

Faden M (2008) "Smart-grid" appliances. Front Ecol Environ 6:62–62

Falkowski PG, Fenchel T, Delong EF (2008) The microbial engines that drive Earth's biogeochemical cycles. Science 320:1034–1039

Fine M, Loya Y (2002) Endolithic algae: an alternative source of photoassimilates during coral bleaching. Proc R Soc Lond B Biol Sci 269:1205–1210

Fine M, Loya Y (2004) Coral bleaching in a temperate sea: from colony physiology to population ecology. In: Rosenberg E, Loya Y (eds) Coral health and disease. Springer, Heidelberg, pp 143–156

Fischer SG, Lerman LS (1983) DNA fragments differing by single base-pair substitutions are separated in denaturing gradient gels: correspondence with melting theory. Proc Natl Acad Sci U S A 80:1579–1583

Garcia GD, Santos EDO, Sousa GV et al (2016) Metaproteomics reveals metabolic transitions between healthy and diseased stony coral *Mussismilia braziliensis*. Mol Ecol 25:4632–4644

Gerner-Smidt P (1992) Ribotyping of the acinetobacter calcoaceticus-*Acinetobacter baumannii* complex. J Clin Microbiol 30:2680–2685

Hadfield MG (2011) Biofilms and marine invertebrate larvae: what bacteria produce that larvae use to choose settlement sites. Annu Rev Mar Sci 3:453–470

Heid C, Stevens J, Livak KJ et al (1996) Real time quantitative PCR. Genome Res 6:986–994

Hills W, Glenn T, Saiki RK, Scharf SJ (1987) 4683195 process for amplifying, detecting, and/or-cloning nucleic acid sequences. Biotechnol Adv 5:313

Huggett MJ, Williamson JE, de Nys R et al (2006) Larval settlement of the common Australian sea urchin *Heliocidaris erythrogramma* in response to bacteria from the surface of coralline algae. Oecologia 149:604–619

Ilboudo SDO, Sombie I, Soubeiga AK et al (2016) Factors influencing the refusal to consult at the health center in the rural region of Western Burkina Faso. Public Health 28:391–397

Joyner J, Wanless D, Sinigalliano CD et al (2014) Use of quantitative real-time PCR for direct detection of Serratia Marcescens in marine and other aquatic environments. Appl Environ Microbiol 80:1679–1683

Kimes NE, Johnson WR, Torralba M et al (2013) The *Montastraea faveolata* microbiome: ecological and temporal influences on a Caribbean reef-building coral in decline. Environ Microbiol 15:2082–2094

Koren O, Rosenberg E (2006) Bacteria associated with mucus and tissues of the coral *Oculina patagonica* in summer and winter. Appl Environ Microbiol 72:5254–5259

Kostman JR, Edlind TD, LiPuma JJ et al (1992) Molecular epidemiology of *Pseudomonas cepacia* determined by polymerase chain reaction ribotyping. J Clin Microbiol 30:2084–2087

Krediet CJ, Ritchie KB, Paul VJ et al (2013) Coral-associated micro-organisms and their roles in promoting coral health and thwarting diseases. Proc R Soc B Biol Sci 280:20122328

Kushmaro A, Kramarsky-Winter E (2004) Bacteria as a source of coral nutrition. In: Rosenberg E, Loya Y (eds) Coral health and disease. Springer, Heidelberg, pp 231–241

Lema KA, Willis BL, Bourneb DG (2012) Corals form characteristic associations with symbiotic nitrogen-fixing bacteria. Appl Environl Microbiol 78:3136–3144

Lesser MP, Falcón LI, Rodríguez-Román A, et al (2007) Nitrogen fixation by symbiotic cyanobacteria provides a source of nitrogen for the scleractinian coral *Montastraea cavernosa*. Mar Eco Progress Ser 346:143–152

Lesser MP, Mazel CH, Gorbunov MY et al (2004) Discovery of symbiotic nitrogen-fixing cyanobacteria in corals. Science 305:997–1000

Little AF, van Oppen MJH, Willis BL (2004) Flexibility in algal endosymbioses shapes growth in reef corals. Science 304:1492–1494

Littman RA, Willis BL, Pfeffer C (2009) Diversities of coral-associated bacteria differ with location, but not species, for three acroporid corals on the Great Barrier Reef. FEMS Microbiol Ecol 68:152–163

Long RA, Azam F (2001) Antagonistic interactions among marine bacteria. Appl Environ Microbiol 67:4875–4983

Mansson M, Gram L, Larsen TTO (2011) Production of bioactive secondary metabolites by marine Vibrionaceae. Mar Drugs 9:1440–1468

Mao-jones J, Ritchie KB, Jones LE et al (2010) How microbial community composition regulates coral disease development? PLoS Biol 8:22–26

Marhaver KL, Vermeij MJA, Rohwer F et al (2013) Janzen-Connell effects in a broadcast-spawning Caribbean coral: distance-dependent survival of larvae and settlers. Ecology 94:146–160

Marquis CP, Baird AH, de Nys R et al (2005) An evaluation of the antimicrobial properties of the eggs of 11 species of scleractinian corals. Coral Reefs 24:248–253

Meron D, Rodolfo-Metalpa R, Cunning R et al (2012) Changes in coral microbial communities in response to a natural pH gradient. ISME J 6:1775–1785

Munn CB, Marchant HK, Moody AJ (2008) Defences against oxidative stress in vibrios associated with corals. FEMS Microbiol Lett 281:58–63

Muyzer G, Dewaal EC, Uitterlinden AG (1993) Profiling of complex microbial populations by denaturing gradient gel electrophoresis analysis of polymerase chain reaction-amplified genes coding for 16S rRNA. Appl Environ Microbiol 59:695–700

Negri AP, Webster NS, Hill RT et al (2001) Metamorphosis of broadcast spawning corals in response to bacteria isolated from crustose algae. Mar Ecol Progress Ser 223:121–131

Olson ND, Lesser MP (2013) Diazotrophic diversity in the Caribbean coral, *Montastraea cavernosa*. Arch Microbiol 195:853–859

Pratte ZA (2013) Microbial functional genes associated with coral health and disease. Dis Aquat Org 107:161–171

Prevost G, Jaulhac B, Piemont Y (1992) DNA fingerprinting by pulsed-field gel electrophoresis is more effective than ribotyping in distinguishing among methicillin-resistant *Staphylococcus aureus* isolates. J Clin Microbiol 30:967–973

Quinn RA, Vermeij MJA, Hartmann AC et al (2016) Metabolomics of reef benthic interactions reveals a bioactive lipid involved in coral defence. Proc Biol Sci R Soc 283:20160469

Raina JB, Tapiolas D, Willis BL et al (2009) Coral-associated bacteria and their role in the biogeochemical cycling of sulfur. Appl Environ Microbiol 75:3492–3501

Raina JB, Tapiolas D, Motti CA et al (2016) Isolation of an antimicrobial compound produced by bacteria associated with reef-building corals. Peer J 4:2275

Ransome E, Munn CB, Halliday N et al (2014a) Diverse profiles of *N*-acyl-homoserine lactone molecules found in cnidarians. FEMS Microbiol Ecol 87:315–329

Ransome E, Rowley SJ, Thomas S et al (2014b) Disturbance to conserved bacterial communities in the cold-water gorgonian coral *Eunicella verrucosa*. FEMS Microbiol Ecol 90:404–416

Reaka-Kudla ML, Wilson DE, Wilson EO (1997) Biodiversity II: understanding and protecting our biological resources. J Insect Conserv 1(1997):247–250

Reshef L, Koren O, Loya Y et al (2006) The coral probiotic hypothesis. Environ Microbiol 8:2068–2073

Ritchie KB (2006) Regulation of microbial populations by coral surface mucus and mucus-associated bacteria. Mar Ecol Progress Ser 322:1–14

Roder C, Arif C, Daniels C et al (2014) Bacterial profiling of White Plague Disease across corals and oceans indicates a conserved and distinct disease microbiome. Mol Ecol 23:965–974

Rohwer F, Seguritan V, Azam F et al (2002) Diversity and distribution of coral-associated bacteria. Mar Eco Progress Ser 243:1–10

Rosenberg E, Falkovitz L (2004) The *Vibrio shiloi*/*Oculina patagonica* model system of coral bleaching. Annu Rev Microbiol 58:143–159

Rosenberg E, Koren O, Reshef L et al (2007) The role of microorganisms in coral health, disease and evolution. Nat Rev Microbiol 5:355–362

Rypien KL, Ward JR, Azam F (2010) Antagonistic interactions among coral-associated bacteria. Environ Microbiol 12(1):28–39

Shikuma NJ, Pilhofer M, Weiss GL et al (2014) Marine tubeworm metamorphosis induced by arrays of bacterial phage tail-like structures. Science 343:529–533

Shnit-Orland M, Kushmaro A (2009) Coral mucus-associated bacteria: a possible first line of defense. FEMS Microbiol Ecol 67:371–380

Smith JE, Shaw M, Edwards RA et al (2006) Indirect effects of algae on coral: Algae-mediated, microbe-induced coral mortality. Ecol Lett 9:835–845

Sneed JM, Sharp KH, Ritchie KB et al (2014) The chemical cue tetrabromopyrrole from a biofilm bacterium induces settlement of multiple Caribbean corals. Proc R Soc 281:1–9

Solanki MK, Singh RK, Srivastava S et al (2014) Isolation and characterization of siderophore producing antagonistic rhizobacteria against *Rhizoctonia solani*. J Basic Microbiol 54:585–596

Spalding M, Ravilious C, Green E (2001) World atlas of coral reefs. University of California Press, Berkeley, CA

Sweet MJ, Croquer A, Bythell JC (2011) Bacterial assemblages differ between compartments within the coral holobiont. Coral Reefs 30:39–52

Tebben J, Tapiolas DM, Motti CA et al (2011) Induction of larval metamorphosis of the coral *Acropora millepora* by tetrabromopyrrole isolated from a *Pseudoalteromonas* bacterium. PLoS One 6:19082

Teplitski M, Ritchie K (2009) How feasible is the biological control of coral diseases? Trends Ecol Evol 24:378–385

Thompson JR, Rivera HE, Closek CJ et al (2014) Microbes in the coral holobiont: partners through evolution, development, and ecological interactions. Front Cell Infect Microbiol 4:176

Thurber RV, Willner-Hall D, Rodriguez-Mueller B et al (2009) Metagenomic analysis of stresses coral holobionts. Environ Microbiol 11:2148–2163

Tout J, Jeffries TC, Webster NS et al (2014) Variability in microbial community composition and function between different niches within a coral reef. Microbiol Ecol 67:540–552

Tremblay P, Grover R, Maguer JF et al (2012) Autotrophic carbon budget in coral tissue: a new 13C-based model of photosynthate translocation. J Exp Biol 215:384–393

Vieira C, Engelen AH, Guentas L et al (2016) Species specificity of bacteria associated to the brown seaweeds Lobophora (Dictyotales, Phaeophyceae) and their potential for induction of rapid coral bleaching in *Acropora muricata*. Front Microbiol 7:1–13

Wegley L, Edwards R, Rodriguez-Brito B et al (2007) Metagenomic analysis of the microbial community associated with the coral *Porites astreoides*. Environ Microbiol 9:2707–2719

Yang SH, Lee STM, Huang CR et al (2016) Prevalence of potential nitrogen-fixing, green sulfur bacteria in the skeleton of reef-building coral *Isopora palifera*. Limnol Oceanogr 61:1078–1086

Index

A
ABC transporter (ABC), 39
Abiotic factors, 40
Abiotic stress
acidification, 125
CRISPR-Cas9 system, 128
genomic information, 128
heat wave, 125
light limitation, 125
light stress, 126
salinity, 127, 128
temperature, 127
Abiotic stresses, 52, 59
Abiotic stressors, 15, 16, 18–20
Acidification, 125
Acutodesmus dimorphus, 262
Acylated homoserine lactones (AHLs), 284
Affinity purification-mass spectrometry (AP-MS), 336
Agarophytic macroalgae, 7
Algicidal bacteria, 288
Alpha-proteobacteria, 227
Alternative electron flows
CO_2 and O_2 conditions, 149
photoprotective roles, 149
photosynthetic carbon assimilation, 149
seagrass systems, 149
Analysis of variance (ANOVA), 72, 73
Anodic stripping voltammetry (ASV), 40
Antioxidant activity, 25, 26
Aquatic environment, 134
Archaeplastids, 280
Ascorbate peroxidase (AP), 38
Ascorbic acid, 256
Astaxanthin, 257, 258, 260
Atlantic and Caribbean coral reefs, 320
ATP-binding cassette (ABC), 40
Autoinducer 1 (AI1), 285, 303
Automated rRNA intergenic spacer analysis (ARISA), 224
Axenic coccolith-bearing strain, 286

B
Bacterioplankton community, 228
Bacteroidetes, 230
Baltic Sea algae, 16
Bangiomorpha pubescens, 18
Benzaldehyde, 101
Betaine lipids, 55, 56
Bioflocculants, 291
Biofuel
and bio-based chemicals, 254
generation of, 254
microalgal, 254
oxidative stress, 262–266
Biological systems
aquatic life history, 169
ATP and NADPH production, 171
bio-optical model, 172, 173
DIC, 172
environmental change sensitivity, 171, 172
filiform pollen grains, *Zostera marina*, 171
high-CO_2, 173
increased $CO_{2(aq)}$, 172
LOQS, 171
photoacclimation, 174
predictive seagrass models, 175–177
Rubisco, balancing redox state, 174

© Springer International Publishing AG 2017
M. Kumar, P. Ralph (eds.), *Systems Biology of Marine Ecosystems*,
DOI 10.1007/978-3-319-62094-7

Biological systems (*cont.*)
taxonomic families, superorder Alismatiflorae, 169
turtlegrass (*Thalassia testudinum*), 169, 170
Zostera marina L. (eelgrass), 174
Bioprospecting
algalomic analyses, 270
carotenogenesis, 270
in cosmetic and pharmaceutical industries, 270
omics projects, 270
Biotic stresses, 52, 86, 89
Braarudosphaera bigelowii, 284

C
Canthaxanthin, 258
Carbon fixation, 244
Carbon reactions
C4 "Kranz" anatomy, 136
carbon-concentrating system, 135
dissolved inorganic carbon pools and atmospheric CO_2, 136
protons extrusion, 136
pulse-chase ^{14}C experiments, 136
seagrass, stomatal differentiation, 135
Carbon-concentrating mechanism (CCM), 148, 153
Carotenogenesis, 259
Carotenoids, 252, 255–261, 266, 270
CCM. *See* Carbon-concentrating mechanism (CCM)
Chlamydomonas reinhardtii, 264, 289
Chlorella sorokiniana C3, 264
Chlorella zofingiensis, 258
Chlorophyta, 105
Chondrus crispus, 5
Chromatographic peak identification, 192–193
Chromophoric dissolved organic matter (CDOM), 243
CO_2
biological CO_2 pump, 236
carbon fixation, 236
concentration, 236
stratospheric ozone, 236
UVR exposure, 236
UVR irradiances, 236
Coccomyxa onubensis, 262
Colony hybridization, 333, 334
Commensalism, 287
Coral bacteriology, 328–333
Coral bleaching, 325–327
Coral ecosystem, 319, 323–327
anthropogenic activities, 318
bacterial diversity, 321–323
coral ecology, 318
coral-associated bacteria
colonization and biofouling, 326
holobiont, 323–326
larva, 325, 326
nutrient supplier, 324, 325
pathogen, 327
DGGE, 330
FISH, 331, 332
membrane-based methods, 333–335
metabolomics, 337
metaproteomics, 336
microarray, 333
next-generation tools, 336
nucleic acid hybridization-based techniques, 331–335
PCR-based methods, 328–330
qRT-PCR, 330
rainforests, 318
and structure, 320, 321
symbiotic relationship, 319
techniques, 327
western blot, 335, 336
Coral holobiont, 319–324, 327, 328, 335, 336
Coral reef ecosystem, 319
Coral reefs, 320
Crustose coralline algae (CCA) surfaces, 325
Cyanobacterium, 231, 324
Cymodocea nodosa, 128
Cystocarp, 306

D
Dehydroascorbate reductase (DHAR) activities, 28
Delisea pulchra, 289, 305
Denaturing gradient gel electrophoresis (DGGE), 321, 328, 330, 336
Desiccation stress
to intertidal zonation patterns, 15, 16
Desiccation stress tolerance, 18, 19
Desiccation-associated proteins, 24–25
Desmarestia anceps (Desmarestiales), 19
Diatoms, 211, 213, 226
Dimethyl sulfide (DMS), 325
Dimethylsulfoniopropionate (DMSP), 103, 105, 325, 330
Dinoflagellates, 319, 321
Dissolved organic matter (DOM), 288
Dunaliella bardawil, 261
Dunaliella salina, 258

Durvillaea antarctica, 42
Dynamic environment research, 141–142
  daily and seasonal fluctuations, 138
  photoacclimation and photoprotection, 138–141
  positive carbon balance and growth, 138
  salinity stress (*see* Salinity stress)

E
Ecosystem engineers
  description, 177
  humans, interactions with, 180, 181
  interaction between seagrasses, 179, 180
  seagrass-sediment interactions, 177, 178
Ectocarpus, 6
Ectocarpus siliculosus, 5
Eel grass. *See* Zostera muelleri (eel grass)
Electron flow
  non-photochemical quenching, 137
  PGR5- and NDH-dependent pathways, 137
  pseudocyclic electron flow/Mehler reaction, 137
  PSII to PSI transportation, 137
  suboptimal growth rates, 137
Emiliania huxleyi, 284
Enteromorpha, 305
Enteromorpha clathrata, 103
Ethylene (ET), 106–108, 125
Ethylene biosynthetic pathway, 106
Eukaryotic microalgae, 211
Eukaryotic transcriptome sequencing project database, 229
Evaporative light scattering detector (ELSD), 64

F
Fatty acid analysis tool (FAAT), 70
Fatty acid methyl esters (FAMEs), 63
Fatty acids (FAs), 56, 57, 63
Flavonoids, 200
Fluorescent in situ hybridization (FISH), 331
Fucus (Fucales, Phaeophyceae), 15
Fucus vesiculosus, 89
Functional genomics, 4–7
  elite variety development, 8
  from context of biofuel, 7–8

G
Gammaproteobacteria, 230, 327
Gas chromatography (GC), 124
Gas chromatography-mass spectrometry (GC-MS), 191, 192
Gelidium arbuscula, 102, 103
Gelidium latifolium, 89
Gene expression, 212, 213, 215
Genomics, 121, 122, 224
  systems biology, 121
Global Ocean sampling expedition (GOS), 224
Glutathione peroxidase (*GPX*), 125
Glyceraldehyde 3-phosphate dehydrogenase (GADPH), 124
Glycerolipids, 55
Glycerophospholipids, 52
Gracilaria conferta, 289
Gracilaria corticata (Gracilariales), 18, 26
Gracilaria tenuistipitata, 42
Grateloupia imbricata, 108, 110
Green fluorescent protein (GFP), 128
Green macroalgae, 305
Gymnodinium catenatum, 283

H
Haematochrom, 258
Haematococcus MT 2877, 261
Haematococcus pluvialis, 258, 259, 265, 268, 269
Halogenated compounds, 101
Haloperoxidases, 104
Heat stress, 142, 143
Heavy metal stress, 39
High-pressure liquid chromatography (HPLC), 60
Hizikia fusiforme, 42
Host-associated bacterial diversity, 306–309
Hydrophilic interactions (HILIC), 64

I
IAA biosynthesis, 286
IAA precursor tryptophan, 290
Interkingdom signalling system, 309
Internal transcribed spacer (ITS), 226
International Census of Marine Microbes (ICoMM), 224
Irradiance, 240

J
Jasmonates
  cell cycle, 105
  ethylene, 106–108
  seaweeds, 103, 106–108
  synthesis, 104
Jasmonic acid (JA), 89, 125

K

Klebsormidium crenulatum (*Charophyta*), 23, 286
Knowledge gaps
  alternative electron flows, 149, 150
  nitrogen metabolisms, 150, 151
  photoreceptors, 146–148
  photosynthesis and respiration, 150

L

Laminaria digitata, 89, 102
Laurencia dendroidea, 7, 102
LC-MS. *See* Liquid chromatography-mass spectrometry (LC-MS)
Lessonia berteroana, 20, 37
Lessonia spicata, 17, 37
Light reactions
  seagrass, 134
Light stress, 126
Light-harvesting
  seagrass, 135
Lipid Metabolites and Pathways Strategy (LIPID MAPS) Consortium, 51
Lipids, 50
Lipoxygenase activity, 107
Lipoxygenases (LOX), 57–59, 85, 89
Liquid chromatography (LC), 124
Liquid chromatography-mass spectrometry (LC-MS), 191, 192, 336
Lithothamnion corallioides, 89
Long-chain aldehydes (LCA), 101
LOQS. *See* Low-oxygen quiescence strategy (LOQS)
Low-oxygen quiescence strategy (LOQS), 171
Lycopene, 257, 258
Lyso-PAF, 337

M

Macroalgae, 109–111
Macroalgal-bacterial interactions
  algal host, 302
  beneficial interaction, 302
  cell-to-cell communication, 303
  green macroalga, 302
  LuxI-LuxR type QS system, 303
  nutrient source and surface, 302
  signalling molecules, 302
Macrophytic blue carbon system, 4
Malondialdehyde (MDA), 260
Marine diazotrophs, 240
Marine ecosystems, 36, 37
  environmental/human origin, 208
  photosynthesis and nutrient cycling, 208
Marine macroalgae, 4
Marine microalgal genomics
  eukaryotic microalgae, 211
  prokaryotic microalgae, 211
  systematic sequencing, 210
Marine microalgal transcriptomics
  laboratory studies, 213
  large genome sizes and mixed cultures, 212
  mRNA sequencing, 212
  proteins, 212
  RNA sequencing, 213, 214
  transcriptomic data, 212
Marine microbes
  evolutionarily diverse organisms, 231
  fingerprinting techniques, 224
  oceanographic processes, 228
Marine microbial metagenomes, 211, 212
Marine microeukaryotes, 211, 214
Marine Strategy Framework Directive (MSFD), 129
Marinobacter, 283
Mass spectrometry (MS), 124
Meristemoid mother cells (MMC), 122
Metabolite enrichment softwares, 73
Metabolite set enrichment analysis (MSEA), 73
Metabolomics, 124
Metabolon's TrueMass™ Complex Lipid Panel technology, 70
Metagenomics, 230
Metal toxicity, 36, 37
  abiotic factors, 40–42
Metallothionein's protein, 267
Metaproteome, 336
Metatranscriptomics, 227, 230
Methionine (M), 304
Methyl alkyl compounds, 103
Methyl jasmonate, 103, 104, 106–109
Mevalonate, 59
Microalgae
  applications, 266, 267
  autotrophic organisms, 252
  bioprospecting, 254
  carotenoid accumulation, 257–261
  eukaryotic, 211
  life cycle of, 284
  lipids and carbohydrates, 254
  long-term interactions, 292
  microalgal species, 253
  microbial engineering, 217
  oxidative stress and biofuel potential, 262–266
  parasites of, 287

photosynthetic mechanism, 254
prokaryotic, 211
simple growth requirements, 252
Microalgal biofactory, 267–270
Microalgal carbohydrates, 253
Microalgal lipid, 262
Microalgal-bacterial interactions
algae, 280
algal-bacterial interactions, 282
applications, 291
causative bioactive compound, 292
commensalism, 287
cyanobacterium, 280
endosymbiotic events, 280
habitat-forming species, 282
mutualism, 282–287
natural environment, 280
parasitism, 287
pathogenicity, 288–291
phytoplankton, 280
research, 292
researchers, 282, 292
small-scale systems, 292
terrestrial environments, 281
Microarray, 331, 333
Microbial ecology, 292
Microbial function
culture-dependent approaches, 224
field-based examples, 224
microbial species, 224
molecular approaches, 224
oceanographers, 224
Microbial gene expression, 227–229
Microbial protein expression, 229–231
Monoraphidium dybowskii, 265, 266
Monostroma oxyspermum, 284
Montreal Protocol, 242
Morphology
Ulva, species, 302
mRNA (gene), 9
Multivariate analysis of variance (MANOVA), 72
Mutualism, 282–287
Mycosporine-like amino acids (MAAs), 244

N

N-acyl homoserine lactone (AHL), 302
Nautella italica, 305
Next-generation sequencing (NGS), 304, 308
Nitrogen-fixing organisms, 241
Nitzschia closterium, 264
Non-photochemical quenching (NPQ), 141
Nostoc, 252
NPQ. *See* Non-photochemical quenching (NPQ)
Nuclear magnetic resonance (NMR), 124
Nucleic acid hybridization-based techniques, 331

O

Ocean acidification (OA), 42–44, 236
atmosphere, 237
$CaCO_3$ solution, 237
carbonate system, 238
chemical processes, 238
CO_2 emission, 238
combined effects, 244–246
epipelagic seawater, 238
equilibrium constants, 237
fossil fuel, 238
Industrial Revolution, 237
photosynthetic organisms, 239, 240, 242
seawater carbonate system, 237
seawater mixing, 237
UV radiation, 244–246
Oceanographic processes
biogeochemical transformations, 226
biotic interactions, 227
Californian coastal waters, 227
fingerprinting techniques, 224
genomic datasets, 227
mesoscale eddies, 226
metagenomics sequencing, 226
microbial community, 226
microbial gene expression, 227–229
microbial protein expression, 229–231
next-generation sequencing, 225
nitrogenase enzyme, 226
oceanographic circulation, 226
16S rRNA gene, 225–226
taxonomic resolution, 226
Omics
CCMs, 153
central carbon and energy metabolisms process, 151
chloroplast-encoded and nuclear-encoded proteins, 153
genome sequences availability, 152
metabolic pathways, 153
metabolomics, 152
transcriptomics and proteomics analysis, 152
Omics platform, 267–270
Operational taxonomic units (OTUs), 224
Ornithine decarboxylase (ODC), 111

Oxidative stress
and biofuel potential, 262–266
in carotenoid accumulation, 257–261
reactive oxygen species, 254–257
Oxylipins, 52, 56–60, 63, 64, 81, 85, 88

P

PAM. *See* Pulse amplitude modulation (PAM) fluorometry
Parasitism, 287
Particulate organic matters (POM), 242–243
Pathogenicity, 288–291
p-coumaric acid (*p*-CA), 290
Pelvetia canaliculata, 16
Permutational multivariate analysis of variance (PERMANOVA), 72
Peroxiredoxin (PRX), 38
Peroxiredoxin Q (*PRXQ*), 125
Phaeocystis globosa, 289
Phlorotannins, 289
Photoacclimation
anthocyanin, role in, 140
anthropogenic activities, 138
CABs and Rubisco activase, 140
chlorophyll b/a ratios, 139
fast chlorophyll fluorescence induction curves, 140
light requirement for seagrass, 138
non-photosynthetic tissues, 139
OJIP analysis, 139
PAM fluorometry, 138
photosynthesis-irradiance characteristics, 139
PSI, role of, 139
UV-B, 141
xanthophyll cycle-related gene upregulation, 140
Photoprotection. *See* Photoacclimation
Photoreceptors
genes, 146
light environment, 146
photoreceptor-mediated physiological responses, terrestrial plants, 146
photosynthetic carbon fixation, 147, 148
PHYA and CRY1 downregulation, 146
phytochromes, 146
plant growth and development, 146
red and far-red wavelengths, 147
transcription factors, 146
Photorespiration
^{14}C pulse-chase experiments, 137
defined, 136
photosynthetic organisms, 137
redox homoeostasis under stress conditions, 137
Rubisco, 136
RuBP oxygenation, 136
temperate seagrasses, 137
Photosynthesis
carbon limitation, 182
DIC, 172
eelgrass stimulation, 173
environmental forcing, 173
metabolic carbon balance determination, 181
organisms, 239, 240, 242
and respiration, 172
UV radiation, 242, 243
β-oxidation, 239
Photosynthesis of seagrass
carbon reactions, 135, 136
electron flow, 137, 138
energy metabolisms, role in, 134
light-harvesting and reactions, 134, 135
photorespiratory activities, 136, 137
Phycobiliproteins, 253
Phycosphere, 282
Phytochelatins, 256
Phytohormones, 286
Phytoplankton, 239, 240
Phytotoxins
antioxidative systems, 144
excessive heavy metals, 144
metalloenzymes inactivator, 145
microbial reduction, sulphate, 145
organic matter load, 145
PCs synthesis, 145
pesticides, herbicides and trace metals, 145
photosynthetic electron transport, 144
PSII and pigment concentration, 144
Picoeukaryotic prasinophytes, 227
Plant-associated bacteria (PAB), 304
Platelet-activating factor (PAF), 337
Polyhydroxyalkanoate (PHA), 285
Polyphenol, 256
Porphyra, 15, 16, 18–21
Porphyra linearis, 42
Porphyra sanjuanensis, 22
Porphyra yezoensis, 8
Precursor ion scans (PIS), 65
Primary polar metabolites, 199
Primary producers
concentration, 237
ocean environment, 247
physiological performance, 236
physiology, 244
Prochlorococcus, 230, 231
Prokaryotic microalgae, 211

Proline, 256
Prorocentrum minimum, 284
Proteomics, 224, 231
Proteomics and metabolomics, 214, 215
PSAL. *See* PSI subunit L (PSAL)
PSI subunit L (PSAL), 135
Pulse amplitude modulation (PAM) fluorometry, 138
Pyropia, 6, 8, 15, 17–21, 26, 27
Pyropia haitanensis, 60
Pyropia leucosticta, 19
Pyropia yezoensis, 19

Q

QS N-acyl-homoserine lactone (AHL) molecules, 286
Quantitative real-time PCR (qRT-PCR), 330
Quorum sensing
  AHL- and DNA-binding domains, 304
  bidirectional interkingdom signalling, 304
  carpospores, 306
  gram-negative bacteria, 304
  green algae, 306
  LuxR regulate virulence, 304–305
  N-acyl side-chains, 305
  plant-associated bacteria, 304
  zoospore, 305
Quorum sensing (QS) systems, 302

R

Reactive oxygen species
  oxidative stress, 254–257
Reactive oxygen species (ROS), 38, 125
Relative electron transport rate (RETR), 126
Ribosomal RNA (rRNA), 224
Ribotyping, 333, 334
Ribulose-1,5-bisphosphate carboxylase oxygenase (Rubisco), 136
RNA sequencing, 213, 214
Rubisco. *See* Ribulose-1,5-bisphosphate carboxylase oxygenase (Rubisco)

S

Saccharina japonica, 5
Saccharolipids, 54, 55, 81
S-adenosylmethionine (SAM), 104
Salinity, 127, 128
Salinity stress, 144–145
  CO_2 enrichment, 143, 144
  cytochrome b559 alpha subunit, 142
  heat stress, 142, 143
  NPQ, 141
  osmoregulatory processes, 141
  oxygen evolution rates, 141
  photosynthetic responses, 141
  phytotoxins (*see* Phytotoxins)
San Pedro Ocean Time-series (SPOT), 226
Sargassum, 6
Scenedesmus acutus, 105
Scenedesmus incrassulatus, 105
Scenedesmus vacuolatus, 267
Seagrasses
  aquatic autotrophic organisms, 190
  carbon traps, 120
  coastal and estuarine sediments, 190
  flowering monocotyledonous plants, 120
  organic carbon, 120
  systems biology, 120
Seaweed, 4–9
  distribution in the intertidal zone, 18–19
  tolerance pathways to desiccation stress, 14–28
Seaweed genetic resource, 8, 9
Seaweed lipidomics, 60–81, 84–86
  allied omics platforms, 86–90
  application to seaweed systems biology, 81
    acclimation strategies to abiotic/biotic stress, 84–86
    elucidation of lipidomic profiling, 81, 84
    identification of novel lipids, 81, 84
  tools and techniques, 61
    analytical platforms for lipidomics, 63–69
    lipidomic data analysis, 69–81
    sampling and lipid extraction, 61, 62
Seaweed lipids, 52
Seaweeds, 4
  ABC transporter proteins, 40
  Buffering Capacities, 17, 18
  CO_2-driven pH, 42
  coastal benthic ecosystems, 37
  community structure, 44
  geographic distribution across intertidal shores, 19, 20
  heavy metal stress, 39
  heavy metal toxicity, 42–44
  human consumption, 43
  inorganic carbon concentration, 43
  marine organisms, 38
  molecular mechanisms of desiccation tolerance, 21–28
  Scytosiphon spp, 38
Secondary metabolites, 190, 193, 199
Siderophore, 283, 324, 327
Silver thiosulphate (STS), 106

Sinorhizobium meliloti (NesR), 304
Sodium hypochlorite (NaClO), 260
Solar UV radiation, 247
SOLiD sequencing, 336
Sphingolipids, 56
Spirulina platensis, 261
Sporulation, 111
Sterols, 59, 60
Stictosiphonia arbuscula (Ceramiales), 25
Stramenopiles-Alveolates-Rhizaria (SAR) group, 280
Sulphated flavonoids, 200
Superoxide dismutase (SOD), 18, 25, 255
Systems biology
  complex network diagrams, 168
  defined, 209
  dynamic environment, 209
  evolutionary theory, 168
  genomics, 121, 122
  high-level computational capacity, 121
  high-throughput techniques, 120
  integration and meta-analysis, 215, 216
  irreducible and interconnected functions, 209
  metabolomics, 124
  molecular system, 209
  nutrient-limited pelagic waters, cyanobacteria, 210
  predictions, 216
  seagrass paradox, 182, 183
  seagrasses (marine angiosperms), 120, 168
  specific organization and intrinsic biological program, 209
  "systems-level" thinking, 209
  transcriptomics, 123, 124
  whole-cell modeling, 217

T
TAG biosynthetic machinery, 270
Taq polymerase, 328
Temperature
  eelgrass distributions, 174
  environmental forcing, 173
  photosynthesis and respiration, 172
  seagrasses, 172
Terminal restriction fragment length polymorphism (T-RFLP), 224
Thalassiosira pseudonana, 228
Tolerance mechanisms
  adaptive mechanism, 40
  external factors, 38
  metal-enriched environments, 38
  O_2 excitation, 38
  Scytosiphon gracilis, 38
  two-stage process, 38
Tolerant marine species, 15
Transcriptomics, 124, 268
  EST sequencing, 123
  genome-guided transcriptome, 123
  high-throughput technologies, 120 (*see* Marine microalgal transcriptomics)
  mediterranean endemic species, 123
  metabolic potential, 229
  metatranscriptomics, 227
  and proteomics, 224
  RNA-seq data, 123
  RuBisCO protein, 124
  Unix-based environment, 123
Tricarboxylic acid (TCA) cycle, 239
Tropodithietic acid (TDA), 285
Two-dimensional gel electrophoresis-mass spectrometry (2DE-MS), 336

U
Ulva linza, 6
Ulva mutabilis, 284
Ulvophyceae, 7
Undaria, 6
Upper mixed layer (UML), 236
UV radiation, 242, 243
UV/PAR radiation, 15

V
Vibrio anguillarum (V. anguillarum), 305
Volatiles
  algal reproduction, 101
  chlorinated compounds, 101
  complexity factors, 102–104
  environmental factors, 100
  experimental factors, 102
  growth regulators, 100
  marine algae, 100
  metabolic origins, 102
  micro- and macroalgae, 101
  organic compounds, 100, 101
  physiological responses, 104, 105
  processed algae, 101
  seagrasses, 102
  seaweeds, 101

W
Western blot, 335, 336

X
Xanthomonas campestris (XccR), 304
Xanthomonas oryzae, 304
Xanthomonas oryzae (OryR), 304
Xanthophylls, 257

Z
Zeaxanthin, 258
Zoospore, 305
Zostera muelleri (eel grass), 121, 126, 191–198
flavone glycoside thalassiolin B, 201
flavonoids, 200
high light and salinity fluctuations, 200
lipid classes, 201
marine flowering plant, 190
metabolites identification
categories, 193
chemicals, 191
chromatographic peak identification, 192–193
GC-MS, 191, 192
GC-MS analysis, 193, 196–198
LC-MS, 191, 192
LC-QTOF analysis, 193–195
nematicidal and antibacterial effects, 190
pharmacological effects, 190
plant defence mechanism, 190
primary polar metabolites, 199
rosmarinic and zosteric acid identification, 193
secondary metabolites, 199
shikimate pathway, 199
targeted and/or non-targeted metabolomic approaches, 193
zosteric acid, 190, 200
Zostera noltii (Alismatales), 22
Zosteric acid, 200

MIX
Papier aus verantwortungsvollen Quellen
Paper from responsible sources
FSC® C105338

If you have any concerns about our products,
you can contact us on
ProductSafety@springernature.com

In case Publisher is established outside the EU,
the EU authorized representative is:
Springer Nature Customer Service Center GmbH
Europaplatz 3, 69115 Heidelberg, Germany

Printed by Libri Plureos GmbH
in Hamburg, Germany